T0260043

Quantum Mechanics

Quantum Mechanics

From Analytical Mechanics to Quantum Mechanics, Simulation, Foundations, and Engineering

Mark Julian Everitt
Department of Physics
Loughborough University

Kieran Niels Bjergstrom
Quantum Technologies
Loughborough University Science and Enterprise Park

Stephen Neil Alexander Duffus
Department of Physics
Loughborough University

This edition first published 2024
© 2024 John Wiley & Sons Ltd.

All rights reserved. No part of this publication may be reproduced, stored in a retrieval system, or transmitted, in any form or by any means, electronic, mechanical, photocopying, recording or otherwise, except as permitted by law. Advice on how to obtain permission to reuse material from this title is available at http://www.wiley.com/go/permissions.

The right of Mark Julian Everitt, Kieran Niels Bjergstrom and Stephen Neil Alexander Duffus to be identified as the author(s) of this work has been asserted in accordance with law.

Registered Offices
John Wiley & Sons, Inc., 111 River Street, Hoboken, NJ 07030, USA
John Wiley & Sons Ltd, The Atrium, Southern Gate, Chichester, West Sussex, PO19 8SQ, UK

For details of our global editorial offices, customer services, and more information about Wiley products visit us at www.wiley.com.

Wiley also publishes its books in a variety of electronic formats and by print-on-demand. Some content that appears in standard print versions of this book may not be available in other formats.

Trademarks: Wiley and the Wiley logo are trademarks or registered trademarks of John Wiley & Sons, Inc. and/or its affiliates in the United States and other countries and may not be used without written permission. All other trademarks are the property of their respective owners. John Wiley & Sons, Inc. is not associated with any product or vendor mentioned in this book.

Limit of Liability/Disclaimer of Warranty
While the publisher and authors have used their best efforts in preparing this work, they make no representations or warranties with respect to the accuracy or completeness of the contents of this work and specifically disclaim all warranties, including without limitation any implied warranties of merchantability or fitness for a particular purpose. No warranty may be created or extended by sales representatives, written sales materials or promotional statements for this work. This work is sold with the understanding that the publisher is not engaged in rendering professional services. The advice and strategies contained herein may not be suitable for your situation. You should consult with a specialist where appropriate. The fact that an organization, website, or product is referred to in this work as a citation and/or potential source of further information does not mean that the publisher and authors endorse the information or services the organization, website, or product may provide or recommendations it may make. Further, readers should be aware that websites listed in this work may have changed or disappeared between when this work was written and when it is read. Neither the publisher nor authors shall be liable for any loss of profit or any other commercial damages, including but not limited to special, incidental, consequential, or other damages.

Library of Congress Cataloging-in-Publication Data applied for:

Paperback ISBN: 9781119829874

Cover design by Wiley
Cover image: © M.J. Everitt

Set in 9.5/12pt STIXTwoText by Straive, Chennai, India
Printed and bound by CPI Group (UK) Ltd, Croydon, CR0 4YY

C9781119829874_110124

Contents

Acronyms

BCH	Baker Campbell Hausdorff
BSIM	Berkeley short-channel IGFET model (IGFET – insulated gate field-effect transistor)
CL	Caldeira–Leggett
CLE	Caldeira–Leggett equation
CSCO	complete set of commuting observables
DC	direct current
DRY	don't repeat yourself
DY	dynodes
FORTRAN	formula translation (language)
GPS	global positioning system
IDE	integrated development environment
ISO	International Standards Organisation
KVN	Koopman–von Neumann
I/O	input–output
JJ	Josephson junction
LC	electrical oscillator circuits with L for inductance and C for capacitance
LCR	damped electrical oscillator circuits with L for inductance, C for capacitance and R for resistance
MA	measurement apparatus
MASER	microwave amplification by stimulation emission of radiation
MBSE	model-based systems engineering
MOT	magneto-optical trap
PDF	probability density function
PMT	photo-multiplier tubes
PNT	positioning, navigation, and timing
RSJ	resistivity shunted junction (model)
SOLID	the principles of: **S**ingle-responsibility; **O**pen–closed; **L**iskov substitution; **I**nterface segregation; **D**ependency inversion
TRL	technology-readiness level
QAA	Quality Assurance Agency
QBM	quantum Brownian motion
QHO	quantum harmonic oscillator

QO	quantum object
QSD	quantum state diffusion
R&D	research and development
RF	radio frequency
SHO	simple harmonic oscillator
SOI	system of interest
SQUID	superconducting quantum interference device
SW	Stratonovich–Weyl
TDD	test-driven development
TDSE	time-dependent Schrödinger equation
TISE	time-independent Schrödinger equation
UTF	unicode transformation format
V&V	verification and validation

About the Authors

MARK JULIAN EVERITT is a senior lecturer and director of studies in the Department of Physics at Loughborough University. His research interests are the foundations of quantum mechanics (such as measurement and the quantum to classical transition, currently with a focus on phase-space methods), open quantum systems, modelling and simulation of quantum systems and devices. Most recently he is leading an effort to establish the disciplines of quantum and low-TRL systems engineering.

KIERAN NIELS BJERGSTROM is director of a technology and strategy consultancy focused on innovative technology translation and commercialisation. He advises organisations globally on the strategic impact of quantum technologies, paths to implementation, and investment priorities. His academic background began in theoretical physics looking at realistically modelling open quantum systems. His PhD in Quantum Systems Engineering was the first on the topic.

STEPHEN NEIL ALEXANDER DUFFUS is a university teacher within the Physics Department at Loughborough University. He has an established reputation of communicating complex ideas in an engaging and accessible fashion. During his PhD, his main area of research was in open quantum systems.

Preface

This book has arisen from my having to write a new set of lecture notes for a second-year module on quantum mechanics at Loughborough University. Considering that very many textbooks on quantum physics already exist, it seems only right to provide some justification as to why there should be yet another. This lecture series is not simply an evolution of my previous teaching of quantum physics, it has been for me a fresh look at how to introduce the subject. The approach I wanted to use is one for which I was surprised not to find a suitable text for the module's reading list. Many who have taught quantum mechanics complain that the first time students typically encounter the Hamiltonian is in this subject. Frustratingly, it makes the subject seem more removed from classical physics than it really is. It also means that students have to wrestle with more new concepts than they should need to. Maybe as a corollary of this historical imperative, most current textbooks are written assuming an audience with no familiarity with Hamiltonian or Lagrangian mechanics. Further, students at this stage of their education tend not to have been provided an appreciation of the role of probability in classical mechanics, such as embodied in Liouville's theorem. As such, they are ill equipped to understand the importance of, and the subtleties in, the probabilistic interpretation of quantum mechanics. This, in turn, makes it difficult to have an open and honest discussion about the metaphysics of the subject. Due to the importance of each of these considerations, I wanted to be able to address all the above and more.

The possibility of making substantial improvements to our undergraduate teaching arose after I became involved in revisions of the regulatory framework for UK physics degrees[1] and at about the same time I became *Director of Studies* for Physics at Loughborough University. I took this opportunity to lead a major review of the Loughborough physics degree. Those of us involved in the review sought to put our provision on the firmest ground possible, leading to the first-year curriculum developing a very strong classical physics foundation. Our first year now covers the analytical mechanics topics referred to above and culminates in presenting the Lagrangian for the electromagnetic field in its covariant form (as well as much-expanded mathematics, laboratory, and computational content). As a result of these changes, we had achieved my ambition to engineer an opportunity to teach quantum physics in a way that makes a strong connection to the framework of classical analytical classical mechanics. Much of this work is the outcome of exploring some of

---

1 Participating in a review of the QAA subject benchmarks statements and Institute of Physics Accreditation.

the various paths that this background enabled. In this respect, the Schrödinger picture, and later the Heisenberg picture, Feynman path integrals, and the Wigner phase-space approach are all introduced and motivated from formal classical mechanics. This is done because each represents an independent pathway into the subject and is best understood in its own right, rather than being derived from each other (although that interlinking is presented subsequently).

Later chapters focus on the consequences of the formalism, some difficulties of quantum theory (such as the measurement problem, interpretations, and the quantum-to-classical transition), applications and problem-solving, including computational methods (the tone of this chapter was set from my experience teaching a first-year computational physics module). We close with a chapter on quantum systems engineering. I started to develop the idea for this subject in 2014, based on a coffee-shop conversation with Vincent Dwyer on reliability engineering (for which I am eternally grateful). With him and many others (such as Kieran Bjergstrom and Michael Henshaw), we have been developing this subject for a number of years. Even though it is still in its infancy, we believe the ideas are mature and valuable enough to deserve a chapter in a book. As with the computational methods chapter, many of the ideas we present here are far more wide-ranging than quantum mechanics, it is just that quantum mechanics brings with it some unique demands that make the subject even more fascinating. This last chapter contains some rather important ideas, even for the theorist. I am becoming increasingly of the view that systems thinking and systems methods can lead to a better way of doing science, not just quantum physics. Making clear the requirements of a model and/or experiment, the acceptance, verification, and validation criteria not only embody the scientific methods but also formalise and systematise the approach, making it less likely that we will overlook assumptions and draw incorrect conclusions from our work.

There are several target audiences for this book:

- To support a first or second course on quantum mechanics following or co-teaching with a course on Hamiltonian and Lagrangian mechanics (Chapters 1–7).
- To support advanced courses on quantum mechanics where coding, open systems, philosophy, or engineering is included.
- To introduce the subject to mathematics majors, who often study the required classical mechanics in courses on dynamical systems or analytical mechanics.
- At master's level (potentially as part of integrated PhD programmes) which often contains advanced conversion course elements (for those switching disciplines) in countries like the UK and US.
- For faculty designing new (or redesigning existing) quantum mechanics courses, where the existence of a text such as this might better facilitate their preferred teaching of the subject over existing texts.
- As a 'reader' in quantum mechanics for the self-motivated advanced student.

Loughborough University   *Mark Julian Everitt*
August, 2023

Acknowledgements

I first and foremost need to thank my co-authors. I was very pleased to be joined in writing this book by two of my former PhD students, Steve and Kieran. It is a source of immense pleasure when one's students go on to be better than one self in subjects that you taught them. The decoherence chapter was led by Steve and based on much of the PhD work of himself and Kieran (my contribution was limited to the narrative and unravellings content). Steve has enjoyed immense success teaching quantum physics to our third-year students over the past few years. We combined our experience to develop the methods and applications chapters. Kieran and I (and others) have been working on developing the subject of quantum systems engineering for some time now. He has a rare talent in his ability to encapsulate and articulate the complexities of this subtle discipline. Working on the last chapter together developed some existing, and stimulated a number of new, insights from us both that we hope the reader will enjoy and will find pleasantly surprising. I also appreciate both of their substantial efforts in editing and proofreading which provided a level of coherence that was well needed in this work. In this last and most challenging of tasks we were very pleased to receive the help of Niels Bjergstrom, whose gargantuan effort resulted in a well-constructed text. For their general input and advice on the first few chapters, I am very grateful to Sasha Balanov and John Samson. Sasha, together with Adam Thompson, Mark Greenaway, and Thomas Steffen, provided very valuable feedback on the Computational Simulation of Quantum Systems chapter (with Adam's professional programmers' eyes being especially useful in ensuring that key topics such as logging were included). We are also very grateful to Finlay Potter for his careful student perspective, careful reading, and valuable observations on the open-systems chapter.

In addition to this, I am deeply grateful to a number of people who have helped deepen my understanding of, and perspectives on, quantum physics in different ways over many years. This includes, but is not limited to: Terry Clark for being my mentor for many years and showing me what commitment to science means; Jason Ralph for many a long conversation on the foundations of quantum mechanics; Todd Tilma, Russell Rundle, John Samson (again), and Ben Davies for adventures in phase space; Tim Spiller, Bill Munro, Kae Nemoto, Ray Bishop, and Alex Zagoskin for always asking good, hard, interesting questions; Vincent Dwyer, Michael Henshaw, Jack Lemon, Laura Justham, Simon Devitt, and Susannah Jones for exploring new ways of engineering quantum systems (as well as my father, who as a professional systems engineer, influenced my world view to naturally accommodate a holistic view); to Pieter Kok, Dan Browne, Derek Raine, and Antje Kohnle for many interesting conversations on how to teach quantum mechanics (especially Pieter's well-voiced issues with unphysical beams in diagrams). Last, but not least, I need to acknowledge the support of my family, Lennie, Alex, and Sophie, who have allowed me the time to work on this text (and who will be seeing much more of me in the future).

Mark Julian Everitt

About the Companion Website

This book is accompanied by the companion website:

http://www.wiley.com/go/everitt/quantum

The website includes: Repository to be discussed in Chapter 8.

Introduction

This book introduces quantum mechanics from the perspective of classical analytical mechanics. The idea of this approach is to highlight more clearly the similarities of quantum and classical mechanics, as well as the differences. As a consequence, we believe that this makes the theoretical framework of the theory more intuitive where this is possible. Our intent is that the actual peculiarities of quantum mechanics are more clearly identified than in treatments where, e.g. the relationship between the dynamics of classical and quantum probability densities is considered.

Chapters 1–4 were written to be studied in order, subject to the various caveats listed below. There is more flexibility in the order in which later material can be studied. The content up to and including Chapter 7 forms the basis of what we consider a first course in quantum mechanics, with the remaining chapters constituting either additional reading or supporting more advanced courses.

In Chapter 1 we present most of the mathematics used in the text, introducing core concepts and notations (such as that of Dirac). We have found that some students can struggle to become comfortable with such notation. Separating the concerns of mathematics from the physics we find improves the understanding of both the mathematics and the physics. Presumably this is because it reduces the number of concepts that students need to think about at any one time. Our approach has the added advantage that we can introduce concepts such as the Heisenberg uncertainty principle in their general mathematical form. Such results have utility beyond quantum mechanics, which this presentation makes clear. In the specific case of the uncertainty principle, the separation of concerns enables us to introduce the result without incorrectly confusing it with ideas of measurement disturbance. This chapter might either be studied entirely on its own before engaging with the rest of the text, or it may be interleaved with the content of the rest of the book. Especially in early chapters, we cross-reference sections back to the prerequisite mathematical material contained within this chapter to enable either approach to be taken.

As some knowledge of Hamiltonian and Lagrangian mechanics is required, Chapter 2 provides a self-contained introduction to the subject. Even if you are familiar with most of this material, it is worth reading, as it contains important material on a formulation of classical mechanics due to Koopman and von Neumann that greatly helps in motivating the Schrödinger equation. Historically, the idea was to make classical mechanics look like quantum mechanics. We have chosen to present the ideas ahistorically, as it is an odd but not conceptually difficult jump to move from the Liouville equation to

Koopman–von Neumann classical mechanics. Once we have Koopman–von Neumann classical mechanics, the Schrödinger equation does not seem anywhere near as surprising as it might do without this context. This chapter finishes with a discussion of the breakdown of classical mechanics and the correspondence principle (some understanding of which is important at this stage, as it allows us to understand the flexibility we have in formulating new theories consistently with existing ones).

In Chapter 3 we introduce the Schrödinger formulation of quantum mechanics. We try to do this in a way that *makes as few assumptions as possible.* This leads to a somewhat lengthy discussion, but it is one that draws out the key assumptions and issues of the Schrödinger picture (such as that the initial state is all that is needed to determine a system's evolution, just like in the Liouville equation). We motivate the Schrödinger equation as being similar in form to the Koopman–von Neumann equation of motion, but where we move to an operator formalism that replaces some Poisson brackets with commutation relations. We make only a few assumptions about the state (i.e. it is a vector in a vector space), and our discussion progresses through measurement axioms before discussing that the representation of a quantum system is a matter of choice and that, e.g. the wave function is just the position representation of a quantum state. This discussion allows us to make clear the axioms associated with dynamics and measurement, and separate these clearly from the ideas of representation.

We then, in Chapter 4, look at some alternative paths into quantum mechanics, the Heisenberg, Wigner phase space, and (a very brief introduction to) Feynman path integral formulations. We do not derive these from the Schrödinger formulation, but instead re-argue from first principles. We do this for two reasons: (i) to show that they are not subordinate to the Schrödinger view, and (ii) because they can be more naturally argued for in a way that helps develop our discussion of the similarities as well as the differences between quantum and classical physics. In fact, there is a strong case for arguing that either of these pictures might be a more natural starting point for developing the subject of quantum physics. We chose to start with the Schrödinger picture, simply because this is the dominant one in most of the textbook and research literature. We think it would be relatively straightforward to mix the Heisenberg picture discussion of Chapter 4 with elements of Chapter 3 to form an alternative opening to the subject.

Chapter 5 introduces vectors and angular momentum and is somewhat unusual in so far as we include an extensive discussion of curvilinear coordinates in quantum mechanics. Our reason for doing this is that many classical mechanics problems become easier to treat by using curvilinear coordinates, if that suits the symmetry of the problem. Such simplification is not seen in quantum mechanics. One of our key aims of this text is to highlight as clearly as possible the similarities and differences between quantum and classical physics. For problems with spherical symmetry, simplification is actually found through an analysis of angular momentum, and we wanted to explain why this is in fact the case. One advantage of this introduction to the subject is that the expression of the kinetic energy operator in terms of radial and angular momentum components does arise naturally (in the usual treatment this is discovered through analysis of the three-dimensional kinetic and angular momentum operators in the coordinate representation which, while effective, lacks elegance). Those not interested in that level of detail can happily skim-read most of Section 5.2. The remainder of the chapter contains the theory needed to understand really important applications, such as the quantum physics of hydrogen. Note that it is often the case that the harmonic oscillator is introduced before angular momentum. We have chosen not to do this, as angular momentum is part of the general theory and the harmonic oscillator is simply a very important example.

The harmonic oscillator discussion in Section 7.3 is not predicated on the content in Chapter 5 and can be studied beforehand.

Chapters 6 and 7 contain methods and applications and important examples such as hydrogen, molecules, and the Jaynes–Cummings model. The order of study can be somewhat flexible.

For the modern physicist, computation has become as important as mathematics for many tasks, for example enabling the solution of problems with no analytic solution. Many texts cover technical aspects of algorithm design pertinent to scientific computing. While there is some literature [93] on good practice in scientific computing, there is a limited amount of textbook resources for physics. In Chapter 8 our focus is on good practice and design principles using quantum physics as an example. In a world where artificial intelligence is getting better at writing routine code, it is these higher-level skills that the physicist will require more and more. This chapter can equally well be used to support either a quantum mechanics module with coding elements, or a coding course where quantum mechanics would provide valuable example applications[1].

In Chapter 9 we provide an introduction to open quantum systems. While there is already substantial literature on this subject, it is a challenging one. Based on our undergraduate teaching and project supervision, our intention is to make this material as accessible as possible, expanding on those areas where our experience has found that students need assistance with the material presented in the existing literature.

Many treatments of quantum mechanics set down a philosophical interpretation of the subject early on. It is one of the main discussion points of the theory that there are multiple interpretations of quantum mechanics and that the subject contains unresolved metaphysical issues. In Chapter 10 we take advantage of much of the preceding content of the book to have an in-depth, open, and honest discussion on some aspects of the foundations of quantum mechanics. Our intent is to stimulate thought and discussion rather than present a single perspective. Interestingly, a discussion of the measurement problem is pertinent to the verification and validation issues that are presented in the final chapter. This provides an interesting link between the very philosophical foundations of quantum mechanics and the very applied goal of engineering quantum systems.

Finally, in Chapter 11 we turn to a general discussion of some challenges associated with the engineering of quantum technologies. We introduce an approach to quantum systems engineering that we believe has value, not just for quantum applications, but to physics as a whole – from the scientific method to technology development[2]. While it has a place in experimental research physics and the scientific method, we avoid extensive discussion of established or modified-for-laboratory systems engineering tools that also apply to

---

1 Much of the conceptual content of this chapter appears in the Loughborough first-year course on computational physics but with classical physics rather than quantum mechanical applications.
2 For this reason, some of this subject matter, as well as some conventional systems engineering, has been strategically introduced into the Loughborough undergraduate curriculum. We find the latter is especially useful in support of laboratory and group project work.

non-quantum technologies. Our discussion centres on the different engineering needs of quantum from non-quantum systems, and why this results in challenges that a quantum physicist might enjoy. This is an emerging field at the forefront of innovation, and we caveat our writings as something of a personal, evolving, perspective

Throughout this document, the following boxes denote different types of materials, as follows:

Normal notes will look like this

Really important notes and observations will look like this

Prerequisite Material: References to needed material or cross-referencing will look like this.

1

Mathematical Preliminaries

"Physics is mathematical not because we know so much about the physical world, but because we know so little; it is only its mathematical properties that we can discover."

Bertrand Russell

"The mathematician plays a game in which he himself invents the rules while the physicist plays a game in which the rules are provided by nature, but as time goes on it becomes increasingly evident that the rules which the mathematician finds interesting are the same as those which nature has chosen."

Paul Dirac

1.1 Introduction

Different people learn in different ways. Some like to study the mathematics they need in some depth before looking at the physics; others prefer to mix and learn both subjects together, a bit at a time. People will also begin their journey into quantum mechanics with different levels of mathematical knowledge. For this reason, the core mathematics needed for the study of quantum mechanics has been organised here in a single chapter, with cross-referencing in later chapters to sections in this chapter. You may wish to study this chapter in depth before proceeding to the Physics that follows. Alternatively, you may wish to skip this chapter and refer back to it as you discover new and unfamiliar mathematics (or skim-read it, just to see what is here). Our aim is to facilitate your study of this material in a way that best suits you.

Quantum mechanics makes use of several branches of mathematics. It does, however, focus on one branch above most others, which centres on a generalisation of vectors, dot products, and linear operators on those vectors. In mathematics, we use the term space to refer to a set (collection of mathematical objects) together with some additional properties. Hence we will study vectors in a vector space where the additional properties are vector addition and multiplication by a scalar. We will also study vectors in an inner product space which is a vector space but with the additional property of a generalisation of the dot product called the inner product. We intend the mathematics we introduce in this chapter to be

Quantum Mechanics: From Analytical Mechanics to Quantum Mechanics, Simulation, Foundations, and Engineering, First Edition. Mark Julian Everitt, Kieran Niels Bjergstrom and Stephen Neil Alexander Duffus.
© 2024 John Wiley & Sons Ltd. Published 2024 by John Wiley & Sons Ltd.
Companion website: www.wiley.com/go/everitt/quantum

sufficiently detailed to enable an understanding of the physics that follows. The treatment is only at this level and, for the sake of brevity, is not a complete treatment of the subject. The language of quantum mechanics is framed in terms of vectors in an extension of inner product space called a Hilbert space. This constrains the space used to describe quantum systems by additional specific mathematical properties[1].

To be consistent with the wider literature, we will often use the term Hilbert space when we only need the properties of an inner product space[2]. Most of our focus will be on the mathematics to do with Hilbert spaces, but some other topics, such as expectation values of random variables, will be briefly introduced. This means that the flow of the text in this chapter will not be as seamless as in more comprehensive treatments. As our approach removes the need to interleave mathematics and physics, better emphasising physics arguments, we hope that you will find this a worthwhile compromise. It is worth noting that while our focus is quantum mechanics, the mathematics introduced in this chapter is applicable to many other subjects. Hilbert spaces are very useful for the study of signals and for gaining a deep understanding of Fourier analysis. A specific benefit of using this approach is that we can introduce the Heisenberg uncertainty principle as a purely mathematical proposition and demonstrate that it has much more general applicability than its usual quantum physics application would imply. The important consequence of doing this is that it extracts from the argument the often incorrectly made assertions about measurement disturbance (we will properly understand what Heisenberg's uncertainty principle actually means for quantum physics when we discuss its realisation in the phase space methods section of Chapter 4). Even if you already have knowledge of vector spaces and functional[3] analysis, this chapter will be of value because it provides some physics context and notation.

1.2   Generalising Vectors

1.2.1   Vector Spaces

Powerful mathematical methods and principles are often developed by identifying shared features of different mathematical objects and exploring the consequences of a generalisation. The advantage of this approach is that we can start with something that is easy to understand (in this case arrows on a plane) and learn the consequences of combined properties of that system (lemmas, propositions, and theorems), which we can then apply to other systems that share those same properties. For instance, the Cartesian co-ordinates (x, y, z), phase space coordinates $\{q_i, p_i\}$, functions (including probability density functions – which shall be introduced later, but are essentially what you would think from the name), sequences, and many other things share a certain set of properties in common with vectors. This means that the theorems that hold for arrows-on-a-plane also hold for all these other

---

1 Such that, for example, sequences defined in an appropriate way converge to elements that are also within the vector space.
2 For those interested in specific details, an excellent introduction can be found in Kreyszig's *Introductory Functional Analysis* text [46] (be warned though – as a pure mathematics text this does use substantially different notation to the one used here).
3 A functional is a particular kind of mapping from a space of mathematical objects into the real or complex numbers.

mathematical objects. We will later see that we use vectors and functions to hold the information that characterises a system. Therefore, generalising vectors will prove to be of central importance to quantum mechanics, and a good understanding now will prepare us for the discussions that follow.

We start by identifying those features and abstracting them into a list of properties (called axioms) that describe the generalised mathematical concept of a vector. Note that, where possible, we have provided a diagrammatic example of each axiom to emphasise the reductionist nature of the approach. In general, vectors $\boldsymbol{u}$, $\boldsymbol{v}$ and $\boldsymbol{w}$ with scalars (complex or real numbers) α and β have the following properties:

- Adding two vectors together results in another vector, i.e. if $\boldsymbol{u}$ and $\boldsymbol{v}$ are vectors, then $\boldsymbol{u} + \boldsymbol{v}$ is also a vector. In terms of arrows, this is the notion of putting the tail of the arrow $\boldsymbol{v}$ on the head of arrow $\boldsymbol{u}$ – the resultant vector connects the tail of $\boldsymbol{u}$ to the head of $\boldsymbol{v}$. The mathematical term for this is that the vector space is closed under addition.

- Multiplying a vector $\boldsymbol{u}$ by a scalar α gives a new vector $\alpha\boldsymbol{u}$ (closed under scalar multiplication).

  - If the scalar is real, then this is just a scale-factor that stretches or shrinks the vector by that amount (i.e. multiplication of the length of the vector).

    Note that a negative scaling will flip the arrow into the opposite direction.
  - If the scalar is complex, then think of this as scaling the length by $|\alpha|$ and adding the phase $\arg(\alpha)$. This can be thought of as a scale and a twist in the complex plan orthogonal to the vector as the direction of the vector does not change.
- Commutativity of vector addition, which states that the resultant vector from adding two vectors together does not change if you change the order of the sum:

  $$\boldsymbol{u} + \boldsymbol{v} = \boldsymbol{v} + \boldsymbol{u}.$$

- Associativity of vector addition, $(\boldsymbol{u} + \boldsymbol{v}) + \boldsymbol{w} = \boldsymbol{u} + (\boldsymbol{v} + \boldsymbol{w})$, which states that it does not matter which order three vectors are added.

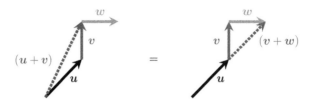

- There is a vector of zero length, commonly notated as $\boldsymbol{0}$, such that $\boldsymbol{u} + \boldsymbol{0} = \boldsymbol{u}$ for any $\boldsymbol{u}$. This is referred to as the null vector or zero vector. Note that we choose not to use $\mathbf{0}$, as we will want to use 0 later for another purpose.
- For any vector $\boldsymbol{u}$, there exists another vector, denoted $-\boldsymbol{u}$ (inverse element), such that the sum $\boldsymbol{u} + (-\boldsymbol{u}) = \boldsymbol{0}$. Note that as only addition is defined (tail of one arrow on the head of another arrow) subtraction does not formally need to exist – but we do use the shorthand $\boldsymbol{u} - \boldsymbol{v} \stackrel{\text{def}}{=} \boldsymbol{u} + (-\boldsymbol{v})$.
- Compatibility is satisfied:

$$\alpha(\beta\boldsymbol{u}) = (\alpha\beta)\boldsymbol{u}$$

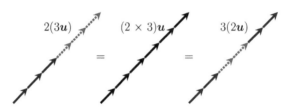

- Multiplication by unity, commonly the number one, leaves a vector unchanged (multiplicative identity):

$$1\boldsymbol{u} = \boldsymbol{u}.$$

- Distributivity across vector sum:

$$\alpha(\boldsymbol{u} + \boldsymbol{v}) = \alpha\boldsymbol{u} + \alpha\boldsymbol{v}.$$

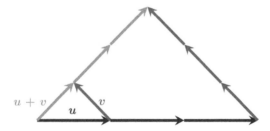

- Distributivity across scalar sum:

$$(\alpha + \beta)\boldsymbol{u} = \alpha\boldsymbol{u} + \beta\boldsymbol{u}.$$

Any example of mathematical objects that satisfies all of the above conditions can be considered a kind of vector. Furthermore, a *set of objects* satisfying the conditions *together with the rules of addition and multiplication by a scalar* is called a vector space. Examples of vector spaces include: complex numbers, vectors (obviously), matrices, co-ordinates [can be seen as $(x_1, y_1, z_1) + (x_2, y_2, z_2) = (x_1 + x_2, y_1 + y_2, z_1 + z_2)$ and $\alpha(x, y, z) = (\alpha x, \alpha y, \alpha z)$], sequences (with elementwise addition), series, linear equations, and last, but by no means least, functions [seen by considering $(f + g)(t) = f(t) + g(t)$ and $(\alpha f)(t) = \alpha f(t)$]. To provide a concrete example: if $f(t) = \sin(n\pi t/L)$ and $g(t) = \cos(n\pi t/L)$, then the sum of these functions is also a function, as is a number times the function. It is this analogy to vectors that explains how functions can act as a basis (with this example pertaining to Fourier analysis).

More importantly, this means that *any theorem we prove for vectors, as defined above, will hold for all other things that satisfy these conditions too*!

The co-ordinates of a physical system in phase space $\{q_i, p_i\}$ are vectors in a vector space.

1.2.2 Inner Product

We are familiar with the idea of the dot product $\boldsymbol{u} \cdot \boldsymbol{v}$. Having now observed that we can generalise the notion of a vector to other things, we can also generalise the notion of the dot product to any vector space.

The dot product has the following important properties:

- $(\boldsymbol{u} + \boldsymbol{v}) \cdot \boldsymbol{w} = \boldsymbol{u} \cdot \boldsymbol{w} + \boldsymbol{v} \cdot \boldsymbol{w}$
- $\boldsymbol{u} \cdot (\alpha \boldsymbol{v}) = \alpha \boldsymbol{u} \cdot \boldsymbol{v}$
- $\boldsymbol{u} \cdot \boldsymbol{v} = (\boldsymbol{v} \cdot \boldsymbol{u})^*$ (where $*$ indicates complex conjugate)
- $\boldsymbol{u} \cdot \boldsymbol{u} \geq 0$, and $\boldsymbol{u} \cdot \boldsymbol{u} = 0$ if and only if $\boldsymbol{u} = \boldsymbol{0}$.

Exercise 1.1 Show that these relations hold for three-element column vectors where $\boldsymbol{u} \cdot \boldsymbol{v} \overset{\text{def}}{=} (\boldsymbol{u}^*)^T \boldsymbol{v}$ (where superscript T means transpose). ∎

The third condition is needed to ensure $\boldsymbol{u} \cdot \boldsymbol{u}$ is real. As such, we can interpret $|\boldsymbol{u}| = \sqrt{\boldsymbol{u} \cdot \boldsymbol{u}}$ as a length in a way that corresponds to the normal notion of Euclidean distance.

The first and third conditions imply:

$$\boldsymbol{u} \cdot (\boldsymbol{v} + \boldsymbol{w}) = \boldsymbol{u} \cdot \boldsymbol{v} + \boldsymbol{u} \cdot \boldsymbol{w},$$

which is sometimes, unnecessarily, listed separately.

From the second and third conditions, $(\alpha \boldsymbol{u}) \cdot \boldsymbol{v} = \alpha^* \boldsymbol{v} \cdot \boldsymbol{u}$.

The generalisation of the dot product is termed an *inner product* or *scalar product* and is denoted by $\langle u|v\rangle$. Other notations for the inner product do exist in the literature, which all look different (and may alter the order of u and v) but mean the same thing. The notation $u \cdot v$ is historically reserved for standard vectors and Cartesian coordinates. We use $\langle u|v\rangle$, as it is the historical origin for a notation developed by Dirac, which we are soon to discuss – and which is needed to be able to properly engage with the literature of quantum mechanics.

As with the dot product, the inner product takes the form of a mapping (strictly speaking a sesquilinear form[4]) of any two vectors (of the same kind but no longer restricted to traditional vectors) to a number, and satisfies the same properties as the dot product:

$$\langle u + v|w\rangle = \langle u|w\rangle + \langle v|w\rangle \tag{1.1}$$

$$\langle u|\alpha v\rangle = \alpha\langle u|v\rangle \tag{1.2}$$

$$\langle u|v\rangle = \langle v|u\rangle^* \tag{1.3}$$

$$\langle u|u\rangle \geq 0 \text{ and } \langle u|u\rangle = 0 \iff u = \theta. \tag{1.4}$$

As with the dot product, the third condition is needed as it ensures $\langle u|u\rangle$ is real. As such, we can interpret $\|u\| = \sqrt{\langle u|u\rangle}$ as a length, which means we can do interesting things such as ask: what is the distance between two functions? We call $\|u\|$ the norm of u. The norm generalises the notion of the modulus of a vector.

The first and third conditions imply that:

$$\langle u|v + w\rangle = \langle u|w\rangle + \langle u|w\rangle.$$

The second and third conditions imply that $\langle \alpha u|v\rangle = \alpha^* \langle u|v\rangle$, just as for the dot product.

Two important examples of the inner product are:

- The usual dot product for standard vectors

$$\langle u|v\rangle = u \cdot v = \sum_i u_i^* v_i \tag{1.5}$$

(so we retain all the good maths that we know to work).
- And, for complex valued functions of a real parameter:

$$\langle f(x)|g(x)\rangle = \int_{-\infty}^{+\infty} dx f^*(x) g(x), \tag{1.6}$$

which means we can apply many ideas that we understand from the geometry of vectors to functions. The lemma (a small theorem) we consider next is an excellent and important

---

4 A sesquilinear form (from Latin: *sesqui*, 'one and a half') is a generalisation of a bilinear form that, in turn, is a generalisation of the concept of the dot product in Euclidean space. A bilinear form is linear in each of its arguments, but a sesquilinear form allows one of the arguments to be 'twisted' in a semilinear manner, thus the name.

example of this, and this will become especially important when we begin to use functions to describe physical systems.

> When integrating, our usual notation is to put the differential of the variable before the integrand. This is a standard notation, but not one you may have seen before. The reason we choose to do this is that it enables us to see which variable we are integrating against from the outset. This can be very useful when things get complicated.

> Introducing an inner product allows us to make further analogies to spatial vectors such as the angle between any two vectors u and u by $\cos\theta = \Re(\langle u|v\rangle)/\|u\|\|v\|$. We will say the two vectors *orthogonal* if $\langle u|v\rangle = 0$.

The Cauchy–Schwarz Inequality states that:

$$\langle u|u\rangle \langle v|v\rangle \geq |\langle u|v\rangle|^2. \tag{1.7}$$

We will later see that the proof of Heisenberg's uncertainty relation relies on this inequality in its derivation, so it is rather important.

For vectors with a dot product, we know that the inner product, $\langle u|v\rangle = u \cdot v$, is just the amount of u in the direction of v, and $\langle u|u\rangle = \|u\|^2$. So this is a generalisation of the geometrical idea that the (square or the length of vector u) times the (square or the length of vector u) is greater than, or equal to, the (square of the amount of u in the direction of v). Specifically, for the dot product, we have $|u|^2|v|^2 \geq ||u||v|\cos(\theta)|^2$ (where θ is the angle between u and v). So the result is perhaps not surprising for traditional vectors as $|\cos(\theta)| \leq 1$ – it is its generalization that makes it interesting.

As a specific example for functions, using the definition of the inner product provided earlier, the result

$$\int_{-\infty}^{+\infty} dx\, |f(x)|^2 \int_{-\infty}^{+\infty} dx\, |g(x)|^2 \geq \left|\int_{-\infty}^{+\infty} dx f^*(x) g(x)\right|^2$$

is perhaps less obvious.

Exercise 1.2   Prove the Cauchy–Schwarz Inequality just using the features (axioms) of the generalised vector and inner product. Hints: where is the inequality in the definition of the inner product? - and consider $w = u + \lambda v$. ∎

1.2.3 Dirac Notation

The way we denoted the inner product introduces the possibility of a new notation for signifying vectors in an inner product space. The approach has the immediate advantage of indicating that such vectors have an inner product associated with them, so we can make use of theorems such as the Cauchy–Schwarz inequality. It is called Dirac notation and comes from noticing that the inner product has a $\langle$**bra**|c|**ket**$\rangle$ shape. If we consider the inner product $\langle u|v\rangle$, we can split it up into $\langle u|\ |v\rangle$ and consider the two sides as entities in their own right: a bra $\langle u|$ and a ket $|v\rangle$. Note that we have dropped the use of bold to denote 'vectors' as this is also convention. Where boldface is used in the following discussion, you can assume it refers to a conventional column vector.

Here, we consider that the ket $|v\rangle$ is just the 'vector' itself. In Dirac notation the axioms (the features that we identified as common in our generalisation of the arithmetic of arrows on a plane) for a vector space now read:

- Commutativity of addition:

$$|u\rangle + |v\rangle = |v\rangle + |u\rangle.$$

- Associativity of addition:

$$(|u\rangle + |v\rangle) + |w\rangle = |u\rangle + (|v\rangle + |w\rangle).$$

- There is a vector of zero length commonly denoted $|0\rangle$ such that $|u\rangle + |0\rangle = |u\rangle$ for any $|u\rangle$.
- For any $|u\rangle$ there exists another vector $-|u\rangle$ (inverse element) such that $|u\rangle + (-|u\rangle) = |0\rangle$ (we use the shorthand $|u\rangle - |v\rangle \stackrel{\text{def}}{=} |u\rangle + (-|v\rangle)$).
- Compatibility is satisfied:

$$\alpha\,(\beta\,|u\rangle) = (\alpha\beta)\,|u\rangle.$$

- Multiplicative identity:

$$1\,|u\rangle = |u\rangle.$$

- Distributivity across a vector sum:

$$\alpha\,(|u\rangle + |v\rangle) = \alpha\,|u\rangle + \alpha\,|v\rangle.$$

- Distributivity across a scalar sum:

$$(\alpha + \beta)\,|u\rangle = \alpha\,|u\rangle + \beta\,|u\rangle.$$

We also can write $\alpha\,|u\rangle = |\alpha u\rangle$ so long as it is clear that u labels the vector and α is a scalar. Please compare these directly with the examples for the properties of 'normal' vectors explained earlier. Convince yourself that we followed exactly the same reasoning here, but added a bracket around the vector instead of using a bold typeface (or putting an arrow over the top) – as *this is all we have done to reach Dirac notation.*

While getting used to Dirac notation it may help to informally think of the bra and ket using the following analogy, if

$$|u\rangle \sim \begin{bmatrix} u_1 \\ u_2 \end{bmatrix} \text{ then } \langle u| \sim \boldsymbol{u}\cdot = (\boldsymbol{u}^*)^T = \begin{bmatrix} u_1^* & u_2^* \end{bmatrix}.$$

In this way we can see that the bra can be thought of as a vector with a dot product waiting to happen. Because of the fact the inner/dot-product is linear the bra behaves just like a vector, so can be thought of as being a vector in its own right (just as $\boldsymbol{u}\cdot$ behaves like a vector). So, the if bra version of one ket vector acts on another ket it will transform that ket into a scalar in just the same way as one can multiply row and column vectors. In other words we make the analogy,

$$\langle u|v \rangle = ((\langle u|)\,|v\rangle) \sim (\boldsymbol{u}\cdot)\boldsymbol{v} = (\boldsymbol{u}^*)^T \boldsymbol{v} = \begin{bmatrix} u_1^* & u_2^* \end{bmatrix} \begin{bmatrix} v_1 \\ v_2 \end{bmatrix}.$$

Alternatively, if we are dealing with a function space, the bra is just like

$$\langle f(x)| \sim \int_{-\infty}^{+\infty} dx\, f(x)^* [\text{'ket' function goes here}].$$

Exercise 1.3 Show that the set of bras also satisfy all the axioms of a vector space and can therefore also be considered vectors in their own right. Hint: The argument for commutativity of addition is to take $\langle u| + \langle v|$ and post multiply by some arbitrary ket $|w\rangle$ and by inner product rules

$$(\langle u| + \langle v|)\,|w\rangle = \langle u|w \rangle + \langle v|w \rangle$$

$$= \langle v|w \rangle + \langle u|w \rangle \text{ (as inner products are just numbers)}$$

$$= (\langle v| + \langle u|)\,|w\rangle$$

as we made no assumptions about $|w\rangle$ so $\langle u| + \langle v| = \langle v| + \langle u|$. ∎

From the preceding exercise we see that to every ket in a vector space we can associate a bra vector through the inner product so that

$$\alpha\,|u\rangle \xleftarrow{\quad\text{Correspondence}\quad} \alpha^*\,\langle u|$$

and we can treat bras just like vectors. In some cases, there may even be more bra than ket vectors. For instance, you may recall that the Dirac delta $\delta(x)$ function not really a function (it's the limit of a distribution) and it is only well-defined under an integral – so there is strictly speaking no ket for $\delta(x)$. But there is a bra vector as the following integral is well-defined:

$$\langle \delta(x)| \sim \int_{-\infty}^{+\infty} dx\, \delta(x)^* [\text{'ket' function goes here}].$$

The set of bras is called the dual space and it may be bigger than the vector space itself.

1.2.4 Basis and Dimension

In this section we generalise the idea of basis vectors such as $\mathbf{i}$, $\mathbf{j}$, and $\mathbf{k}$. Just as with conventional vectors, any vector $|u\rangle$ can be represented in terms of a basis set of vectors $\{|b_i\rangle\}$ as a linear combination

$$|u\rangle = \sum_i c_i\,|b_i\rangle,$$

where the c_i are suitable coefficients. A basis is said to be *complete* if all possible vectors (in the vector space) can be represented by that basis. It is said to be *orthogonal* if each basis vector is perpendicular to every other basis vector, and *orthonormal* if each basis vector is also of length one. Most bases you will encounter in this book are complete and orthonormal. Note that continuous bases are also possible, and in such cases the series is replaced by an integral.

The minimum number of basis vectors needed to represent any vector in a space is the dimension of the space, which may in some cases be infinite. An example of an infinite basis is the sines and cosines in the Fourier expansion of a function.

As with Cartesian systems, the situation is greatly simplified if the vectors in the basis are orthogonal (just like $\mathbf{i}$, $\mathbf{j}$, and $\mathbf{k}$), and each basis vector is of length one. For orthonormal bases, we often use the standard notation:

$$|u\rangle = \sum_n u_n |e_n\rangle \text{ where } \langle e_n|e_m\rangle = \delta_{nm}$$

(in the continuous case, the Kronecker delta is replaced by a Dirac delta function). The vector $\mathbf{u}$ with elements $u_n = \langle e_n|u\rangle$ is a representation of $|u\rangle$ in that basis (we will consider this point in much more detail in Section 1.4).

While we can make an analogy of vector space basis vectors to $\{\mathbf{i}, \mathbf{j}, \mathbf{k}\}$, spatial basis vectors in quantum mechanics are a different kind of vector and the two should not be confused. When we later consider angular momentum in quantum mechanics, we will discuss Cartesian and spherical polar coordinates as examples of spatial bases. For non-Cartesian coordinate systems, quantum-classical analogies are not as simple as one might expect. For example, defining an (unnormalised) spatial basis vector for the coordinate q_i as

$$\mathbf{e}_i = \frac{\partial \mathbf{r}}{\partial q_i}$$

has some subtle difficulties if its application is attempted in quantum mechanics (if you have not seen this notation before it might help to note, e.g. that $\frac{\partial(x,y,z)}{\partial x} = (1,0,0) = \mathbf{e}_x = \mathbf{i}$ as expected).

1.3 Linear Operators

1.3.1 Definition and Some Key Properties of Linear Operators

Recall that if a matrix acts on a vector it creates another vector,

$$\mathbf{Au} = \mathbf{v}.$$

In this way, a matrix encodes the operations of rotating and scaling a vector. In the same way we generalised vectors and the dot product, we may also generalise the notion of a matrix. Such a generalisation is termed a (linear) operator, and its definition is motivated by the arithmetic of matrices as follows (the 'hat' denotes that an object is an operator):

- Operators transform vectors to other vectors (of the same kind),

$$|\hat{X}u\rangle = |u'\rangle.$$

- Linearity means that for vectors $|u\rangle$ and $|v\rangle$ and a scalar α,

$$|\hat{X}(u+v)\rangle = |\hat{X}u\rangle + |\hat{X}v\rangle$$

and

$$|\hat{X}\alpha u\rangle = \alpha|\hat{X}u\rangle.$$

- The operators $\hat{X}$ and $\hat{Y}$ are equivalent operators if, for all $|u\rangle$, the following holds:

$$|\hat{X}u\rangle = |\hat{Y}u\rangle.$$

This is true even if $\hat{X}$ and $\hat{Y}$ *look* different – operators are defined by their *effect* on vectors.
- As with matrices, operators are
  - commutative under addition:

$$\hat{X} + \hat{Y} = \hat{Y} + \hat{X}$$

  - have associativity of addition:

$$(\hat{X} + \hat{Y}) + \hat{Z} = \hat{X} + (\hat{Y} + \hat{Z})$$

  - have associativity under multiplication (e.g. composition of rotation matrices):

$$(\hat{X}\hat{Y})\hat{Z} = \hat{X}(\hat{Y}\hat{Z}).$$

- The null operator $\hat{O}$ maps all vectors to the zero vector
- The identity operator $\hat{I}$ (or $\hat{1}$) maps all vectors to themselves

The notation $|\hat{X}u\rangle$ does not look like the matrix-vector notation $\mathbf{Au}$. In physics, to emphasise the vector nature of the ket, we often use the alternative notation,

$$|\hat{X}u\rangle \stackrel{\text{def}}{=} \hat{X}|u\rangle,$$

where it is understood that $\hat{X}$ acts to the right, just as a matrix acts on a vector. Most of the time this is not confusing, but sometimes it is. In these cases it may help to return to the $|\hat{X}u\rangle$ notation.

When operators appear in an inner product between two vectors

$$\langle u|\hat{X}v\rangle = \langle u|\hat{X}|v\rangle \stackrel{\text{def}}{=} \hat{X}_{uv}$$

is said to be a matrix element of the operator as $\hat{X}_{uv}$ has two indices just like the element of a matrix. Recall that if $\{|e_n\rangle\}$ is a basis, a vector $\mathbf{u}$ with elements $u_n = \langle e_n|u\rangle$ is a representation of $|u\rangle$ in that basis. In the same way, the matrix with elements $\hat{X}_{nm} = \langle e_n|\hat{X}|e_m\rangle$ represents $\hat{X}$ as an actual matrix in that basis. The action of $\hat{X}|u\rangle$ will be the same as $\mathbf{Xu}$ (again, we will consider this more deeply in Section 1.4).

For the case where the same vector appears on both sides,

$$\langle v|\hat{X}v\rangle = \langle v|\hat{X}|v\rangle, \tag{1.8}$$

we term this an expectation value of the operator (the reason for this will be explained later). This is sometimes given the shorthand $\langle\hat{X}\rangle_v$, or $\langle\hat{X}\rangle$ if the actual vector is not relevant to the discussion.

1.3.2 Expectation Value of Random Variables

We briefly digress from our discussion of operators to that of random variables, specifically to justify the term expectation value introduced above. Consider some random variable (traditionally labelled using a capital letter) X, representing the outcome of an experiment. First, let us assume that X has a finite number, or countably infinite number, of outcomes $\{x_1, \ldots, x_k\}$, and the probability of each outcome occurring is $\{p_1, \ldots, p_k\}$. If we performed many experiments and averaged the results, we would get the expected value of the experiment:

$$E(X) = \sum_i p_i x_i.$$

Some care needs to be taken when dealing with expectation values – they don't necessarily tell you what outcome you would get (the expected outcome of a random string of zeros and ones with equal weight would be half – not an outcome that is ever measured). Since many distributions in physics are Gaussian, or in some other way localised, the expectation value can be a very useful concept (especially when coupled with higher-order moment or cumulant analysis – a subject not covered here).

Generalisation to a continuous case is, as one would expect,

$$E(X) = \int dx\, p(x)\, x.$$

and further generalisations to sets of outcomes that depend on multiple outcomes, such as the position on the (x, y) plane, is again a natural extension of this idea

$$E(X, Y) = \int dx \int dy\, p(x, y)\, (x, y).$$

Simply, this adds up all the values of the random variable times their outcome probabilities to get the mean of all possible outcomes.

If, as in the above example, we don't actually know the values x of X or y of Y – if we have some function of the random variables X and Y, $g(X, Y)$ say – but we do know the probability density $p(x, y)$, we can invoke *the law of the unconscious statistician*, which states:

$$E(g(X, Y)) = \int dx \int dy\, p(x, y)\, g(x, y)$$

(the two-dimensional example is given, for reference, in Section 2.3).

With this definition of expectation in mind, let us lay some groundwork that links random variables to Dirac notation, and therein quantum mechanics. This may seem purely notational, but it also puts down the foundations for an interesting representation of classical mechanics by Koopman and von Neumann, which looks very much like quantum mechanics (and is briefly discussed at the end of Chapter 2).

Consider some arbitrary probability density function $p(x_i)$ over some set of random variables X_i. Then there will exist some functions $\psi(x_i)$ such that $p(x_i) = \psi^*(x_i)\psi(x_i)$ (which, for reasons that will become apparent when we study quantum mechanics, we allow to be complex). We restrict $\psi(x_i)$ in the space of square integrable functions (the integral of the modules squared exists and is finite) and so it is also a vector in that space. The space has the inner product given in Eq. (1.6), so we can write the function as a ket $|\psi(x_i)\rangle$ and define:

$$p(x_i) = |\psi(x_i)\rangle \langle \psi(x_i)|.$$

By doing this, the expectation value of any function of random variables can be written as:

$$E(g(X_i)) = \langle \psi(x_i) | g(x_i) | \psi(x_i) \rangle .$$

This is why we consider Eq. (1.8) the expectation value of an operator. We will give a more physical argument for this terminology in Section 2.5.

1.3.3 Inverse of Operators

Not all operators have an *inverse*, but if one exists it is defined as follows: for an operator $\hat{A}$ its inverse is denoted $\hat{A}^{-1}$ and must obey:

$$\hat{A}\hat{A}^{-1} = \hat{A}^{-1}\hat{A} = \hat{\mathbb{1}}.$$

1.3.4 Hermitian Adjoint Operators

Consider the matrix-vector equation

$$\boldsymbol{u} \cdot (\mathbf{A}\boldsymbol{v}) = \alpha,$$

where α is some scalar. The *Hermitian adjoint* (or Hermitian conjugate) of $\mathbf{A}$ is a matrix denoted $\mathbf{A}^\dagger$ that satisfies:

$$\left(\mathbf{A}^\dagger \boldsymbol{u}\right) \cdot \boldsymbol{v} = \alpha,$$

or, more succinctly, $\mathbf{A}^\dagger$ is the matrix associated with $\mathbf{A}$ that satisfies

$$\boldsymbol{u} \cdot (\mathbf{A}\boldsymbol{v}) \overset{\text{def}}{=} \left(\mathbf{A}^\dagger \boldsymbol{u}\right) \cdot \boldsymbol{v}.$$

Exercise 1.4   Use $\boldsymbol{v}$, written in the form of a column vector, along with this equation

$$[\boldsymbol{u}^T]^*(\mathbf{A}\boldsymbol{v}) = \alpha,$$

to show that $\mathbf{A}^\dagger = (\mathbf{A}^T)^*$.   ∎

As an example to help you with the above exercise, let us consider the matrix

$$\mathbf{A} = \begin{pmatrix} 1 & 2 \\ 3i & 4 \end{pmatrix}$$

then,

$$\boldsymbol{u} \cdot (\mathbf{A}\boldsymbol{v}) = \begin{pmatrix} u_1^* & u_2^* \end{pmatrix} \begin{pmatrix} 1 & 2 \\ 3i & 4 \end{pmatrix} \begin{pmatrix} v_1 \\ v_2 \end{pmatrix}$$

$$= \begin{pmatrix} u_1^* & u_2^* \end{pmatrix} \begin{pmatrix} v_1 + 2v_2 \\ 3iv_1 + 4v_2 \end{pmatrix}$$

$$= u_1^*(v_1 + 2v_2) + u_2^*(3iv_1 + 4v_2).$$

Further,

$$\mathbf{A}^\dagger = \begin{pmatrix} 1 & -3i \\ 2 & 4 \end{pmatrix}.$$

You can now show that

$$\left(\mathbf{A}^\dagger \mathbf{u}\right) \cdot \mathbf{v} = \left\{ \begin{pmatrix} 1 & -3i \\ 2 & 4 \end{pmatrix} \begin{pmatrix} u_1 \\ u_2 \end{pmatrix} \right\} \cdot \begin{pmatrix} v_1 \\ v_2 \end{pmatrix}$$

$$= \left[\left\{ \begin{pmatrix} 1 & -3i \\ 2 & 4 \end{pmatrix} \begin{pmatrix} u_1 \\ u_2 \end{pmatrix} \right\}^* \right]^T \begin{pmatrix} v_1 \\ v_2 \end{pmatrix},$$

which equals the same thing.

> *Hermitian adjoint* and *adjoint* mean two different things – here we are talking about the former, also known as the *conjugate transpose* or *Hermitian transpose*. The latter meaning, which we do not use, refers to the adjugate that is the transpose of the cofactor matrix, used to find the inverse of a square matrix.

Note that the Hermitian adjoint is defined with respect to the dot product. As before, this leads us to generalise the concept of a Hermitian adjoint to operators and a given inner product.

> The *Hermitian adjoint* of an operator $\hat{X}$ is denoted $\hat{X}^\dagger$ and is the operator that satisfies:
>
> $$\left\langle u \middle| \hat{X}v \right\rangle = \left\langle \hat{X}^\dagger u \middle| v \right\rangle$$
>
> for every possible u and v. Note that the Hermitian adjoint is a property shared by the operator and the inner product together. Change the definition of the inner product, and the Hermitian adjoint of a given operator will change with it.

> An operator is said to be Hermitian if $\hat{X} = \hat{X}^\dagger$ (this is the operator analogy to a real number). This generalises the idea of a matrix whose transpose complex conjugate is exactly the same as the original matrix, and so behaves in the same way irrespective of the side of an inner product to which it is applied.

Exercise 1.5 Show that the expectation value of an Hermitian operator is real. It may help to just look at Hermitian matrices first and then see if you can generalise the argument. ∎

> An operator is said to be anti-Hermitian if $\hat{X} = -\hat{X}^\dagger$ (analogous to an imaginary number). For matrices, the transpose complex conjugate will be the negated version of the original matrix.

Exercise 1.6 Show that the expectation value of an anti-Hermitian operator is imaginary. The argument for this is nearly identical to the one you used in the previous exercise, with a few sign changes. ∎

Exercise 1.7 Show $(X^\dagger)_{vu} = (X_{uv})^*$ (from which we conclude that the matrix representing the Hermitian adjoint of an operator is the transpose complex conjugate of the matrix representing the operator). Hint: Start by looking for the inner product axiom involving the complex conjugate. ∎

Note that, with the exception of unitary operators, very many of the operators seen in quantum mechanics are Hermitian. Those that are not can be made from Hermitian operators according to $\hat{X} = \hat{A} + i\hat{B}$ where both $\hat{A}$ and $\hat{B}$ are Hermitian, $\hat{A} = \frac{1}{2}(\hat{X} + \hat{X}^\dagger)$ and $\hat{B} = \frac{i}{2}(\hat{X}^\dagger - \hat{X})$. You will see examples of this when we consider the harmonic oscillator and angular momentum.

1.3.5 Unitary Operators

Unitary operators are those that satisfy the property

$$\hat{U}\hat{U}^\dagger = \hat{U}^\dagger\hat{U} = \hat{\mathbb{1}},$$

so that $\hat{U}^\dagger = \hat{U}^{-1}$. It can be shown that the eigenvalues of such operators lie on the unit circle – which it why they are termed *unitary*.

Unitary operators preserve the length of the vectors they act on, and thus also preserve inner products. This makes them useful for transforming one orthonormal basis into another (such as rotating **i**, **j**, and **k**, for the study of a rotating system). In addition to transformations, unitary operators will become important when we consider the dynamics of quantum systems (the evolution operator), symmetry considerations such as permutations (which lead to the exclusion principle), and will also be used in understanding open quantum systems (those coupled to an external environment).

1.3.6 Commutators

Just as with matrices, operators do not generally commute: $\hat{X}\hat{Y} \neq \hat{Y}\hat{X}$; order matters. In the case that $\hat{X}\hat{Y} = \hat{Y}\hat{X}$, then the operators $\hat{X}$ and $\hat{Y}$ are said to commute. The commutator is another operator (used throughout quantum mechanics) that expresses how much two operators do, or do not, commute:

$$[\hat{X}, \hat{Y}] \overset{\text{def}}{=} \hat{X}\hat{Y} - \hat{Y}\hat{X}, \tag{1.9}$$

which is the null operator if $\hat{X}$ and $\hat{Y}$ commute.

Exercise 1.8 Consider matrices

$$\sigma_x = \begin{bmatrix} 0 & 1 \\ 1 & 0 \end{bmatrix}$$

$$\sigma_y = \begin{bmatrix} 0 & -i \\ i & 0 \end{bmatrix}$$

$$\sigma_z = \begin{bmatrix} 1 & 0 \\ 0 & -1 \end{bmatrix}$$

Show $[\sigma_x, \sigma_y] = 2i\sigma_z$. ∎

As another example, consider the operator on functions of a variable x that consist of (multiply by x) and (differentiate w.r.t x). This would have the commutator:

$$[(\text{multiply by} x), (\text{differentiate w.r.t } x)] = x\frac{d}{dx} - \frac{d}{dx}x.$$

This can be evaluated if we act on some test function $f(x)$, as

$$\left(x\frac{d}{dx} - \frac{d}{dx}x\right)f(x) = xf'(x) - \frac{d}{dx}xf(x)$$

$$= xf'(x) - [f(x) + xf'(x)]$$

$$= -f(x).$$

Hence, we can see that the commutator is

$$[(\text{multiply by} x), (\text{differentiate w.r.t } x)] = -1,$$

since its effect is the same as multiplying by -1. We will later see that commutations relations are central to understanding the essence of quantum mechanics.

We will often consider a commutator of the form:

$$[\hat{A}, \hat{B}] = \zeta\hat{\mathbb{I}},$$

where ζ is a scalar. In these circumstances it is standard to drop the identity operator and use the shorthand

$$[\hat{A}, \hat{B}] = \zeta,$$

leaving the identity implied.

Exercise 1.9 Show that if $\hat{X}\hat{Y} = \hat{\mathbb{I}}$ then $\hat{X}$ and $\hat{Y}$ commute. Hint: start by writing out the commutator and then see if you can use some trick to show it is zero. ∎

Exercise 1.10 Show that commutators of complex valued matrices satisfy the following algebraic properties

- $[\hat{A}, \hat{B}] = -[\hat{B}, \hat{A}]$
- $[\hat{A}, (\hat{B} + \hat{C})] = [\hat{A}, \hat{B}] + [\hat{A}, \hat{C}]$
- $[\hat{A}, \hat{B}\hat{C}] = [\hat{A}, \hat{B}]\hat{C} + \hat{B}[\hat{A}, \hat{C}]$
- $[\hat{A}, [\hat{B}, \hat{C}]] + [\hat{B}, [\hat{C}, \hat{A}]] + [\hat{C}, [\hat{A}, \hat{B}]] = 0$
- $[\hat{A}, \hat{B}]^{\dagger} = [\hat{B}^{\dagger}, \hat{A}^{\dagger}]$. ∎

Note that the anti-commutator is defined as

$$\{\hat{X}, \hat{Y}\}_{+} \stackrel{\text{def}}{=} \hat{X}\hat{Y} + \hat{Y}\hat{X}. \tag{1.10}$$

Here we are using the standard notation of curly braces with, for us, the addition of a + subscript to distinguish it from the Poisson bracket.

There are a set of key commutation relations as follows:

$$[\hat{A}, \hat{B}] = -[\hat{B}, \hat{A}] \tag{1.11}$$

$$[\hat{A}, \hat{B} + \hat{C}] = [\hat{A}, \hat{B}] + [\hat{A}, \hat{C}] \tag{1.12}$$

$$[\hat{A}, \hat{B}\hat{C}] = [\hat{A}, \hat{B}] \, \hat{C} + \hat{B} \, [\hat{A}, \hat{C}] \tag{1.13}$$

$$0 = [\hat{A}, [\hat{B}, \hat{C}]] + [\hat{C}, [\hat{A}, \hat{B}]] + [\hat{B}, [\hat{C}, \hat{A}]] \tag{1.14}$$

$$[\hat{A}, \hat{B}]^{\dagger} = [\hat{B}^{\dagger}, \hat{A}^{\dagger}]. \tag{1.15}$$

Note that relation Eq. (1.13) also works for anti-commutators nested within commutators $[\hat{A}, \{\hat{B}, \hat{C}\}_{+}]$; this is important to the study of Quantum Brownian Motion and open quantum systems later in the book.

Exercise 1.11   Prove each of the above commutation relations. Which, if any, also hold for the anti-commutator? ∎

1.3.7   Eigenvectors and Eigenvalues

Just as with matrices and column vectors, we can define eigenvectors and eigenvalues of operators. If a vector $|u\rangle$ satisfies

$$\hat{A}\,|u\rangle = \lambda\,|u\rangle,$$

it is said to be an eigenvector of $\hat{A}$ with eigenvalue λ.

If more than one vector shares the same eigenvalue,

$$\hat{A}\,|u^{(i)}\rangle = \lambda\,|u^{(i)}\rangle,$$

then the eigenvectors are said to be degenerate and the degree of degeneracy is the total number of orthogonal eigenvectors that share this property. Eigenvalues and eigenvectors are important when we consider representations of quantum systems and the postulates of quantum mechanics associated to measurement.

Exercise 1.12   Show that the eigenvalues of a Hermitian operator are real, and that any two non-degenerate eigenvectors are orthogonal (degenerate eigenvectors are those that share the same eigenvalues). Hint: (i) for proving eigenvalues are real you will at some point use that $\langle u|u\rangle$ is always real (ii) for orthogonality use $\langle n|\hat{X}|m\rangle = \langle m|\hat{X}|n\rangle^{*}$ if $\hat{X} = \hat{X}^{\dagger}$. ∎

Exercise 1.13   Consider two operators $\hat{n}$ and $\hat{x}$ whose commutator is

$$[\hat{n}, \hat{x}] = \alpha\hat{x},$$

where α is a scalar. If $\hat{n}$ has eigenvalues n with corresponding eigenvectors $|n\rangle$ (i.e. $\hat{n}\,|n\rangle = n\,|n\rangle$) show that $\hat{x}\,|n\rangle$ is also an eigenvector of $\hat{n}$ with eigenvalue $n + \alpha$.

Further, show that if $\hat{n}$ is Hermitian, $[\hat{n}, \hat{x}^{\dagger}] = -\alpha\hat{x}^{\dagger}$ and conclude that $\hat{x}^{\dagger}\,|n\rangle$ is also an eigenvector of $\hat{n}$ with eigenvalue $n - \alpha$. If α is a positive real number, explain why $\hat{x}$ might be referred to as a *raising* or creation operator and $\hat{x}^{\dagger}$ a *lowering* or annihilation operator. Why might the pair $\hat{x}$ and $\hat{x}^{\dagger}$ be termed ladder operators? Hint: consider recursive application of $\hat{x}$ and $\hat{x}^{\dagger}$. ∎

1.3.8   Eigenvectors of Commuting Operators

If operators commute, they share eigenvectors; if the commutator of two arbitrary operators $\hat{A}$ and $\hat{B}$ is the null operator, then the eigenvectors of $\hat{A}$ are the same as those for $\hat{B}$.

To see this, let $|a\rangle$ be an eigenvector of $\hat{A}$ with eigenvalue a (and for now assume that this is non-degenerate). Then

$$\hat{A}\hat{B}\,|a\rangle = \hat{B}\hat{A}\,|a\rangle = a\hat{B}\,|a\rangle,$$

so $\hat{B}\,|a\rangle$ is also an eigenvector of $\hat{A}$ with the same eigenvalue a. Remember that if two operators commute, then the order in which they act can be changed without consequence. We will find commuting operators are important when we wish to fully label the basis of a system in a physically meaningful way. Some important examples will be given in Section 1.4.3 and other will be discussed when we later consider the simple harmonic oscillator and angular momentum.

1.3.9   Functions of Operators

The first example of a function of an operator is that of taking an integer power. Unsurprisingly, this can be written as:

$$\hat{A}^n \overset{\text{def}}{=} \hat{A}\hat{A}\hat{A}\ldots \text{ for } n \text{ times.}$$

Exercise 1.14   Consider the commutator of two operators $\hat{A}$ and $\hat{B}$

$$\left[\hat{A}, \hat{B}\right] = \zeta,$$

where ζ is a scalar. Using this relation, show that

$$\left[\hat{A}, \hat{B}^2\right] = 2\zeta\hat{B}.$$

∎

Exercise 1.15   Further, using the same commutation relation as in the previous exercise, show that

$$\left[\hat{A}, \hat{B}^n\right] = n\zeta\hat{B}^{(n-1)}.$$

Hint 1: use proof by induction. Induction is the method of mathematical reasoning that states if (i) some mathematical statement for an integer n implies the equivalent statement is true for $n + 1$ and (ii) that the statement is know to be true for a given integer N then it must also hold for all integers greater than N. Hint 2: if the first hint is not enough help, try looking at the special cases for $n = 3$ and 4. ∎

Functions of operators can be defined in terms of their power series expansion. So if, for example, we have the function

$$f(x) = \sum_n c_n x^n,$$

then the operator form would be

$$f(\hat{A}) \overset{\text{def}}{=} \sum_n c_n \hat{A}^n.$$

Exercise 1.16 Using the commutation relation $\left[\hat{A}, \hat{B}^n\right] = n\zeta\hat{B}^{(n-1)}$, where ζ is a scalar, show

$$[\hat{A}, f(\hat{B})] = \zeta f'(\hat{B}).$$

This result is known as McCoy's theorem. ∎

A very important example of a power series is:

$$\exp \hat{A} = \sum_{n=0}^{\infty} \frac{1}{n!} \hat{A}^n.$$

Note that while you may see the notation:

$$\exp \hat{A} = e^{\hat{A}}$$

do not think of this as Euler's number raised to a power of an operator, as this is undefined – it is simply a shorthand for the function exp(). That said, it is convenient shorthand, and many of the standard rules for exponentials will still work.

1.3.10 Differentiation of Operators

In the previous section we looked at using functions with operators as arguments. It is also possible to have operators that take a scalar as an argument, for example, $\hat{X}(t) = \hat{A}\sin(t)$. As with normal functions of a scalar, we may want to differentiate or ingrate the operator over that scalar. Differentiation of operators is defined in the way one would expect:

$$\frac{d\hat{A}(t)}{dt} = \lim_{\delta t \to 0} \frac{\hat{A}(t + \delta t) - \hat{A}(t)}{\delta t}.$$

There are more choices for defining integration (Riemann, Lebesgue, etc.) but their generalisation is equally straightforward. Rules such as the sum and product rule are recovered accordingly:

$$\frac{d}{dt}\left(\hat{A}(t) + \hat{B}(t)\right) = \frac{d\hat{A}(t)}{dt} + \frac{d\hat{B}(t)}{dt}$$

and

$$\frac{d}{dt}\hat{A}(t)\hat{B}(t) = \frac{d\hat{A}(t)}{dt}\hat{B}(t) + \hat{A}(t)\frac{d\hat{B}(t)}{dt},$$

with the exception that, in the last expression, order is important as $\hat{A}$ and $\hat{B}$ may not commute.

1.3.11 Baker Campbell Hausdorff, Zassenhaus Formulae, and Hadamard Lemma

The following contains three useful formulae for dealing with the consequences of non-commuting operators that are given without derivations. The first is the Baker Campbell Hausdorff formula

$$e^{\hat{A}} e^{\hat{B}} = e^{\hat{C}}, \tag{1.16}$$

where

$$\hat{C} = \hat{A} + \hat{B} + \frac{1}{2}\left[\hat{A},\hat{B}\right] + \frac{1}{12}\left[\hat{A},\left[\hat{A},\hat{B}\right]\right]$$
$$- \frac{1}{12}\left[\hat{B},\left[\hat{A},\hat{B}\right]\right] + \dots. \tag{1.17}$$

This series does not necessarily terminate quickly; computer algebra system can be usefully employed to help evaluate it. The next is the Zassenhaus formula:

$$e^{t(\hat{A}+\hat{B})} = e^{t\hat{A}}\,e^{t\hat{B}}\,e^{-\frac{t^2}{2}[\hat{A},\hat{B}]}\,e^{\frac{t^3}{6}(2[\hat{B},[\hat{A},\hat{B}]]+[\hat{A},[\hat{A},\hat{B}]])} \tag{1.18}$$

$$e^{\frac{-t^4}{24}([[[\hat{A},\hat{B}],\hat{A}],\hat{A}]+3[[[\hat{A},\hat{B}],\hat{A}],\hat{B}]+3[[[\hat{A},\hat{B}],\hat{B}],\hat{B}])}\dots \tag{1.19}$$

The final one is the Hadamard lemma:

$$e^{\hat{A}}\hat{B}e^{-\hat{A}} = \hat{B} + \left[\hat{A},\hat{B}\right] + \frac{1}{2!}\left[\hat{A},\left[\hat{A},\hat{B}\right]\right] + \dots \tag{1.20}$$

All three will be of use, as expressions of this form regularly occur in quantum mechanics. Note that if the commutator $\left[\hat{A},\hat{B}\right]$ commutes with $\hat{A}$ and $\hat{B}$, as is often the case, then these formulae become much simpler.

1.3.12 Operators and Basis State – Resolutions of Identity

1.3.12.1 Outer Product and Projection

As we have already discussed, in the same way that the transpose complex conjugate of a column vector can be thought of as 'half of a dot product waiting to happen' (i.e. $(\mathbf{v}\cdot)$), the bra vector can be thought of as 'half of an inner product waiting to happen'. To illustrate this, if

$$|u\rangle \sim \begin{bmatrix} u_1 \\ u_2 \end{bmatrix} \text{ then, } \langle u| \sim \mathbf{u}\cdot \sim (\mathbf{u}^*)^T = \begin{bmatrix} u_1^* & u_2^* \end{bmatrix}.$$

So, the if bra version of one ket vector acts on another ket, it will transform that ket into a scalar. In this sense, the bra is not an operator, as its action does not result in another vector. But if we were to pre-multiply a bra $\langle v|$ with a ket $|u\rangle$ then we get

$$|u\rangle\langle v|,$$

which is the generalisation of $\mathbf{u}(\mathbf{v}\cdot)$. We call this the outer product of the two vectors $|u\rangle$ and $|v\rangle$. It is an operator, and we can see its effect by its action on a third vector $|w\rangle$:

$$(|u\rangle\langle v|)|w\rangle = |u\rangle\langle v|w\rangle = (\langle v|w\rangle)|u\rangle.$$

This, in terms of our usual idea of vectors, looks like $\mathbf{u}(\mathbf{v}\cdot)\mathbf{w} = (\mathbf{v}\cdot\mathbf{w})\mathbf{u}$. The operator $|u\rangle\langle v|$ acts on $|w\rangle$ to produce a vector $|u\rangle$ scaled by $\langle v|w\rangle$, and acquiring any complex phase associated with the inner product. We say that $|u\rangle\langle v|$ *projects* $|w\rangle$ onto $|u\rangle$.

Exercise 1.17 By mapping a vector to another vector we see that $|u\rangle\langle v|$ is indeed an operator. Check that this outer product satisfies all the axioms of a linear operator. ∎

1.3.12.2 Resolutions of Identity

It is useful to note that some combinations of operator outer products can be equal to the identity operator. For example, let us consider the $\mathbf{i}, \mathbf{j}$, and $\mathbf{k}$ basis of Cartesian coordinates and the operator

$$\left(\mathbf{i}\mathbf{i}\cdot + \mathbf{j}\mathbf{j}\cdot + \mathbf{k}\mathbf{k}\cdot\right).$$

If this acts upon a vector we see that the effect is:

$$\left(\mathbf{i}\mathbf{i}\cdot + \mathbf{j}\mathbf{j}\cdot + \mathbf{k}\mathbf{k}\cdot\right)\left(3\mathbf{i} + 2\mathbf{j} + 7\mathbf{k}\right) = \left(3\mathbf{i} + 2\mathbf{j} + 7\mathbf{k}\right),$$

it does not change the vector it acts on. If we were to write this operator in Dirac notation, it would look like:

$$|i\rangle\langle i| + |j\rangle\langle j| + |k\rangle\langle k| = \hat{\mathbb{1}},$$

equal to the identity operator.

Exercise 1.18 Show that the operator leaves the second term unchanged in the expression below.

$$(|i\rangle\langle i| + |j\rangle\langle j| + |k\rangle\langle k|)(3|i\rangle + 2|j\rangle + 7|k\rangle).$$ ∎

It is the case that for any orthonormal basis $\{|e_i\rangle\}$ that

$$\hat{\mathbb{1}} = \sum_i |e_i\rangle\langle e_i|,$$

which is termed a resolution of the identity. Note that for a continuous basis the same can be done, but we replace the sum with an integral.

We have explained that the outer product projects a vector onto another vector; thus, we can also use it to change basis. If one wishes to transform a vector from one $\{|e_i\rangle\}$ basis to another basis $\{|e'_j\rangle\}$, we simply multiply the vector in the first basis by a resolution of the identity in the second:

$$|u\rangle = \sum_i c_i |e_i\rangle$$

$$= \sum_i c_i \hat{\mathbb{1}} |e_i\rangle$$

$$= \sum_i c_i \sum_j |e'_j\rangle\langle e'_j||e_i\rangle$$

$$= \sum_j \left(\sum_i c_i \langle e'_j|e_i\rangle\right) |e'_j\rangle$$

$$= \sum_j c'_j |e'_j\rangle,$$

where

$$c'_j = \sum_i c_i \langle e'_j|e_i\rangle,$$

which is evidently a sum of scalars. As the basis state are normalised (scaled to unit length), it is handy to recall that $\langle e_j' | e_i \rangle$ can be interpreted as the amount of $|e_j'\rangle$ in the direction of $|e_i\rangle$; directly analogous to unit vectors and the dot product.

1.4 Representing Kets as Vectors, and Operators as Matrices and Traces

If we consider the operator $(\mathbf{i}\,\mathbf{i}\cdot)$ on its own and apply it to any vector on its own such as $(3\mathbf{i} + 2\mathbf{j} + 7\mathbf{k})$ we obtain the ith component of the vector, in this instance $3\mathbf{i}$. If we take each basis vector in turn and construct a column vector, for an arbitrary vector $\boldsymbol{u}$, by applying the projection $(\boldsymbol{e}_i \cdot)$ to the ith row we find that

$$\boldsymbol{u} \to \begin{pmatrix} \boldsymbol{e}_1 \cdot \boldsymbol{u} \\ \boldsymbol{e}_2 \cdot \boldsymbol{u} \\ \boldsymbol{e}_3 \cdot \boldsymbol{u} \end{pmatrix} \to \begin{pmatrix} \mathbf{i} \cdot \boldsymbol{u} \\ \mathbf{j} \cdot \boldsymbol{u} \\ \mathbf{k} \cdot \boldsymbol{u} \end{pmatrix} = \begin{pmatrix} u_x \\ u_y \\ u_z \end{pmatrix},$$

recovering the usual component form of a vector.

We can generalise this notion to an arbitrary vector space, so long as we have a basis. We will now see what this looks like in Dirac notation.

The **column vector representation** of a vector $|u\rangle$ with respect to a basis $\{|e_i\rangle\}$ is

$$|u\rangle \to \begin{pmatrix} \langle e_1 | \, |u\rangle \\ \langle e_2 | \, |u\rangle \\ \vdots \end{pmatrix} = \begin{pmatrix} \langle e_1 | u \rangle \\ \langle e_2 | u \rangle \\ \vdots \end{pmatrix} = \begin{pmatrix} u_1 \\ u_2 \\ \vdots \end{pmatrix}.$$

Exercise 1.19 (optional – for the mathematically interested). Think about the sines and cosines in a Fourier series as basis vectors. What does the vector representation of a function look like in this basis? ∎

Exercise 1.20 (optional – for the mathematically interested). There are other possible sets of functions that can be used as a basis, such as the Hermite polynomials (which we will need to understand the quantum harmonic oscillator). How might you convert from one vector representation to another and can you express this in Dirac notation? ∎

Note that if the basis is infinite, then so is the dimension of the vector representation of any vector in that basis. Also note that if the basis is continuous, as might be represented by $\{|\alpha\rangle\}$ where α is some real or complex number, then $|u\rangle \to (\langle \alpha | u \rangle)$ is a vector with a continuous index, more commonly known as a function. Namely, $|u\rangle \to (\langle \alpha | u \rangle) = u(\alpha)$. Therein, it is perhaps unsurprising that functions can be thought of as vectors in a vector space; functions are just vectors with a continuous label.

If the basis is continuous, with a continuous index α, then the elements of the vector representation of $|u\rangle$ are $\langle \alpha | u \rangle$, which is simply a function $u(\alpha)$. That is to say,

$$|u\rangle \rightarrow \begin{pmatrix} \langle \alpha | \; | u \rangle \\ \downarrow \end{pmatrix} = \begin{pmatrix} \langle \alpha | u \rangle \\ \downarrow \end{pmatrix} \equiv u(\alpha).$$

Here, the downward pointing arrow attempts to denote the continuous nature of α. The arguments presented here hold regardless of the nature of the basis (so long as it is complete and orthonormal).

If we take a ket and make it look like a column vector, or a function, by choosing different bases, the underlying mathematical object is still the same and we expect it to have the same behaviour. Although it is mathematically unnecessary, we can think of the ket notation as a shorthand that delays the choice of how we represent a vector until we have chosen a basis (we do not need Dirac notation to do this, so a good mathematician may, and probably should, object to this abusive language – but for a physicist the idea is of utility).

If we can represent a ket as a vector/function by specifying a basis, the next logical topic to explore is that of representing operators as matrices. We do this by following an extension of the above argument. That is, we map $\hat{A}$ to a matrix whose elements A_{ij} are $\langle e_i | \hat{A} | e_j \rangle$. We may examine this to see that it is intuitive, if

$$\hat{A} |u\rangle = |v\rangle,$$

then pre-multiplying both sides by $\langle e_i |$ and inserting an identity operator gives

$$\langle e_i | \hat{A} \hat{\mathbb{1}} | u \rangle = \langle e_i | v \rangle.$$

But as the $\{|e_i\rangle\}$ form a basis (by definition), we can write $\hat{\mathbb{1}} = \sum_j |e_j\rangle \langle e_j|$ as a resolution of identity,

$$\langle e_i | \hat{A} \sum_j |e_j\rangle \langle e_j| |u\rangle = \langle e_i | v \rangle.$$

This can be expanded to

$$\sum_j \left(\langle e_i | \hat{A} | e_j \rangle \right) \left(\langle e_j | u \rangle \right) = \langle e_i | v \rangle,$$

which is

$$\sum_j A_{ij} u_j = v_i.$$

This is exactly the component-wise form of a matrix, multiplied by a column vector. The terminology we previously adopted of matrix element for $(\mathbf{X})_{vu} = X_{vu} = \langle v | \hat{X} | u \rangle$ is therefore well justified.

1.4.1 Trace

Just as the trace of a matrix is the sum of its diagonal elements, the trace of an operator is defined as

$$\text{Tr}[\hat{A}] = \sum_i \left\langle e_i \left| \hat{A} \right| e_i \right\rangle \tag{1.21}$$

for any basis $\{|e_i\rangle\}$.

Exercise 1.21 Show that the trace of an operator is invariant after a change of basis (i.e. if that basis is transformed by a unitary operator). ∎

Exercise 1.22 Show that the trace of the product of an arbitrary number of operators is invariant over cyclic permutations of those operators, i.e.

$$\text{Tr}[\hat{A}\hat{B}\hat{C}] = \text{Tr}[\hat{C}\hat{A}\hat{B}].$$

∎

Exercise 1.23 Show

$$\text{Tr}[\hat{A}\,|u\rangle\langle u|] = \left\langle u \left| \hat{A} \right| u \right\rangle$$

for any vector $|u\rangle$. This will be important later so well worth the effort of solving! ∎

1.4.2 Basis, Representation, and Inner Products

Here we show, by example, that the choice of basis and the rules of vector arithmetic (specifically the properties of an inner product) are sufficient to define the specific form that an inner product takes.

For a discrete orthonormal basis, $\{|e_n\rangle\}$

$$\langle u|v\rangle = \left\langle u \left| \hat{\mathbb{1}} \right| v \right\rangle = \left\langle u \left| \sum_n |e_n\rangle\langle e_n| \right| v \right\rangle = \sum_n \langle u|e_n\rangle \langle e_n|v\rangle = \sum_n u_n^* v_n,$$

which is the same as the vector dot product $\boldsymbol{u} \cdot \boldsymbol{v} = (\boldsymbol{u}^T)^* \boldsymbol{v}$;
and, for a continuous orthonormal basis, $\{|\alpha\rangle\}$

$$\langle f|g\rangle = \left\langle f \left| \hat{\mathbb{1}} \right| g \right\rangle = \left\langle f \left| \int d\alpha\, |\alpha\rangle\langle\alpha| \right| v \right\rangle = \int d\alpha\, \langle f|\alpha\rangle \langle\alpha|g\rangle = \int d\alpha f^*(\alpha)g(\alpha),$$

which is the same as the inner product for functions.

Exercise 1.24 Identify exactly which axioms from vector and inner product spaces were used in the above two examples. ∎

1.4.3 Observables

The term observable is used to mean any Hermitian operator whose eigenvectors form a complete orthonormal basis. We already know that eigenvectors of Hermitian operators with differing eigenvalues are orthogonal. Since eigenvectors can be arbitrarily scaled, imposing normalisation is trivial. Furthermore, if there is some degeneracy, the process of Gram–Schmidt orthonormalisation[5] can be used to construct the orthonormal condition for all eigenvectors. For this reason, the only extra condition we are adding to the notion of a Hermitian operator is that an observable's eigenbasis must be complete (i.e. it comprises a set of eigenvectors such that any vector can be written using it).

Exercise 1.25   Show that an observable satisfying

$$\hat{A}\,|a\rangle = a\,|a\rangle$$

can be written in its eigenvector expansion

$$\hat{A} = \sum_n a\,|a\rangle\,\langle a|$$

in the case that the eigenvectors are non-degenerate. What if there is degeneracy of the eigenvectors? ∎

Exercise 1.26   If $f(x)$ has a power series expansion, and we know the eigenvalues and eigenvectors for an observable $\hat{A}$ satisfy

$$\hat{A}\,|a\rangle = a\,|a\rangle\,,$$

with non-degenerate eigenvectors, show

$$f(\hat{A}) = \sum_n f(a)\,|a\rangle\,\langle a|\,.$$

What if there is degeneracy of the eigenvectors? ∎

1.4.4 Labelling Vectors – Complete Sets of Commuting Observables – CSCO

If an observable has non-degenerate eigenvectors, then it is possible to label its eigenvectors by the eigenvalues, as exemplified by

$$\hat{A}\,|a\rangle = a\,|a\rangle\,,$$

as $\{|a\rangle\}$ forms a complete orthonormal basis. This means we can proceed as described in Section 1.4 and make a matrix representation of vectors and operators in this basis – this is something that is often done in quantum mechanics, so it is an important concept to grasp.

If we have degenerate eigenvalues, we cannot immediately do this, and must first find a way to uniquely label all eigenvectors. To start, let us consider the eigenequation for $\hat{A}$

$$\hat{A}\,\big|e_n^{(i)}\big\rangle = a_n\,\big|e_n^{(i)}\big\rangle,$$

---

5 The Gram–Schmidt process is a way of constructing orthonormal vectors from some set of vectors - as we do not use it in this book, we have not presented any detail. See a good linear algebra text for a description.

where the superscript indicates the additional labels needed to distinguish the eigenvectors of $\hat{A}$.

If we can find another observable $\hat{B}$ that commutes with $\hat{A}$ then, as we have seen in Section 1.3.8, they share a common eigenbasis (which by construction must also be degenerate – otherwise we could label the eigenvectors by the eigenvalues of $\hat{B}$). Two observables that commute are said to be compatible. So we now also have

$$\hat{B}\left|e_n^{(j)}\right\rangle = b_n \left|e_n^{(j)}\right\rangle.$$

If a_n and b_n can be used to distinguish all the eigenvector, we are done. If not, we need another operator $\hat{C}$ *etc.* until each set of labels a_n, b_n, c_n, ... describes one, and only one, eigenvector $\left|e_n\right\rangle$. We then write this as:

$$\left|e_n\right\rangle = \left|a_n, b_n, c_n, ...\right\rangle$$

(sometimes instead of commas we may use semicolons, other delimiters, or even a mix if we want to make some distinction between indices). The set of operators $\left\{\hat{A}, \hat{B}, \hat{C}, ...\right\}$ is termed a complete set of commuting observables (CSCO). *Complete* because the eigenvalues of the operators can be used to label a complete basis; *Set* because it may take more than one operator to provide the labels; *Commuting* because each operator must commute with every other operator in the set to share common eigenvectors; and *Observables* as Hermitian operators are guaranteed to have the properties we want (real eigenvalues and eigenvectors that can form an orthonormal basis).

1.5 Tensor Product

This subject is essential for understating composite quantum systems and the phenomenon of entanglement. The physics is somewhat subtle, and the full implications that the mathematics has for quantum physics can take a while to become comfortable with. That said, the mathematics itself comprises a set of straightforward rules. While notation may be unfamiliar, it requires little practice performing the operations of tensor product spaces to become competent at using the mathematics.

1.5.1 Setting the Scene: The Cartesian Product

In the mathematical description of classical systems we often like to think of space as the product of one-dimensional spaces (this is what we are doing when we draw the three perpendicular axes). We formalise this by defining the Cartesian product of sets as the ordered tuple

$$A \times B \times C \times ... = \{(a, b, c, ...) \ : \ a \in A, b \in B, c \in C, ...\}.$$

Two-dimensional space is then the Cartesian product

$$\mathbb{R}^2 \stackrel{\text{def}}{=} \mathbb{R} \times \mathbb{R} = \{(x, y) \ : \ x \in \mathbb{R} \text{ and } y \in \mathbb{R}\}.$$

The product is termed the Cartesian product precisely because it is motivated by Cartesian coordinates. For this reason, addition and multiplication by a scalar take the expected forms of $(a, b, c, \ldots) + (a', b', c', \ldots) = (a + a', b + b', c + c', \ldots)$ and $\lambda(a, b, c, \ldots) = (\lambda a, \lambda b, \lambda c, \ldots)$. We can extend the idea to arbitrary dimensions and composite systems in the obvious way (e.g. the position of two particles in a plane would be given by $\mathbb{R}^2 \times \mathbb{R}^2 = \{(x_1, y_1, x_2, y_2) : x_i \in \mathbb{R}$ and $y_i \in \mathbb{R}\}$). It is important to note that the axis for each particle lives in different product spaces. As we often use Cartesian coordinates that share the same origin and orientation, this can be easy to forget and to think that they all live in the same space (visualisations and animations of systems of particles can further reinforce this misconception as we tend to plot spaces on top of one another). The axis for each particle does not need to even use the same system - just contain enough generalised coordinates to specify the position of each particle.

In the preceding section, we discussed labelling vectors in a vector space using a CSCO. We also know that any vector, in a vector space, can be made from a linear combination (superposition) of the basis state labelled by that CSCO. Let us, for a moment, consider two operators, $\hat{X}$ and $\hat{Y}$, that form a CSCO. We can label the basis state according to the eigenvalues: $\hat{X} |x, y\rangle = x |x, y\rangle$ and $\hat{Y} |x, y\rangle = y |x, y\rangle$ where, in this example, we will assume the eigenvalues x and y are the continuum of real numbers. We have done this to make clear the connection to, and differences from, the motivating example of the coordinate (x, y).

Just as in Eq. (1.23) any vector $|\psi\rangle$ in this space can be written in terms of this basis:

$$|\psi\rangle = \int dx \int dz \, \psi(x, y) |x, y\rangle, \tag{1.22}$$

where $\psi(x, y) = \langle x, y | \psi\rangle$. The important difference from labelling a particle's position is that in Eq. (1.22) $|x, y\rangle$ uniquely labelled a vector in terms of the eigenvector of two operators $\hat{X}$ and $\hat{Y}$ that live in the *same* Hilbert space. We see that the index/label space for this composite system is still the Cartesian product of the label spaces of the component systems – in this case it is the same as two-dimensional space.

1.5.2 The Tensor Product

As we have already stated, for the Cartesian product we can think of two-dimensional space as the product of two one-dimensional spaces, $(x, y) = (x) \times (y)$. This motivates us to ask if it is possible to write $|x, y\rangle$ as the product of two vectors $|x\rangle$ (an eigenvector of $\hat{X}$) and $|y\rangle$ (an eigenvector of $\hat{Y}$). Importantly, here we now make the distinction from the preceding discussion that $\hat{X}$ and $|x\rangle$ should now live in different vector spaces from $\hat{Y}$ and $|y\rangle$, say X and Y. This is analogous to the way that the coordinates x and y live in different sets (one for each axis).

This leads us to make the following definition: If X and Y are two vector spaces, we denote their tensor product by $X \otimes Y$. The resultant tensor product space is spanned by vectors of the form $|x\rangle \otimes |y\rangle$, or simply $|x\rangle |y\rangle$. These tensor products are subject to the following rules:

$$(|x_1\rangle + |x_2\rangle) \otimes |y\rangle = |x_1\rangle \otimes |y\rangle + |x_2\rangle \otimes |y\rangle$$
$$|x\rangle \otimes (|y_1\rangle + |y_2\rangle) = |x\rangle \otimes |y_1\rangle + |x\rangle \otimes |y_2\rangle$$
$$\alpha(|x\rangle \otimes |y\rangle) = (\alpha |x\rangle) \otimes |y\rangle = |x\rangle \otimes (\alpha |y\rangle).$$

The fact that $|x\rangle \otimes |y\rangle$ is an ordered pair means we can associate to this product a vector $|x, y\rangle \stackrel{\text{def}}{=} |x\rangle \otimes |y\rangle \in X \otimes Y$ that we can show satisfies all the axioms of a vector, and hence that $X \otimes Y$ must be a vector space. Note that operators are only allowed to act on vectors in their own space. So, if $\hat{A} \in X$ then $\hat{A} |x, y\rangle = (\hat{A} |x\rangle) \otimes |y\rangle$. This operator can be *extended* into the tensor product space according to $\hat{A} \otimes \hat{\mathbb{I}}_Y$ where $\hat{\mathbb{I}}_Y$ is the identity operator in space Y. This process looks superficially like the Cartesian product and $|x, y\rangle$ reminds us of the notation from CSCO, but there are some important differences which we will now explore.

To see how this relates to our previous comments on Cartesian products and CSCOs, let us, for the sake of simplicity, assume that $\{\hat{X}\}$ is a CSCO for space X and $\{\hat{Y}\}$ is a CSCO for space Y. The eigenvectors are labelled

$$\hat{X} |x\rangle = x |x\rangle \text{ where } |x\rangle \in X$$
$$\hat{Y} |y\rangle = y |y\rangle \text{ where } |y\rangle \in Y.$$

As each operator is a CSCO, we know that $\{|x\rangle\}$ forms an orthonormal basis of some vector space, as does $\{|y\rangle\}$ for another. The extended operators $\{\hat{X} \otimes \hat{\mathbb{I}}_Y, \hat{\mathbb{I}}_X \otimes \hat{Y}\}$ form a CSCO in $X \otimes Y$. The set $\left\{ |x, y\rangle \overset{\text{def}}{=} |x\rangle \otimes |y\rangle \right\}$ will form an orthonormal basis of $X \otimes Y$. We also know that any vector can be written as a superposition of these basis state,

$$|\psi\rangle = \int dx \int dy \, \psi(x, y) |x, y\rangle = \int dx \int dy \, \psi(x, y) |x\rangle \otimes |y\rangle.$$

Therefore, this means that vectors in the tensor product space are made up of superpositions of basis vectors, indexed over the set of all possible Cartesian products of the basis state.

Let X and Y be two vector spaces, then the tensor product space is the vector space $X \otimes Y$, spanned by a basis that is the Cartesian product of the basis vectors of the component spaces. So if $\{|x_i\rangle\}$ is a basis for X and $\{|y_i\rangle\}$ for Y then $\{|x_i\rangle \otimes |y_i\rangle\}$ is a basis of $X \otimes Y$.

Exercise 1.27 Consider two operators $\{\hat{n}_x, \hat{m}_y\}$ forming a CSCO. Convince yourself that any vector can be written as a superposition of these basis state as

$$|\psi\rangle = \sum_{n_x=0}^{\infty} \sum_{m_y=0}^{\infty} \psi_{n_x m_y} |n_x, m_y\rangle,$$

where $|n_x, m_y\rangle = |n_x\rangle \otimes |m_y\rangle$. If only the first three states of the system are important, we can truncate the basis to

$$|\psi\rangle = \sum_{n_x=0}^{2} \sum_{m_y=0}^{2} \psi_{n_x m_y} |n_x, m_y\rangle.$$

How many basis state are there in total? How do these scale if we require more observables to form a CSCO (think carefully, this is a trick question)? How does the number of total states scale if we take the tensor product of some integer number of systems? Show that there exists some states that can *not* be written in the form

$$|\psi_x\rangle \otimes |\phi_y\rangle = \left(\sum_{n_x=0}^{2} \psi_{n_x} |n_x\rangle \right) \otimes \left(\sum_{m_y=0}^{2} \phi_{m_y} |m_y\rangle \right).$$

States that cannot be written in this form in any basis formed from a natural CSCO are said to be *entangled*. ∎

Exercise 1.28 In the above discussion we made a basis of a tensor product space, where the CSCO for each space was a single operator. Given the method for extending an operator into a tensor product space, clearly operators that are extensions of those from different spaces will always commute with each other. The CSCO for the tensor product space is therefore just the set of operators used for determining the basis of each individual space. Is it possible to reverse this process and decompose a CSCO into tensor product sub-spaces? ∎

Exercise 1.29 Consider the example given above for a single particle in a two-dimensional space. Compare and contrast this mathematically and conceptually with that of two one-dimensional particles. ∎

Exercise 1.30 Look up the Kronecker product *of matrices* (maybe choose your favourite open source of information!). Consider two systems, X and Y, with discrete basis state, $\{|n_X\rangle\}$ and $\{|n_Y\rangle\}$, and the matrix representation of vectors $|\psi\rangle_X$ and $|\phi\rangle_Y$ given by

$$\psi_X = \begin{pmatrix} \langle 0_X|\psi\rangle_X \\ \langle 1_X|\psi\rangle_X \\ \vdots \end{pmatrix} = \begin{pmatrix} \psi_0 \\ \psi_1 \\ \vdots \end{pmatrix} \text{ and } \phi_Y = \begin{pmatrix} \langle 0_Y|\phi\rangle_Y \\ \langle 1_Y|\phi\rangle_Y \\ \vdots \end{pmatrix} = \begin{pmatrix} \phi_0 \\ \phi_1 \\ \vdots \end{pmatrix}$$

and operators $\hat{A}_X$ and $\hat{B}_Y$ with matrix representation

$$A_X = \begin{pmatrix} \langle 0_X|\hat{A}_X|0_X\rangle & \langle 0_X|\hat{A}_X|1_X\rangle & \cdots \\ \langle 1_X|\hat{A}_X|0_X\rangle & \langle 1_X|\hat{A}_X|1_X\rangle & \cdots \\ \vdots & \vdots & \ddots \end{pmatrix} = \begin{pmatrix} a_{00} & a_{01} & \cdots \\ a_{10} & a_{11} & \cdots \\ \vdots & \vdots & \ddots \end{pmatrix}$$

and

$$B_Y = \begin{pmatrix} \langle 0_Y|\hat{B}_Y|0_Y\rangle & \langle 0_Y|\hat{B}_Y|1_Y\rangle & \cdots \\ \langle 1_Y|\hat{B}_Y|0_Y\rangle & \langle 1_Y|\hat{B}_Y|1_Y\rangle & \cdots \\ \vdots & \vdots & \ddots \end{pmatrix} = \begin{pmatrix} b_{00} & b_{01} & \cdots \\ b_{10} & b_{11} & \cdots \\ \vdots & \vdots & \ddots \end{pmatrix}.$$

Show that the matrix representation of (a) $|\psi\rangle_X \otimes |\phi\rangle_Y$ is the Kronecker product of ψ_X & ϕ_Y and (b) $\hat{A}_X \otimes \hat{B}_Y$ is the Kronecker product of A_X & B_Y. ∎

1.6 The Heisenberg Uncertainty Relation

The Heisenberg uncertainty relation is rather famous, and there is a fair chance you already know of it, even if not in any specific detail. You may be somewhat surprised to see this section; you might reasonably expect to see this topic in a section on quantum measurement. This presentation has been carefully selected to remove some of the standard confusion that surrounds this neat bit of mathematics. As we shall see, the Heisenberg uncertainty relation says *nothing about measurement*. Later, when we consider measuring quantum states, the uncertainty relation does tell us some important information about the nature of quantum states – the details of which will become clear when we describe the phase space formulation of quantum mechanics due to Wigner in Section 4.3. Regardless, the Heisenberg uncertainty relation *has absolutely nothing to do with measurement disturbance*.

There are three, and only three, mathematical ingredients needed to understand the Heisenberg uncertainty relation:

- a vector space (we will assume $|u\rangle$, $|v\rangle$ and $|w\rangle$ are elements of this space),
- a couple of *Hermitian* operators, say $\hat{A}$ and $\hat{B}$, and
- an 'inner product'.

As such, this is a *mathematical proof*[6] and holds for (and is important to) any system where such quantities are relevant, from signal processing to quantum mechanics.

As a shorthand notation we will use

$$\langle w|\hat{A}|w\rangle \overset{\text{def}}{=} \langle \hat{A}\rangle_w \overset{\text{def}}{=} \langle \hat{A}\rangle,$$

where, if we get really lazy, we sometimes just drop the subscript if the same vector is being used unambiguously throughout an equation. This quantity is called the *expectation value* of $\hat{A}$. And if $\hat{A}$ is Hermitian, the expectation value is real (see Exercise: 1.5).

We also define

$$\delta_w \hat{A} \overset{\text{def}}{=} \hat{A} - \langle \hat{A}\rangle_w,$$

so that this operator is like $\hat{A}$ but with $\langle \delta_w \hat{A}\rangle_w = 0$. You may think of this as a new operator, but with the 'mean' centred around zero.

Exercise 1.31 Show that the following holds if $\hat{A}$ is Hermitian:

$$\left\langle \left(\delta\hat{A}\right)^2\right\rangle = \left\langle \hat{A}^2\right\rangle - \langle \hat{A}\rangle^2,$$

where we have dropped the use of the subscript w. ∎

Now consider

$$|u\rangle = \delta_w \hat{A}\,|w\rangle \text{ and } |v\rangle = \delta_w \hat{B}\,|w\rangle.$$

As $\hat{A}$ and $\hat{B}$ are Hermitian,

- $\langle u|u\rangle = \left\langle \left(\delta\hat{A}\right)^2\right\rangle$
- $\langle v|v\rangle = \left\langle \left(\delta\hat{B}\right)^2\right\rangle$
- $\langle u|v\rangle = \left\langle \left(\delta\hat{A}\right)\left(\delta\hat{B}\right)\right\rangle,$

---

6 The presentation which we give here follows standard derivations such as those found in [34, 77] (note some steps have been deliberately left as problems here, but intermediate steps can be found in [34]).

so that by the Cauchy–Schwarz inequality,

$$\langle u|u\rangle \langle v|v\rangle \geq |\langle u|v\rangle|^2.$$

We have, by substitution,

$$\left\langle (\delta \hat{A})^2 \right\rangle \left\langle (\delta \hat{B})^2 \right\rangle \geq \left| \langle (\delta \hat{A}) (\delta \hat{B})\rangle \right|^2.$$

Note: we have dropped the subscript w, as it is the only vector in these equations.

Now we shall use some mathematical trickery;

$$\delta \hat{A} \, \delta \hat{B} = \frac{1}{2}\delta \hat{A} \, \delta \hat{B} + \frac{1}{2}\delta \hat{A} \, \delta \hat{B},$$

$$= \frac{1}{2}\delta \hat{A} \, \delta \hat{B} + \underbrace{\frac{1}{2}\delta \hat{B} \, \delta \hat{A} - \frac{1}{2}\delta \hat{B} \, \delta \hat{A}}_{=0} + \frac{1}{2}\delta \hat{A} \, \delta \hat{B},$$

$$= \frac{1}{2}\{\delta \hat{A}, \delta \hat{B}\}_+ + \frac{1}{2}[\delta \hat{A}, \delta \hat{B}].$$

The trick was to use $1 = (1/2) + 0 + (1/2)$, with the curly braces indicating an anti-commutator.

Exercise 1.32 Show the following is true

$$[\delta \hat{A}, \delta \hat{B}] = [\hat{A}, \hat{B}].$$

∎

Exercise 1.33 Show that the following is Hermitian

$$\{\delta \hat{A}, \delta \hat{B}\}_+$$

∎

Exercise 1.34 Show that the following is anti-Hermitian

$$[\hat{A}, \hat{B}]$$

∎

You should already have shown in Section 1.3.4 that the expectation value of a Hermitian operator is real, and that the expectation value of an anti-Hermitian operator is imaginary.

Hence, in

$$\langle \delta \hat{A} \, \delta \hat{B}\rangle = \frac{1}{2}\langle \{\delta \hat{A}, \delta \hat{B}\}_+\rangle + \frac{1}{2}\langle [\delta \hat{A}, \delta \hat{B}]\rangle,$$

the first term following the equals sign is purely real, and the second purely imaginary. Noting that we can write $|z|^2 = \Re(z)^2 + \Im(z)^2$,

$$\left| \langle \delta \hat{A} \, \delta \hat{B}\rangle \right|^2 = \frac{1}{4}\left| \langle \{\delta \hat{A}, \delta \hat{B}\}_+\rangle \right|^2 + \frac{1}{4}\left| \langle [\delta \hat{A}, \delta \hat{B}]\rangle \right|^2,$$

$$\geq \frac{1}{4}\left| \langle [\delta \hat{A}, \delta \hat{B}]\rangle \right|^2.$$

So combining with our application of the Cauchy–Schwarz inequality

$$\left\langle (\delta \hat{A})^2 \right\rangle \left\langle (\delta \hat{B})^2 \right\rangle \geq \left| \langle \delta \hat{A} \, \delta \hat{B}\rangle \right|^2,$$

$$\geq \frac{1}{4}\left| \langle [\delta \hat{A}, \delta \hat{B}]\rangle \right|^2. \tag{1.23}$$

Recall that you have recently shown

$$[\delta\hat{A}, \delta\hat{B}] = [\hat{A}, \hat{B}],$$

so it is possible to write

$$\left\langle \left(\delta\hat{A}\right)^2 \right\rangle \left\langle \left(\delta\hat{B}\right)^2 \right\rangle \geq \frac{1}{4}\left|\left\langle [\hat{A}, \hat{B}]\right\rangle\right|^2. \tag{1.24}$$

Finally, we can define

$$\Delta\hat{A} \overset{\text{def}}{=} \sqrt{\left\langle \left(\delta\hat{A}\right)^2 \right\rangle} = \sqrt{\left\langle \hat{A}^2 \right\rangle - \left\langle \hat{A} \right\rangle^2}.$$

For a given vector $|w\rangle$, $\Delta_w\hat{A}$ quantifies the magnitude of deviation that is obtained from acting with the operator $\hat{A}$ on $|w\rangle$. That is, how much the action of $\hat{A}$ on $|w\rangle$ changes that vector – or, alternatively, this quantifies how much $|w\rangle$ is *not* an eigenvector of $\hat{A}$. Hence, we refer to $\Delta_w\hat{A}$ as the *dispersion* of $\hat{A}$ with respect to the vector $|w\rangle$. If you want, you might like to consider this as an operator generalisation of the root mean square, or standard deviation, of a distribution.

Square rooting equation (1.23) and substituting this into the dispersion yields the Heisenberg uncertainty relation:

$$\Delta\hat{A}\,\Delta\hat{B} \geq \frac{1}{2}\left|\left\langle [\hat{A}, \hat{B}]\right\rangle\right| \tag{1.25}$$

which relates the dispersions of two Hermitian operators, with respect to a given vector, to the expectation value of the commutator of those operators. Thus, it can be thought of as representing how much two given operators do, or do not, commute.

Exercise 1.35 Recall the operators (multiply by x) and (differentiate by x) used in the commutator example in Section 1.3.6. Show that (multiply by x) is Hermitian. Next show that $i\frac{d}{dx}$ is Hermitian [Hint: first show that (multiply by i) is anti-Hermitian, and then show that (differentiate by x) is anti-Hermitian (use integration by parts and the correct inner product definition)]. Now apply the Heisenberg uncertainty relation to show:

$$\Delta x \Delta\left(i\frac{d}{dx}\right) \geq \frac{1}{2}.$$

∎

Uncertainty relations are important, not only for quantum mechanics. They can, for example, provide an understanding of the limitations of time frequency analysis, some details of which can be found in [57].

If any two Hermitian operators commute, then they minimise the Heisenberg uncertainty relation – this is why such operators are said to be compatible. If not, they are said to be incompatible.

1.7 Concluding Remarks

The Dirac notation of quantum mechanics can confuse many that are new to this subject, and annoy some mathematicians that may see it as unnecessary. Once you are used to working with it, you will find that it makes it easier to spot some patterns and to perform some mathematical tricks. Specifically, the notation makes the methods of calculation and reasoning in quantum mechanics more visually clear.

If you struggle with the concepts from this chapter, you can always use the analogies that kets behave like column vectors, bras like their transpose complex conjugate, operators like matrices, etc. It may also help to think of a function as a column vector with a continuous index. There are some subtleties to the functional analysis of quantum mechanics which means such analogies do not always hold – but unless you plan to become a practitioner in the subject these are rarely an issue.

It is by trying, and making mistakes, and learning from those mistakes, that we gain understanding and eventually some degree of mastery of the discipline. So the key to making progress will be to practice, and solve problems, until you are comfortable with the mathematics. That way you can later focus on the physics without being distracted by struggling with the mathematics at the same time.

Before we embark on a description and explanation of quantum mechanics, we will next summarise the Hamiltonian formulation of classical physics. This will enable us to make clear what aspects of quantum mechanics are specifically different from classical physics and which are not. Our intent is to remove some of the potential for confusion that can arise when introducing quantum mechanics without taking classical Hamiltonian physics into account.

2

Notes on Classical Mechanics

"The chief law of physics, the pinnacle of the whole system is, in my opinion, the principle of least action."

Max Planck/Public Domain

2.1   Introduction

Both classical and quantum mechanics can be studied from different theoretical perspectives. Each has its own advantages and brings different insights. Maybe because of historical imperative, the teaching of a first course in undergraduate classical mechanics seems almost universally to be given in the Newtonian framework. There are two other approaches to classical mechanics due to Lagrange and Hamilton that are closely related to each other. They differ from each other in a similar way to the integral versus point (differential) forms of Maxwell's equations. The Lagrangian formulation is like the former, insofar as it provides a global understanding of the system, whilst that of Hamilton view is more like the second, providing local insight at the particle level. They make use of one of two quantities, the Lagrangian (the difference between kinetic and potential energies) or the Hamiltonian (usually, but not always, the sum of kinetic and potential energies) to determine the dynamical behaviour of mechanical systems. This way of understanding classical physics is predominately, and we think incorrectly, viewed as too advanced for first-year undergraduates. Anyone studying their first course in quantum mechanics early in these studies will encounter the Hamiltonian. That which was deemed too hard for students to deal with only a semester before now has to be dealt with, but additionally within the context of also trying to understand a brand-new and apparently different theory. This approach makes quantum mechanics seem far more different to classical mechanics than it actually is. Moreover, it also obfuscates some of the ways in which quantum mechanics is specifically and shockingly different to classical mechanics. A good grounding in classical Lagrangian and Hamiltonian mechanics will enable us to introduce quantum mechanics in a far more natural way than we can without it.

In this chapter, we provide an introduction to classical Lagrangian and Hamiltonian mechanics. We will start by detailing how a system of particles can be described by generalised coordinates. We will then introduce something called *the action*. This is

Quantum Mechanics: From Analytical Mechanics to Quantum Mechanics, Simulation, Foundations, and Engineering, First Edition. Mark Julian Everitt, Kieran Niels Bjergstrom and Stephen Neil Alexander Duffus.
© 2024 John Wiley & Sons Ltd. Published 2024 by John Wiley & Sons Ltd.
Companion website: www.wiley.com/go/everitt/quantum

simply a function that takes the coordinates and velocities needed to describe the state of the system and maps them onto a numerical value describing how the physical system changes over time. We will then use one of the most fundamental concepts in physics, the principle of least action, to derive the Euler–Lagrange equations. These equations provide a mechanism for obtaining the equations of motion of position and velocity from the Lagrangian. Noting that from the Lagrangian we find that a quantity which is conserved for systems with translational symmetry exists, we introduce the momentum. It turns out that in most formulations of quantum mechanics, momentum is more fundamental than velocity. The canonical momentum in many circumstances reduces to that with which we are familiar from Newtonian mechanics, but in some cases it is different. A change of variables from coordinates and velocities to coordinates and conjugate momenta enables us to introduce the Hamiltonian. We then derive Hamilton's equations, which give the dynamics of the system in terms of position and momentum. This enables us to introduce a quantity that we will find of fundamental importance in understanding quantum mechanics, the *Poisson bracket*. Please pay particular attention to how the Poisson bracket and Hamiltonian together define classical dynamics as a flow in phase space. We see that the Poisson bracket is the thing that determines classical evolution – or we could say that the definition of the Poisson bracket also defines classical physics. In particular, the Poisson brackets for position and momentum, as well as those for angular momentum, should be well noted, as they will turn out to be of great importance for developing our understanding of quantum physics. This quote from Dirac's Varenna Lectures puts in historical and scientific context the importance of this topic for our future discussions on quantum mechanics:

'I went back to Cambridge at the beginning of October 1925, and resumed my previous style of life, intense thinking about these problems during the week and relaxing on Sunday, going for a long walk in the country alone. The main purpose of these long walks was to have a rest so that I would start refreshed on the following Monday.

It was during one of the Sunday walks in October 1925, when I was thinking about this $(uv - vu)$, in spite of my intention to relax, that I thought about Poisson brackets. I remembered something which I had read up previously, and from what I could remember, there seemed to be a close similarity between a Poisson bracket of two quantities and the commutator. The idea came in a flash, I suppose, and provided of course some excitement, and then came the reaction 'No, this is probably wrong'. I did not remember very well the precise formula for a Poisson bracket, and only had some vague recollections. But there were exciting possibilities there, and I thought that I might be getting to some big idea. It was really a very disturbing situation, and it became imperative for me to brush up on my knowledge of Poisson brackets. Of course, I could not do that when I was right out in the countryside. I just had to hurry home and see what I could find about Poisson brackets.

I looked through my lecture notes, the notes that I had taken at various lectures, and there was no reference there anywhere to Poisson brackets. The textbooks which I had at home were all too elementary to mention them. There was nothing I could do, because it was Sunday evening then and the libraries were all closed. I just had to wait impatiently through that night without knowing whether this idea was really any good or not, but I still think that my confidence gradually grew during the course of the night.

Note: Dirac *'did not remember very well the precise formula for a Poisson bracket'*. Even a very great contributor like Dirac had 'vague recollections' of the subject matter and was anxious to check his idea against textbooks he did not have access to at the time. His narrative well exemplifies that physics is created by people who have glimmers of ideas and do not know everything, but then work hard to see where those ideas lead. Hopefully, this bit of history is something you can relate to compared with the idealised arguments presented in textbooks like this one. Our arguments have been optimised over many years of post-hoc thinking and teaching – it is not how the subject was created.

One of the most popular interpretations of quantum mechanics is probabilistic in its presentation. We may wish to properly understand the implications of that interpretation and how they differ or do not differ from the predictions of classical mechanics. Our discussion therefore moves on to considerations of probability in classical mechanics. The deterministic nature of classical mechanics may mean that the role of probability in experimental empiricism is overlooked. Given that any experimental apparatus has finite precision and accuracy, the predictive nature of any theory needs to be understood within that context. We explore this topic in sufficient depth to enable us to have well-informed discussions of the possible role of probability in quantum mechanics.

We then introduce the outline of an unusual presentation of classical mechanics due to Koopman and von Neumann. This is ahistorical since the idea was put forward to make classical mechanics look very much like one of the main presentations of quantum theory due to Schrödinger. We choose to present it in advance, as its discussion naturally follows on from the previous discussion of probability in classical mechanics. We believe that if this is understood, the transition from quantum to classical theory becomes much easier to understand and make. At the very least, quantum mechanics should not look so radically different from a formulation of classical mechanics as it would without discussion of this topic. A further advantage of understanding Koopman-von Neumann classical mechanics is that it will make it much easier for us to discuss the ways in which quantum mechanics is fundamentally different from any presentation of classical mechanics.

We finally outline some problems with classical mechanics and note that it is a theory that cannot explain certain observed physical phenomena. For this reason, we need a new theory, not only capable of explaining those phenomena, but preferably also able to reproduce the predictions of classical mechanics. Consequently, we shall briefly discuss the correspondence principle, providing a formal statement as to how new theories can be developed whilst preserving empirical findings of existing theories, where these have proven to be effective. In this way, we set the scene for introducing quantum mechanics in a manner that seeks to highlight the similarities rather than the differences between classical and quantum mechanics.

Our discussion is constrained to containing only that which is needed to support our later discussion on quantum mechanics. For a full treatment, there are many good textbooks on classical mechanics that you may turn to if you find our presentation too brief. Of particular note is [36] which is a standard first course on the subject, and a personal favourite [37]. A more gentle introduction can be found in [84]. A text which really comes into its own once you already have a grasp of the subject is [47]. Its clarity, depth of argument, and insight make it a valuable read. Another noteworthy mention is Arnold [87], but unless you have the required mathematics, it is a very demanding presentation. This chapter brings in ideas from each of those texts, but is also original in its form.

2.2 A Brief Revision of Classical Mechanics

Prerequisite Material: In this chapter we assume familiarity with mathematics typically encountered within the first year of a physics degree. Specifically, you will need to have experience with: partial and total derivatives and be clear in your mind of the differences between the two, set notation, exponentials with real and complex arguments, and solutions to simple partial differential equations.

2.2.1 Lagrangian Mechanics

Generalised coordinates, $q_i(t)$, are a set of numbers, with index i, that completely describe the state of a system at any one point in time, t. There will be one generalised coordinate per degree of freedom within any system. The total time-derivative of each coordinate, $\dot{q}_i(t)$, provides the associated generalised velocities. In three-dimensional space, the state of an N-particle system, and sufficient information to predict all future states, is given by a set of generalised coordinates and their velocities

$$\left\{ q_i(t), \dot{q}_i(t) \, | \, i = 1 \dots 3N \right\}. \tag{2.1}$$

The set of all possible values is the phase space of the system in terms of position and velocity. To a given system we can describe an action which is a quantity that characterises, as a number, how the system changes over time. In other words, the action, S, is some functional that maps the generalised coordinates, their velocities, and time onto a real number in a way that characterises the system's dynamics. This can be expressed as

$$S \stackrel{\text{def}}{=} \int_{t_i}^{t_f} dt \, \mathscr{L}(q_i(t), \dot{q}_i(t), t), \tag{2.2}$$

where $\mathscr{L}(q_i(t), \dot{q}_i(t), t)$ is a function called the Lagrangian. Hamilton's principle essentially states that where the action does not change for (is not sensitive to) small changes in the trajectory (stationary points), we can determine the equations of motion for the system. That is, Dynamics is given by the principle of 'least' action, so that the path taken by the system of particles is between two points in time, t_i (initial) and t_f (final), and is the one for which the action is stationary (usually minimised), $\delta S = 0$, to first order. Note in the derivation that follows that we do not specifically use the form of the Lagrangian. The result is quite general. We consider problems where the initial $q_i(t_1)$ and final $q_i(t_2)$ sets of coordinates are known (it is a boundary value problem). Associated with this will be one or more trajectories $q_i(t)$ associated with that 'least' action, which gives the physically allowable trajectories of the particle(s), the system's dynamics. In other words, the principle of least action tells us how the system navigates its way between the two.

There may be an infinite number of possible paths between $q_i(t_1)$ and $q_i(t_2)$, but we assume we know the end points, which remain fixed. This is why we consider the Lagrangian formulation a global rather than a local view. Our task is now to minimise that action (strictly speaking, we seek to find stationary points). Let us denote by $\overline{q}_i(t)$ the path that 'minimises' the action. As the initial and final points are fixed, we have $\overline{q}_i(t_1) = q_i(t_1)$ and $\overline{q}_i(t_2) = q_i(t_2)$. Even if we vary the path from the stationary-action path by an amount $\delta q(t)$, these points are not allowed to change. That is, we assert that

$$\delta q(t_1) = \delta q(t_2) = 0. \tag{2.3}$$

We have required that $\delta \mathcal{S} = 0$, to first order. This means that $\mathcal{S}\left[\overline{q}(t) + \delta q(t)\right] - \mathcal{S}\left[\overline{q}(t)\right] = 0$.

We are now going to use an argument from the calculus of variations. This is only used at this particular point in the book, so do not worry too much if the subject is unfamiliar. That said, it is a powerful method and well worth spending time mastering. The calculus of variations is used to minimise expressions, such as Eq. (2.2), by considering what would happen by making small changes to the function in the integrand. In this example, we will be changing the trajectories $q_i(t)$ to determine which one 'minimises' $\mathcal{S}$. Importantly, note that the variation in the velocity is determined by the variation in the coordinates, as the velocity depends entirely on the change in coordinates for a given increment in time (this may seem obvious once it is pointed out, but from experience teaching computational implementations of variational calculus, this point appears to be often misunderstood). See Figure 2.1 for an illustration of this method.

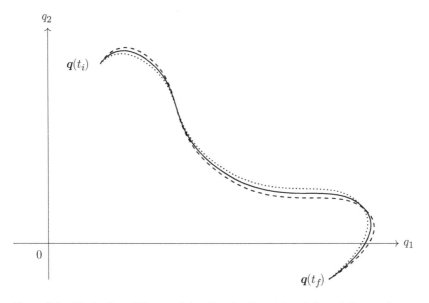

Figure 2.1 Illustration of the principle of least action determining a trajectory for some non-trivial potential. In this picture, assume that the dark line is the actual dynamical trajectory of the system. Changing the trajectory to either the dotted or dashed line results in a different set of coordinates and velocities that in turn result in an increase in the action. The principle of least action states that those trajectories will not be followed.

By the calculus of variations, from Eq. (2.2), we have

$$
\mathcal{S}\left[q(t)+\delta q(t)\right]=\int_{t_i}^{t_f}\mathrm{d}t\,\mathcal{L}\left(q_i(t)+\delta q_i(t),\dot{q}_i(t)+\delta\dot{q}_i(t),t\right) \tag{2.4}
$$

$$
=\int_{t_i}^{t_f}\mathrm{d}t\left[\mathcal{L}\left(q_i(t),\dot{q}_i(t),t\right)+\delta q(t)\frac{\partial\mathcal{L}}{\partial q}+\delta\dot{q}(t)\frac{\partial\mathcal{L}}{\partial\dot{q}}\right] \tag{2.5}
$$

$$
=\mathcal{S}\left[q(t)\right]+\int_{t_i}^{t_f}\mathrm{d}t\left[\delta q(t)\frac{\partial\mathcal{L}}{\partial q}+\delta\dot{q}(t)\frac{\partial\mathcal{L}}{\partial\dot{q}},\right] \tag{2.6}
$$

so we have the variation of the action

$$
\delta\mathcal{S}=\mathcal{S}\left[\overline{q}(t)+\delta q(t)\right]-\mathcal{S}\left[\overline{q}(t)\right]=\int_{t_i}^{t_f}\mathrm{d}t\left[\delta q(t)\frac{\partial\mathcal{L}}{\partial q}+\delta\dot{q}(t)\frac{\partial\mathcal{L}}{\partial\dot{q}}\right].
$$

Integrating by parts yields

$$
\delta\mathcal{S}=\left[\delta q\left(\frac{\partial\mathcal{L}}{\partial\dot{q}}\right)\right]_{t_1}^{t_2}-\int_{t_i}^{t_f}\mathrm{d}t\,\delta q(t)\left[\frac{\mathrm{d}}{\mathrm{d}t}\left(\frac{\partial\mathcal{L}}{\partial\dot{q}}\right)-\frac{\partial\mathcal{L}}{\partial q}\right].
$$

Noting Eq. (2.3) means the first term vanishes, and in order for the action to be stationary we obtain the Euler–Lagrange equations:

$$
\frac{\mathrm{d}}{\mathrm{d}t}\frac{\partial\mathcal{L}}{\partial\dot{q}_i}-\frac{\partial\mathcal{L}}{\partial q_i}=0, \tag{2.7}
$$

which, in turn, give the equations of motion for each of the system's coordinates.

Now we need to determine a Lagrangian that can be used to derive the equations of motion. Consider the simplest of examples, a free particle in space. Assuming space is both isotropic and homogeneous (so the action is to be the same in different inertial frames of reference), the Lagrangian cannot depend on position or time, nor can it depend upon which direction in space the particle is travelling. It must therefore be a function of the speed of the particle, and one such example is $\dot{q}^2$. An immediate corollary of the Euler–Lagrange Eq. (2.7) is the law of inertia: that a free particle travels at a constant velocity in any inertial frame. Such observations in turn lead to Galilean relativity. Following, e.g. [47] one can show that a Lagrangian of the form

$$
\mathcal{L}=\frac{1}{2}m\dot{q}^2 \tag{2.8}
$$

satisfies the Galilean relativity principle. From this we understand that the kinetic energy, which we will denote by $\mathcal{T}$, is a Lagrangian for a free particle and the notion of mass arises as a constant of proportionality. The contribution to a suitable Lagrangian that is not due to the free motion, but has the same units as $\mathcal{T}$, we term the Potential energy and denote this by $\mathcal{V}$ (which, by Galilean relativity, should usually be a function of only coordinates and time). The usual Lagrangian of classical mechanics is then defined as

$$
\mathcal{L}\left(q_i(t),\dot{q}_i(t),t\right)\stackrel{\mathrm{def}}{=}\mathcal{T}-\mathcal{V}, \tag{2.9}
$$

which describes the trade-off between kinetic and potential energies at any instant. As such, it makes a good candidate as a functional for characterising the system's capacity for change (in conservative systems, the total energy is conserved, which means that we *cannot* use that as a Lagrangian for applying the principle of least action).

The approach of Lagrangian mechanics is strikingly powerful, but also formulaic. In order to find the equations of motion for any system, all we have to do is to figure out any convenient set of generalised coordinates, express the kinetic energy and the potential energy using these coordinates, subtract them to make the Lagrangian, substitute them into the above equation, and *voilà*, we are done. For any complicated system, this is a much simpler process than applying Newton's laws. The next exercise looks at one of the most important concepts in physics and is worth spending some time solving.

Exercise 2.1   Use the Euler–Lagrange equations to derive the equation of motion for the simple harmonic oscillator. You may use $\mathcal{T} = \frac{1}{2}m\dot{q}^2$ and $\mathcal{V} = \frac{1}{2}m\omega^2 q^2$ where m is mass and ω is the resonant angular frequency.   ■

Exercise 2.2   The Euler–Lagrange equations also lead us to an understanding of Noether's theorem. This states that any conservative system with a continuous symmetry will have an associated conservation law. Show that for any system with the translational symmetry $q_i \rightarrow q_i + \delta q_i$ the quantity $\frac{\partial \mathcal{L}}{\partial \dot{q}_i}$ is conserved. Specific example systems will be those systems where the potential between particles is a function of the difference between coordinates.   ■

Exercise 2.3   For the simple harmonic oscillator, show that $\frac{\partial \mathcal{L}}{\partial \dot{q}_i} = m\dot{q}$. Is this a conserved quantity?   ■

For our later discussion on the path-integral formulation of quantum mechanics, note that in the above discussion we have assumed (i) there is one definite start and one definite end point to the particle's trajectory; (ii) at all times, the state of the system is described by the positions and velocities of a set of particles. This is one point in the system's phase space; and (iii) there is only one trajectory - the particle follows the one that makes the action stationary. It will be some time before our discussion gets us to the path-integral formulation of quantum mechanics, but as you read the intervening material, reflect on how the above assumption conflicts with the development of our quantum mechanical argument and how they might be reconciled. We find that quantum mechanic's path-integral formulation will remove some of these assumptions, but recover what they say as a limiting case.

2.2.2   Hamiltonian Mechanics

An alternative to Lagrangian mechanics is that of Hamiltonian mechanics. We might wonder 'why bother, after all, isn't Lagrangian Mechanics enough?' Given that we have just stated that a path-integral formulation of quantum mechanics exists, which is close to the principle of least action and reasonably intuitive to understand, this question needs to be answered. The first point is, that while the path-integral formulation of quantum mechanics is elegant, very few problems can easily be solved using that approach. The next argument is that we found it hard to argue for the full framework of quantum mechanics, especially those arguments related to measurement, from the path-integral approach alone. For this reason, we continue developing classical mechanics to a point that enables us to naturally introduce standard quantum theory. We find that Hamiltonian mechanics provides a geometric and algebraic interpretation of classical mechanics that is missing from the Lagrangian formulation. It is from this geometric view that the standard Schrödinger view of quantum mechanics can be clearly motivated. The following discussion is designed to set the necessary scene from which to do this.

We start by looking at Exercise 2.2 and the quantity that is conserved for all those systems with translational symmetry. Exercise 2.3 tells us this conserved quantity looks like the momentum of Newtonian mechanics (and, by Noether's theorem we understand when and why, i.e. the symmetry, for which it is conserved). This leads us to define the canonically conjugate momenta to the coordinates

$$p_i \overset{\text{def}}{=} \frac{\partial \mathcal{L}}{\partial \dot{q}_i}. \tag{2.10}$$

It is worth observing that from a global (integral) view, namely least action, of the system, we have derived a local (differential) set of equations describing the dynamics of the system (the Euler–Lagrange equations). These two views are quite interchangeable. Our definition of momentum through a derivative is a local concept, as is the material that follows (hence our analogy in the introduction to the integral versus point form of Maxwell's equations).

Substituting the definition of momentum into the Euler–Lagrange Eq. (2.7), we get

$$\frac{\mathrm{d}}{\mathrm{d}t}p_i - \frac{\partial \mathcal{L}}{\partial q_i} = 0 \implies \dot{p}_i = \frac{\partial \mathcal{L}}{\partial q_i}. \tag{2.11}$$

A Lagrangian is a function of coordinates and velocities. Now that we have defined momentum, we can ask if it is possible to transform the Lagrangian into a function of coordinates and their canonically conjugate momenta. Let us assume this is true and that the Hamiltonian is a function $\mathcal{H}(q_i, p_i)$. The product rule for the total differential is:

$$\mathrm{d}\mathcal{H} = \sum_i \left(\frac{\partial \mathcal{H}}{\partial q_i} \mathrm{d}q_i + \frac{\partial \mathcal{H}}{\partial p_i} \mathrm{d}p_i \right) + \frac{\partial \mathcal{H}}{\partial t} \mathrm{d}t. \tag{2.12}$$

With the benefit of hindsight, let us see what happens if we define the Hamiltonian as:

$$\mathcal{H} \overset{\text{def}}{=} \sum_i \dot{q}_i p_i - \mathcal{L}. \tag{2.13}$$

As our discussion progresses, observe that the Hamiltonian will be implicitly shown to represent the mechanical energy of a system. We can see what the total derivative of this quantity is and compare it to the previous equation:

$$
\begin{aligned}
\mathrm{d}\mathcal{H} &= \sum_i \left(\dot{q}_i \, \mathrm{d}p_i + p_i \, \mathrm{d}\dot{q}_i \right) - \left[\sum_i \left(\frac{\partial \mathcal{L}}{\partial q_i} \mathrm{d}q_i + \frac{\partial \mathcal{L}}{\partial \dot{q}_i} \mathrm{d}\dot{q}_i \right) + \frac{\partial \mathcal{L}}{\partial t} \mathrm{d}t \right] \\
&= \sum_i \left(\dot{q}_i \, \mathrm{d}p_i + p_i \, \mathrm{d}\dot{q}_i - \dot{p}_i \mathrm{d}q_i - p_i \mathrm{d}\dot{q}_i \right) - \frac{\partial \mathcal{L}}{\partial t} \mathrm{d}t \\
&= \sum_i \left(\dot{q}_i \, \mathrm{d}p_i - \dot{p}_i \, \mathrm{d}q_i \right) - \frac{\partial \mathcal{L}}{\partial t} \mathrm{d}t. \tag{2.14}
\end{aligned}
$$

Here we substituted in Eqs. (2.10) and (2.11). By comparing the two total derivatives, we see that this definition of the Hamiltonian works very nicely. It is guaranteed to be a function of coordinates and momentum, and equating coefficients of differentials yields

$$\frac{dq_i}{dt} = \frac{\partial \mathcal{H}}{\partial p_i}, \quad \frac{dp_i}{dt} = -\frac{\partial \mathcal{H}}{\partial q_i} \quad \text{and} \quad \frac{\partial \mathcal{H}}{\partial t} = -\frac{\partial \mathcal{L}}{\partial t}. \tag{2.15}$$

The first two of these equations are Hamilton's equations and, just like the Euler–Lagrange equations, they provide the dynamical equations of motion for the system.

When $\mathcal{V}$ is a function of the coordinates q_i alone (known of as Scleronomous constraints) the Hamiltonian reduces to the total energy:

$$\mathcal{H} = \mathcal{T} + \mathcal{V} \tag{2.16}$$

As most potentials considered in quantum mechanics are scleronomous, you may see the Hamiltonian being incorrectly defined in this way. There are some situations where the potential is velocity dependent and it important to derive the momentum and Hamiltonian properly.

Let us continue to develop our use of total time-derivatives, and now note that the dynamics of any quantity $A(q_i(t), \dot{q}_i(t), t)$ is then given by

$$\frac{dA}{dt} = \frac{\partial A}{\partial t} + \sum_i \frac{\partial A}{\partial q_i} \frac{dq_i}{dt} + \frac{\partial A}{\partial p_i} \frac{dp_i}{dt}, \tag{2.17}$$

which is just the chain rule for the total derivative of any quantity. We can now add the constraint that this dynamics obeys the laws of classical physics (for undamped systems) use by substituting Eqs. (2.15) to obtain

$$\frac{dA}{dt} = \frac{\partial A}{\partial t} + \sum_i \frac{\partial A}{\partial q_i} \frac{\partial \mathcal{H}}{\partial p_i} - \frac{\partial A}{\partial p_i} \frac{\partial \mathcal{H}}{\partial q_i}. \tag{2.18}$$

This links the dynamics of A to the principle of least action through Hamilton's equations. This link is very important and we can emphasise the role of the Hamiltonian by writing this equation using the standard notation of the Poisson bracket,

$$\{A, B\} \stackrel{\text{def}}{=} \sum_i \frac{\partial A}{\partial q_i} \frac{\partial B}{\partial p_i} - \frac{\partial A}{\partial p_i} \frac{\partial B}{\partial q_i} \tag{2.19}$$

so that

$$\frac{dA}{dt} = \frac{\partial A}{\partial t} + \{A, \mathcal{H}\} \tag{2.20}$$

specifies the dynamics of *any* quantity.

It is the Hamiltonian that specifies the system and the initial conditions of its state. Together these determine the system's dynamics. It is, most importantly, the *Poisson bracket* that defines the way in which the evolution happens as

$$dp_i = \{p_i, \mathcal{H}\} \, dt \quad \text{and} \quad dq_i = \{q_i, \mathcal{H}\} \, dt. \tag{2.21}$$

The Poisson bracket thus defines the way that the Hamiltonian of the system increments the state of the system from one point in time to another and provides the structure of classical mechanics.

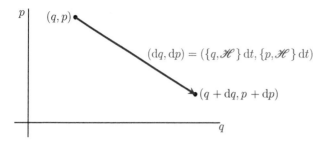

Figure 2.2 An exaggerated illustration of how Poisson bracket and Hamiltonian define a flow in phase space by mapping one point to another in a time increment dt.

To potentially over-labour the point: the Poisson bracket specifies the way in which the system's state transforms incrementally in time for a given Hamiltonian as a flow in phase space. For example, in a time increment dt the coordinates

$$(q_i, p_i) \rightarrow (q_i + \{q_i, \mathcal{H}\} \, dt, p_i + \{p_i, \mathcal{H}\} \, dt), \tag{2.22}$$

and we can see how the Hamiltonian defines a 'flow' in phase space (Figure 2.2 illustrates this in one-dimension).

We now point out a few properties of Hamiltonian mechanics and Poisson brackets that are important for our later discussion on quantum mechanics. There is rich physics beyond this that deserves more study than we can accommodate here, but can be found in the textbooks recommended in the introduction.

We begin by observing, as Figure 2.2 indicates, just like we could with positions and velocities, that each position and momentum can be ascribed to a Cartesian coordinate system and form a phase space. Let us consider the case where $A(q_i, p_i, t)$, $B(q_i, p_i, t)$, and $C(q_i, p_i, t)$ are any phase space functions. If $\{A, B\} = 0$, we say that A and B commute. If $dA/dt = 0$ then it is a conserved quantity and termed a constant of the motion. A constant of the motion that does not depend explicitly on time is called an integral of the motion. If a function A commutes with the Hamiltonian and is not explicitly dependent on time, then it is an integral of the motion. Next we consider two sets of results for Poisson brackets that as we will later see are crucially important for understanding Dirac's approach to quantisation. The first arises from the definition of the Poisson bracket, that for all canonical coordinate systems:

$$\{q_i, q_j\} = 0 \tag{2.23}$$

$$\{p_i, p_j\} = 0 \tag{2.24}$$

$$\{q_i, p_j\} = \delta_{ij}, \tag{2.25}$$

where δ_{ij} is the Kronecker delta which is 1 if $i = j$ and zero otherwise. The second is:

$$\{A, A\} = 0 \tag{2.26}$$

$$\{A, B\} = -\{B, A\} \tag{2.27}$$

$$\{aA + bB, C\} = a\{A, C\} + b\{B, C\} \tag{2.28}$$

$$\{AB, C\} = \{A, B\}C + A\{B, C\} \tag{2.29}$$

$$0 = \{\{A, B\}, C\} + \{\{C, A\}, B\} + \{\{B, C\}, A\}, \tag{2.30}$$

where a and b real numbers. This last set of relations is so important that they form the basis for a branch of mathematics called Lie algebra which generalises the idea of infinitesimal symmetry motions such as spatial rotations.

Exercise 2.4   Prove each of these relations. Then show from two of these equations that $\{h, af + bg\} = a\{h, f\} + b\{h, g\}$. ∎

Exercise 2.5   Show, in one line, that $\mathcal{H}$ for a closed system is an integral of the motion. ∎

Now we introduce some notation that will help us better introduce the Schrödinger picture of quantum mechanics. Observe that by slightly rearranging the Poisson bracket for the Hamiltonian with a function that we have yet to define (denoted missing by the dot in the bracket) we have a quantity

$$\{\mathcal{H}, \cdot\} = \frac{\partial \mathcal{H}}{\partial q} \frac{\partial}{\partial p} - \frac{\partial \mathcal{H}}{\partial p} \frac{\partial}{\partial q} \tag{2.31}$$

that takes the form of a differential operator. We can define another operator, called the Liouvillian as:

$$\hat{L} \overset{\text{def}}{=} -\mathrm{i}\{\mathcal{H}, \cdot\} \tag{2.32}$$

so that we conclude our introduction to Hamiltonian mechanics by presenting it in a form that begins to look like our introduction to quantum theory

$$\frac{\mathrm{d}A}{\mathrm{d}t} = \mathrm{i}\,\hat{L}A + \frac{\partial A}{\partial t}. \tag{2.33}$$

To make the connection between quantum and classical mechanics even clearer, we next turn to a discussion of probability in classical mechanics. Note that the introduction of an apparently unnecessary 'i' will become evident when we discuss quantum mechanics.

2.3   On Probability in Classical Mechanics

2.3.1   The Liouville Equation

Because of its apparent deterministic nature, the usual presentation of classical mechanics is in terms of definite trajectories of a system of particles (or fields) that follow from some set of perfect initial conditions. It is not possible to measure a given physical system to indefinite accuracy and precision. As an example, consider the simple school textbook problem of determining if a ladder leaning against a wall will slip. In the Platonist ideal (the imagined mathematically perfect version of reality), for a given ladder and coefficient of friction, there will be an exact angle at which slipping will begin. Because of the empirical nature of science, the physics starts when we then use this model to predict when actual ladders will slip; but that means measuring length, weight, and centre of mass of the not-quite-uniform ladder, its position against the imperfect wall to estimate the angle of inclination and the coefficient of friction, which may also vary with position. Each of these quantities comes with a degree of uncertainty. The empirical physicist will then quantify these uncertainties and use the underlying deterministic mechanics to form a probabilistic model calculating the chance of a ladder slipping. As Max Born pointed out in his 1954 Nobel Prize Lecture:

> *'Ordinary mechanics must also be statistically formulated: the determinism of classical physics turns out to be an illusion, it is an idol, not an ideal in scientific research.'*

For any complex system the situation is even more complicated than this as there is a reasonable chance that we will get our model wrong. For example, we might not identify all the contributions to the system's kinetic and potential energies. In systems that are non-linear, this could be a real issue. It is a problem that we will not consider here, but modelling and simulation in such situations is something to think about. In classical and quantum electronics, for example, there will be parasitical inductances and capacitances that might alter a system's Hamiltonian. We will limit the present discussion to the situation where our model is understood to be correct but we have some uncertainty in establishing the system's initial state. Not only is this the simplest first step; it is all that we need to motivate our discussion of quantum mechanics.

The deterministic Hamiltonian view describing the initial state of a system of particles is a set of coordinates and conjugate momenta in phase space. Uncertainty will smear these points in phase space into a probability distribution of its possible initial conditions. The shape of that smearing will depend on the nature of the uncertainty but it will usually take a form approximating a Gaussian distribution (the width of which may be different for different degrees of freedom). The probability density function over the phase space is denoted $\rho(q_i, p_i)$. For a given area/volume/hypervolume V of phase space,

$$P(V) = \int_V dV \, \rho(q_i, p_i)$$

is the probability of finding the system in this volume of phase space. Note that $P(V = [\text{all of phase space}]) = 1$, since the system must always be in some configuration. If the position and momentum were known exactly, at (Q_i, P_i) say, then the distribution would take the form of a Dirac delta functioncentred at one point in phase space $\rho(q_i, p_i) = \prod_i \delta(q_i - Q_i, p_i - P_i)$. We will not consider this situation here, but those interested in how the following arguments develop consistently with the Hamiltonian formulation of classical mechanics might like to investigate, e.g. the Klimontovich equation used in plasma physics. As our focus is understanding classical mechanics and probability density functions in a way that makes quantum physics seem more natural, we will discuss only the case where $\rho(q_i, p_i)$ is a smeared distribution.

Also note that the probability density function can be used to study not only a system of particles whose initial state is not well known, but also an ensemble where each copy starts off in a different configuration. For example, we could consider a statistical mixture represented by starting off one half of an experiment at one point in phase space and one half at another (an odd example that will later be seen to be relevant for two-state quantum systems). Importantly, this approach allows us to express mathematically what happens to a statistical ensemble of experiments, and generalises the notion of the 'state' of a system from a single configuration to something more general (and more empirical - especially for non-linear systems). A statistical mixture defines a probability density according to:

$$\rho = \sum w_i \rho_i, \tag{2.34}$$

where w_i represents the weight of each contribution. Note that this sum may become an integral if there is a continuum of contributions to the statistical mixture.

One of the nice things about Hamiltonian mechanics is that we have the dynamics of any quantity from Eq. (2.20). We can therefore apply this directly to the probability density so that

$$\frac{d\rho(q_i, p_i, t)}{dt} = \{\rho(q_i, p_i, t), \mathscr{H}\} + \frac{\partial \rho(q_i, p_i, t)}{\partial t}. \tag{2.35}$$

We will find that we can make an important simplification to this equation. To understand this, we start by making an analogy with fluid dynamics or electromagnetism and consider ρ as some density distribution (like mass density or charge). By analogy, we can associate a probability current density

$$\boldsymbol{j}(q_i, p_i, t) \triangleq \rho(q_i, p_i, t)\boldsymbol{v}(q_i, p_i, t)$$
$$= \left(\rho\frac{dq_1}{dt}, \rho\frac{dq_2}{dt}, \dots, \rho\frac{dp_1}{dt}, \rho\frac{dp_2}{dt}, \dots, \right)$$

with the probability density.

Now note that

$$\nabla \cdot \boldsymbol{j}(q_i, p_i, t) = \frac{\partial \rho}{\partial q_1}\frac{dq_1}{dt} + \frac{\partial \rho}{\partial q_2}\frac{dq_2}{dt} + \dots$$
$$\dots + \frac{\partial \rho}{\partial p_1}\frac{dp_1}{dt} + \frac{\partial \rho}{\partial p_2}\frac{dp_2}{dt} \dots$$

where $\nabla = (\frac{\partial}{\partial q_1}, \frac{\partial}{\partial q_2}, \dots, \frac{\partial}{\partial p_1}, \frac{\partial}{\partial p_2}, \dots)$. Now we use Hamilton's equations (2.15) to obtain

$$\nabla \cdot \boldsymbol{j}(q_i, p_i, t) = \frac{\partial \rho}{\partial q_1}\frac{\partial H}{\partial p_1} + \frac{\partial \rho}{\partial q_2}\frac{\partial H}{\partial p_2} + \dots$$
$$\dots - \frac{\partial \rho}{\partial p_1}\frac{\partial H}{\partial q_1} - \frac{\partial \rho}{\partial p_2}\frac{\partial H}{\partial q_2} - \dots$$

By inspection we see that the right-hand side of this equation is just the Poisson bracket of the probability density with the Hamiltonian, so

$$\nabla \cdot \boldsymbol{j}(q_i, p_i, t) = \{\rho(q_i, p_i, t), \mathscr{H}\}. \tag{2.36}$$

Substitution into Eq. (2.35) yields

$$\frac{d\rho(q_i, p_i, t)}{dt} = \nabla \cdot \boldsymbol{j}(q_i, p_i, t) + \frac{\partial \rho(q_i, p_i, t)}{\partial t}. \tag{2.37}$$

As previously stated, we know that the system must always be somewhere in its phase space, which gives

$$\int_V dV \rho(q_i, p_i, t) = 1,$$

and the fact that this probability must be conserved is expressed by

$$\frac{d\rho(q_i, p_i, t)}{dt} = 0.$$

This again means that

$$\frac{\partial \rho(q_i, p_i, t)}{\partial t} = -\nabla \cdot \boldsymbol{j}(q_i, p_i, t),$$

which is just like the continuity equation for an incompressible fluid or electric charge. When written in terms of the Poisson bracket we have:

$$\frac{\partial \rho(q_i, p_i, t)}{\partial t} = - \{\rho(q_i, p_i, t), \mathcal{H}\},$$

which is simply Eq. (2.35), noting that $\frac{d\rho}{dt} = 0$.

If $\rho(q_i, p_i)$ is some probability density function over phase space, then, as with any phase space function, its evolution is determined by Eq. (2.20) which yields the Liouville equation

$$\frac{\partial \rho(q_i, p_i)}{\partial t} = - \{\rho(q_i, p_i), \mathcal{H}\}, \tag{2.38}$$

as $\rho(q_i, p_i)$ has no explicit time dependence. This, importantly, shows that *Hamiltonian mechanics can naturally describe the dynamics of a probability distribution of possible states of its phase space*. More pedantically we can say that, in the absence of damping, the Poisson bracket embodies the principle of least action as a divergenceless flow of the system's probability density function in phase space. Moreover, *any formally correct real-world application of classical mechanics necessitates the use of probability densities* in this way.

Perhaps unsurprisingly from the nomenclature, the Liouville equation can be written, using the Liouvillian, as

$$\hat{L}\rho = i \frac{\partial}{\partial t} \rho. \tag{2.39}$$

This takes us one step closer to presenting classical mechanics in the way that makes it look most similar to quantum mechanics. Before taking that last step, we will take two asides. The first is a very brief consideration of damped systems, and the next is a discussion of expectation values. We will then complete our arguments with a description of Koopman–von Neumann theory, followed by some key failures of classical mechanics.

2.3.2 Expectation Values

Prerequisite Material: It would be helpful to read this in conjunction with Section 1.3.2: Expectation Value of Random Variables (and have some superficial understanding of the preceding mathematics).

For one degree of freedom the expected value of a quantity, $A(q, p)$, is

$$\langle A \rangle = \int dq \int dp \, A(q, p) \rho(q, p). \tag{2.40}$$

The expected value can be thought of as the continuous extension of the sum of the (probability of the system being in that state) times (the weight of the thing at [such as energy] at the point in phase space). Expectation values can be used to calculate other interesting quantities such as the Shannon entropy $\langle \log \rho^{-1} \rangle$ which represents the total amount of information[1] in the probability distribution. Generalisation to higher degrees of freedom is the natural extension of this formula, integrating over the entire phase space. For systems of many particles it may be useful to consider a reduced probability density function that results from integrating a system's full probability density function over all the coordinates and conjugate momenta except those components we are interested in,

$$\rho_{\text{reduced}} = \int_{\text{degrees of freedom excluding those of the system(s) of interest}} \rho. \tag{2.41}$$

Much of the time we would expect to do this to find the reduced probability density function for a single 'particle' to try to understand how its behaviour is effected by the rest of the system. For linear systems, such as the simple harmonic oscillator, any uncertainty in knowledge of the initial state of the system makes little difference in our ability to determine long-term behaviour. For the simple harmonic oscillator for example, if the initial state is a Gaussian centred around any point other than the origin, then the dynamics of the system will remain this Gaussian undergoing circular motion. Here, the expected position and momentum will be the centre of the distribution and its dynamics will be the same as the dynamics of a particle described by the particle's equations of motion. Note for this example of the harmonic oscillator, any Gaussian centred at the origin[2] does not change over time. This is an example of a stationary state that arises whenever $\{\rho, \mathcal{H}\} = 0$, as this implies $\frac{\partial \rho}{\partial t} = 0$. In quantum physics we will find that the notion of a stationary state also exists but takes a somewhat different form.

In other circumstances, some care may be needed to understand the information that the expectation value provides. As soon as we introduce any non-conservative or non-linear behaviour, such as found in the double pendulum or the Duffing oscillator, this situation changes drastically. We will consider the second example in some detail in the next section. For now, we note that the Duffing oscillator comprises a double-well system. It is straightforward to arrange it so that the probability density function for such systems can be composed of two disjoint pieces. In this case the system could be in one part of phase space or the other but nowhere else. The expected value of, e.g. position under such circumstances may not actually be a physically realisable point (in the similar way that the average of sampling a uniformly random selection of zeros and ones is a half).

Even though there may be some subtlety in interpreting the expectation value, an important result for later comparison with quantum theory results is that we can take Eq. (2.20) and from it calculate the dynamics of expectation values according to

$$\left\langle \frac{dA}{dt} \right\rangle = \langle \{A, \mathcal{H}\} \rangle + \left\langle \frac{\partial A}{\partial t} \right\rangle. \tag{2.42}$$

---

1 In this formulation we see that unlikely events can be considered information. It may help intuition to note that the maximum of the weighted information $\rho \log \rho^{-1}$ of a two-dimensional Gaussian lies in an annulus centred at the mean with a radius between one and two standard deviations.
2 Including the delta function as a limiting case of a particle whose position and momentum is known with absolute certainty.

For future reference, note that as

$$\frac{\partial A\rho}{\partial t} = \frac{\partial A}{\partial t}\rho + A\frac{\partial \rho}{\partial t}$$

and, using Leibniz rule,

$$\frac{d}{dt}\int dq\,dp A\rho = \int dq\,dp\frac{\partial A\rho}{\partial t} = \int dq\,dp\frac{\partial A}{\partial t}\rho + \int dq\,dp A\frac{\partial \rho}{\partial t}$$

so that

$$\frac{d}{dt}\langle A\rangle = \left\langle \frac{\partial A}{\partial t}\right\rangle + \int dq\,dp\left[A\frac{\partial\rho}{\partial t}\right] \tag{2.43}$$

and we see that $\frac{d}{dt}\langle A\rangle \neq \left\langle \frac{\partial A}{\partial t}\right\rangle$ in general. We will later use this observation in a critical analysis of the Schrödinger picture of quantum mechanics.

2.4 Damping

In classical mechanics we are familiar with the non-conservative forces such as friction. Friction isn't just important in understanding and modelling simple damping phenomena, its presence enables an important class of dynamics seen in many systems – dissipative chaos. Some systems only exhibit chaos for certain values of damping. Under such circumstances, systems will be constrained to dynamics in phase space that have a fractal structure – so-called strange attractors. Quantum mechanics was initially developed to understand the physics of systems such as atoms and molecules where there is no friction. We will shortly discuss the correspondence principle, from which we will see that there will be a requirement, if it is a 'good' theory, for quantum mechanics to predict phenomena such as chaos.

In Lagrangian and Hamiltonian mechanics, damping can be modelled in more than one way. A full treatment of damping and dissipation is beyond the scope of this book. For this reason, we will focus our attention on just one approach. However, we will later consider damped (open) quantum systems in some detail. Our discussion will be framed in terms of the quantum versions of the Liouville equation and the probability density function. To set that narrative in context, we here provide a short overview of the treatment of this topic within Hamiltonian mechanics. We aim to enable an understanding of how quantum damping and dissipation is different from their classical counterparts.

A system can experience two kinds of forces: those due to a potential $F = -\nabla\mathcal{V}$ and those not due to a potential, such as friction. Friction is dependent on the velocities of the coordinates (a stationary object does not experience friction). Although it is possible to derive a more general case, we will restrict this discussion to those forces that linearly depend on velocity. This means that the friction force will take the form

$$f_i = \sum_j a_{ij}\dot{q}_i.$$

Just as with normal forces, we would like to be able to view frictional forces as a gradient of some function like the potential but this time differentiating with respect to velocities rather than coordinates. That is, we also want $f_i = \partial \mathcal{F}/\partial \dot{q}_i$ for some dissipative function, $\mathcal{F}$. This, by inspection, we see is satisfied by

$$\mathcal{F} = \sum_j a_{ij} \dot{q}_i \dot{q}_j. \tag{2.44}$$

We can interpret this as the rate of dissipation of energy in the system. By dimensional analysis we see that each term in the Euler–Lagrange equations has units of force, so it is not unreasonable to expect

$$\frac{d}{dt} \frac{\partial \mathcal{L}}{\partial \dot{q}_i} - \frac{\partial \mathcal{L}}{\partial q_i} = -\frac{\partial \mathcal{F}}{\partial \dot{q}_i}. \tag{2.45}$$

Exercise 2.6 Look up Euler's theorem for Homogeneous functions and use it to prove

$$\frac{d\mathcal{H}}{dt} = -2\mathcal{F}.$$

Under what circumstances does $\mathcal{F}$ represent a loss of energy? ∎

It is possible to show that the above treatment and a repetition of our previous analysis leads to a modified version of Hamilton's equations:

$$\frac{dq_i}{dt} = \frac{\partial \mathcal{H}}{\partial p_i}, \quad \frac{dp_i}{dt} = -\left(\frac{\partial \mathcal{H}}{\partial q_i} + \frac{\partial \mathcal{F}}{\partial \dot{q}_i} \right). \tag{2.46}$$

We are not aware of any success in modifying the Liouville equation in a similar way. There are alternatives to the Liouville equation for damped systems, e.g. the Fokker–Planck equation. Their consideration is beyond the scope of this text. Interestingly, in the quantum case we will be able to extend the quantum Liouville equation to take dissipation and damping into account. Even without a Liouville equation for damped systems, we can still calculate the dynamics of probability density functions numerically. We do this by taking the modified Hamilton's equations and solve them for a very large number of trajectories whose initial conditions are sampled from a probability density function of initial values (the so-called Monte Carlo approach).

As a specific example to sum up all the preceding discussions on probability density functions, let us consider the dynamics of the Duffing oscillator which has a double well potential and a Hamiltonian of the form

$$\mathcal{H} = \frac{p^2}{2m} + \frac{\alpha}{4} q^4 - \frac{\beta}{2} q^2 + q \cos(t). \tag{2.47}$$

This example has been chosen because it exhibits dissipative chaos. In Figure 2.3 we show several snapshots of the probability density function for the Duffing oscillator described by this Hamiltonian, but where we have also introduced loss through a friction term. In the top figure, we show the initial condition, which is a Gaussian distribution around a somewhat arbitrary point in phase space. The next three figures give the PDF after one, three, and nine drive periods, respectively. We see that after one drive period the phase has changed and the Gaussian has begun to smear a little bit. By the third period, the distribution has changed considerably and is no longer localised. By the ninth period, we clearly see a complex structure that is a signature of the non-linear dynamics of dissipated classical chaos.

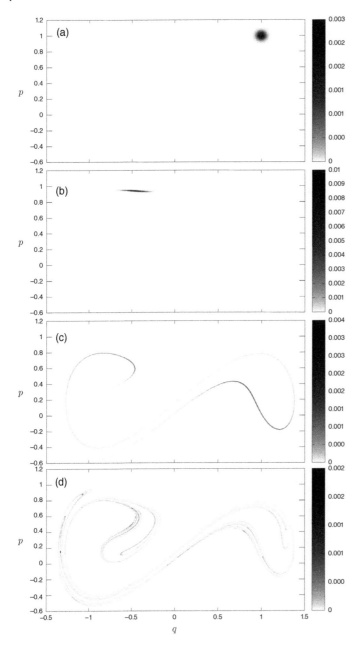

Figure 2.3 Evolution of a phase space distribution for the damped driven Duffing oscillator. (a) Initial probability density function (a Gaussian); (b) PDF after one drive period (showing the Gaussian smearing); (c) PDF after three drive periods (showing the beginning of chaotic like folding of phase-space); (d) PDF after nine drive periods (even stronger signatures of chaos).

We will find that quantum mechanics is a linear theory. If quantum mechanics is indeed a true extension of classical mechanics, its application to the Duffing oscillator should be able to reproduce an arbitrarily good approximation to this chaotic behaviour in a suitable limit. As quantum mechanics is a linear theory, both Hamiltonian and dissipative chaos present core challenges for understanding the connection between quantum and classical physics. We will need to do a reasonable amount of work before we can discuss this topic fully in chapter 10.

2.5 Koopman–von Neumann (KvN) Classical Mechanics

Prerequisite Material: One of the reasons we include this topic is to present classical mechanics within the same mathematical language as the standard presentation of quantum mechanics. In particular, we will make use of Dirac notation – demonstrating that this view is relevant to classical mechanics and not just to quantum mechanics. The minimum advised pre-reading for this section is thus: Sections 1.2, 1.3.1, and 1.3.2. We have written this section so that those already familiar with Dirac notation should not need to refer back to this material. While this results in some duplication of material, it does provide a clearer narrative.

In this section, we present a reformulation of classical Hamiltonian mechanics due to von Neumann, extending some earlier work by Koopman. Our approach is ahistorical in that their motivation was to see how close they could make classical mechanics look like quantum mechanics. However, no quantum mechanics is needed to understand the approach that they took. As such, we believe it is a worthwhile exercise to summarise some elements of their approach, as it will make introducing quantum mechanics more intuitive (or, at least, much less counter-intuitive). The topic that KvN theory addresses could be phrased as: how much can we make this formulation of classical mechanics look like linear algebra? Doing so adds value as it would, for example, allow us to apply standard algebraic results such as the Cauchy–Schwarz inequality (see page 7) or Heisenberg's uncertainty principle (Section 1.6) in a classical context. An alternative way of saying this is that our challenge is to reformulate classical mechanics in Dirac notation.

Our starting point is the Liouville equation of motion, Eq. (2.39),

$$\hat{L}\rho = i\frac{\partial}{\partial t}\rho$$

where we recall that the Liouvillian operator is defined to be

$$\hat{L} = -i\,\{\mathscr{H},\cdot\}.$$

Writing the Poisson bracket of the Hamiltonian with an unknown function in this way:

$$\{\mathscr{H},\cdot\} = \sum_i \frac{\partial\mathscr{H}}{\partial q_i}\frac{\partial}{\partial p_i} - \frac{\partial\mathscr{H}}{\partial p_i}\frac{\partial}{\partial q_i}$$

makes it clear that this is a linear operator on a function space. Our initial thought might be to consider the probability density as vectors in a vector space and introduce Dirac notation on them (this is certainly mathematically allowable). However, recall that the expectation value of a physical quantity is given by

$$\langle A \rangle = \int dq \int dp\, A(q, p) \rho(q, p),$$

which does not look like an expectation value in Dirac notation.

Dirac notation and expectation values assumes the existence of an inner product. We already know one inner product for function spaces:

$$\langle f | g \rangle = \int f^* g,$$

where this integral is the definite (Lebesgue) integral over the entire parameter space. The leads us to the following speculation: what happens if we write the probability density in terms of its own square root $\rho = \psi^* \psi$? To be general, we allow ψ being complex valued (maybe we have been inspired by the form of the inner product). That is

$$|\psi\rangle \sim \psi \text{ and } \langle \psi | \sim \int \psi^* \cdot$$

where the dot indicates that the bra vector is waiting for a ket to act on. From this, we immediately see that

$$\langle \psi | \psi \rangle = \int \psi^* \psi = \int \rho \stackrel{\text{def}}{=} 1.$$

Hence, all $|\psi\rangle$ are unit vectors in the function space. We can thus interpret classical mechanics in terms of 'ray vectors' where the state of the system is given by the 'direction' of the vector $|\psi\rangle$. For this reason, we will call it the (classical) state vector. Expectation values now take the form

$$\langle \hat{A} \rangle = \left\langle \psi \left| \hat{A} \right| \psi \right\rangle = \int \psi^*(q, p) A(q, p) \psi(q, p)$$

in agreement with our discussion in Section 1.3.2.

The above argument looks like a potentially valid approach for representing the classical probability density as a state vector. The next question we address is to see if we can link this to Liouville's equation. Let us return to our original definition of ψ, some function that satisfies $\rho = \psi^* \psi$. Then, by the product rule, we have

$$\frac{\partial}{\partial t} \rho = \frac{\partial}{\partial t} \psi \psi^* = \psi \frac{\partial \psi^*}{\partial t} + \psi^* \frac{\partial \psi}{\partial t},$$

so Liouville's equation reads

$$i \psi \frac{\partial \psi^*}{\partial t} + i \psi^* \frac{\partial \psi}{\partial t} = \hat{L} \psi^* \psi = \psi^* \hat{L} \psi + \psi \hat{L} \psi^*,$$

where we have used the fact that the Liouvillian is a differential operator and thus must also follow the product rule. We see that this holds so long as

$$i \frac{\partial}{\partial t} \psi = \hat{L} \psi \quad \text{and} \quad i \frac{\partial}{\partial t} \psi^* = \hat{L} \psi^*.$$

As the second equation is the complex conjugate of the first, this holds trivially (in hindsight it adds partial justification for the '–i' prefactor in the Liouvillian as $\hat{L}^* = -\hat{L}$). In direct notation, the first of these equations reads:

$$\mathrm{i}\frac{\partial}{\partial t}\,|\psi\rangle = \hat{L}\,|\psi\rangle,\tag{2.48}$$

which, we will later find, looks the same as the equation for the dynamics of a quantum state – the Schrödinger equation.

> **Prerequisite Material:** One of the advantages of Dirac notation is the ability to delay the choice of the representation as a vector. You may wish to skip this section on a first read and return to it after reading the next chapter. In order to understand the augment that follows, you will need to have an understanding of the material in Section 1.4

With the benefit of hindsight, we start by asking if the arguments of ψ at a given time, the coordinates and their conjugate momenta, could be the eigenvalues of some operators in a CSCO. Specifically, we will postulate that $\{\hat{q}_i, \hat{p}_i\}$ exist, where

$$\hat{q}_i\,|q_j, p_j\rangle = q_i\,|q_j, p_j\rangle \quad \text{and} \quad \hat{p}_i\,|q_j, p_j\rangle = p_i\,|q_j, p_j\rangle$$

and where $\{|q_j, p_j\rangle\}$ forms a complete orthonormal basis and j runs over all degrees of freedom so $|q_j, p_j\rangle \equiv |q_1, p_1, q_2, p_2, \ldots\rangle$. The representation of $|\psi\rangle$ in this basis is $\langle q_i, p_i|\psi\rangle = \psi(q_i, p_i, t)$.

Now that we have these operators for position and momentum, our first task is to determine their action on $|\psi\rangle$ in this representation. If

$$|\phi\rangle = \hat{q}_i\,|\psi\rangle \quad \text{and} \quad |\varphi\rangle = \hat{p}_i\,|\psi\rangle,$$

the phase space representation of these state vectors is found by premultiplying both sides by $\langle q_j, p_j|$

$$\phi(q_j, p_j) \overset{\text{def}}{=} \langle q_j, p_j|\phi\rangle = \langle q_j, p_j|\hat{q}_i|\psi\rangle = \langle q_j, p_j|q_i|\psi\rangle = q_i\langle q_j, p_j|\psi\rangle = q_i\psi(q_j, p_j)$$

and likewise for $\hat{p}_i$. We abbreviate this to

$$\hat{q}_i\psi(q_j, p_j) = q_i\psi(q_j, p_j) \quad \text{and} \quad \hat{p}_i\psi(q_j, p_j) = p_i\psi(q_j, p_j)$$

so we find the action of $\hat{q}_i$ (or $\hat{p}_i$) to be the same operation as multiplication by q_i (and p_i). Again, with the benefit of hindsight, we can consider introducing operators whose effect is

$$\hat{Q}_i \sim -\mathrm{i}\frac{\partial}{\partial q_i} \quad \text{and} \quad \hat{P}_i \sim -\mathrm{i}\frac{\partial}{\partial p_i}.$$

We will later see that this is equivalent to defining the operators by imposing the previous orthonormality condition together with $[\hat{Q}_i, \hat{q}_i] = \mathrm{i}$ and $[\hat{P}_i, \hat{p}_i] = \mathrm{i}$.

The important thing to note is that doing this means that we can construct a model of classical mechanics entirely within an abstract Hilbert space. Liouville's equation including the Hamiltonian can be written entirely in terms of the above operators and time. The state of the system is represented by a unit vector in a state space. Finally, the usual form of Liouville's equation is then recovered by choosing a CSCO of all the operators q_i and p_i and imposing the condition that the set of eigenvalues equate to all of phase space. It would be well worth revisiting this section after reading the next chapter, as the potential similarities and differences between quantum and classical mechanics will then become much more apparent.

2.6 Some Big Problems with Classical Physics

One of the first examples of an issue with classical physics is that electromagnetism predicted, and empirical observation verified, that the speed of light is the same in all reference frames. A fundamental cornerstone of physics is the principle of relativity. It states that the laws of physics must take the same mathematical form in all different frames of reference. The constancy of the speed of light is inconsistent with the Galilean transform and the addition of velocities – both of which need to be abandoned. The resulting physics is special relativity. General relativity, a most beautiful theory of gravity, follows by further imposing the relativity principle to all frames of reference – but that is not the subject of this story. Our focus will be other phenomena not explainable by classical physics.

2.6.1 Atoms and Polarisers

In quantum mechanics, our path is set by some other key observations. Rutherford's experiments on scattering in atoms led us to abandon the so-called plum pudding model. The best guess at the structure of atoms at the time was of electrons as point particles orbiting a nucleus (they cannot stay still because they are experiencing a centripetal force due to the Coulomb interaction). Once more, the issue is made apparent from our understanding of electromagnetism. From Maxwell's equations and observation, we know that accelerating charges radiate light. Since circular motion is accelerated motion, the implication is that electrons in atoms would lose all their Coulomb energy as electromagnetic radiation. This means that matter would not be stable, which is incompatible with the universe as we see it (an issue that the plum pudding model did not suffer from). In the early 1900s, explaining the structure of atoms was one of the biggest challenges that faced physics, and classical mechanics is not up to the task.

Another oddity arises when describing the action of a series of polarisers. For high intensities of light, polarisers behave in a way that is entirely consistent with Maxwell's equations. However, if one attenuates the intensity of light down sufficiently, the effect that one sees on a detection screen (such as photographic plates or an array of photodetectors) is that the screen is developed a bit at a time. What is even more peculiar is that each measurement of light records only one of two possible polarisation states, not the continuum one would expect from Maxwell's equations. We now go on to discuss an experiment that very well highlights this odd behaviour.

2.6.2 The Stern–Gerlach Experiment

One of the most experimentally straightforward examples of quantum physics is the Stern–Gerlach experiment, which looks at the measurement of the magnetic moments of silver atoms. A schematic of the experiment is shown in Figure 2.4. From left to right: we

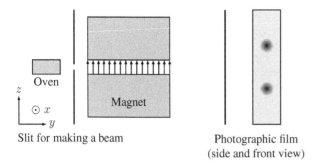

Figure 2.4 Set-up of the Stern–Gerlach experiment. Hot silver atoms leave the oven and we see events on the screen as the film developed by whatever process has happened. We should be careful not to think in terms of classical ideas such as beams of atoms. For this reason, such common visualisation 'aides' are deliberately omitted from this schematic.

have an oven that serves to heat up atoms (historically Silver) to make them energetic; this is followed by a screen which embodies the idea of collimating whatever comes out of the oven (if what comes out of the oven is a wave, we would expect diffraction effects – while it is beyond the scope of the present discussion, in a full treatment of this experiment this will need to be accounted for). The collimator is followed by a magnetic field that will affect atoms with different orientations of magnetic moments in different ways; and finally we have a detector, which in this case is photographic film.

Exercise 2.7 Convince yourself that, for an oven containing only one type of atoms with a given magnetic moment, the expected pattern on the film after very many events would always be a single Gaussian of exposed film. ∎

Surprisingly, for different atomic sources, we get different patterns on the film. We focus our discussion on those sources that result in two distinct lumps. An idealised representation is shown in Figure 2.4 (for interesting reasons actual experimental data is not as clean as this. A reasonably strong amount of quantum mechanics is needed to begin to understand why this is the case and we postpone the discussion of that part of the topic until later). As the atoms are always measured in one of two possible states, they behave in the outcome of experiments just like polarisation experiments using highly attenuated light. This may make us suspect that a single physical principle might underlie such behaviour.

Our first explanation, applying Occam's razor, might be that there are two species of this kind of atom. We should call them 'top' atoms and 'bottom' atoms. In Figure 2.5 we show a stylised set of diagrams of a set of experiments that test this hypothesis. In Figure 2.5a we start by adding a screen that blocks all the 'bottom' atoms. We can see by the lack of appearance on the screen that this is effective. Someone who doubts the theory might object and claim that the splitting into lumps on the screen has only something to do with the magnetic field. In Figure 2.5b we check that this kind of objection is unfounded by verifying that adding another magnetic field after the screen has no effect on the number of lumps we see on the photographic film. Just to check that we now have only one kind of atoms, the top kind, in Figure 2.5c we change the orientation of the second magnet. This is now getting very peculiar, as it appears that there are 'left' and 'right' atoms as well. Along with

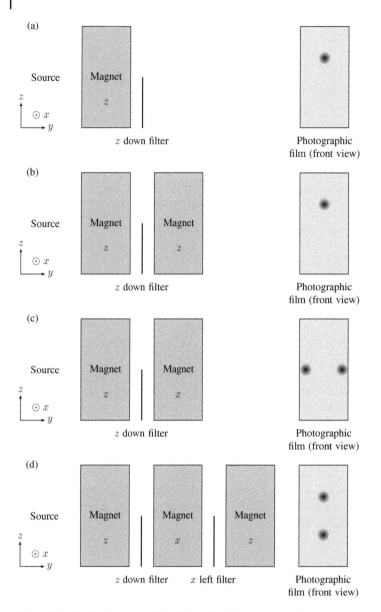

Figure 2.5 Stylised set-up of the Stern–Gerlach experiment.

the principle of relativity, the principle that physical laws should be independent of coordinate system, lies at the heart of physics. If this were to be the real Physics, it would therefore be deeply unsettling, as it might imply that the universe had some specific preferred axis for different, coordinate-system-dependent, flavours of atoms. We can block left and right flavours, just like we did for top and bottom. In Figure 2.5d we show the outcome of filtering out all bottom atoms and all left atoms. After filtering out the left atoms, we then apply one more z-oriented magnetic field. We see on the photographic film that we recover the original distribution of top and bottom outcomes. This means that we cannot with our first z filter have filtered out particles that have an intrinsic bottom characteristic. It also means that something else must be going on, which is independent of the specific configuration of our filters and the orientations of the magnets. There is no explanation of this phenomenon within classical mechanics. This means that we need to seek a new theory capable of providing insight into what is going on here. If that theory also could explain the structure of matter, then we would have a high degree of confidence that it was a very good theory. But this is not the only requirement for a new theory to be good, as the following, and final, section of this chapter should make clear.

The filtering arrangement of the above experiment does exhibit some key features that we will use to formulate quantum mechanics. The first is that the initial state (as set up by an arrangement of filters) determines the final outcome of each experiment. This is something we have seen before in the Liouville equation, where the dynamics of the probability density function depends only on the initial state and the form of the Hamiltonian (i.e. we do not need to specify $\dot{\rho}$ as an initial condition). As such, this implies that any equation of motion should have at most a first order time-derivative of the state. The second key observation is that the behaviour of this apparatus, and that of a series of polarisers, is completely compatible with the principle of superposition. That is, if $\psi_1(t)$ and $\psi_2(t)$ are allowable states, then so is $\psi_1(t) + \psi_2(t)$. We will use the fact that the principle of superposition is a property of all linear systems.

2.6.3 The Correspondence Principle – What It Is and What It Is Not

We have already mentioned some fundamental principles of physics, such as relativity, where the mathematical expression of a theory must be independent of the reference-frame. We have also mentioned that our theories should be independent of the coordinate system. In addition, we have briefly touched on Noether's very important observation, that continuous symmetries of physical systems lead to conservation laws. In this section, we consider another fundamental principle of physics that lays down what it means for a new theory to be 'good'. This principle comes into play when we cease to be satisfied with existing theories and seek to either extend or unify them. A key concern – especially from a unification point of view – is that it is important that in introducing a new theory, we do not somehow lose all the good bits of the preceding theories. This idea is formalised within the correspondence principle. In developing quantum mechanics, we seek to fix some known problems in classical mechanics, but the way the theory develops can make it hard to apply this correspondence principle. As such, this is a topic that needs to be treated with some care. Here, we will consider enough of a general statement that we can use when discussing and comparing the axiomatic foundations of classical and quantum mechanics.

Some care needs to be put into the specific phrasing of the correspondence principle. Many expressions are not sufficiently empirical in nature to make it clear to which elements of physics this principle must apply. We phrase the principle as follows:

For any new theory to be 'good', it *must reproduce the predictions* of established existing theories (and all relevant theories if on the unification rather than extension hunt) in the domains where the existing theory(ies) were 'good'. Here, good simply means that the existing theory(ies) made predictions that were in 'agreement' with experimental observation and tolerances.

Note that there is no constraint on ontology in the above – the theories may look very different. It is only their predictions that must agree, and only then where the old theory was 'good'. In subjects like mechanics and relativity, this statement is quite straightforward to apply, as their ontological construction/metaphysics are very similar (one arrives at relativity simply by dropping axioms). In quantum mechanics the situation, we will find, is more subtle.

The above statement suffers from some vagaries, such as 'what do you mean by agreement?', 'what is really meant by a prediction?', 'how is agreement with experiment determined?', or 'does the ontology constrain the comparisons that can be made?' Such questions, coupled with the fact that quantum mechanics has a good deal of different ontology/interpretations/metaphysics, cause one to often enter into discussions about the philosophy of science (see works by, e.g. Popper, Lakatos, and Feyerabend). We would argue that a lack of mastery of the actual technical nature of quantum physics, as well as a proper understanding of the actual similarities with, and differences to, classical physics (as, for example, is provided by Koopman–von Neumann theory), may render interesting and plausible-sounding philosophical arguments inadequate. One of the reasons that we have chosen to present our discussion of quantum mechanics in the way that we have, is to facilitate engagement with the metaphysical aspects of quantum theory from a firm foundation of classical and quantum theory. It is natural to have an interest in the philosophy of science, and relevant texts can contain a lot of good and bad ideas and are never the final word in any discussion. For example, the following text:

> 'The consistency condition which demands that new hypotheses agree with accepted theories is unreasonable because it preserves the older theory, and not the better theory. Hypotheses contradicting well-confirmed theories give us evidence that cannot be obtained in any other way. Proliferation of theories is beneficial for science, while uniformity impairs its critical power. Uniformity also endangers the free development of the individual.'
>
> Paul Feyerabend, Against Method

In our view, this demonstrates a misunderstanding of a well-expressed version of the correspondence principle and what it means (it may be apt for poorly expressed versions). It also overstates how much of a shift exists between different theories. Progress is far more incremental than it might at first appear. By including our discussion on Koopman–von Neumann theory, we hope that we have made it possible for us to clearly show that quantum and classical physics have more in common than they do not. The correspondence principle as stated above may not be without ambiguity but it is, we believe, robust against Feyerabend's criticism.

Of particular note for quantum mechanics is that many formulations of the theory can seem very different, to the extent that they might be considered different theories. However, all have been shown to be mathematically equivalent and therefore do not differ in their predictions of experimental outcomes. Unless a formulation or interpretation of a theory contains distinct statements that are testable and falsifiable by observation, it needs to be considered as having equal merit to all of its alternatives.

In terms of testable hypotheses, one of the areas of quantum mechanics that will turn out to be interesting, is the recovery of classical chaos. As we observed at the end of the previous section, experiments clearly showing quantum behaviour also seemed to indicate that nature obeys the principle of superposition. This means that our quantum theory will be linear in nature. Even without damping, Hamiltonian systems, such as the famous three-body problem, exhibit the extremely nonlinear phenomena of chaos. As we will discuss later, satisfying the correspondence principle for such systems in a completely satisfactory way is still an open problem.

The metaphysics of quantum mechanics is a very interesting area of study in and of itself. The subject can, will, and often does, inspire enthusiastic conversation. We will try to make our discussion as neutral as possible with regard to the different interpretations of quantum theory. When we do talk about interpretations, we do so to motivate aspects of the formulations of quantum theory, where we can think of no other way of presenting the material (and where possible we will try to draw analogies with classical physics). Please be aware that there may be a number of different ways of understanding the emergence of the same model from other arguments. With this in mind, it is now time to move to the main subject of this book, quantum theory.

3

The Schrödinger View/Picture

"From the point of view of basic physics, the most interesting phenomena are, of course, in the new places, the places where the rules do not work - not the places where they do work! That is the way in which we discover new rules."

Richard Feynman

3.1 Introduction

In the preceding chapter, we reviewed classical mechanics and provided a couple of examples of its incompleteness. It is tempting to introduce the development of quantum mechanics from a historical perspective. In arriving at the modern formulation(s) of the theory many avenues were taken and it is interesting to study the way, and thought processes, that led the scientific community to a solution. That presentation has the disadvantage that it obfuscates some of the similarities between classical and quantum mechanics (it is also found in many other texts on the subject). In this and the follow-ing chapter, we chose alternative justifications for quantum theory that emphasise its commonalities with classical mechanics. It is worth noting that none of what follows constitutes a proof or derivation. Rather, we try to make clear a set of observations that lead to a set of postulates, which in turn form the theory. From these, predictions can be made and tested against reality in experiment. The 'better' the theory, the better the agreement between its predictions and observed reality.

This chapter is concerned with the most common presentation of quantum mechanics at the undergraduate level, the Schrödinger picture (and it is for this reason alone that it comes first). More conceptually efficient ways of introducing alternative presentations of quantum theory are covered in the following chapter.

First, an overview of what we will cover in this chapter. We will start by highlighting some key observations from experiments that we wish to impose on the 'new theory' (of quantum mechanics). We try to keep the observations and assumptions made to a minimum. A *key outcome* of our observations will be that we arrive at the idea that the state of a system can be encoded in a vector within a vector space – the state vector. This approach is much like that of the Koopman–von Neumann presentation of classical mechanics, but this time more general, as we do not assume what sort of vector this is.

Quantum Mechanics: From Analytical Mechanics to Quantum Mechanics, Simulation, Foundations, and Engineering, First Edition. Mark Julian Everitt, Kieran Niels Bjergstrom and Stephen Neil Alexander Duffus.
© 2024 John Wiley & Sons Ltd. Published 2024 by John Wiley & Sons Ltd.
Companion website: www.wiley.com/go/everitt/quantum

Our next step will be setting up the equation that govern a system's dynamics – the Schrödinger equation. We will then use a corollary of this equation due to Ehrenfest. This takes the form of a dynamical equation, and looks very similar to classical dynamics, but replaces the Poisson bracket with another mathematical object called a commutator. The idea that the Poisson bracket of classical mechanics needs to be replaced by something else is a *key observation*. Some similarities between quantum mechanics and classical mechanics will be quite apparent when comparing the dynamics of 'probability densities' between the classical Liouville and its counterpart, the quantum von Neumann equations.

The theme central to all of our discussions will be that, in moving from classical to quantum physics, the Poisson bracket is replaced by another mathematical operation – usually a commutation relation. This is the *key outcome* that this introduction to quantum mechanics tries to make clear (although, as we shall see, this is not without its difficulties).

Next, we use an examination of the energy of the system that arises from deriving the time-independent Schrödinger equation, to motivate a discussion of what it means to make a measurement in quantum mechanics. We discuss this in comparison to the classical analogy of measurement in statistical physics. The measurement process is the third *key outcome* of this introduction.

Up until this point, a lot of the mathematical presentation may have appeared rather abstract. At some point we need to solve problems and make predictions. So we now give some specific examples of quantum states in terms of position, spin, and position and spin. The derivations here are long, but have been chosen as they make clear *how few* assumptions are needed to go from an abstract theory to some equations we can solve. We do not advise you try to remember all these long derivations – the key point is to understand how the few assumptions made at the beginning lead to the destination reached at the end of each section (such as the form of wave-function and the Schrödinger equation in position representation).

We will close this chapter by presenting (i) a technical summary of what we have done, and (ii) a number of possible axiomatic foundations of quantum mechanics. These are important, as they *are the theory*.

3.2 Motivating the Schrödinger Equation

Prerequisite Material: Section 1.2: Generalising Vectors, Section 1.2.1: Vector Spaces, Section 1.2.2: Inner Product Section 1.2.3: Dirac Notation

In developing a new theory, we do not want to abandon the best parts of an old theory. In the following discussion, we try to emphasise how much of classical physics is in fact retained in quantum physics, which is far less of a revolutionary theory than many people think. Note that arguments, such as the one that follows, are always speculative and based on having to reject some ideas thought evidently correct, and replace those with new ones that might look odd.

This can be an uncomfortable experience. There is no single correct argument leading to the final answer, and the process is not one of derivation. There are historical arguments that go a long way 'around the houses' to reach the answer, designed to help one understand the process that was followed but at the penalty of unnecessary steps. There are others, such as the one we present below, that attempt to take a shorter path and serve a different purpose. Ours is highlighting the similarities between quantum and classical physics.

Let us start with a conclusion based on the behaviour of chains of polarisers, or the Stern–Gerlach apparatus.

> Observation 1: We know that it is *only the initial state that is needed* to specify all future states of quantum systems. An immediate corollary of this is that the dynamical equations that give *a system's evolution must at most contain only first-order time-derivatives of the systems 'state'.*

Note that this condition is met, e.g. by Liouville's equation, as well as those of Koopman–von Neumann, so such a constraint is not unique to quantum mechanics. As such, this is an axiom we are keeping rather than replacing, and we choose to keep it as it satisfies observations of quantum phenomena.

Although this statement is consistent with the Hamiltonian formulation of classical mechanics, it does present problems for the Euler–Lagrange equations. The need to specify velocities in a system's initial condition is in conflict with the above observation (as these equations will contain accelerations). So there is a subtle distinction between Hamiltonian and Lagrangian mechanics that you might like to ponder on (think about the proper definition of momentum, least action, and local versus global theories).

Let us keep things as general as we can for the moment and just assume that the state is encoded in some vector in a vector space (as this includes functions, coordinates as used in classical mechanics, plus other mathematical objects, so it is the easiest generalisation to make).

> Assumption 1: The initial state is some vector in a vector space – we will use the new notation $|\psi\rangle$ to denote these vectors. We will not yet decide what sort of vectors they are (column vectors, functions, etc. are all possible choices at this point). Suffice it to say, this covers a vast range of possibilities, including one significant one that is important to classical physics: the probability density function.

So Observation 1 and Assumption 1 together imply that the equation of motion for the 'state of the system' must take the form:

$$\frac{d|\psi\rangle}{dt} = \text{some function of } |\psi\rangle, \tag{3.1}$$

which is nothing new compared with classical mechanics, as this looks equivalent to Koopman–von Neumann equations.

> Observation 2: Quantum systems also seem to satisfy the principle of superposition – so, much like in wave equations you have seen in classical electrodynamics, any dynamical property must act in a linear way (as non-linearity always breaks superposition). This means that
>
> $$\frac{d|\psi\rangle}{dt} \propto \hat{H}|\psi\rangle,$$
>
> where $\hat{H}$ is some as yet undetermined linear operator.

Since the Liouvillian is also a linear operator, we still have nothing new compared with classical mechanics (as expressed by Koopman–von Neumann). But we do have the option here to add some *new physics* by asking if there are operators that could take the place of the Liouvillian.

Let us look at this equation in some more detail, and start off by constraining $\hat{H}$ and the constant of proportionality, $c = a + i\,b$, so that the solutions of

$$\frac{d|\psi\rangle}{dt} = c\hat{H}|\psi\rangle$$

are in some way well-behaved. Let us consider what happens to the normalisation of $|\psi\rangle$ (we will use the fact that $\|\psi\|^2 = \langle\psi|\psi\rangle$). For this theory to be worth investigating, its solutions must not be ill-behaved (i.e. vanish to zero or diverge to infinity, as it is not clear what this would mean for our notion of a 'state').

The square of the norm of the state evolves according to

$$\frac{d}{dt}\langle\psi|\psi\rangle = \left\{\frac{d}{dt}\langle\psi|\right\}|\psi\rangle + \langle\psi|\left\{\frac{d}{dt}|\psi\rangle\right\}$$
$$= \left\{\langle\psi|c^*\hat{H}^\dagger\right\}|\psi\rangle + \langle\psi|\left\{\hat{H}|\psi\rangle\right\}$$
$$= \left\langle\psi\left|c\hat{H} + c^*\hat{H}^\dagger\right|\psi\right\rangle$$
$$= a\left\langle\psi\left|\hat{H} + \hat{H}^\dagger\right|\psi\right\rangle + i\,b\left\langle\psi\left|\hat{H} - \hat{H}^\dagger\right|\psi\right\rangle.$$

If the right-hand side of this equation contains a non-zero real component, the norm of $|\psi\rangle$ will have an amplitude that grows or decays exponentially (which does not seem physically sensible), and any non-zero imaginary component would result in a complex norm, which is not allowed. This hints at two possibilities. We can guarantee that solutions of this equation are well-behaved (bounded) if either:

1. c is purely real (so $b = 0$) and $\hat{H}$ is anti-Hermitian or
2. c is purely imaginary (so $a = 0$) and $\hat{H}$ is Hermitian.

This is because either case leads to $(\partial/\partial t)\langle\psi|\psi\rangle = 0$ and conservation of the norm of $|\psi\rangle$ follows.

Conversely, if (a) c is purely real (so $b = 0$) and $\hat{H}$ is Hermitian, or (b) c is purely imaginary (so $a = 0$) and $\hat{H}$ is anti-Hermitian, then the system would be badly behaved with divergent solutions. So we will rule out those possibilities.

If we consider various combinations of c that are complex (neither purely real nor purely imaginary), or a case where $\hat{H}$ is neither Hermitian nor anti-Hermitian, then it is hard to say how the dynamics of the system would behave. A lot of activity went into considering the case where c is purely imaginary (so $a = 0$) and $\hat{H}$ is neither Hermitian nor anti-Hermitian to try to model phenomena such as loss and damping in quantum systems. While interesting, all such approaches contain some problematic elements.

Our ideas are now beginning to nucleate around some simple core speculations and, if this were a truly original theory, the details would need to be hammered out by trial and error until some nice (in the archaic sense of the word) structure would finally be forthcoming. We will shortcut the dead ends that are followed in the development of any new theory by applying liberal amounts of hindsight. The trouble is – like in a good jigsaw – the order and the way everything fits together does not truly fall into place until the picture is nearly complete. In the context of our derivation, this may lead to logic that sounds circular at first, or confusions as to why some assumptions are made. For example, until our discussion of measurement in quantum mechanics, it will not become properly clear that the assumptions we have just made: c being purely imaginary (so $a = 0$), and $\hat{H}$ being Hermitian, are well justified. For now, please accept that it gives mathematically well-behaved solutions and that other choices lead to substantial difficulties. With this in mind, our next key assumption is:

Assumption 2: With the benefit of historical hindsight, and in order to ensure that the solutions of our equation are well-behaved, we write

$$i\hbar\frac{d|\psi\rangle}{dt} = \hat{H}|\psi\rangle, \tag{3.2}$$

where $\hbar$ is a real constant of proportionality (the value of which is to be determined later by experiment – once the theory is well enough developed to make predictions). This is Schrödinger's equation in its most general form [81].

With the assumption that $\hat{H}$ is Hermitian, we also have

$$\frac{d\langle\psi|}{dt} = -\frac{1}{i\hbar}\langle\psi|\hat{H} \tag{3.3}$$

(as $i \to -i$ when moving from kets to bras).

As a result, we now have some *dynamical equations* for the state of a system. But how does this help us? In classical physics we talk about coordinates (and their velocities or canonically conjugate momenta) or the value of a quantity (such as kinetic energy). This is a very Pythagorean world view of attaching a set of changing numbers to a property, or configuration, of a system. For a theory to be properly predictive, it needs to be quantifiable – that is, we need to be able to extract numbers which we can then compare in some meaningful way to the outcome of a measurement (be it the deflection of a needle of a Galvanometer, clicks from a Geiger counter, or something else). For now, let us just focus on trying to extract numbers from our theory, and postpone the discussion of relating them to measurement for a little while longer. Our new theory has only states and operators; how might we make the connection from a given operator and state, to a number that might later be related to a measurement? One way is through the expectation value:

$$\left\langle\psi\left|\hat{A}\right|\psi\right\rangle. \tag{3.4}$$

3.2.1 Ehrenfest's Theorem, Poisson Brackets, and Commutation Relations

> **Prerequisite Material:** Section 1.3.2: Expectation Value of Random Variables, Section 1.3.6: Commutators, Section 1.6: The Heisenberg Uncertainty Relation.

In this section, we are going to present an argument for the fundamentally quantum mechanical world view that Poisson brackets should be replaced by commutation relations. There is a much more elegant argument than the one presented here, which we will explain in the next chapter. Our reason for presenting both arguments is that there are some issues with the Schrödinger picture that are best understood by comparing and contrasting the two different approaches.

Previously, we considered the dynamics of the norm of a state vector. Let us now see what we can say about the rate of change of the expectation value, $\langle \hat{A} \rangle = \left\langle \psi \middle| \hat{A} \middle| \psi \right\rangle$, and where this leads us. By application of the product rule, we find

$$\frac{d}{dt} \langle \hat{A} \rangle = \frac{d}{dt} \left\langle \psi \middle| \hat{A} \middle| \psi \right\rangle = \frac{d\langle\psi|}{dt} \hat{A} |\psi\rangle + \left\langle \psi \middle| \frac{d\hat{A}}{dt} \middle| \psi \right\rangle + \langle\psi| \hat{A} \frac{d|\psi\rangle}{dt}$$

$$= \left\langle \psi \middle| \frac{1}{i\hbar} \hat{A}\hat{H} - \frac{1}{i\hbar} \hat{H}\hat{A} + \frac{d\hat{A}}{dt} \middle| \psi \right\rangle$$

$$= \left\langle \psi \middle| \frac{1}{i\hbar} [\hat{A}, \hat{H}] + \frac{d\hat{A}}{dt} \middle| \psi \right\rangle$$

$$= \frac{1}{i\hbar} \langle [\hat{A}, \hat{H}] \rangle + \left\langle \frac{d\hat{A}}{dt} \right\rangle, \tag{3.5}$$

where $[\hat{A}, \hat{H}]$ is the commutator of $\hat{A}$ with $\hat{H}$. We recall from Eq. (2.42) page 49 we that classically

$$\left\langle \frac{dA}{dt} \right\rangle = \langle \{A, \mathcal{H}\} \rangle + \left\langle \frac{\partial A}{\partial t} \right\rangle,$$

which is close, but not exactly the same as the expression in our new 'quantum' theory:

$$\frac{d}{dt} \langle \hat{A} \rangle = \frac{1}{i\hbar} \langle [\hat{A}, \hat{H}] \rangle + \left\langle \frac{d\hat{A}}{dt} \right\rangle. \tag{3.6}$$

Note that not only do these two equations not quite agree, we cannot make them any closer by rearranging them. We could insist that operators in this new quantum theory have *no implicit time dependence* so that

$$\frac{d\hat{A}}{dt} = \frac{\partial \hat{A}}{\partial t}$$

and

$$\frac{d}{dt} \langle \hat{A} \rangle = \frac{1}{i\hbar} \langle [\hat{A}, \hat{H}] \rangle + \left\langle \frac{\partial \hat{A}}{\partial t} \right\rangle, \tag{3.7}$$

which is called *Ehrenfest's theorem* [21]. Please always remember that in deriving this equation, while we have made no assumptions on $\hat{A}$ beyond it being a linear operator with no implicit time dependence, we have imposed that $\hat{H}$ *must be Hermitian* – and if it is not, the above equation does not hold.

While there are some issues with the above argument (which we will discuss at the end of the section), the similarity between the two equations leads us, by comparison, to *our main observation* of the shift needed *to move from classical to quantum physics*. Your eyes may have already scanned this page and given the game away, but if not, can you guess what we do?

The new theory, we are developing (quantum mechanics), looks very similar to our old theory (classical mechanics) if we *replace any physical quantity by an operator counterpart* and apply the following rule: for two classical quantities, A and B, their quantum corresponding operators, $\hat{A}$ and $\hat{B}$, are defined according to

$$[\hat{A}, \hat{B}] = i\,\hbar\,\{A, B\}$$

Note that the operators here are defined by a commutator and a Poisson bracket, so this must be done by referencing one quantity against another.

The above proposal was put forward by Dirac [18], but $[\cdot, \cdot] = i\,\hbar\,\{\cdot, \cdot\}$ does not always hold. For polynomials above a certain order, one can find Poisson brackets of different functions that evaluate to the same thing, but where a commutation relation cannot be found *that also agrees* for all lower-order commutation relations. Seeking a resolution to this difficulty will later lead us to a discussion of operator ordering, Groenewold's theorem and a phase space description of quantum mechanics. *Until then, we will restrict our discussion to the cases where we can write $[\cdot, \cdot] = i\,\hbar\,\{\cdot, \cdot\}$ without ambiguity* (most examples in quantum physics textbooks stay firmly in this domain, so it is not as much of a restriction as one might fear).

Exercise 3.1 Start by assuming $[\hat{q}, \hat{p}] = i\,\hbar\,\{q, p\} = i\,\hbar$. Show that the Poisson bracket $\frac{1}{3}\{q^3, p^3\} = \{q^2 p, q p^2\}$ but $\frac{1}{3}[\hat{q}^3, \hat{p}^3] \neq [\hat{q}^2 \hat{p}, \hat{q} \hat{p}^2]$. ∎

An immediate consequence of replacing the Poisson bracket with a commutator is that it brings in the Heisenberg uncertainty relation

$$\Delta\hat{A}\,\Delta\hat{B} \geq \frac{1}{2}\left|\langle[\hat{A}, \hat{B}]\rangle\right| = \frac{\hbar}{2}\left|\langle\{A, B\}\rangle\right|,$$

as $[\hat{A}, \hat{B}] = i\,\hbar\{\hat{A}, \hat{B}\}$ (the curly braces here denote the Poisson bracket and not the anti-commutator). For the specific, standard, example of position and momentum, this yields $\Delta\hat{q}\,\Delta\hat{p} \geq \hbar/2$. Note that we have made *no discussion of measurement*, which is very deliberate. The connection between the Heisenberg uncertainty relation, the nature of quantum states, and thus the implication for measurement is often overstated. We will delay a detailed discussion of this topic until it can be properly formulated, following the introduction of the phase space description of quantum mechanics due to Wigner and others in the next chapter.

3.2.2 The Main Proposition

To extract the main finding from the preceding discussion means that we have the key equations for quantum mechanics, the (time-dependent) *Schrödinger Equation*:

$$\hat{H}\,|\psi\rangle = i\hbar\frac{d|\psi\rangle}{dt}, \tag{3.8}$$

where we now interpret $\hat{H}$ as a quantum analogue of the classical Hamiltonian, which gives the dynamics of the state vector $|\psi\rangle$.

Let us make this more clear by way of the example of the simple harmonic oscillator. The classical Hamiltonian is:

$$\mathcal{H} = \frac{p^2}{2m} + \frac{1}{2}m\omega^2 q^2$$

with $\{p, q\} = 1$. The quantum Hamiltonian should therefore be, by our rule,

$$\hat{H} = \frac{\hat{p}^2}{2m} + \frac{1}{2}m\omega^2\hat{q}^2$$

with $[p, q] = i\hbar$.

From this, we can now use *Ehrenfest's theorem*

$$\frac{d}{dt}\langle\hat{A}\rangle = \frac{1}{i\hbar}\langle[\hat{A},\hat{H}]\rangle + \left\langle\frac{\partial\hat{A}}{\partial t}\right\rangle \tag{3.9}$$

to see if our theory so far is reasonable.

Exercise 3.2   Apply Ehrenfest's theorem to find $d\langle\hat{q}\rangle/dt$ and $d\langle\hat{p}\rangle/dt$ for the simple harmonic oscillator – how do these equations compare to those for the classical coordinate and momenta? We will discuss what you find here and its significance in detail later. What would happen for a more general potential $V(\hat{q})$? What would Newton think of this?   ∎

Before proceeding to the next section, let us just review what we have 'determined' so far: we represent states by vectors in vector spaces. As a corollary, the dynamics are given by the Schrödinger equation, from which we derive Ehrenfest's theorem. By comparing this with classical mechanics, we proposed to replace Poisson brackets by commutators according to the rule given, and classical quantities by their operator counterpart, the only additional restrictions being that the Hamiltonian is Hermitian and that operators have no implicit time dependence. In essence, this is all we have done, and it is nearly enough to lead us to all modern quantum mechanics.

3.2.2.1 Summarising an Issue with the Above Argument

In this section, we argued that we can replace our classical equation of motion

$$\frac{d}{dt}A(t) = \{A(t), \mathcal{H}(t)\} + \frac{\partial A(t)}{\partial t}$$

with a linear equation of motion called the Schrödinger equation for a vector in a vector space that represents the system's state (a concept not too different to the probability density function (PDF)). As part of this process, we presented the well-known result of Erhenfest's theorem. We used this to argue that the Poisson bracket of classical mechanics should be replaced by a commutation relation. You may have noticed that the argument is not entirely convincing as

$$\frac{\mathrm{d}}{\mathrm{d}t} \langle \text{Something} \rangle \neq \left\langle \frac{\mathrm{d}}{\mathrm{d}t} \text{Something} \right\rangle$$

in general, and worse still, in our new quantum theory we equated

$$\frac{\mathrm{d}}{\mathrm{d}t}\hat{A} = \frac{\partial}{\partial t}\hat{A} \implies \left\langle \frac{\mathrm{d}}{\mathrm{d}t}\hat{A} \right\rangle = \left\langle \frac{\partial}{\partial t}\hat{A} \right\rangle$$

by imposing that operators in the new theory cannot have any implicit time dependence. This is at odds with classical physics and, quite rightly, might make you feel uncomfortable.

Happily, there is a way to argue for a quantum equation that takes exactly the same form as the classical equation of motion with the exception that we map $\{\cdot, \cdot\} \to [\cdot, \cdot] / \mathrm{i}\,\hbar$:

$$\frac{\mathrm{d}}{\mathrm{d}t}\hat{A}_H(t) = \frac{1}{\mathrm{i}\,\hbar} \left[\hat{A}_H(t), \hat{H}(t)\right] + \frac{\partial \hat{A}_H(t)}{\partial t},$$

which argues more strongly and elegantly for substituting Poisson brackets with commutators (called the Heisenberg picture). Unlike the Schrödinger picture, here the operators do have explicit time dependence and the state has none (and we distinguish the pictures with a subscript H).

We delay the discussion of this until later since the Ehrenfest's theorem provides sufficient motivation to justify replacing Poisson brackets with commutators, and the Schrödinger picture of quantum mechanics is a more natural one for constructing many of the arguments that follow. We will return to both the Heisenberg picture and Ehrenfest's theorem in Section 4.2 (the interested reader may wish to read ahead).

3.3 Measurement

3.3.1 Introducing Measurement

The fact that our new theory in some ways looks similar to classical mechanics is very promising. One big question remains before we can test it against real-world experiments: how do we find out what the state of the system is? In other words – how do we make measurements?

In classical mechanics we think we have a notion of what a particle is, and that we can simply measure its properties, e.g. its position. The fact that we can see physical objects, such as a pendulum, leads to this view of definiteness. But even here, things are not as clear as they seem – the 'position' of a particle is its centre of mass, which we can only determine by approximation. Other properties, like the velocity of a particle, the electric field at a point, or the current flowing through a wire, all require some form of interaction[1] (from light bouncing off an object, to changing the state of a more complex piece of apparatus). From this, the actual state of a system is inferred.

---

1 From which one obtains a classical measurement disturbance relation, which is often forgotten in discussions of quantum measurement disturbance, and then wrongly linked to the Heisenberg uncertainty principle.

Measuring electrical current is a good example of the subtlety of measurement, even in classical physics. Before we even begin to talk about the connection between measurement and the uncertainty principle[2] applied to quantum mechanics, let us briefly discuss measurement disturbance in all settings. A simple (at least in theory) physical way of measuring current is using the moving coil galvanometer. Here, a circuit with low impedance allows a current to flow through a coil placed around a permanent magnet, to which a needle restrained by a light spring is attached. The force that acts on the needle is the Lorentz force, and the resulting displacement of the needle allows us to infer the current flowing through the wire. However, this is not a lossless process, and the act of measuring the current will change the current itself (in a very good galvanometer this will not be by much, but it will always be by something). This is the notion of measurement disturbance, and it is as relevant to classical systems as it is to quantum ones.

We also have the fact that any measurement carries with it some uncertainty (even if it is ± half the smallest division of some apparatus that is perfect in all other respects). This implies that all classical models should be probabilistic to reflect this uncertainty (so in this respect Liouville's equation is more appropriate than Hamilton's - something that is most apparent for non-linear systems). We can think of Liouville's equation as describing the average behaviour of a collection of experiments performed independently, as much as we can think of it describing the behaviour of an ensemble. So, whilst the PDF can evolve into something quite complicated, we still think that any one experiment is in a specific 'coherent' state (delta function) representing a particular deterministic 'reality'. We simply do not know what that state is unless we measure the system. If we are to measure the state of the system at distinct points in time, then the process is as follows: we measure the initial conditions with some error (which must, therefore, be a distribution). Liouville's equation now predicts the evolution of the distribution. This carries on until we measure the system again, which 'fixes' the state again within experimental uncertainty (the new PDF will depend on what has been measured. If, for example, we have only measured position, we have learnt nothing about momentum.). We could call this process of determining the current PDF by measurement from the PDF estimated by Liouville's equation 'collapsing' the PDF to a new one, representing our updated knowledge of the system state. Note that the subject of continuous measurement is a very interesting one that we shall postpone for now; suffice it to say that whereas we are familiar with the process of continuously measuring classical systems, this intuition does not carry over so clearly into the quantum domain.

Let us now consider the measurement of a quantum (i.e. not classical) system. An example apparatus is the photomultiplier tube or Geiger counter, which operates by different but similar principles. This measurement process is often described as starting when an incoming electron or particle excites an electron off a conducting plate in an environment containing a high-voltage difference between some electrodes. This lone electron, accelerated by the high-voltage difference, 'impacts' the next electrode which frees a number of other electrons, which in turn does the same again and again until a current pulse of sufficient magnitude is produced that an 'event' (such as the click of a speaker) may be recorded. Numerous good questions can be asked about such devices,

---

2 which, as already noted, we will not do in this chapter as we need to make use of later-developed phase space methods for a proper discussion of the topic.

such as: 'What happens if more than one thing enters the device, separated only by a short interval?'; 'How often does the device fail to trigger when it should?'; or 'Can the device trigger even when nothing enters it?'. Really, the big question here is: 'When, in fact, does the measurement take place? Was it at the freeing of the first electron (an undetectable event in its own right), was it at any of the many electron-electrode events in between (each a quantum process), or was it at the 'click'?[3]'

We note from the above discussion that measurement, in any context, is potentially more subtle than we might at first think. But what is not clear in our new theory, is what the quantities we measure are. While they may at first seem unrelated to measurement, the next two subsections seek to set up the answer to this question within standard quantum theory.

3.3.2 On the Possible Connection Between the State Vector and Probabilities

Prerequisite Material: Section 2.3: On Probability in Classical Mechanics, discussion on the naming of the expectation value on page 12. Section 2.5: Koopman–von Neumann (KvN) Classical Mechanics

In this section, we discuss the probabilistic interpretation of the quantum state by direct analogy to classical mechanics. We start by recalling that, for the classical PDF, we have conservation of probability leading to $\frac{d\rho}{dt} = 0$, which in turn leads to

$$\frac{\partial \rho}{\partial t} = -\{\rho, \mathscr{H}\}. \tag{3.10}$$

If we look at the above equation and take the expectation of each side, we obtain

$$\left\langle \frac{\partial \rho}{\partial t} \right\rangle = -\langle \{\rho, \mathscr{H}\} \rangle. \tag{3.11}$$

We might ask if there is some quantum analogue of this equation, which would need some quantum operator, $\hat{\rho}$, to exist – an analogue to the PDF ρ. We take inspiration from Ehrenfest's theorem,

$$\frac{d}{dt} \langle \hat{\rho} \rangle = \frac{1}{i\hbar} \langle [\hat{\rho}, \hat{H}] \rangle + \left\langle \frac{\partial \hat{\rho}}{\partial t} \right\rangle, \tag{3.12}$$

and, analogous to $\frac{d\rho}{dt} = 0$, we require a similar conservation of probability to hold. Using Ehrenfest as a basis for the analogy, this would then take the form $\frac{d}{dt} \langle \hat{\rho} \rangle = 0$. Before proceeding, let us note that if

$$\frac{\partial \hat{\rho}}{\partial t} = -\frac{1}{i\hbar} [\hat{\rho}, \hat{H}], \tag{3.13}$$

then $\frac{d}{dt} \langle \hat{\rho} \rangle = 0$ would immediately follow.

---

3 Given that the click could be recorded and played back many years later, we will not bring musings about mind-conciseness playing a role here – there are too many counterarguments for us to take that view seriously.

Recall that we have already asserted on page 66 that the state of the system should be normalised, so $\langle \psi(t)|\psi(t)\rangle$ is constant (to ensure the well-behavedness of solutions to the Schrödinger equation). This means that its square is also constant, so

$$\frac{\partial}{\partial t}\langle \psi(t)|\psi(t)\rangle^2 = \frac{\partial}{\partial t}\langle \psi(t)|\psi(t)\rangle\langle \psi(t)|\psi(t)\rangle = \frac{\partial}{\partial t}\langle \psi(t)|\left[|\psi(t)\rangle\langle \psi(t)|\right]|\psi(t)\rangle = 0.$$

Exercise 3.3 Using the chain rule and Eqs. (3.2, 3.4) show that

$$\frac{\partial}{\partial t}|\psi\rangle\langle \psi| = -\frac{1}{i\hbar}\left[|\psi\rangle\langle \psi|, \hat{H}\right].$$

∎

Therefore, we see that an operator of the form

$$\hat{\rho}(t) = |\psi(t)\rangle\langle \psi(t)|$$

has the required properties of a quantum analogue to the classical PDF. We see that this is strikingly similar to the Koopman–von Neumann idea of setting $\rho = \psi^*\psi$, the main difference being that for these bra and ket vectors, the specific inner product has yet to be defined. We can thus think of $|\psi\rangle$ as describing the quantum state of the system in a similar way to the KVN ψ, and $\hat{\rho}$ as an analogue to the classical probability density ρ. This leads us to the term $\hat{\rho}$, the density operator, whose dynamics are given by the Liouville–von Neumann equation

$$\frac{\partial}{\partial t}\hat{\rho}(t) = -\frac{1}{i\hbar}\left[\hat{\rho}(t), \hat{H}\right], \tag{3.14}$$

which is of a similar form to Liouville's equation, but with the standard replacement of the Poisson bracket $[\hat{\rho}, \hat{H}] = i\hbar\{\hat{\rho}, \hat{H}\}$.

In the next two exercises, we further show how $\hat{\rho}$ and $|\psi\rangle$ are related to probability through expectation values.

Exercise 3.4 If $\hat{\rho} = |\psi\rangle\langle \psi|$ and $\hat{X}|n\rangle = x_n|n\rangle$ for some observable $\hat{X}$ (so $\{|n\rangle\}$ forms a complete basis with discrete eigenvalues x_n), using Eq. (1.21) show that

$$\text{Tr}\left[\hat{X}\hat{\rho}\right] = \sum_n p_n x_n,$$

where $p_n = |\psi_n|^2$ and $\psi_n = \langle n|\psi\rangle$.

∎

Exercise 3.5 Show that if instead of a discrete set of eigenvalues in the above question one assumes a continuous spectrum for $\hat{X}$ where $\hat{X}|x\rangle = x|x\rangle$ then

$$\text{Tr}\left[\hat{X}\hat{\rho}\right] = \int dx\, p(x)\, x,$$

where $p(x) = |\psi(x)|^2$ and $\psi(x) = \langle x|\psi\rangle$.

∎

The above sum and integral are of the same form as expectation values, so we can assert $\langle \hat{X}\rangle = \text{Tr}\left[\hat{X}\hat{\rho}\right]$. With a bit of linear algebra we can also show that if $\hat{\rho} = |\psi\rangle\langle \psi|$ then $\langle \hat{X}\rangle = \left\langle \psi\left|\hat{X}\right|\psi\right\rangle$, just as with KVN classical mechanics.

This argument leads us to impose the condition that $\langle \psi|\psi\rangle = \|\psi\|^2 = 1$, which represents that the probability of a particle being in some state is always 1, which is equivalent to the notion of $\text{Tr}[\hat{\rho}] = 1$. We often just say that $|\psi\rangle$ is normalised to one as $\||\psi\|\| = 1$. Our previous arguments motivating the Hermiticity of the Hamiltonian also guarantee that this condition is always met as an initial condition, due to conservation of the norm[4].

Note that if we have a set of possible states $\hat{\rho}_n(t) = |\psi_n(t)\rangle \langle\psi_n(t)|$, each of which are solutions to the Schrödinger equation, then the statistical mixture

$$\hat{\rho}(t) = \sum_n P_n \hat{\rho}_n = \sum_n P_n |\psi_n(t)\rangle \langle\psi_n(t)| \text{ where } P_n \in \mathbb{R} \text{ and } \sum_n P_n = 1$$

will still satisfy the von Neumann equation – even though it may not be possible to write a state $|\psi(t)\rangle$ that satisfies the Schrödinger equation. If $\hat{\rho}(t)$ can be written as $|\psi(t)\rangle \langle\psi(t)|$ then it is referred to as a *pure state*, otherwise as a *mixed state*. In this way, we see that the density operator generalises the notion of the quantum state of a system, similarly to how the PDF generalises the notion of the classical state.

Exercise 3.6 If $\hat{\rho}$ is a pure state, show that $\sum_n P_n^2 = 1$. ∎

3.3.3 The Time-independent Schrödinger Equation

> **Prerequisite Material:** Section 1.3.7: Eigenvectors and Eigenvalues, Section 1.2.4: Basis and Dimension

Here we look at the special case of systems with no time dependence in the Hamiltonian. We will use these results in the next section to argue for a quantum theory of measurement. Recall that the time-dependent Schrödinger equation is:

$$\hat{H}(t) |\psi(t)\rangle = i\hbar \frac{d|\psi(t)\rangle}{dt}.$$

We know from classical mechanics that there are many interesting systems to study where the Hamiltonian does not depend on time. Let us examine what this means for quantum mechanics.

Taking a separation of variables approach, let us assume that if the Hamiltonian is time independent, solutions to the Schrödinger equation exist that can be written in the form:

$$|\psi(t)\rangle = f(t) |\phi\rangle, \tag{3.15}$$

where $|\phi\rangle$ has no time dependence, and $f(t)$ is just a complex-valued function of time. Recall, from the previous section, that we argued that $\langle\psi|\psi\rangle = 1$, and note that

$$\langle\psi|\psi\rangle = \langle\phi|f^*(t)f(t)|\phi\rangle = |f(t)|^2 \langle\phi|\phi\rangle.$$

Therefore, we can ensure, without loss of generality, that $\langle\phi|\phi\rangle = 1$ if $f(t)$ lies on the unit circle. This is the case if $f(t) = \exp[ig(t)]$ where $g(t)$ is some real-valued function of time.

Substituting Eq. (3.15) into the Schrödinger equation yields

$$\hat{H}f(t) |\phi\rangle = i\hbar \frac{df(t)|\phi\rangle}{dt}. \tag{3.16}$$

---

4 We note that non-probabilistic interpretations of the state vector, $|\psi\rangle$, are possible and we will return to this in later discussions.

Now $\hat{H}(t)f(t)\,|\phi\rangle = f(t)\hat{H}(t)\,|\phi\rangle$, as $f(t)$ is just a function, and

$$\frac{\mathrm{d}f(t)\,|\phi\rangle}{\mathrm{d}t} = |\phi\rangle\,\frac{\mathrm{d}f(t)}{\mathrm{d}t},$$

since $|\phi\rangle$ has no time dependence, so

$$f(t)\hat{H}\,|\phi\rangle = \mathrm{i}\,\hbar\,|\phi\rangle\,\frac{\mathrm{d}f(t)}{\mathrm{d}t}. \tag{3.17}$$

Pre-multiplying by $\langle\phi|$, we obtain

$$f(t)\left\langle\phi\left|\hat{H}\right|\phi\right\rangle = \mathrm{i}\,\hbar\,\langle\phi|\phi\rangle\,\frac{\mathrm{d}f(t)}{\mathrm{d}t}. \tag{3.18}$$

Using $\langle\phi|\phi\rangle = 1$ and noting that $\langle\phi|\hat{H}|\phi\rangle$ is time-independent and must be some real number (as $\hat{H}$ is Hermitian), which we shall call E, we have

$$f(t)E = \mathrm{i}\,\hbar\frac{\mathrm{d}f(t)}{\mathrm{d}t}. \tag{3.19}$$

We recognise that this has the solution

$$f(t) = \exp\left[-\frac{\mathrm{i}E}{\hbar}t\right]. \tag{3.20}$$

Substituting back into Eq. (3.17) yields:

$$\hat{H}\,|\phi\rangle = E\,|\phi\rangle, \tag{3.21}$$

which we recognise as an eigenequation.

For systems where the Hamiltonian is independent of time,

$$\hat{H}\,|\phi\rangle = E\,|\phi\rangle$$

is known of as the *Time-Independent Schrödinger Equation* (TISE), and the associated solutions to the Time-Dependent Schrödinger Equation (TDSE) take the form

$$|\psi(t)\rangle = \exp\left[-\frac{\mathrm{i}E}{\hbar}t\right]|\phi\rangle,$$

which form a complete basis. Here E is termed the *energy-eigenvalue* and $|\phi\rangle$ the *energy-eigenvector*.

Note that as $\hat{H}$ is Hermitian, its eigenvectors form an orthonormal basis, so any state can be written in the form

$$|\psi\rangle = \sum_i \psi_n\,|\phi_n\rangle.$$

Hence, the dynamics of the state are simply

$$|\psi(t)\rangle = \sum_i \psi_n \exp\left[-\frac{\mathrm{i}E_n}{\hbar}t\right]|\phi_n\rangle,$$

which is very much like a normal mode expansion in harmonic analysis. Note that the basis may be continuous or a mixture of continuous and discrete, and that the above can be generalised in the natural way by using integrals, or sums and integrals, as required.

3.3.4 Measurement Outcomes

Prerequisite Material: Discussion of expectation value on page 68

When we study the TISE

$$\hat{H}\,|\phi\rangle = E\,|\phi\rangle$$

we see that the eigenvalue of the Hamiltonian, E, is a number that will have units of energy. We also note that any good measurement takes the form of a number (or set of numbers). Moreover, we note that

$$\left\langle \phi \left| \hat{H} \right| \phi \right\rangle = E\,\langle\phi|\phi\rangle = E,$$

as $\langle\phi|\phi\rangle = 1$, and that $\langle\phi|\hat{H}|\phi\rangle = \mathrm{Tr}[\hat{A}\,|\phi\rangle\,\langle\phi|]$ looks like an expectation value in classical mechanics (at least in the KVN formulation). Maybe it will be the case that when we measure the energy of a system, the outcome of that measurement is an eigenvalue of the Hamiltonian. This is a testable hypothesis, and experiments agree with it.

We can only measure real numbers, so maybe the only things that are directly measurable are those with eigenvalues that are guaranteed to be real – Hermitian operators. We may generalise this idea to define that an operator can be an *observable* if it is a Hermitian operator, and if this represents a quantity that can be measured. The outcome of such a measurement must always be an eigenvalue of the related observable.

So, we could then postulate that the dynamics of a quantum system is given by any one of the dynamical equations (Schrödinger, Heisenberg, or Ehrenfest) already presented and that the outcome of any measurement is given as above. But this *does not work* on its own; if the state of the system is not an eigenstate of the observable before measurement, we find that the quantum dynamics of the system is no longer represented by these equations (an eigenstate is eigenvector of an operator that can represent the state of a system).

From examples of sequential chains of Stern–Gerlach experiments, what seems to be the case is that, directly after a measurement, the system is going to be in an eigenstate of the observable associated with the eigenvalue just measured (and if the eigenvalue is non-degenerate, then the state after the measurement is uniquely determined). The quantum dynamics of the system is then given, again after the measurement, by the quantum dynamical equations (Schrödinger, Heisenberg, or Ehrenfest), but now using that eigenstate as the post-measurement initial condition. This is not so unusual, since it fits with our initial discussion of 'realistic' measurement and evolution of classical systems through the Liouville equation (as in the introduction to our discussion of measurement).

> The recipe that appears to work is that for some observable $\hat{A}$ and some state $|\psi\rangle$ the mean or expected value from measuring a quantity is $\langle\psi|\hat{A}|\psi\rangle$, but on any one measurement the outcome will be an eigenvalue α of $\hat{A}$, and immediately on measurement the state of the system is projected into a normalised eigenvector associated with α. Moreover, the probability of measuring a given eigenvalue is the sum of the expectation values of the current state of the system with all the eigenvectors of that eigenvalue.

Note that the normalisation of the measured state is historically imposed because of the probabilistic interpretation of the state vector and density operator, as after measurement the system is known to be in that state with probability 1. It is possible to re-work quantum mechanics without this constraint, but the mathematics begin to look quite messy.

At first sight, this recipe for modelling measurement seems to include a fundamental measurement disturbance effect that always changes the state of a measured system. Do note, however, that if a system is already in an eigenstate of the observable when one makes the measurement, then nothing happens – the system stays in the same state. If sufficiently 'strong' repeated measurements are taken, the dynamics of the system is essentially frozen and this behaviour is termed the quantum Zeno effect (after Zeno's arrow paradox).

Please note that we have taken great care to avoid discussions on subjects like wave-particle duality, as much of the text written on this subject is at best confusing and at worst incorrect.

Note also that in some ways measurement in quantum mechanics is 'better' defined than in classical physics, as it is phrased completely in terms of eigenstates of observables. In classical mechanics (of PDFs) measurement simply acts as projecting/reshaping the PDF as a result of the measurement made. But the ideas are still similar: if we measure a position degree of freedom either classically or quantum mechanically, we gain knowledge of that degree of freedom. The difference is that classically the momentum distribution would, post measurement, reflect the classical probability distribution inherited from the original PDF. But, in the quantum case, the quantum mechanical momentum distribution would be that of the eigenstate of the position operator[5].

3.4 Representation of Quantum Systems

We have had a lot of abstract discussion so far, and still do not know what these quantum states, $|\psi\rangle$, actually look like. In this section we will show how to move from our core ansatz of Poisson brackets being replaced by commutation relations, and our assumptions regarding the measurable values of observables, to construct what the state must look like based on this.

3.4.1 The Position and Momentum Representation

> **Prerequisite Material:** Section 1.3.7: Eigenvectors and Eigenvalues, Section 1.3.9: Functions of Operators, Section 1.3.12: Operators and Basis State – Resolutions of Identity, Section 1.3.6: Commutators, Section 1.4.3: Observables

3.4.1.1 The One-dimensional Case

Let us start by discussing the so-called *position representation* in one dimension. There are two important observations from classical physics:

---

5 Of course, we would also need to consider how to model instrument error, which can be done either with density matrices or Wigner functions, but this is a discussion for another time.

1. Our experience is that if we record the position of a particle that, in one dimension, one unique real number gives us that position.
2. $\{q, p\} = 1$

From this we construct two important corollaries for the quantum mechanical position operator $\hat{q}$ and its canonically conjugate momentum operator $\hat{p}$:

1. Applying our postulates of measurement, $\hat{q}$ is an observable and its eigenequation is

    $$\hat{q} |q\rangle = q |q\rangle,$$

    where the eigenvalues q must be continuous (like the classical position, as this is what is measured) and non-degenerate so $\langle q|q'\rangle = \delta(q - q')$ as $\{|q\rangle\}$ form an orthonormal continuous basis (this form of the delta function assumes rectilinear coordinates).
2. By our previous discussion on commutators, we know we should replace the Poisson bracket with $[\hat{q}, \hat{p}] = i\hbar$.

Let us see where this leads. If all we are considering is position, then we know that $\hat{q}$ is a complete set of commuting observables all on its own, and we can ask how the position representation, usually termed the *wave-function*, of a state

$$\psi(q) = \langle q|\psi\rangle$$

behaves. By this, we mean: what do the actions of various operators on $|\psi\rangle$ look like, allowing us to explore the system's dynamics using, for example, the Schrödinger equation?

The easiest thing to consider is the action of the position operator on $|\psi\rangle$ in this representation. If

$$|\phi\rangle = \hat{q} |\psi\rangle,$$

then the position representation of this is found by premultiplying both sides by $\langle q|$

$$\phi(q) = \langle q|\phi\rangle = \langle q |\hat{q}| \psi\rangle = \langle q |q| \psi\rangle = q \langle q|\psi\rangle = q\psi(q),$$

which is traditionally written in the abbreviated form

$$\hat{q}\psi(q) = q\psi(q).$$

Let us now proceed to discuss the position representation of the momentum operator acting on an arbitrary state $|\psi\rangle$. If we now consider

$$|\phi\rangle = \hat{p} |\psi\rangle,$$

then what is

$$\phi(q) = \langle q|\phi\rangle = \langle q |\hat{p}| \psi\rangle?$$

The derivation that follows is a little long-winded, but it has been chosen as it makes perfectly clear that the form of the position representation of the momentum operator is dependent only on two factors: the eigenvalues and vectors of the position operator, and the commutation relation between the position and momentum operators. So, in the same way Poisson brackets define how classical mechanics works, it is clear that replacing these with commutation relations is needed to make the transition to quantum mechanics. This is one of those 'proofs' that one would never be expected to guess from the outset, and is developed *a posteriori*[6].

We start by defining the operator

$$\hat{D}(Q) = \exp\left[-\frac{i}{\hbar}Q\hat{p}\right],$$

where Q is a real number. Clearly,

$$\hat{D}^\dagger(Q) = \hat{D}^{-1}(Q) = \hat{D}(-Q)$$

and

$$\hat{D}(Q')\hat{D}(Q'') = \hat{D}(Q' + Q'').$$

Using operator algebra for operators whose commutation relation is a scalar[7] we can apply McCoy's theorem, $[A, B] = \alpha \implies [\hat{A}, f(\hat{B})] = \alpha f'(\hat{B})$, so we have

$$[\hat{q}, \hat{D}(Q)] = i\hbar\frac{\partial}{\partial\hat{p}}\hat{D}(Q)$$

$$= i\hbar\left(\frac{-iQ}{\hbar}\right)\hat{D}(Q)$$

$$= Q\hat{D}(Q).$$

Expanding the commutator and rearranging, we get:

$$\hat{q}\hat{D}(Q) - \hat{D}(Q)\hat{q} = Q\hat{D}(Q)$$

$$\hat{q}\hat{D}(Q) = Q\hat{D}(Q) + \hat{D}(Q)\hat{q}$$

$$= \hat{D}(Q)(\hat{q} + Q).$$

It follows that

$$\hat{q}\hat{D}(Q)\,|q\rangle = \hat{D}(Q)(\hat{q} + Q)\,|q\rangle$$

$$= \hat{D}(Q)(\hat{q}\,|q\rangle + Q\,|q\rangle)$$

$$= \hat{D}(Q)(q\,|q\rangle + Q\,|q\rangle)$$

$$= \hat{D}(Q)(q + Q)\,|q\rangle$$

$$= (q + Q)\hat{D}(Q)\,|q\rangle,$$

so

$$\hat{q}\left[\hat{D}(Q)\,|q\rangle\right] = (q + Q)\left[\hat{D}(Q)\,|q\rangle\right].$$

Thus we see that $\hat{D}(Q)$ acting on an eigenstate of $\hat{q}$ with eigenvalue q produces another eigenstate of $\hat{q}$ with eigenvalue $q + Q$. We have assumed that eigenvalues are non-degenerate, but this can only fix an eigenstate up to some arbitrary phase. So let us use one state $|0\rangle$ as a reference, and define $|q\rangle := \hat{D}(q)\,|0\rangle$. We then also have:

$$\hat{D}(Q)\,|q\rangle = \hat{D}(Q)\hat{D}(q)\,|0\rangle$$

---

6 This presentation is close to that given in [12].
7 Note that in the following expression defining exponential can be tricky for unbounded operators, and McCoy's theorem relies on the existence of an underlying power series expansion - that one does not always exist, can cause issues. We will return to this when we discuss curvilinear coordinates in chapter 5.

$$= \hat{D}(Q + q) |0\rangle$$

$$= |Q + q\rangle .$$

For this reason, $\hat{D}(q)$ is termed the displacement operator, and we can see that the exponentiated form of the momentum is what generates displacement. Also note that the displacement operator is a continuous ladder operator as considered in Exercise 1.13, and its form could have been constructed from the requirement $[\hat{q}, \hat{D}] = Q\hat{D}$.

Let us now consider the effect of an infinitesimal displacement by $-\varepsilon$

$$\psi (q + \varepsilon) = \left\langle q \left| \hat{D}(-\varepsilon) \right| \psi \right\rangle = \left\langle q \left| \hat{\mathbb{1}} + \frac{i\varepsilon\hat{p}}{\hbar} + \mathcal{O}(\varepsilon^2) \right| \psi \right\rangle$$

$$= \left\langle q \left| \hat{\mathbb{1}} \right| \psi \right\rangle + \frac{i\varepsilon}{\hbar} \langle q |\hat{p}| \psi \rangle + \mathcal{O}(\varepsilon^2)$$

$$= \psi(q) + \frac{i\varepsilon}{\hbar} \langle q |\hat{p}| \psi \rangle + \mathcal{O}(\varepsilon^2).$$

Rearranging gives us

$$\langle q |\hat{p}| \psi \rangle = -i\hbar \frac{\psi(q + \varepsilon) - \psi(q)}{\varepsilon} + \mathcal{O}(\varepsilon),$$

which in the limit $\varepsilon \to 0$ yields

$$\langle q |\hat{p}| \psi \rangle = -i\hbar \frac{\partial \psi(q)}{\partial q}. \tag{3.22}$$

This again is often written in the abbreviated form

$$\hat{p}\psi(q) = -i\hbar \frac{\partial \psi(q)}{\partial q}.$$

In the position representation:

$$\hat{q}\,\psi(q) = q\psi(q)$$

and

$$\hat{p}\psi(q) = -i\hbar \frac{\partial \psi(q)}{\partial q}.$$

Now to the Schrödinger equation,

$$\frac{d}{dt} |\psi(t)\rangle = \hat{H} |\psi(t)\rangle ,$$

in this representation. We start off by projecting on the left by $\langle q|$

$$\langle q| \frac{d}{dt} |\psi\rangle = \langle q| \hat{H} |\psi\rangle .$$

On the right-hand side we can use similar arguments to those above, $\langle q| \hat{H} |\psi\rangle = \left\langle q \left| \hat{H}\hat{\mathbb{1}} \right| \psi \right\rangle$, to map

$$\hat{H}(\hat{q}, \hat{p}, t) \to \hat{H} \left(q, -i\hbar \frac{\partial}{\partial q}, t \right).$$

With the left-hand side we make use of the fact that eigenstates, $|q\rangle$, of the position operator have no time dependence, so neither will the eigenvalue labels q and so

$$\langle q| \frac{d}{dt} |\psi(t)\rangle = \frac{d}{dt} \langle q|\psi(t)\rangle$$
$$= \frac{d}{dt} \psi(q, t)$$
$$= \frac{\partial}{\partial t} \psi(q, t),$$

where we have used the assumption that in the Schrödinger picture, operators do not implicitly depend on time. As an example, consider the one-dimensional simple harmonic oscillator. The Schrödinger equation

$$\frac{d}{dt} |\psi(t)\rangle = \left(\frac{\hat{p}^2}{2m} + \frac{1}{2} m\omega^2 \hat{q}^2 \right) |\psi(t)\rangle$$

becomes in the position representation

$$\frac{\partial}{\partial t} \psi(q, t) = \left(-\frac{\hbar^2}{2m} \frac{\partial^2}{\partial q^2} + \frac{1}{2} m\omega^2 q^2 \right) \psi(q, t),$$

which is a partial differential (wave) equation we can solve. Note that there are other representation of the Harmonic oscillator that look very different to this and can be more useful in calculations – we will postpone discussion of this point. For now, it is sufficient to realise that the above PDE is one of many possible representations of the Schrödinger equation for the example system. It is also very important to realise how few assumptions we have made to get to this equation.

Exercise 3.7   Repeat the above argument but with the role of position and momentum reversed to show that in the momentum representation

$$\hat{q} \psi(p) = i\hbar \frac{\partial \psi(p)}{\partial p}$$

and

$$\hat{p} \psi(p) = p\psi(p).$$
■

Exercise 3.8   Show that the Schrödinger equation for the simple harmonic oscillator in the momentum representation is

$$\hat{H}(q) = \frac{p^2}{2m} - \frac{1}{2} m\omega^2 \hbar^2 \frac{\partial^2}{\partial p^2}.$$
■

Let us now return to Eq. (3.22), but write $\psi(q)$ as $\langle q|\psi\rangle$ and let the state under consideration be some eigenstate of momentum $|\psi\rangle = |p\rangle$, so that we have

$$\langle q |\hat{p}| p \rangle = -i\hbar \frac{\partial \langle q|p\rangle}{\partial q},$$

$$\langle q |p| p \rangle = -i\hbar \frac{\partial \langle q|p\rangle}{\partial q},$$

$$p \langle q|p \rangle = -i \hbar \frac{\partial \langle q|p \rangle}{\partial q},$$

or

$$\frac{\partial \langle q|p \rangle}{\partial q} = \frac{i p}{\hbar} \langle q|p \rangle.$$

Exercise 3.9 Show that the solution to this equation is the plane wave:

$$\langle q|p \rangle = \frac{1}{\sqrt{2\pi\hbar}} \exp \left(\frac{i p q}{\hbar} \right).$$

Hint: use $\langle q|q' \rangle = \delta(q - q')$ and a resolution of identity in the momentum representation. You may also need to refer to one of the definitions of the Dirac delta function. ■

Exercise 3.10 Show that the eigenfunctions of position in the position representation, and momentum in the momentum representation, are delta functions (think Fourier analysis). How is this similar and different to the probability density function of a particle in classical physics? ■

Exercise 3.11 By inserting the appropriate resolutions of identity and using the above result show

$$\psi(q) = \frac{1}{\sqrt{2\pi\hbar}} \int_{-\infty}^{\infty} dp \, \exp \left[\frac{i p q}{\hbar} \right] \psi(p),$$

and

$$\psi(p) = \frac{1}{\sqrt{2\pi\hbar}} \int_{-\infty}^{\infty} dq \, \exp \left[-\frac{i p q}{\hbar} \right] \psi(q),$$

so that the position and momentum representation can be seen as Fourier transforms of each other. ■

Exercise 3.12 You might like to verify that these two representations of the Hamiltonian are Fourier transforms of each other – note for the simple harmonic oscillator how the form remains unchanged even though the coefficients exchange roles – this is an important symmetry of the harmonic oscillator. ■

3.4.1.2 Three Dimensions

What about higher dimensions? We now consider three dimensions; two dimensions, which we do not consider here, is simply a cut-down version of this argument – they are equally easy to make. Let us start by what we know classically in Cartesian coordinates:

1. Our experience is that if we record the position of a particle, one unique real number for each axis q_x, q_y, and q_z, gives us that position.

2. $\{q_i, q_j\} = 0$, $\{p_i, p_j\} = 0$, and $\{q_i, p_j\} = \delta_{ij}$.

As before, we construct quantum mechanical position operators $\hat{q}_i$ ($i \in \{x, y, z\}$) and their canonically conjugate momenta $\hat{p}_j$ according to:

1. For the position representation, each $\hat{q}$ is an observable and its eigenequation is

$$\hat{q}_i \left| q_x, q_y, q_z \right\rangle = q_i \left| q_x, q_y, q_z \right\rangle$$

with[8]

$$\left\langle q_x, q_y, q_z | q'_x, q'_y, q'_z \right\rangle = \delta(q_x - q'_x)\delta(q_y - q'_y)\delta(q_z - q'_z).$$

Importantly, note that q_i for a given x, y, or z no longer uniquely specifies the eigen-state – just as in classical physics we need to measure all three simultaneously to uniquely define the position state of the system.

2. $\left[\hat{q}_i, \hat{q}_j\right] = 0$, $\left[\hat{p}_i, \hat{p}_j\right] = 0$, and $\left[\hat{q}_i, \hat{p}_j\right] = i\hbar\left\{\hat{q}_i, \hat{p}_j\right\} = i\hbar\delta_{ij}$.

In this case. as they all commute, the $\{\hat{q}_i\}$ forms a CSCO (they are complete by the con-struction of the problem). So we can label the basis state as $\left| q_x, q_y, q_z \right\rangle$ (note we *choose* the ordering of the label to follow the conventional ordering of Cartesian coordinates – we do not *have to* do this). In fact, we could even choose combinations such as $\{q_x, p_y, q_z\}$ as our CSCO, and there are occasions where this might be a sensible thing to do (you should be able to see how this would pan out by adapting the following discussion). Following the same logic as in the preceding section, we define the wave-function for any state $|\psi\rangle$ as

$$\psi(q_x, q_y, q_z) = \left\langle q_x, q_y, q_z | \psi \right\rangle$$

and find elementwise

$$\left\langle q_x, q_y, q_z \left| \hat{q}_i \right| \psi \right\rangle = q_i \left\langle q_x, q_y, q_z | \psi \right\rangle$$

and

$$\left\langle q_x, q_y, q_z \left| \hat{p}_i \right| \psi \right\rangle = -i\hbar\frac{\partial \left\langle q_x, q_y, q_z | \psi \right\rangle}{\partial q_i}$$

or in our shorthand

$$\hat{q}_i\psi(q_x, q_y, q_z) = q_i\psi(q_x, q_y, q_z)$$

and

$$\hat{p}_i\psi(q_x, q_y, q_z) = -i\hbar\frac{\partial\psi(q_x, q_y, q_z)}{\partial q_i}.$$

We can simplify this notationally, as we can take the direct analogy of representing clas-sical coordinates as a vector:

$$(q_x, q_y, q_z) \rightarrow \mathbf{q} = q_x\mathbf{i} + q_y\mathbf{j} + q_z\mathbf{k} = \begin{pmatrix} q_x \\ q_y \\ q_z \end{pmatrix}$$

to define a quantum equivalent

$$\hat{\mathbf{q}} = \hat{q}_x\mathbf{i} + \hat{q}_y\mathbf{j} + \hat{q}_z\mathbf{k} = \begin{pmatrix} \hat{q}_x \\ \hat{q}_y \\ \hat{q}_z \end{pmatrix}$$

---

8 This assumes rectilinear coordinates.

and

$$\hat{\mathbf{p}} = \hat{p}_x \mathbf{i} + \hat{p}_y \mathbf{j} + \hat{p}_z \mathbf{k} = \begin{pmatrix} \hat{p}_x \\ \hat{p}_y \\ \hat{p}_z \end{pmatrix} = -i\hbar \begin{pmatrix} \partial/\partial q_x \\ \partial/\partial q_y \\ \partial/\partial q_z \end{pmatrix}.$$

> Thus we can write that
>
> $$\hat{\mathbf{q}}\psi(\mathbf{q}) = \mathbf{q}\psi(\mathbf{q}) \text{ and } \hat{\mathbf{p}}\psi(\mathbf{q}) = -i\hbar\nabla\psi(\mathbf{q}).$$

For the momentum representation we would repeat this argument in terms of the eigenvalues of the momentum operator, and also note that we could *use a mix* of position and momentum representations in different directions as they too form a CSCO (so it is perfectly possible to end up with a wave-function $\psi(q_x, p_y, q_z)$ which, for some physical systems, may be of use).

Exercise 3.13 Satisfy yourself that

$$\hat{\mathbf{q}}\psi(\mathbf{p}) = i\hbar\nabla\psi(\mathbf{p}) \text{ and } \hat{\mathbf{p}}\psi(\mathbf{p}) = \mathbf{p}\psi(\mathbf{p}).$$

∎

Exercise 3.14 Satisfy yourself that the Hamiltonian for the three-dimensional harmonic oscillator in the position representation is:

$$\frac{\partial}{\partial t}\psi(\mathbf{q}, t) = \left(-\frac{\hbar^2}{2m}\nabla^2 + \frac{1}{2}m\omega^2\mathbf{q}^2 \right) \psi(\mathbf{q}, t).$$

What is it in the momentum representation? How about if the x and y components are taken in the position representation and the z component in the momentum representation? ∎

3.4.2 Spin

> **Prerequisite Material:** Section 2.6.2: The Stern–Gerlach Experiment. While not an actual prerequisite, some understanding of quaternions and the application of quaternions to rotations in classical physics will aid your intuition towards the findings in this section. The non-commutativity and matrix representation are of particular interest. Even outside quantum mechanics, quaternions are useful things to know about; they have broad relevance. For example, they often find application in the video-games industry as they avoid an issue that other methods of coding rotations have.

Let us consider the Stern–Gerlach experiment for a particle with no net orbital angular momentum but which retains a magnetic moment, such as silver (this is due to the spin of one electron). We can make the following observations:

1. The measurement of the intrinsic angular momentum, which we term 'spin' in the Stern–Gerlach experiment, only ever yields one of two results for a z-oriented magnet – therefore, it must be represented by an operator with two distinct eigenvalues of equal and opposite magnitude.

2. Repeats of experiments with the same and other set-ups indicate that these eigenvalues should not be degenerate.
3. Classically, angular momentum satisfies the Poisson bracket $\{L_x, L_y\} = L_z$ (and cyclic permutations), so for a quantum mechanical equivalent we are looking for three operators that satisfy $[\hat{S}_x, \hat{S}_y] = i\hbar\hat{S}_z$ (and cyclic permutations). Note, we are using the letter $\hat{S}$ instead of $\hat{L}$ as we will want to use $\hat{L}$ to represent orbital angular momentum in later discussions – $\hat{S}$ is what we will use for spin[9].

Note the similarity with the beginning of the discussion of the previous section. Let's see how far these observations get us in determining what these spin operators look like.

We have the eigenvalue equation,

$$\hat{S}_z |m\rangle = m\hbar |m\rangle,$$

where $\langle m|n\rangle = \delta_{mn}$, as $\hat{S}_z$ is an observable. We include a factor of $\hbar$ here to make the mathematics that follows tidier (an $\hbar$ factor in each of the $\hat{S}_i$ sorts out the lone $\hbar$ in the commutation relation, and as $\hbar$ has units of angular momentum, this ensures that the above equation is dimensionally correct if m is just a number).

As we have done nothing more than set some commutation relations and an eigenvalue equation, we are actually free to fix m – whatever follows will set the scale of $\hat{S}_z$. Let us *choose* $m = \pm 1/2$, as this sets the energies associated with each eigenstate exactly $\hbar$ apart (alternatively one can look at this the other way – from the development of the theoretical framework that follows we will determine this constant of proportionality and once this is fixed, we have a framework we can test by experiment).

In the previous section we had a continuous basis, so working in wave-functions was sensible. For this system the basis is discrete so we will work in a matrix formulation. We will order our basis state so

$$\psi \rightarrow \begin{pmatrix} \langle +\frac{1}{2}|\psi\rangle \\ \langle -\frac{1}{2}|\psi\rangle \end{pmatrix}$$

and

$$\hat{A} \rightarrow \begin{pmatrix} \langle +\frac{1}{2}|\hat{A}|+\frac{1}{2}\rangle & \langle +\frac{1}{2}|\hat{A}|-\frac{1}{2}\rangle \\ \langle -\frac{1}{2}|\hat{A}|+\frac{1}{2}\rangle & \langle -\frac{1}{2}|\hat{A}|-\frac{1}{2}\rangle \end{pmatrix}.$$

As we have chosen the eigenbasis of the S_z operator, we simply have

$$\hat{S}_z \rightarrow \mathbf{S}_z = \frac{\hbar}{2}\begin{pmatrix} 1 & 0 \\ 0 & -1 \end{pmatrix},$$

$$\left|+\frac{1}{2}\right\rangle \rightarrow \begin{pmatrix} 1 \\ 0 \end{pmatrix},$$

$$\left|-\frac{1}{2}\right\rangle \rightarrow \begin{pmatrix} 0 \\ 1 \end{pmatrix}.$$

---

9 A general discussion of angular momentum in quantum mechanics, and the reason we do not choose spherical coordinates and conjugate momenta, has sufficient subtleties to necessitate postponing a comprehensive study until a later chapter devoted to the subject. For now, we shall only consider this two-state example.

We now proceed to determine the matrix representation of $\hat{S}_x$ and $\hat{S}_y$. Without loss of generality we can write

$$S_x = \frac{\hbar}{2}\begin{pmatrix} a & c \\ c^* & b \end{pmatrix} \text{ and } S_y = \frac{\hbar}{2}\begin{pmatrix} \alpha & \gamma \\ \gamma^* & \beta \end{pmatrix},$$

where a, b, α, β, are real numbers, c and γ are complex numbers, and the symmetry of the off-diagonal elements is a requirement of the Hermiticity of observables. Now,

$$\left\langle m\left|\hat{S}_y\right|n\right\rangle = -\frac{i}{\hbar}\left\langle m\left|[\hat{S}_z, \hat{S}_x]\right|n\right\rangle$$

$$= -\frac{i}{\hbar}\left\langle m\left|\hat{S}_z\hat{S}_x - \hat{S}_x\hat{S}_z\right|n\right\rangle$$

$$= -\frac{i}{\hbar}\left\langle m\left|m\hbar\hat{S}_x - \hat{S}_x n\hbar\right|n\right\rangle$$

$$= i(n-m)\left\langle m\left|\hat{S}_y\right|n\right\rangle.$$

From this we can immediately deduce that

$$\alpha = \beta = \left\langle\pm\frac{1}{2}\left|\hat{S}_y\right|\pm\frac{1}{2}\right\rangle = 0$$

and

$$\left\langle\pm\frac{1}{2}\left|\hat{S}_y\right|\mp\frac{1}{2}\right\rangle = \mp i\left\langle\pm\frac{1}{2}\left|\hat{S}_y\right|\mp\frac{1}{2}\right\rangle$$

(an identical argument using $[\hat{S}_y, \hat{S}_z] = i\,\hbar\hat{S}_x$ can be used to show that $a = b = 0$), so that

$$S_x = \frac{\hbar}{2}\begin{pmatrix} 0 & c \\ c^* & 0 \end{pmatrix} \text{ and } S_y = \frac{\hbar}{2}\begin{pmatrix} 0 & -ic \\ ic^* & 0 \end{pmatrix}. \tag{3.23}$$

Now let us use the last of the three angular momentum commutation relations $[S_x, S_y] = i\,\hbar S_z$ which in matrix form is

$$\frac{\hbar^2}{4}\left[\begin{pmatrix} 0 & c \\ c^* & 0 \end{pmatrix}\begin{pmatrix} 0 & -ic \\ ic^* & 0 \end{pmatrix} - \begin{pmatrix} 0 & -ic \\ ic^* & 0 \end{pmatrix}\begin{pmatrix} 0 & c \\ c^* & 0 \end{pmatrix}\right] = i\frac{\hbar^2}{2}\begin{pmatrix} 1 & 0 \\ 0 & -1 \end{pmatrix}. \tag{3.24}$$

This simplifies to

$$\begin{pmatrix} i|c|^2 & 0 \\ 0 & -i|c|^2 \end{pmatrix} = \begin{pmatrix} i & 0 \\ 0 & -i \end{pmatrix}, \tag{3.25}$$

which means that c must be a complex number of length one, i.e. of the form $e^{i\theta}$ for some θ so that

$$S_x = \frac{\hbar}{2}\begin{pmatrix} 0 & e^{i\theta} \\ e^{-i\theta} & 0 \end{pmatrix} \text{ and } S_y = \frac{\hbar}{2}\begin{pmatrix} 0 & -ie^{i\theta} \\ ie^{-i\theta} & 0 \end{pmatrix}. \tag{3.26}$$

But how do we fix this phase? If we fix the phase to zero, each of these matrices looks like the matrix representation of quaternions multiplied by $\hbar/2$. Indeed, a non-zero phase would correspond to rotating the xy plane around the z-axis (so this is something we are free to do). Setting the phase to zero connects these spin operators to the notion of rotation in space. To make this connection more explicit, let us define

$$\sigma_x = \begin{pmatrix} 0 & 1 \\ 1 & 0 \end{pmatrix}, \sigma_y = \begin{pmatrix} 0 & -i \\ i & 0 \end{pmatrix}, \sigma_z = \begin{pmatrix} 1 & 0 \\ 0 & -1 \end{pmatrix}.$$

These are referred to as the Pauli matrices. For completeness, let us note that $\hat{\mathbb{1}}$, $\hat{\sigma}_x$, $\hat{\sigma}_y$, and $\hat{\sigma}_z$, have exactly the same algebraic properties as the quaternions 1, i, j, and k. We can also consider these as matrix representations of some operators $\hat{\sigma}_x$, $\hat{\sigma}_y$, $\hat{\sigma}_z$ in the $\hat{S}_z$ basis – known as the Pauli operators – whose commutation relation is $[\hat{\sigma}_i, \hat{\sigma}_j] = 2i\epsilon_{ijk}\hat{\sigma}_k$ (just like the quaternions).

To summarise, we have shown that

$$S_x = \frac{\hbar}{2}\sigma_x, \quad S_y = \frac{\hbar}{2}\sigma_y, \quad S_z = \frac{\hbar}{2}\sigma_z,$$

which together with the identity operator is enough to completely describe any operation on a single spin, and is directly connected to spatial rotations as described by quaternions. So, in the same way that we saw momentum operators as generators of displacement, we can view angular momentum operators as generators of rotation.

The quantum mechanical description of spin arises as a consequence of a few observations: an eigenvalue equation with two orthonormal eigenvectors, the fact that observables are represented by Hermitian operators, and the three angular momentum commutators. It is this commutation relation, inherited from the Poisson bracket, that leads to the connection with quaternions and spatial rotations – a profoundly elegant set of connections.

3.4.3 Spin and Position – The Spinor

Our previous discussion about spin was motivated by the Stern–Gerlach experiment, but here we only considered measuring the spin degree of freedom. If we wanted to better describe the state of the atom, we should at least consider its spatial (or momentum) degrees of freedom. Unlike orbital angular momentum, spin is found to be intrinsic and, therefore, does not depend on position and momentum of a particle in any way. For this reason, the three spin operators will commute with position and momentum and not share an eigenbasis. Even though atoms have more complex internal degrees of freedom, this is of value for situations where those are not important. For particles like the electron that, to the best of our knowledge, have no internal structure, spin and position (or momentum) comprise a CSCO that can be used to completely describe the state of the particle.

If we choose a position eigenbasis of

$$\hat{q}\,|q, \ldots\rangle = q\,|q, \ldots\rangle$$

and, keeping our convention of a z reference, a spin basis[10] of

$$\hat{S}_z\,|m, \ldots\rangle = m\hbar\,|m, \ldots\rangle,$$

$\{\hat{q}, \hat{S}_z\}$ form a CSCO, so we can represent a state with basis state $|q, m\rangle$ in this space by:

$$\psi(q) = \begin{pmatrix} \left\langle q, +\frac{1}{2} | \psi \right\rangle \\ \left\langle q, -\frac{1}{2} | \psi \right\rangle \end{pmatrix} \triangleq \begin{pmatrix} \psi_+(q) \\ \psi_-(q) \end{pmatrix}.$$

---

10 Where the dots indicate that the eigenvalue alone is not enough to uniquely determine each state – for example, if we know the position of a particle it tells us nothing about its spin.

This object is termed a *spinor*. We use spinors to describe particles with spin-half such as electrons or, should we want to do a more comprehensive study than is conventionally presented, silver atoms in the Stern–Gerlach experiment. We will come back to this in a later problem.

Exercise 3.15   What is the effect of the different spin, position, and momentum operators on this state?   ∎

3.5   Closing Remarks and the Axioms of Quantum Mechanics

If we return to the position representation discussion for a moment, we see that only if the potential is a function of position and time, can we write Schrödinger's equation as

$$\left[-\frac{\hbar^2}{2m}\frac{\partial^2}{\partial q^2} + V(q,t) \right] \psi(q,t) = i\hbar\frac{\partial}{\partial t}\psi(q,t),$$

which, in this form, is the historical beginnings of modern quantum mechanics. We can see that it takes the form of a wave equation. This explains the historical notions of wave mechanics (a discussion we have carefully avoided, as it is not needed and can produce some fundamental confusion with discrete bases, and regarding the problematic concept of wave-particle duality).

It is traditional in quantum mechanics texts to list the axioms of quantum mechanics. Oddly, they don't always match up. The reason for this is that the same theory (as a mathematical framework) can be reached through a number of different paths. So, before looking at the axioms themselves, let us summarise the arguments made to get to this point:

1. We observed that the initial state of a system could be encoded in a vector in a vector space – generalising notions of a state in classical mechanics (so this captures notions such as coordinates, PDFs, Cartesian vectors, etc., as well as the extensions we need for quantum mechanics). As such, this is *just a generalisation of what we do in classical mechanics*.

2. We argued that it is only this initial state vector that is needed to determine the system's future evolution; the dynamical equation for any model in this framework must, at most, be of first order in time. This is no different from Hamilton's equations or the Liouville equation in classical mechanics. Again, this is *not really new* (you might like to ponder on why this makes an important distinction between Hamiltonian and Lagrangian mechanics, considering that they represent the same physics).

3. We used the principle of superposition to argue that the other side of the dynamical equation, for the rate of change of the state vector, must take the form of a linear operator acting on that vector, leading to the Schrödinger equation up to a constant of proportionality (but this looks like Koopman–von Neumann classical mechanics, so again not a new assumption).

4. We then argued for the solutions to be mathematically well-behaved (in terms of the rate of change of the norm, which we wanted to be bounded and non-vanishing), and

invoked some historical imperative to prevent us investigating dead ends (such as for non-Hermitian Hamiltonians[11]). This gave us this form of the Schrödinger equation:

$$\hat{H} |\psi\rangle = i\hbar \frac{\partial}{\partial t} |\psi\rangle.$$

The next phase of the argument focused on understanding what this might mean, and we noted again that this looks like Koopman–von Neumann classical mechanics but with an extra constant of proportionality (that, if set to $\hat{L} = \hat{H}/\hbar$, it would look equivalent).

5. Looking at the rate of change of expectation values,

$$\frac{d}{dt} \langle \hat{A} \rangle = \frac{1}{i\hbar} \langle [\hat{A}, \hat{H}] \rangle + \left\langle \frac{\partial \hat{A}}{\partial t} \right\rangle,$$

and comparing with classical mechanics,

$$\frac{dA}{dt} = \{A, \mathcal{H}\} + \frac{\partial A}{\partial t},$$

we at last arrived at some *non-classical physics*. We postulated that we replace any physical quantity by its operator counterpart, such that: for two classical quantities, A and B, their quantum corresponding operators, $\hat{A}$ and $\hat{B}$, are defined by *replacing Poisson Brackets with commutators* according to:

$$\{A, B\} \rightarrow \frac{1}{i\hbar} [\hat{A}, \hat{B}],$$

where the operators have no implicit time dependence. We also *impose that the eigenvalues of the operators agree with the measured values of that quantity* and, therefore, that observables must be Hermitian operators. These two ideas complete our argument for the form of the Schrödinger equation and the behaviour of operators. So we now understand that $\hat{H}$ is the quantum version of the classical Hamiltonian, and takes the form of an operator.

6. We next investigated the similarity between Liouville's theorem

$$\frac{d\rho}{dt} = \{\rho, \mathcal{H}\}$$

and the von Neumann equation

$$\frac{d\hat{\rho}}{dt} = \frac{1}{i\hbar} [\hat{\rho}, \hat{H}],$$

where $\hat{\rho} = |\psi\rangle \langle\psi|$, and we have made use of our rule for replacing Poisson brackets with commutators. We used this to motivate interpreting the quantum state in a probabilistic framework (it is not the only interpretation, but it works for our current argument).

---

11 Non-Hermitian Hamiltonians are still an active area of investigation and should not be completely ruled out. One word of caution is that there is a currently active topic of research that goes under the title *PT*-symmetry quantum mechanics, where some works give the impression that this is in some way non-Hermitian quantum mechanics. With careful investigation, one can observe that the inner product structure may be changed from the conventional inner product (based on the spectral decomposition of the Hamiltonian – which is somewhat tautological) to actually make the Hamiltonian Hermitian (as Hermiticity is defined with respect to the inner product). For time-dependent Hamiltonians, this could have all kinds of odd effects, as the inner product structure would be time dependent).

7. Studying time-independent Hamiltonians and the Time-Independent Schrödinger Equation, we speculated that it is the eigenvalues of operators, corresponding to classical physical quantities (defined by replacing Poisson brackets with commutators), that are the numbers observed in a given experiment. We discussed a number of possibilities and in the end arrived at the following prescription: *for some observable $\hat{A}$ and some state $|\psi\rangle$, the mean or expected value of measuring its associated quantity is $\left\langle \psi \left| \hat{A} \right| \psi \right\rangle$ – but for any* single *measurement the outcome will be an eigenvalue α of $\hat{A}$. Immediately on measurement, the state of the system is projected into a normalised eigenvector associated with α.* The probability of measuring this eigenvalue is the amplitude of the state in the direction of the associated eigenvector $|\langle \alpha|\psi \rangle|^2$ (at least in the non-degenerate case). This is the Copenhagen view, and our presentation is a simplification of the original argument, which took a number of years to formulate. That said, we also noted that the idea of a measurement of the classical PDF (of the ψ in KVN theory) follows a similar recipe (one that is not usually articulated in this way for classical mechanics).

8. This finalised our formulation of quantum mechanics and, in *The Schrödinger View/Picture* we applied this theory to a number of physical systems, starting with the position operator and deriving Schrödinger's wave equation. It is important to note that there were no new assumptions in this section – our observations were consequences of the theory we had developed in the preceding sections.

Let us record the assumptions that we have made to get here, since these form our axioms of quantum mechanics:

A1 That the complete information of a system is contained within a normalised state vector, $|\psi\rangle$.

A2 That we replace any classical measurable quantity with its operator counterpart, which must be an observable with no implicit time dependence.

A3 Classical quantities are, when there is no ambiguity, mapped to quantum operators by replacing the Poisson bracket with a commutator relation:

$$[\cdot, \cdot] = i\hbar \{\cdot, \cdot\} .$$

A4 Dynamics of the state are given by the Schrödinger equation:

$$\hat{H} |\psi\rangle = i\hbar \frac{\partial}{\partial t} |\psi\rangle .$$

This is followed by some further **measurement axioms**:

M1 When an observable is measured, the only possible outcome of that measurement is an eigenvalue of the observable.

M2 The probability of measuring a given eigenvalue is the expectation value of the current state of the system with the projection operator onto the subspace spanned by all the eigenvectors of that eigenvalue.

M3 On measurement, the state of the system is projected into a normalised eigenstate of the system, associated with the measured eigenvalue. Whilst this 'breaks' Schrödinger dynamics, system evolution post-measurement continues according to the Schrödinger equation.

Here, M1 and M2 are known of the *Born rule*, and M3 is the projective measurement/collapse of the wave-function postulate. As a final step, we must *choose a representation*. This is needed, not from a physics point of view, but in order to write down equations we can actually apply in order to model the system at hand and solve problems. We:

R1 Choose a complete set of commuting observables as a representation of the state-space.
R2 Use our understanding of the classical counterparts to, or experimentally observed behaviour of, each of the operators in the CSCO to set the domain and inner product for the vector space (which will have important consequences, such as providing the resolution of identity).
R3 Apply the above to make a representation of the vectors and operators in the state-space in terms of this 'basis'.

We can then, for example, solve Schrödinger's equation or Ehrenfest's theorem in this representation and compare predictions against experimental results.

In any theory, the set of axioms is the collection of assumptions that are used to arrive at that theory – however, as already noted, there may be multiple routes to the same destination. In our discussion so far, you may have noticed alternative sets of assumptions that lead to the same mathematical framework. We will return to this discussion when we look at the Heisenberg, phase space, and path integral pictures of quantum mechanics. We will also discuss measurement in much more detail once we have developed a deeper understanding of the framework.

4

Other Formulations of Quantum Mechanics

"For a brief period at the beginning of 1926, it looked as though there were, suddenly, two self-contained but quite distinct systems of explanation extant: matrix mechanics and wave mechanics. But Schrödinger himself soon demonstrated their complete equivalence."

Max Born, Nobel Lecture (1954)[1]

4.1 Introduction

We have already seen that there are a number of formulations of classical physics, such as Lagrangian, Hamiltonian, Liouvillian, and Koopman–von Neumann, that may look very different from each other. Each has its own strengths and weaknesses, and we tend to use the formulation that best serves our needs at a given time. It is therefore unsurprising that there also exist a number of different formulations of quantum mechanics that look very different from each other. In this chapter, we introduce three of them. They are the *Heisenberg, Wigner phase space*, and *Feynman Path integral* pictures. The first two we discuss in some detail, but to the last we only give a very gentle introduction (we refer the reader to [28] for a full treatment). We do not follow the usual approach of deriving the different formulations from one another. Instead, we continue the spirit of the previous chapter and motivate each case directly from classical physics, emphasising the similarities as well as the differences between quantum and classical physics. This approach allows us to make clear the way in which each formalism is an extension of classical mechanics. It is interesting to note that while these arguments are different in their formulation, they share the same common destination (even though the mathematical formulations may look different, there is an equivalence between each of them).

In the previous chapter, we argued for the Schrödinger picture as an ahistorical extension of Koopman–von Neumann theory. One could argue that each of the formulations presented in this chapter is in some ways more natural than the Schrödinger view. In the Heisenberg picture, the dynamical equations are closer to those of Hamiltonian mechanics, and expressed in terms of operators rather than the dynamics of a state vector (because

---

1 Actually, as we shall later see, there are some differences.

Quantum Mechanics: From Analytical Mechanics to Quantum Mechanics, Simulation, Foundations, and Engineering, First Edition. Mark Julian Everitt, Kieran Niels Bjergstrom and Stephen Neil Alexander Duffus.
© 2024 John Wiley & Sons Ltd. Published 2024 by John Wiley & Sons Ltd.
Companion website: www.wiley.com/go/everitt/quantum

of this, interpreting quantum measurement is less of an issue, as unitary evolution of the state is not assumed). The Heisenberg picture sits conceptually between an extension of Koopman–von Neumann and Hamiltonian mechanics. In both the Wigner phase space and Feynman path integral formalism, the connection to classical physics is even more clear, as each seeks to retain as much of the conceptual framework of classical mechanics as is possible. The Wigner phase space formalism, by generalising the Poisson bracket, is a direct extension of Hamiltonian/Liouvillian mechanics. The Feynman path integral formulation, by relaxing the least action condition, is a generalisation of Lagrangian mechanics.

One might quite rightly wonder why we did not begin our discussion with one of these other frameworks. Our preferences for starting with the Schrödinger picture were twofold. The first was that we are used to thinking of the state of a system evolving dynamically. This makes the Heisenberg picture, which is in all other respects more elegant, a less natural staring point. The second was that we needed at some point to introduce the standard axioms of quantum measurement. This is more challenging for the Wigner phase space and Feynman path integral formulations. Now that we have that framework in place, it acts as a useful point of reference for the discussion in this chapter, aiding the comparison between quantum and classical physics. We have chosen to present the alternative formulations of quantum mechanics, progressing from the one most similar to the Schrödinger picture to the least similar. Each one of these formulations of quantum mechanics could have been used to develop the theory. In fact, one could argue that they become increasingly conceptually easier to understand (but not necessarily to use).

4.2 The Heisenberg Picture

4.2.1 Background

It has become standard in quantum mechanics texts to derive the so-called Heisenberg picture from the Schrödinger equation. Heisenberg matrix mechanics is a view of quantum theory that actually predates the Schrödinger picture. Moreover, the Heisenberg picture contains some important physical insights, and some aspects of this perspective can be more readily related to classical Hamiltonian physics than the Schrödinger picture. For this reason, we will motivate this formulation from first principles and only then connect it to the Schrödinger equation. It is interesting to read Max Born's Nobel lecture, as it not only gives a perspective on the historical development at a summary level, but also provides some useful insights into the struggles and triumphs of the early pioneers of quantum theory (Nobel lectures are available from the NobelPrize.org website). In a very rough summary, his argument developed along these lines: Niels Bohr discovered that '*an atomic system cannot exist in all mechanically possible states, forming a continuum, but in a series of discrete « stationary » states*'. The energy differences or frequencies between these states can be laid out as a table and the formula for the elements $v_{nm} = (E_n - E_m)/h$ was known. Heisenberg then made the leap of '*Instead of describing the motion by giving a coordinate as a function of time, x(t), an array of transition amplitudes should be determined x_{mn}*'. It is interesting that the use of matrices was not at all common at this time and it was when Heisenberg passed a draft paper to Born, who was aware of matrices, that the penny dropped. Here, Born

...applied this rule to Heisenberg's quantum condition and found that this agreed in the diagonal terms. It was easy to guess what the remaining quantities must be, namely, zero; and at once there stood before me the peculiar formula

$$pq - qp = h/2\pi i$$

This meant that coordinates q and momenta p cannot be represented by figure values but by symbols, the product of which depends upon the order of multiplication – they are said to be « non-commuting ».

I was as excited by this result as a sailor would be who, after a long voyage, sees from afar, the longed-for land, and I felt regret that Heisenberg was not there. I was convinced from the start that we had stumbled on the right path. Even so, a great part was only guess-work, in particular, the disappearance of the non-diagonal elements in the above-mentioned expression.

Born then went on with Jordan to work these ideas into a full matrix mechanics that was later claimed by Schrödinger to be equivalent (there are actually some subtle differences). If these early pioneers had been aware of the theory of representations, it might have been quite natural for them that this could be done. But they were not, and so the link between matrices & states and operators & wave functions as two different manifestations of the same thing was rather surprising for them.

We will follow a different argument from that presented in Born's lecture. We instead follow the same logic as we used when motivating the Schrödinger equation. We do this to once more make clear the connection of quantum physics to classical Hamiltonian mechanics and to highlight the similarities rather than differences between the two theories. We will again use the three key observations that (i) nature (at least regarding quantum phenomena) is linear, as experiments indicate that the principle of superposition holds (at least until we do a measurement); (ii) the initial state of the system is all that is needed to describe the dynamics of the evolution (but this time we will *not explicitly use* the assumption that the rate of change of the state is not needed); and (iii) that the initial state of the system is described by a vector in a vector space $|\psi(0)\rangle$. Unlike in our previous argument, we will keep front and centre only classical Hamiltonian mechanics. We will specifically focus on the Poisson bracket, as it plays a key role in understanding the symmetries and conservation laws of any given system. We will seek to preserve as much of this formal structure as we can, and in doing so follow a process known of as canonical quantisation. In this way, the presentation of the Heisenberg picture might be argued to more elegantly arrive at certain aspects of quantum physics than the one presented in the previous chapter. The fact that the state does not evolve in this view in-between measurements is a subtlety that is quite different from classical mechanics and complicates the argument to the point that we felt it preferable to introduce the Schrödinger picture first. In any case, let us now proceed to make that argument.

4.2.2 Motivating the Heisenberg Equation of Motion

Recall that the equation for the dynamics of any classical quantity A is

$$\frac{dA}{dt} = \{A, \mathscr{H}\} + \frac{\partial A}{\partial t}. \tag{4.1}$$

One of the key assumptions we have required of our quantum theory is that it is linear (observation (1) in Section 3.2). The physics encoded in the above equation does not meet this requirement. The non-linear element of classical mechanics arises from the fact that the dynamics of the classical state (coordinates and conjugate momenta) of the system are described by a closed but interdependent set of equations (e.g. to determine $\frac{dq}{dt}$ we will also need p and to get $\frac{dp}{dt}$ we need q). As we noted in our discussion of the correspondence principle, even simple systems such as the three-body problem are very non-linear in nature and exhibit Hamiltonian chaos. From this, we postulate that the above model cannot stand as it is. However, if we could find a way to modify this expression so that the principle of superposition holds, maybe that would provide a path to a new theory. We know that the Liouville equation, which describes the state of the system in a single probability density function, is linear in this way. However, it does not model quantum phenomena, so we need something else. In the previous chapter, we invoked the Koopman–von Neumann theory as a way to achieve this end and argued for the Schrödinger equation. But if we did not want to do that, what other options are there? The alternative 'obvious' thing to do is to see if we can replace the Poisson bracket with something else (that, like the Poisson bracket, is linear in its first and second arguments). There is more than one choice for doing this, and later in this chapter we discuss a phase space method that explores an alternative to the one we consider in this section.

Let us begin by reviewing the properties of the Poisson bracket, so we can see the mathematical properties of that which we are trying to generalise. Specifically, recall that

$$\{u, v\} = -\{v, u\} \tag{4.2}$$

$$\{au_1 + bu_2, v\} = a\{u_1, v\} + b\{u_2, v\} \tag{4.3}$$

$$\{u_1 u_2, v\} = \{u_1, v\}u_2 + u_1\{u_2, v\} \tag{4.4}$$

$$\{u, v_1 v_2\} = \{u, v_1\}v_2 + v_1\{u, v_2\} \tag{4.5}$$

$$\{\{u, v\}, w\} + \{\{w, u\}, v\} + \{\{v, w\}, u\} = 0, \tag{4.6}$$

where a and b are just numbers (we note that linearity in the second argument follows from the first two equations, so this is not included as a specific property). Dirac argued for a quantum Poisson bracket which satisfies all the same above algebraic relations of the Poisson bracket [18]. However, motivated by experiments such as the Stern–Gerlach experiment and light in polarisers, where the ordering in which we do things matters, Dirac also imposed from the outset that the ordering of the quantities we use matters. To some extent, the argument that is developed in this way presupposes the solution that Dirac finds. We present a modified argument to make the non-commutativity of quantum mechanics become exposed as a natural part of the thought process of the argument itself. In any case, our discussion remains only motivational, as either argument is still subject to the flaw that it is not always possible to find a quantum Poisson bracket on non-commuting quantities that is matched by a classical Poisson bracket (this is Groenewold's theorem, a point we will expand later). The narrative will, importantly, also set the scene for introducing phase space quantum mechanics, which we will discuss later in this chapter. We start from the same point as in Dirac's argument and examine $\{u_1 u_2, v_1 v_2\}$ subject to the above assumptions and see where this leads us. We first apply Eq. (4.4) to obtain:

$$\{u_1 u_2, v_1 v_2\} = \{u_1, v_1 v_2\}u_2 + u_1\{u_2, v_1 v_2\} \tag{4.7}$$

and then Eq. (4.5) to obtain:

$$\{u_1 u_2, v_1 v_2\} = \{u_1 u_2, v_1\} v_2 + v_1 \{u_1 u_2, v_2\}. \tag{4.8}$$

Equating these yields

$$\{u_1, v_1 v_2\} u_2 + u_1 \{u_2, v_1 v_2\} = \{u_1 u_2, v_1\} v_2 + v_1 \{u_1 u_2, v_2\}. \tag{4.9}$$

Making use of the other relations and simplifying, we find:

$$\{u_1, v_1\} \left[u_2, v_2\right] = \left[u_1, v_1\right] \{u_2, v_2\}, \tag{4.10}$$

where $\left[a, b\right] \stackrel{\text{def}}{=} ab - ba$ is, as in the previous chapter, the commutator. In classical mechanics Eq. (4.10) holds trivially because $\left[u_1, v_1\right] = \left[u_2, v_2\right] = 0$. However, as we are seeking to generalise the Poisson bracket, we can ask ourselves what happens if we relax this condition and allow ourselves to replace classical quantities with something that does not commute - what then?

Exercise 4.1 Show that Eq. (4.10) is indeed correct. ∎

First we note that commutators satisfy the same algebra as the Poisson brackets – this is good as it means we retain some important structure that was essential in defining classical mechanics. As this is very important, let us state the identities explicitly:

$$[u, v] = - [v, u] \tag{4.11}$$

$$\left[au_1 + bu_2, v\right] = a \left[u_1, v\right] + b \left[u_2, v\right] \tag{4.12}$$

$$\left[u_1 u_2, v\right] = \left[u_1, v\right] u_2 + u_1 \left[u_2, v\right] \tag{4.13}$$

$$\left[u, v_1 v_2\right] = \left[u, v_1\right] v_2 + v_1 \left[u, v_2\right] \tag{4.14}$$

$$[[u, v], w] + [[w, u], v] + [[v, w], u] = 0. \tag{4.15}$$

Specifically (and don't worry if you do not get this bit), both the Poisson bracket and the commutator are examples of a Lie bracket. This mathematical structure links both classical and quantum mechanics to the mathematics of Lie groups and algebras. This branch of mathematics is concerned with continuous symmetries in systems and the behaviour of infinitesimal transformations within them. That canonical quantisation, which seeks to preserve classical symmetries in the quantisation process, is so closely connected to Lie algebra through the Poisson bracket is quite natural.

Now, since u_1 and v_1 may be chosen independently of u_2 and v_2, we can guarantee that Eq. (4.10) holds only if

$$\left[u_1, v_1\right] = \alpha \{u_1, v_1\} \text{ and } \left[u_2, v_2\right] = \alpha \{u_2, v_2\}, \tag{4.16}$$

where α is some constant independent of $\{u_1, v_1, u_2, v_2\}$. The idea is to obtain a new theory that is 'compatible' with Hamiltonian mechanics by replacing the Poisson bracket with the commutator accordingly. If the above equality were to always hold, then the commutator divided by α would evaluate to the same thing as Poisson brackets, and Eq. (4.1) would always hold. As we later discuss, according to Groenewold's theorem this equality does not always hold. We will end up applying the rule $[u, v] = \alpha \{u, v\}$ only to coordinates and their conjugate momenta. These will be the canonical commutation relations which form the basis of quantum mechanics. For now, we posit that for any two classical quantities A and B we have

$$\left[\hat{A}_H, \hat{B}_H\right] = \alpha \left\{\hat{A}_H, \hat{B}_H\right\}, \tag{4.17}$$

where we have used the hats to indicate that these are non-commuting quantities that have replaced the usual classical variables. The subscript H indicates that we are working in the Heisenberg picture. We need to distinguish these quantities from the operators we used in the Schrödinger picture (we will later find that while they are indeed different, they are closely relatable). Observe that the commutator in the above expression arises naturally from the argument that: anything that replaces the Poisson bracket must have the same algebraic properties as the Poisson bracket. In this way, we note once more, the development of the Heisenberg picture is arguably cleaner than the Schrödinger picture. By changing the Lie bracket structure, we are in essence changing the underlying symmetries of the system. If we consider Noether's theorem, that to any continuous symmetry there is a conservation law, we expect that replacing Poisson brackets with commutators, and thus changing the underlying symmetry, means that the conservation laws in quantum mechanics will be different to those in classical physics. When we later find new conserved quantities in quantum mechanics, it is worth remembering that it is the commutation relation and its associated symmetries that are their root cause.

By substitution, Eq. (4.17) and Eq. (4.1) yield

$$\frac{\mathrm{d}\hat{A}_H}{\mathrm{d}t} = \frac{1}{\alpha}\left[\hat{A}_H, \hat{H}_H\right] + \frac{\partial \hat{A}_H}{\partial t}. \tag{4.18}$$

Just as we did in our development of the Schrödinger equation, we now explore some implications of our assumptions and argue to constrain the form of the constant α.

Consider two quantities whose properties we know well in classical mechanics, namely position and momentum, and whose Poisson bracket is $\{q, p\} = 1$. Our ansatz Eq. (4.17) leads to

$$\left[\hat{q}_H, \hat{p}_H\right] = \alpha.$$

We know from commutator algebra[2] that if $\left[\hat{A}, \hat{B}\right] = \alpha$ then $\left[\hat{A}, f\left(\hat{B}\right)\right] = \alpha f'\left(\hat{B}\right)$. As an immediate corollary, $\left[\hat{p}_H, V\left(\hat{q}_H\right)\right] = \alpha V'\left(\hat{q}_H\right)$. As our core assumption is to keep as much classical physics as we can, we will assume that the Hamiltonian in the new theory will be the same as the classical Hamiltonian but replacing the coordinates and conjugate momenta with their non-commuting counterparts. For any classical Hamiltonian of the form $\mathcal{H} = \frac{p^2}{2m} + V(q)$ we have

$$\hat{H}_H = \frac{\hat{p}_H^2}{2m} + V(\hat{q}_H).$$

From Eq. (4.18) we find the equations of motion for the non-commuting quantities $\hat{q}_H$ and $\hat{p}_H$:

$$\frac{\mathrm{d}\hat{q}_H}{\mathrm{d}t} = \frac{\hat{p}_H}{m}$$

---

2 To be more precise, if one looks at the proof of this statement, we see it can be generalised to: for any Lie bracket $\{\cdot, \cdot\}$ where $\{A, B\} = \alpha$ for any function f expandable as a power series, we have $\{A, f(B)\} = \alpha f'(B)$. So the result does not only apply in quantum physics. As such, the result that follows is actually not at all surprising.

and

$$\frac{d\hat{p}_H}{dt} = -V'\left(\hat{q}_H\right).$$

These equations are of exactly the same form as the equations of motion for the commuting classical counterparts of $\hat{q}_H$ and $\hat{p}_H$, q and p (because, like the Poisson bracket, the commutator is also a Lie bracket).

Exercise 4.2 Complete the argument above to derive the equations of motion given above for $\hat{q}_H$ and $\hat{p}_H$ from Eq. (4.18). ∎

Note that as α cancels out, it is not present in the equations of motion. As $\alpha = 0$ would impose that the commutator is zero, we might expect $\alpha \to 0$ to be a classical limit for a quantum-classical correspondence principle. As we will discuss later, the correspondence principle in quantum mechanics is somewhat more subtle than taking such a simple limiting case would imply. Nevertheless, we would expect α to be present only for dynamical equations of motion that differ from their classical counterparts. As these equations of motion arise for the (Lie) algebraic structure alone, the fact that α does not contribute to the equations of motion for systems of the type considered above, is expected and not a concern. What it does mean is that we shall need alternative approaches to determining this constant.

To get a better understanding of α, we now turn our attention to what these non-commuting quantities might be. Matrices and, more generally, linear operators, satisfy all the right properties for non-commuting things like $\hat{q}_H$ and $\hat{p}_H$. In classical mechanics, physical quantities such as position and momentum are represented by real numbers. The analogous restriction for operators is that they be Hermitian. So let us impose the constraint that the non-commuting quantities in this new theory be linear operators and that those corresponding to real classical quantities also be Hermitian. The commutator of two Hermitian operators is anti-Hermitian. Looking once more at the coordinate and its conjugate momenta, where $\left[\hat{q}_H, \hat{p}_H\right] = \alpha$ we see that

$$\left[\hat{q}_H, \hat{p}_H\right]^\dagger = \left[\hat{p}_H^\dagger, \hat{q}_H^\dagger\right] = \left[\hat{p}_H, \hat{q}_H\right] = -\left[\hat{q}_H, \hat{p}_H\right] = -\alpha.$$

This means that $\alpha^* = -\alpha$, and that this must then be purely imaginary or zero. We can thus write $\alpha = i\hbar$ where $\hbar$ is a real constant to be determined by some, yet to be designed, experiment.

In our discussion of the Schrödinger picture, we noted that $[\cdot, \cdot] = i\hbar \{\cdot, \cdot\}$ cannot always hold, as this equality cannot always be made (Groenewold's theorem). This difficulty is just as much a problem in the Heisenberg picture. The resolution of this issue leads to a phase space representation of quantum mechanics due to Wigner, Weyl, Groenewold, and others, which we will discuss later. In the meantime, we posit that since

$$\left[\hat{A}_H, \hat{B}_H\right] = i\hbar \left\{\hat{A}_H, \hat{B}_H\right\} \tag{4.19}$$

cannot always be satisfied (Groenewold's theorem), then our ansatz is that for any coordinate and its conjugate momenta:

$$\left[\hat{q}_H, \hat{p}_H\right] = i\hbar \tag{4.20}$$

and the dynamics of any quantity $\hat{A}_H$ is given by

$$\frac{d\hat{A}_H}{dt} = \frac{1}{i\hbar} [\hat{A}_H, \hat{H}_H] + \frac{\partial \hat{A}_H}{\partial t}. \tag{4.21}$$

This is historically known as the Heisenberg equation. One important consideration is missing from the above framework: we have not accounted for the initial state at all. This is quite profound, as it really makes clear that it is the commutation relation that defines a system's dynamical structure via the Hamiltonian in a general sense. We will consider a specific example of solving the Heisenberg equation and then return to the matter of the state.

4.2.3 A Specific Example: the One-dimensional Harmonic Oscillator

Starting from the ansatz that the Hamiltonian in the Heisenberg picture is simply the classical Hamiltonian, but replacing variables with operators, we have

$$\hat{H}_H = \frac{\hat{p}_H^2}{2m} + \frac{1}{2} m\omega^2 \hat{q}_H^2.$$

We next impose the commutation relation on the coordinates and their canonically conjugate momenta and apply the above scheme to yield:

$$[\hat{q}_H(t), \hat{p}_H(t)] = i\hbar \{\hat{q}_H(t), \hat{p}_H(t)\} = i\hbar. \tag{4.22}$$

Substitution into Eq. (4.21) and using the commutation arithmetic, we have

$$\begin{aligned}
\frac{d}{dt}\hat{q}_H(t) &= \frac{1}{i\hbar} [\hat{q}_H(t), \hat{H}_H] + \left(\frac{\partial \hat{q}_H(t)}{\partial t}\right) \\
&= \frac{1}{i\hbar} \left(\frac{1}{2m} [\hat{q}_H(t), \hat{p}_H^2(t)]\right) \\
&= \frac{\hat{p}_H(t)}{m}.
\end{aligned} \tag{4.23}$$

By similar arguments (or alternatively $[\hat{q}_H, \hat{p}_H] = i\hbar$ implies $[\hat{q}_H, f(\hat{p})_H] = i\hbar f'(\hat{p})_H$), we have

$$\frac{d}{dt}\hat{p}_H(t) = -m\omega^2 \hat{q}_H(t). \tag{4.24}$$

As expected these are of exactly the same form as the equations of motion for classical physics with the exception that $\hat{q}_H(t)$ and $\hat{p}_H(t)$ no longer commute.

Exercise 4.3 Show that

$$\hat{q}_H(t) = \hat{q}_H(0) \cos(\omega t) + \frac{1}{m\omega} \hat{p}_H(0) \sin(\omega t)$$

$$\hat{p}_H(t) = \hat{p}_H(0) \cos(\omega t) - m\omega \hat{q}_H(0) \sin(\omega t)$$

is a solution of the Heisenberg equations of motion for the SHO. Show that $[\hat{q}_H(t), \hat{p}_H(t)] = i\hbar$ for all t (which confirms that these solutions remain consistent with the scheme $[\hat{q}_H(t), \hat{p}_H(t)] = i\hbar \{\hat{q}_H(t), \hat{p}_H(t)\}$). ∎

4.2.4 The State, Representation, and Dynamics

For inspiration as to how we treat the state of the system, we return once more to Koopman–von Neumann theory (Section 2.5). That is, we assume that the initial state, $|\psi(0)\rangle$, is some vector in a vector space.

Now, we may like to think about the dynamics of a system in terms of an evolving state of the system. This point of view is a very classical one, based on human experience. It is, however, not necessarily the best way to think about physical models for which it is not possible to make continuous measurements of the system's total state or probability density (in classical mechanics we assume that such a thing is possible). The empirical process is to start a system in some known initial state or probability density, and then ask what the likely outcome of some future measurement will be. That is, experimentally, we do not ever know the dynamics of the state. We only know the state at $t = 0$, and our model only predicts the expected outcome of a future measurement. To reconstruct the state takes many experiments, just as it does to reconstruct the probability density function in classical physics. In the Heisenberg picture, we have seen that it is the operators that are the dynamical quantities – independent of any state – and determined by Heisenberg's equations. From the initial state, the dynamics of expectation values of any quantity is then given by:

$$\left\langle \hat{A}_H \right\rangle (t) = \left\langle \psi(0) \left| \hat{A}_H(t) \right| \psi(0) \right\rangle .$$

The fact that the dynamics of a system actually only depend on its initial state is thus most clearly shown in this picture.

If we conduct experiments on systems, such as the Stern–Gerlach experiment, we will find that the above model works in terms of the dynamics of expectation values but that the results or measurements may never take that specific value (just like the average of a random string of zeros and ones). Observations like this lead us to once more posit that measurement in quantum mechanics takes the form of the measurement axioms discussed in the previous chapter. From this position we can bring in the theory or representation and then, e.g. show from the commutation relations that $\hat{q}_H(0)$ and $\hat{p}_H(0)$ are exactly the same operators as $\hat{q}$ and $\hat{p}$ in the Schrödinger picture (with the same differential form in the position representation).

4.2.5 Axioms of Quantum Mechanics Revisited

A few, but important, assumptions have been made in the above discussion that differ from those axioms presented at the end of the previous chapter. To make clear the differences, let us present an alternative set of axioms. We denote by a subscript H those axioms that have changed to make clear that some remain the same. The main axioms of quantum mechanics in the Heisenberg picture are

A1$_H$ That the complete information about a system's *initial state* is contained within a normalised state vector, $|\psi(0)\rangle$.

A2 That we replace any classical measurable quantity with its operator counterpart.

A3 Classical quantities are, when there is no ambiguity, mapped to quantum operators by replacing the Poisson bracket with a commutator relation:

$$[\cdot, \cdot] = i \hbar \{\cdot, \cdot\} .$$

A4$_H$ Dynamics of *operators are given by the Heisenberg equation*:

$$\frac{d\hat{A}_H}{dt} = \frac{1}{i\hbar}\left[\hat{A}_H, \hat{H}_H\right] + \frac{\partial \hat{A}_H}{\partial t}$$

and their expectation value by $\left\langle \hat{A}_H \right\rangle (t) = \left\langle \psi(0) \left| \hat{A}_H(t) \right| \psi(0) \right\rangle.$

The measurement axioms contain an important alteration to the third axiom, removing one of the main conceptual difficulties in interpreting measurement in the Schrödinger picture. The first two are unaltered, and the full list reads:

M1 When an observable is measured, the only possible outcome of that measurement is an eigenvalue of the observable.

M2 The probability of measuring a given eigenvalue is the expectation value of the current state of the system with the projection operator onto the subspace spanned by all the eigenvectors of that eigenvalue.

M3$_H$ On measurement, the state of the system is projected onto a normalised eigenstate of the system, associated with the measured eigenvalue.

For the last axiom, recall that projection of the state was not consistent with the dynamics of the Schrödinger equation. In the Schrödinger picture, this leads to the conceptual difficulty that measurement process appears to be in conflict with dynamics. In the Heisenberg picture, this issue does not arise. The state vector corresponds to specifying an initial state – measurement simply corresponds to setting a new initial state and fits naturally with the Heisenberg axiom A1$_H$. The dynamics, A4$_H$, are unaffected by measurement.

The final 'axioms' of representation only need one slight change:

R1$_H$ Choose a complete set of commuting observables at $t = 0$ as a representation of the state-space.

R2 Use our understanding of the classical counterparts to, or experimentally observed behaviour of, each of the operators in the CSCO to set the domain and inner product for the vector space.

R3 Apply the above to make a representation of the vectors and operators in the state-space in terms of this 'basis'.

4.2.6 The Evolution Operator

We can connect the Heisenberg picture with the Schrödinger picture by taking the time-derivative of both sides of Heisenberg's equation and noting that $|\psi(0)\rangle$ is constant so that

$$\underbrace{\left\langle \frac{d\hat{A}_H}{dt} \right\rangle = \overbrace{\left\langle \frac{1}{i\hbar}\left[\hat{A}_H, \hat{H}_H\right] \right\rangle + \left\langle \frac{\partial \hat{A}_H}{\partial t} \right\rangle}^{\text{Ehrenfest's theorem}} = \frac{d}{dt}\left\langle \hat{A}_H \right\rangle.}_{\text{Same form as classical Hamiltonian physics}} \qquad (4.25)$$

This leads to an expression that is in agreement both with the dynamics of expectation values in classical Hamiltonian dynamics and with Ehrenfest's theorem (which takes the same form in the Schrödinger and Heisenberg pictures). In this way we see that the Heisenberg picture is again in some ways nicer than the Schrödinger picture, as the analogy with classical physics is more direct.

However, we can do better than this in connecting the two pictures. To do this, we will borrow a concept from geometry used to change sets of coordinates, namely the *similarity transformation*. This is used for moving (conformal mapping) matrices between different rotated reference frames and takes the form $A' = B^{-1}AB$, where A' and A are called similar matrices. For a given operator in the Heisenberg picture, we can ask if it can be written in the form:

$$\hat{A}_H(t) = \hat{U}^{-1}(t)\hat{A}\hat{U}(t), \tag{4.26}$$

where $\hat{A}$ has no implicit time dependence. The idea is to extract all the time dependence from $\hat{A}_H(t)$ and put it into an operator $\hat{U}(t)$ (like a reference frame rotating with the earth would remove time dependence from the location of an object at rest on the earth's surface). One important property of similarity transformations is that they do not change key properties, such as the eigenvalues of an operator. In this way we see that such a transformation will preserve the measurable properties of an observable and both versions of the operator should be compatible with the Copenhagen approach to measurement. The following exercise should reinforce the point that such a similarity transformation will preserve the core physics of quantum mechanics.

Exercise 4.4 Show that $\hat{A}_H\hat{B}_H = (\hat{A}\hat{B})_H$. Hint: use $\hat{I} = \hat{U}^{-1}(t)\hat{U}(t)$. Also show that $[\hat{A}_H, \hat{B}_H] = ([\hat{A}, \hat{B}])_H$. As a corollary, and starting from $[\hat{q}_H, \hat{p}_H] = i\hbar$, what is $[q, p]$? ∎

Now, expectation values can be written in the Heisenberg picture as

$$\left\langle \psi(0) \middle| \hat{A}_H(t) \middle| \psi(0) \right\rangle = \left\langle \psi(0) \middle| \hat{U}^{-1}(t)\hat{A}\hat{U}(t) \middle| \psi(0) \right\rangle. \tag{4.27}$$

We see that if we add the condition that $\hat{U}(t)$ is unitary so that $\hat{U}^\dagger(t) = \hat{U}^{-1}(t)$, then we can define a time-dependent state

$$|\psi(t)\rangle = \hat{U}(t)|\psi(0)\rangle. \tag{4.28}$$

Because this operator generates the dynamics of the state from its initial conditions, $\hat{U}(t)$ is termed the evolution operator. This operator is sufficiently important to be worth a small digression from our main discussion to examine it in more detail.

As well as through the above argument, the evolution operator can be introduced by the principle of superposition alone. While this is a little repetitious, the alternative perspective adds value. This argument starts with one of our main empirical observations: that any dynamical representation of a quantum state must obey the principle of superposition.

This also means that any dynamical evolution of the state mapping

$$|\psi(t_0)\rangle = a|\psi_1(t_0)\rangle + b|\psi_2(t_0)\rangle$$

to

$$|\psi(t)\rangle = a|\psi_1(t)\rangle + b|\psi_2(t)\rangle$$

is a linear operation. As such, it can be represented by a single linear operator:

$$|\psi(t)\rangle = \hat{U}(t, t_0)|\psi(t_0)\rangle,$$

where $\hat{U}(t, t_0)$ is, just as before, termed the evolution operator. There are some properties of the evolution operator that we can deduce must be satisfied from simple reasoning:

- $\hat{U}(t, t) = \hat{I}$ which follows from the fact that evolution over zero time does not change the state.
- $\hat{U}(t, t')\hat{U}(t', t'') = \hat{U}(t, t'')$, which states that evolution over two consecutive time steps is the same as evolution over the total time. This can be extended to any number of intermediate time steps.
- As evolving forward in time and then back in time to where one started does nothing, we have $\hat{U}(t, t')\hat{U}(t', t) = \hat{I} = \hat{U}(t', t)\hat{U}(t, t')$, which implies $\hat{U}(t', t) = \hat{U}^{-1}(t, t')$.
- Finally, from normalisation, since $1 = \langle\psi(t)|\psi(t)\rangle = \left\langle\psi(t_0)\left|\hat{U}^\dagger(t, t_0)\hat{U}(t, t_0)\right|\psi(t_0)\right\rangle$, $\hat{U}(t', t) = \hat{U}^\dagger(t, t') = \hat{U}^{-1}(t, t')$ and the evolution operator is unitary (as the TDSE preserves norm, and this operator represents Schrödinger evolution, this to be expected).

In addition to using the evolution operator, we can find the dynamics of the state from the Schrödinger equation

$$\hat{H}(t)|\psi(t)\rangle = i\hbar\frac{d|\psi(t)\rangle}{dt},$$

and we can substitute $|\psi(t)\rangle = \hat{U}(t, t_0)|\psi(t_0)\rangle$ to write this as

$$\hat{H}(t)\hat{U}(t, 0)|\psi(0)\rangle = i\hbar\frac{d\hat{U}(t, 0)|\psi(0)\rangle}{dt}.$$

As this equation must hold for any initial state $|\psi(0)\rangle$, we can omit the state and write an operator-only equation

$$\hat{H}(t)\hat{U}(t, 0) = i\hbar\frac{d\hat{U}(t, 0)}{dt}.$$

We therefore see that the evolution operator is the operator solution of the TDSE. The general solution is

$$\hat{U}(t, t_0) = \hat{I} - \frac{i}{\hbar}\int_{t_0}^{t} dt'\, \hat{H}(t')\hat{U}(t', t_0)$$

or

$$\hat{U}(t + dt, t) = \hat{I} - \frac{i}{\hbar}\hat{H}(t)\, dt.$$

If the Hamiltonian is time-independent, the evolution takes on this simple form

$$\hat{U}(t, t_0) = \exp\left(-\frac{i}{\hbar}\hat{H}[t - t_0]\right). \tag{4.29}$$

If $\hat{H}$ is time-dependent but commutes with itself at different times $[\hat{H}(t), \hat{H}(t')] = 0$,

$$\hat{U}(t, t_0) = \exp\left(-\frac{i}{\hbar}\int_{t_0}^{t} dt'\, \hat{H}(t')\right). \tag{4.30}$$

Finally, we note a more, but not completely, general case:

$$\hat{U}(t, t_0) = \hat{1} + \sum_n \left(-\frac{i}{\hbar}\right)^n \int_{t_0}^t dt_1 \int_{t_1}^t dt_2 \ldots$$

$$\int_{t_0}^{t_{n-1}} dt_n \hat{H}(t_1)\hat{H}(t_2) \ldots \hat{H}(t_n), \tag{4.31}$$

where $t_1 > t_2 > \ldots > t_n$. This is known as the Dyson series.

4.2.7 Connection to the Schrödinger Picture, and Revisiting Issues with Ehrenfest's Theorem

We start by considering Ehrenfest's theorem from Eq. (3.7) in the Schrödinger picture:

$$\frac{d}{dt}\left\langle \psi(t)\left|\hat{A}(t)\right|\psi(t)\right\rangle = \left\langle \psi(t)\left|\frac{1}{i\hbar}[\hat{A}(t), \hat{H}(t)] + \frac{\partial\hat{A}(t)}{\partial t}\right|\psi(t)\right\rangle$$

and make use of the evolution operator to write $|\psi(t)\rangle = \hat{U}(t)|\psi(0)\rangle$ and by substitution

$$\frac{d}{dt}\left\langle \psi(0)\left|\hat{U}^\dagger(t)\hat{A}(t)\hat{U}(t)\right|\psi(0)\right\rangle = \left\langle \psi(0)\left|\hat{U}^\dagger(t)\frac{\partial\hat{A}(t)}{\partial t}\hat{U}(t)\right|\psi(0)\right\rangle +$$

$$\frac{1}{i\hbar}\left\langle \psi(0)\left|\hat{U}^\dagger(t)[\hat{A}(t), \hat{H}(t)]\hat{U}(t)\right|\hat{U}(t)\psi(0)\right\rangle.$$

Now note that

$$\hat{U}^\dagger[\hat{A}, \hat{H}]\hat{U} = \hat{U}^\dagger[\hat{A}\hat{H} - \hat{H}\hat{A}]\hat{U}$$

$$= \hat{U}^\dagger\hat{A}(\hat{1})\hat{H}\hat{U} - \hat{U}^\dagger\hat{H}(\hat{1})\hat{A}\hat{U}$$

$$= \hat{U}^\dagger\hat{A}(\hat{U}\hat{U}^\dagger)\hat{H}\hat{U} - \hat{U}^\dagger\hat{H}(\hat{U}\hat{U}^\dagger)\hat{A}\hat{U}$$

$$= (\hat{U}^\dagger\hat{A}\hat{U})(\hat{U}^\dagger\hat{H}\hat{U}) - (\hat{U}^\dagger\hat{H}\hat{U})(\hat{U}^\dagger\hat{A}\hat{U})$$

$$= [\hat{U}^\dagger\hat{A}\hat{U}, \hat{U}^\dagger\hat{H}\hat{U}].$$

So if we define the following operators (and the subscript H hints at where we are going):

$$\hat{H}_H \stackrel{\text{def}}{=} \hat{U}^\dagger\hat{H}\hat{U}$$

$$\hat{A}_H \stackrel{\text{def}}{=} \hat{U}^\dagger\hat{A}\hat{U}$$

$$\left(\frac{\partial\hat{A}}{\partial t}\right)_H \stackrel{\text{def}}{=} \hat{U}^\dagger\frac{\partial\hat{A}}{\partial t}\hat{U}$$

Ehrenfest's theorem can be written as

$$\frac{d}{dt}\left\langle \psi(0)\left|\hat{A}_H\right|\psi(0)\right\rangle = \left\langle \psi(0)\left|\frac{1}{i\hbar}[\hat{A}_H, \hat{H}_H] + \left(\frac{\partial\hat{A}}{\partial t}\right)_H\right|\psi(0)\right\rangle$$

and as $|\psi(0)\rangle$ has no time dependence,

$$\left\langle \psi(0)\left|\frac{d}{dt}\hat{A}_H\right|\psi(0)\right\rangle = \left\langle \psi(0)\left|\frac{1}{i\hbar}[\hat{A}_H, \hat{H}_H] + \left(\frac{\partial\hat{A}}{\partial t}\right)_H\right|\psi(0)\right\rangle.$$

Since this equation must hold for any initial state (as we have made no assumptions about $|\psi(0)\rangle$), we can omit the state to read:

$$\frac{\mathrm{d}}{\mathrm{d}t}\hat{A}_H = \frac{1}{\mathrm{i}\,\hbar}\,[\hat{A}_H, \hat{H}_H] + \left(\frac{\partial \hat{A}}{\partial t}\right)_H,\tag{4.32}$$

which is *nearly* the Heisenberg equation of motion, Eq. (4.21). The one difficulty here is that

$$\left(\frac{\partial \hat{A}}{\partial t}\right)_H \neq \frac{\partial \hat{A}_H}{\partial t}.\tag{4.33}$$

This discrepancy causes little practical difficulty, but the two pictures are nevertheless slightly different. This issue links directly to the issues with Ehrenfest's theorem that we discussed in the previous chapter.

4.3 Wigner's Phase-space Quantum Mechanics

4.3.1 Background

There is perhaps little better introduction to motivating this subject than that presented by Moyal in his seminal work *'Quantum Mechanics as a Statistical Theory'*:

> Statistical concepts play an ambiguous role in quantum theory. The critique of acts; of observation, leading to Heisenberg's 'principle of uncertainty' and to the necessity for considering dynamical parameters as statistical variates, not only for large aggregates, as in classical kinetic theory, but also for isolated atomic systems, is quite fundamental in justifying the basic principles of quantum theory; yet paradoxically, the expression of the latter in terms of operations in an abstract space of 'state' vectors is essentially independent of any statistical ideas. These are only introduced as a post hoc interpretation, the accepted one being that the probability of a state is equal to the square of the modulus of the vector representing it; other and less satisfactory statistical interpretations have also been suggested (cf. Dirac(l)).
>
> One is led to wonder whether this formalism does not disguise what is an essentially statistical theory, and whether a reformulation of the principles of quantum mechanics in purely statistical terms would not be worthwhile in affording us a deeper insight into the meaning of the theory. From this point of view, the fundamental entities would be the statistical variates representing the dynamical parameters of each system; the operators, matrices and wave functions of quantum theory would no longer be considered as having an intrinsic meaning, but would appear rather as aids to the calculation of statistical averages and distributions. Yet there are serious difficulties in effecting such a reformulation. Classical statistical mechanics is a 'crypto-deterministic' theory, where each element of the probability distribution of the dynamical variables specifying a given system evolves with time according to deterministic laws of motion; the whole uncertainty is contained in the form of the initial distributions. A theory based on such concepts could not give a satisfactory

account of such non-deterministic effects as radioactive decay or spontaneous emission (cf. Whittaker (2)). Classical statistical mechanics is, however, only a special case in the general theory of dynamical statistical (stochastic) processes. In the general case, there is the possibility of 'diffusion' of the probability 'fluid', so that the transformation with time of the probability distribution need not be deterministic in the classical sense. In this paper, we shall attempt to interpret quantum mechanics as a form of such a general statistical dynamics.

While the motivation of this section is somewhat sympathetic with Moyal's, we will take an alternative approach that, with some historical irony, borrows from Dirac's motivation of the Heisenberg picture (Dirac was apparently unaware of some of the mathematics that made phase space methods a complete quantum theory and disliked the approach).

One other key application of phase space quantum mechanics is to gain intuition through visualisation of the nature of quantum states, and when, how, and if they are different to classical probability density functions. The phase space representation of the quantum state that we consider here is the Wigner function. To understand this motivation, you may want to look ahead to some example figures: Figure 4.1 on page 115 and Figure 4.2 on page 119.

4.3.2 Motivating the Phase-space Equation of Motion

Once more we return to the classical equation for the dynamics of any classical quantity A as our starting point:

$$\frac{dA}{dt} = \{A, \mathcal{H}\} + \frac{\partial A}{\partial t}. \tag{4.34}$$

We recall that if $A(t)$ is the probability density function $\rho(\boldsymbol{q}, \boldsymbol{p}, t)$ then, as conservation of probability implies $\frac{d\rho}{dt} = 0$, we obtain the Liouville equation:

$$\frac{\partial \rho(\boldsymbol{q}, \boldsymbol{p}, t)}{\partial t} = \{\mathcal{H}, \rho(\boldsymbol{q}, \boldsymbol{p}, t)\}.$$

This time, however, we will not invoke Koopman–von Neumann theory. We will instead try to keep as much of the apparatus of classical mechanics as possible. We will seek to keep coordinates and momenta, the classical Hamiltonian and, importantly, the notion of a probability density function.

Beyond the motivation presented by Moyal quoted above, we will also seek insight into two practical difficulties with the theoretical framework which we have briefly mentioned. The first is how to reconcile the Schrödinger and Heisenberg pictures with classical mechanics. In other words, can we formulate (linear) quantum mechanics in a way that clearly satisfies the correspondence principle with (non-linear) classical mechanics? The second is that operator ordering seems to be some additional set of rules that have no nice physical origin, and $[\cdot, \cdot] = i\hbar\{\cdot, \cdot\}$ cannot always be satisfied (Groenewold's theorem).

As we did with the Heisenberg picture, let us begin by reviewing the properties of the Poisson bracket:

$$\{u, v\} = -\{v, u\} \tag{4.35}$$

$$\{au_1 + bu_2, v\} = a\{u_1, v\} + b\{u_2, v\} \tag{4.36}$$

$$\{u_1 u_2, v\} = \{u_1, v\} u_2 + u_1 \{u_2, v\} \tag{4.37}$$

$$\{u, v_1 v_2\} = \{u, v_1\} v_2 + v_1 \{u, v_2\} \tag{4.38}$$

$$\{\{u, v\}, w\} + \{\{w, u\}, v\} + \{\{v, w\}, u\} = 0, \tag{4.39}$$

where a and b are just numbers. As we have previously stated, the above set of equations defines a Lie algebra – the rules that infinitesimal transformations, such as rotation or translation, must satisfy. Given the constraints of wanting to retain all the mathematical objects of classical mechanics, it must be the Poisson bracket we look to change. An important argument for retaining the Lie algebraic structure is that any dynamical theory for state evolving from time t to $t + \delta t$ must take the form of an infinitesimal transformation of some kind.

4.3.3 Quantising Phase Space

We once more adapt an argument from Dirac [18]: let us consider the Poisson bracket and modify it with care. Recall that Dirac argued for a quantum Poisson Bracket which satisfies all the same algebraic relations of the Poisson bracket above, i.e. to keep a Lie algebra. But now let us do something similar but different. We will formulate a new 'quantum Poisson Bracket' denoted by $\{\{\cdot, \cdot\}\}$ and termed the Moyal Bracket. Just as the Poisson bracket, it must satisfy all the Lie algebraic relations above but we will also seek to impose that

$$\lim_{\hbar \to 0} \{\{\cdot, \cdot\}\} = \{\cdot, \cdot\} \tag{4.40}$$

so that the Moyal Bracket continuously tends towards the Poisson bracket in the limit of vanishing Planks constant (the quantifier of 'quantumness'). This approach has several important implications:

- We cannot solve the problem by introducing non-commuting operators as we would never be able to get this limit - in other words, the conceptual framework of a phase space must remain.
- The Poisson bracket can be considered as defining a 'flow' in phase space. We retain the idea of dynamics as a flow in phase space, but by introducing a Moyal bracket, we are redefining that which is an *allowable* flow.
- The resulting theory can be considered as a continuous deformation of phase space as a function of $\hbar$. The intention is to guarantee a correspondence principle. Limits can be funny things, so it is not clear that Eq. (4.40) is always well-defined. In fact, it turns out that the Moyal bracket is a continuous version of the commutator, and issues related to ordering remain for $\hbar \neq 0$.

As this algebra is of central importance, we will be explicit (if a bit repetitive) and note that the Moyal bracket must also satisfy the Lie algebra conditions:

$$\{\{u, v\}\} = -\{\{v, u\}\} \tag{4.41}$$

$$\{\{au_1 + bu_2, v\}\} = a\{\{u_1, v\}\} + b\{\{u_2, v\}\} \tag{4.42}$$

$$\{\{u_1 u_2, v\}\} = \{\{u_1, v\}\} u_2 + u_1 \{\{u_2, v\}\} \tag{4.43}$$

$$\{\{u, v_1 v_2\}\} = \{\{u, v_1\}\} v_2 + v_1 \{\{u, v_2\}\} \tag{4.44}$$

$$\{\{\{\{u, v\}\}, w\}\} + \{\{\{\{w, u\}\}, v\}\} + \{\{\{\{v, w\}\}, u\}\} = 0, \tag{4.45}$$

over the *same phase space* as the Poisson bracket.

We know from both the Schrödinger and Heisenberg pictures that quantum mechanics must be non-commutative in nature. We are now insisting that this be built into the phase space itself, rather than separated into operators acting in a vector space. This means that multiplication of phase space distributions themselves must be reviewed and replaced with something else (there is no other choice). We will denote this new multiplication as $f \star g$ – the *star* or *Moyal* product. Inspired by commutation relations, we might guess that $\{\{f, g\}\} \propto f \star g - g \star f$ would be a good candidate to replace $\{f, g\}$. Invoking the correspondence principle to require a classical limit implies that this proposed Moyal bracket must reproduce the classical Poisson bracket structure in the limit $\hbar \to 0$. The implication of this is that any new type of multiplication rule must, as $\hbar \to 0$, recover the usual way of multiplying functions. We therefore propose a new product of phase space functions of the form

$$f \star g \stackrel{\text{def}}{=} \sum_{n=0}^{\infty} \hbar^n \Xi_n(f, g), \tag{4.46}$$

where $\Xi_n(f, g)$ is some satisfactory function of the phase space distributions $f(\boldsymbol{q}, \boldsymbol{p})$ and $g(\boldsymbol{q}, \boldsymbol{p})$. If we set the first term in this expression as $\Xi_0(f, g) = fg$ then we have

$$\lim_{\hbar \to 0} f \star g = fg$$

and, as $\hbar \to 0$, we recover the usual multiplication of functions as required. Now we recall that the Poisson bracket is

$$\{f, g\} \stackrel{\text{def}}{=} \sum_i \frac{\partial f}{\partial q_i} \frac{\partial g}{\partial p_i} - \frac{\partial f}{\partial p_i} \frac{\partial g}{\partial q_i}. \tag{4.47}$$

Note for future reference that we can write this equation in terms of left and right derivatives as

$$\{f, g\} \stackrel{\text{def}}{=} f \left[\sum_i \left(\frac{\overleftarrow{\partial}}{\partial q_i} \frac{\overrightarrow{\partial}}{\partial p_i} - \frac{\overleftarrow{\partial}}{\partial p_i} \frac{\overrightarrow{\partial}}{\partial q_i} \right) \right] g, \tag{4.48}$$

where we can consider the differential operators in the brackets as forming what we will term a 'Poisson operator'. Let us now add another term of $\mathcal{O}(\hbar)$ to $f \star g$, so that

$$f \star g = fg + i\hbar \sum_{i=1}^{N} \frac{\partial f}{\partial q_i} \frac{\partial g}{\partial p_i} + \mathcal{O}(\hbar^2), \tag{4.49}$$

where N is just the number of degrees of freedom. We have made the constant of proportionality $i\hbar$ and not just $\hbar$ with the benefit of hindsight. Adding this term to $f \star g$ removes commutativity of function multiplication. Recalling our anzats that $\{\{f, g\}\} \propto f \star g - g \star f$, we define the Moyal bracket as:

$$\{\{f, g\}\} \stackrel{\text{def}}{=} \frac{1}{i\hbar} \left[f \star g - g \star f \right] \tag{4.50}$$

so that we satisfy the correspondence limit

$$\{f, g\} = \lim_{\hbar \to 0} \{\{f, g\}\}. \tag{4.51}$$

Because of this limit we propose that phase space quantum mechanics is achieved by directly replacing the Poisson bracket with some Moyal bracket. By redefining the multiplication of phase space distribution with a star product and replacing the Poisson bracket with the Moyal bracket, our new physical theory puts the rules of quantum mechanics into the geometry of phase space itself. It is 'better' than our previous approaches in so far as it is conceptually connected to classical physics in the limit $\hbar \to 0$. The equations for the dynamics of any phase space distribution then become

$$\frac{dA}{dt} = \frac{1}{i\hbar} [A \star \mathcal{H} - \mathcal{H} \star A] + \frac{\partial A}{\partial t}$$
$$= \{\{A, \mathcal{H}\}\} + \frac{\partial A}{\partial t}. \tag{4.52}$$

By construction, this equation is equivalent in form to Hamiltonian dynamics, but we also see the same observations hold true for the Heisenberg picture,

$$\frac{d\hat{A}_H(t)}{dt} = \frac{1}{i\hbar} [\hat{A}_H(t), \hat{H}] + \frac{\partial \hat{A}_H(t)}{\partial t}.$$

We can follow arguments similar to those we followed when deriving the Liouville equation for the probability density $\rho(\boldsymbol{q}, \boldsymbol{p}, t)$. We will introduce a new function, $W(\boldsymbol{q}, \boldsymbol{p}, t)$, called the Wigner function that serves that same purpose – a quantum probability density function. As such, we will require it to be normalised to one and that its dynamics are conserved, so:

$$\int W(\boldsymbol{q}, \boldsymbol{p}, t) = 1 \text{ and } \frac{dW(\boldsymbol{q}, \boldsymbol{p}, t)}{dt} = 0.$$

We will later find that this function is not positive definite. As we have become used to complex probability amplitudes of wave functions, a small amount of negativity in the Wigner function will not cause us any issue so long as the marginals of the distribution behave sensibly. That is, e.g., that integrating over the momentum degrees of freedom produces a probability density function in the coordinates that agrees with the predictions of the Schrödinger picture position representation $|\psi(\boldsymbol{q}, t)|^2$. By using the same reasoning as we followed with the derivation of the Liouville equation, we find that

$$\frac{\partial W(\boldsymbol{q}, \boldsymbol{p}, t)}{\partial t} = \frac{1}{i\hbar} [\mathcal{H} \star W(\boldsymbol{q}, \boldsymbol{p}, t) - W(\boldsymbol{q}, \boldsymbol{p}, t) \star \mathcal{H}]$$
$$= \{\{\mathcal{H}, W(\boldsymbol{q}, \boldsymbol{p}, t)\}\}. \tag{4.53}$$

It is very encouraging to see that this equation looks both similar to the classical Liouville equation and the Schrödinger picture von Neumann equation

$$\frac{\partial \hat{\rho}(t)}{\partial t} = \frac{1}{i\hbar} [\hat{H}, \hat{\rho}(t)].$$

The approach of replacing the Poisson bracket with the Moyal bracket is guaranteed to satisfy all the nice results above, no matter how we define $\Xi_n(f, g)$ in the star product. The problem that needs to be solved is then: what should the higher-order terms in $\Xi_n(f, g)$ be? This problem was solved by Groenewold and embeds the Weyl ordering of operators within its formulation[3].

---

3 In 'conventional' operator-based quantum mechanics, Weyl ordering looks like yet another axiom bolted on to fix a bit of a messy theory while in phase space quantum mechanics, we can see that it is built into the

To get a feel for where this is going, consider what the next term (those of $\mathcal{O}\left(\hbar^2\right)$) might look like. The zeroth-order term was just usual function multiplication. The next term involved differentiating with respect to position and momentum, and enabled us to be able to reconstruct the Poisson bracket. It seems like a good guess that the following term will involve differentials at the next higher order. To simplify notation for a moment, let us consider only systems comprising one coordinate and its conjugate momentum (generalisation to N degrees of freedom follows exactly the same procedure as in classical physics). The possible candidates for $\mathcal{O}\left(\hbar^2\right)$ derivatives of f and g are terms of the form:

$$\frac{\partial^2}{\partial q^2}f\frac{\partial}{\partial p}g, \quad \frac{\partial}{\partial p}\frac{\partial}{\partial q}f\frac{\partial}{\partial p}\frac{\partial}{\partial q}g \quad \text{and} \quad \frac{\partial}{\partial q}f\frac{\partial^2}{\partial p^2}g.$$

These must be combined in a way that preserves the Lie algebra structure and produces dynamics consistent with the quantum behaviour we see in nature. It turns out that the full product is given by

$$f \star g = \sum_{i=1}^{N}\frac{1}{n!}\left(\frac{i\hbar}{2}\right)^n \Pi_n(f,g), \tag{4.54}$$

where the previously defined $\Xi_n(f,g) = \frac{1}{n!}\left(\frac{i}{2}\right)^n \Pi_n(f,g)$ and

$$\Pi_n(f,g) = \sum_{k=0}^{n}(-1)^k \binom{n}{k}\left[\frac{\partial^k}{\partial p^k}\frac{\partial^{(n-k)}}{\partial q^{(n-k)}}f\right]\left[\frac{\partial^{(n-k)}}{\partial p^{(n-k)}}\frac{\partial^k}{\partial q^k}g\right]. \tag{4.55}$$

With some work, the star-product can be more compactly written as

$$f \star g \triangleq f\sum_{i=1}^{N}\exp\left[i\hbar\left(\frac{\overleftarrow{\partial}}{\partial q_i}\frac{\overrightarrow{\partial}}{\partial p_i} - \frac{\overleftarrow{\partial}}{\partial p_i}\frac{\overrightarrow{\partial}}{\partial q_i}\right)\right]g, \tag{4.56}$$

where the arrows indicate the direction in which the derivative is to act. The Moyal bracket takes on the following simplified expression:

$$\{\{f,g\}\} = \frac{1}{i\hbar}\left(f \star g - g \star f\right) \tag{4.57}$$

$$= \frac{2}{\hbar}\sum_{i=1}^{N}f\sin\left[\frac{\hbar}{2}\left(\frac{\overleftarrow{\partial}}{\partial q_i}\frac{\overrightarrow{\partial}}{\partial p_i} - \frac{\overleftarrow{\partial}}{\partial q_i}\frac{\overrightarrow{\partial}}{\partial p_i}\right)\right]g. \tag{4.58}$$

We see that the star product is essentially an exponentiation of the Poisson operator we introduced in Eq. (4.48). Both of the above expressions are rather cumbersome. It is a worthwhile exercise to compute the star product and Moyal bracket of some simple functions to get a feel for how they work. In terms of actually doing calculations, it is useful to note that:

$$f(q,p) \star g(q,p) = f\left(q + \frac{i\hbar}{2}\overrightarrow{\partial_p}, p - \frac{i\hbar}{2}\overrightarrow{\partial_q}\right)g(q,p) \tag{4.59}$$

$$= f(q,p)g\left(q - \frac{i\hbar}{2}\overleftarrow{\partial_p}, p + \frac{i\hbar}{2}\overleftarrow{\partial_q}\right) \tag{4.60}$$

---

very structure of the quantum world - it is a defining part of the algebra. Different orderings result in different physical theories, and it is the Weyl ordering that appears to work for our reality (at least in flat spacetime).

$$= f\left(q + \frac{i\hbar}{2}\overrightarrow{\partial_p}, p\right) g\left(q - \frac{i\hbar}{2}\overleftarrow{\partial_p}, p\right) \tag{4.61}$$

$$= f\left(q, p - \frac{i\hbar}{2}\overrightarrow{\partial_q}\right) g\left(q, p + \frac{i\hbar}{2}\overleftarrow{\partial_q}\right). \tag{4.62}$$

So, for example:

$$\{\{\mathcal{H}, W\}\} = \frac{1}{i\hbar}(\mathcal{H} \star W - W \star \mathcal{H})$$

$$= \frac{1}{i\hbar}\left[\mathcal{H}\left(q + \frac{i\hbar}{2}\overrightarrow{\partial_p}, p - \frac{i\hbar}{2}\overrightarrow{\partial_q}\right) W(q, p)\right.$$

$$\left. - W(q, p)\mathcal{H}\left(q - \frac{i\hbar}{2}\overleftarrow{\partial_p}, p + \frac{i\hbar}{2}\overleftarrow{\partial_q}\right)\right], \tag{4.63}$$

which can be used to greatly simplify calculations.

To summarise, phase space quantum mechanics provides a framework that, except for a non-local star product and associated Moyal bracket, preserves classical ontology, and where the dynamics of any quantity is given by:

$$\frac{\mathrm{d}A}{\mathrm{d}t} = \frac{1}{i\hbar}[A \star \mathcal{H} - \mathcal{H} \star A] + \frac{\partial A}{\partial t}$$

$$= \{\{A, \mathcal{H}\}\} + \frac{\partial A}{\partial t}. \tag{4.64}$$

This framework allows us to understand how quantum concepts such as the discrete eigenvalues of observables (use to explain atomic emission and absorption spectra) can be connected to their classical counterpart. As in classical physics, stationary states will occur when $\{\{W, \mathcal{H}\}\} = 0$ and $\frac{\partial W}{\partial t} = 0$ as this implies $\frac{\mathrm{d}A}{\mathrm{d}t} = 0$ (a direct quantum analogue of classical integrals of the motion and stationary states in statistical physics). This will happen when

$$\mathcal{H} \star W = EW = W \star \mathcal{H},$$

which is the phase space version of the time-independent Schrödinger equation. The Wigner function solutions to these equations directly correspond to the eigenstates of the Hamiltonian in operator-based quantum mechanics and, as we will later discuss, it is possible to directly convert between the state-space and phase space representations.

In comparing classical physics and quantum physics, we can now see that the idea of stationary states is common to both theories. In classical physics, we have $\{\rho, \mathcal{H}\} = 0$, so $\frac{\partial \rho}{\partial t} = 0$ since $\frac{\mathrm{d}\rho}{\mathrm{d}t} = 0$. This, for example, in the harmonic oscillator will be satisfied by Gaussians centred at the origin (including the delta function as a limiting case of a particle whose position and momentum are known with absolute certainty). Other states with the same symmetry are also possible stationary states, but are less often realised. In quantum physics, we have $\{\{W, \mathcal{H}\}\} = 0$, so $\frac{\partial W}{\partial t} = 0$ as $\frac{\mathrm{d}W}{\mathrm{d}t} = 0$. The only change between classical and quantum physics is the (continuous) deformation of the Poisson bracket with the Moyal bracket. The Moyal bracket, due to its non-local nature, allows for a wider variety of stationary states. So, while Gaussian states centred at the origin may still be stationary - as with classical physics - there are some important differences. The first is that the delta function is no longer an acceptable limit (except mathematically when $\hbar \to 0$) – which as we will see in the next section is a consequence of understanding the Heisenberg uncertainty relation in phase space. The other is that non-Gaussian stationary states are possible, but these will not be positive valued distributions. These 'negative probabilities' are what we get by deforming the Poisson bracket structure; they represent non-local quantum correlations. While negative probabilities may seem unusual at first, we have in state vectors been dealing with complex amplitudes already and, as Dirac observed, 'negative energies and probabilities should not be considered as nonsense. They are well-defined concepts mathematically, like a negative of money'[17]. The important thing is that the predictions of the theory remain reasonable and in line with observation.

One more important observation is that the 'quantumness' encoded in the star product averages out over phase space. This is expressed by:

$$\int f \star g = \int g \star f = \int fg.$$

Here the integral is taken over all of phase space. As with classical statistical physics, the expectation value of any phase space quantity A is then

$$\langle A \rangle = \int A \star W = \int W \star A = \int AW.$$

We see from the above that the expectation value is the same expression for quantum and classical distributions (in the limit $\hbar \to 0$ we have $W \to \rho$).

4.3.4 Joining the Dots

We have asserted that the phase space formulation of quantum mechanics is equivalent to the previous operator formalisms. At least this should be the case for systems where there is a clear classical representation in terms of canonical coordinates and conjugate momentum (we have made no argument for systems where this is not natural, such as spin). We should therefore be able to create a mapping between the state vector or density operator and the Wigner function. The mapping for state vectors may include an arbitrary phase factor, as this will not affect any observable properties of the state. Since density operators will not have the phase-factor ambiguity of state vectors, let us start here. Because quantum mechanics is a linear theory, we will look for a mapping that is a linear transformation. Specifically, we will search for a transformation of the form:

$$W_{\hat{A}}(q,p) = \mathrm{Tr}\left[\hat{A}\,\hat{\Pi}(q,p)\right] \tag{4.65}$$

$$\hat{A} = \int dq\,dp\,W_{\hat{A}}(q,p)\hat{\Pi}(q,p), \tag{4.66}$$

where, once more, we have restricted our analysis to one degree of freedom to simplify notation (generalisation to more degrees of freedom takes the usual extension of the argument). Note that we are looking for a trace in the first example, as we are thinking that the Wigner function will be the expectation value of some observable $\hat{\Pi}(q,p)$. The second equation expresses the fact that we want the inverse transformation to be the symmetric equivalent mapping to trace (integrating over all of phase space is like tracing the Wigner function). Várilly termed this mapping between phase space distributions and quantum mechanical operators the Stratonovich–Weyl correspondence[4]. The conditions, in terms of

---

4 After [83] and the form presented here adapted from Ref. [86].

some arbitrary parametrisation of phase space that we label Ω (as that is the tradition in much of the historical literature), are as follows:

SW 1 The mappings $W_{\hat{A}}(\Omega) = \text{Tr}\left[\hat{A}\,\hat{\Pi}(\Omega)\right]$ and $\hat{A} = \int_\Omega W_{\hat{A}}(\Omega)\hat{\Pi}(\Omega)d\Omega$ exist and are informationally complete. Simply put, we can fully reconstruct $\hat{A}$ from $W_{\hat{A}}(\Omega)$ and vice versa.

SW 2 When $\hat{A}$ is Hermitian, $W_{\hat{A}}(\Omega)$ is real valued. This means that $\hat{\Pi}(\Omega)$ must be Hermitian.

SW 3 $W_{\hat{A}}(\Omega)$ is 'standardised' so that the definite integral over all space $\int_\Omega W_{\hat{A}}(\Omega)d\Omega = \text{Tr}\left[\hat{A}\right]$ exists and $\int_\Omega \hat{\Pi}(\Omega)d\Omega = \hat{1}$.

SW 4 Unique to Wigner functions over other phase space representations, $W_{\hat{A}}(\Omega)$ is self-conjugate (this means both Eq. (4.65) and Eq. (4.66) are satisfied by the same $\hat{\Pi}(\Omega)$); the definite integral $\int_\Omega W_{\hat{A}'}(\Omega)W_{\hat{A}''}(\Omega)d\Omega = \text{Tr}\left[\hat{A}'\hat{A}''\right]$ exists.

SW 5 Covariance: mathematically, any Wigner function generated by 'rotated' operators $\hat{\Pi}(\Omega')$ (by some unitary transformation V) must be equivalent to 'rotated' Wigner functions generated from the original operator ($\hat{\Pi}(\Omega') \equiv V\hat{\Pi}(\Omega)V^\dagger$), i.e. if $\hat{A}$ is invariant under global unitary operations, then so is $W_{\hat{A}}(\Omega)$.

As a note on terminology, such trace transformations (noting that the integral is a specific kind of trace), mapping one thing to another, the transforming quantity, here $\hat{\Pi}(q,p)$, is termed the kernel of the transformation.

Before continuing our discussion, we first need to do a little bit of groundwork (for physical context you may wish to skim read section 7.3). We begin by introducing an operator and its Hermitian adjoint,

$$\hat{a} = \frac{\hat{q} + i\hat{p}}{\sqrt{2\hbar}} \quad \text{and} \quad \hat{a}^\dagger = \frac{\hat{q} - i\hat{p}}{\sqrt{2\hbar}}, \tag{4.67}$$

which are termed the annihilation operator and the creation operator (our reason for doing so, and naming both, will become more clear after we have studied the quantum simple harmonic oscillator). We note that they have the important commutation relation

$$\left[\hat{a}, \hat{a}^\dagger\right] = \hat{1}. \tag{4.68}$$

The annihilation operator has well-defined eigenstates, but the creation operator has none. We denote the eigenstates of the annihilation operator according to

$$\hat{a}\,|\alpha\rangle = \alpha\,|\alpha\rangle. \tag{4.69}$$

These states are termed (canonical) coherent states [35]. By solving Eq. (4.69) in the position representation,

$$\frac{1}{\sqrt{2\hbar}}\left(q + \hbar\frac{\partial}{\partial q}\right)\psi_\alpha(q) = \alpha\psi_\alpha(q), \tag{4.70}$$

we see that their position-representation wave function takes the form of a Gaussian

$$\psi_\alpha(q) = \left(\frac{1}{\pi\hbar}\right)^{1/4}\exp\left(-\frac{1}{2\hbar}\left(q - \sqrt{2\hbar}\,\Re(\alpha)\right)^2 + i\sqrt{\frac{2}{\hbar}}\,\Im(\alpha)q\right) \tag{4.71}$$

up to some arbitrary phase factor. With some calculation, one can show that $\langle\hat{q}\rangle = \sqrt{2\hbar}\,\Re[\alpha]$ and $\langle\hat{p}\rangle = \sqrt{2\hbar}\,\Im[\alpha]$. We see that this distribution is 'centred' around the eigenvalue

$$\alpha = \frac{\langle \hat{q} \rangle + i \langle \hat{p} \rangle}{\sqrt{2\hbar}} \text{ or, in abbreviated form, } \frac{q + i p}{\sqrt{2\hbar}}. \tag{4.72}$$

We therefore see that $\psi_{\alpha=0}(q)$ is an eigenstate with eigenvalue zero and this leads us to define the so-called vacuum state, $|0\rangle$, by

$$\hat{a} \, |\alpha = 0\rangle \overset{\text{def}}{=} \hat{a} \, |0\rangle = 0. \tag{4.73}$$

The vacuum state is a Gaussian centred around the origin (this state corresponds to the lowest energy eigenstate of the harmonic oscillator which is why it bears this name). Note that $|0\rangle$ is not the null vector $|\emptyset\rangle$ and it, like all states, has unit norm. Wigner functions of two example states are shown in Figure 4.1.

We can define an operator $\hat{D}$ by

$$\hat{D}(\alpha) \overset{\text{def}}{=} \exp\left[\alpha \hat{a}^{\dagger} - \alpha^* \hat{a}\right] = \exp\left[\frac{i}{\hbar}(p\hat{q} - q\hat{p})\right] \overset{\text{def}}{=} \hat{D}(q, p). \tag{4.74}$$

This operator has some very interesting properties: it is unitary and $\hat{D}(-\alpha) = \hat{D}^{\dagger}(\alpha) = \hat{D}^{-1}(\alpha)$. For us, most importantly

$$\hat{D}(\alpha) \, |0\rangle = |\alpha\rangle, \tag{4.75}$$

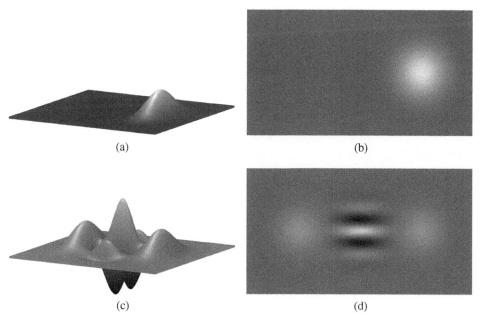

(a) (b)

(c) (d)

Figure 4.1 Examples based on the displaced vacuum 'coherent state'. (c,d) show an example of a Schrödinger cat. The interface terms between the Gaussians demonstrate quantum coherence of the superposition – these would not be present in a classical statistical mixture of $|\alpha = \pm 2\rangle$. (a) Wigner function of the coherent state $|\alpha = 2\rangle$. For reference with (c,d) note that $|\alpha = -2\rangle$ is just a Gaussian of the same shape biased to the other side of the origin. (b) Top-down view of (a). Here, q is the horizontal axis, p the vertical axis. (c) Wigner function of the single-mode Schrödinger cat state $|\alpha = 2\rangle + |\alpha = -2\rangle$. (d) Top-down view of (c). Here, q is the horizontal, p the vertical axis.

so that this operator $\hat{D}(\alpha)$ creates any coherent state $|\alpha\rangle$ by displacing the vacuum state to some new origin α (we will prove this is in Section 7.3.7.3). For this reason $\hat{D}(\alpha) = \hat{D}(q, p)$ is termed the displacement operator, generalising the one defined in Section 3.4 to include displacements in both position and momentum[5].

Exercise 4.5 Show, using commutator algebra, that coherent states minimise the Heisenberg uncertainty relation. ∎

Exercise 4.6 Show

$$\hat{D}(\alpha) = e^{-|\alpha|^2/2} \exp[\alpha\hat{a}^\dagger] \exp -[\alpha^*\hat{a}] \tag{4.76}$$

$$\hat{D}(\alpha) = e^{|\alpha|^2/2} \exp[-\alpha^*\hat{a}] \exp[\alpha\hat{a}^\dagger]. \tag{4.77}$$

Also derive similar expressions in terms of $\hat{D}(q, p)$, $\hat{q}$, and $\hat{p}$. ∎

In addition to the displacement operator, we shall need another operator, the parity operator, $\hat{\Pi}$. This is defined by

$$\hat{\Pi}|q\rangle = |-q\rangle \text{ and } \hat{\Pi}|p\rangle = |-p\rangle, \tag{4.78}$$

so one can show that $\hat{\Pi}^2(q, p) = \hat{I}$ and $\hat{\Pi}^\dagger(q, p) = \hat{\Pi}(q, p)$. The parity operator is unitary and its own inverse; its eigenvalues are ± 1. Eigenstates with eigenvalue $+1$ are said to be even and those with eigenvalue -1, odd (the position or momentum representation of even and odd eigenstates will be even and odd functions). The action of the parity operator on a coherent state is

$$\hat{\Pi}|\alpha(q, p)\rangle = |\alpha(-q, -p)\rangle = |-\alpha(q, p)\rangle.$$

That is, the act of the parity operator on a coherent state is to produce, as one would expect, the input state reflected about the origin.

Now let us return to the problem at hand of using the Stratonovich–Weyl correspondence to find a kernel that allows us to move between phase space and operator quantum mechanics. There is more than one kernel that can move from operator quantum mechanics to a phase space representation. There are also numerous kernels that map back to operator quantum mechanics from that of phase space. There is, however, only one kernel which works for both mappings. It is also the only kernel that produces the phase space representation of the Wigner function. It can be shown that this kernel is

$$\hat{\Pi}(q, p) = \hat{D}(q, p)\hat{\Pi}\hat{D}^\dagger(q, p). \tag{4.79}$$

We understand $\hat{\Pi}(q, p)$ as a displaced parity operator, as it acts like a parity operator about an origin shifted to (p, q). See [5, 39, 62] for details.

Exercise 4.7 Using $W_{\hat{\rho}}(q, p) = \text{Tr}\left[\hat{\rho}\,\hat{\Pi}(q, p)\right]$, and by taking the trace in the position representation $\text{Tr}\left[\hat{A}\right] = \int dq' \left\langle q' \left| \hat{A} \right| q' \right\rangle$, show that

$$W_{\hat{\rho}}(q, p) = \int d\zeta \left\langle q - \frac{1}{2}\zeta \left| \hat{\rho} \right| q + \frac{1}{2}\zeta \right\rangle \exp\left[\frac{ip\zeta}{\hbar}\right].$$

---

5 Note that for some coherent state $|\beta\rangle$, $\hat{D}(\alpha)|\beta\rangle = \exp[(\alpha\beta^* - \alpha^*\beta)/2]|\alpha + \beta\rangle$ which acts as a displacement of $|\beta\rangle$ by α but also gains an extra physically irrelevant phase term. The fact that $|0\rangle$ does not pick up a phase, makes its combination with $\hat{D}(\alpha)$ somewhat special.

This is the expression that Wigner magically pulled out of the air in the original paper 'On the Quantum Correction For Thermodynamic Equilibrium' [92] (note: this expression is the usual staring point for introducing the Wigner function). ■

Exercise 4.8 Show that $W_{\hat{q}}(q, p) = q$ and $W_{\hat{p}}(q, p) = p$. Hint: Use the Hadamard lemma together with the fact that trace is invariant under cyclic permutations, so

$$\mathrm{Tr}\left[\hat{A}\hat{D}(q, p)\hat{\Pi}\hat{D}^{\dagger}(q, p)\right] = \mathrm{Tr}\left[\hat{D}^{\dagger}(q, p)\hat{A}\hat{D}(q, p)\hat{\Pi}\right], \tag{4.80}$$

and an appropriate basis for each trace. ■

Exercise 4.9 Does $W_{\hat{q}\hat{p}}(q, p) = W_{\hat{p}\hat{q}}(q, p)$? ■

As the operator coordinates and momenta of N different degrees of freedom commute with each other, the kernel transformation between phase space and operator quantum mechanics is simply the tensor product of each component displaced parity operator:

$$\hat{\Pi}(\boldsymbol{q}, \boldsymbol{p}) = \bigotimes_{i=1}^{N}\hat{\Pi}(q_i, p_i). \tag{4.81}$$

This observation will be important later when we consider the phase space representation of composite particles such as atoms and molecules.

4.3.5 The Heisenberg Uncertainty Relation

Recall that the Heisenberg uncertainty relation quantifies how much two Hermitian operators, say $\hat{A}$ and $\hat{B}$, do not commute, and is given by

$$\Delta\hat{A}\,\Delta\hat{B} \geq \frac{1}{2}\left|\left\langle\left[\hat{A}, \hat{B}\right]\right\rangle\right|.$$

For historical reasons, this expression is often confused with measurement disturbance. The phase space formulation of quantum mechanics can provide some insight as to why such a misunderstanding can persist. We will focus this discussion on the specific example:

$$\Delta\hat{q}\,\Delta\hat{p} \geq \frac{1}{2}\left|\left\langle\left[\hat{q}, \hat{p}\right]\right\rangle\right| = \frac{\hbar}{2}.$$

In phase space, we do not have operators, but distributions associated with those operators. In conventional quantum physics, Hermitian operators represent observable quantities. In phase space, we will make Stratonovich–Weyl correspondence and use the distributions of the corresponding classical quantity (so q instead of $\hat{q}$ and p instead of $\hat{p}$ in this example). The only thing, we must be mindful of when seeking to apply the uncertainty principle, is that the classical quantity must be one whose equivalent quantum operators (in Heisenberg or Schrödinger pictures) are Hermitian.

We have already noted that we can map concepts such as expectation values onto quantum phase space in the natural way. The mathematical formulation of uncertainty in terms of probability density functions is the standard deviation $\Delta A = \sigma_A$ given by the usual expression,

$$\Delta A = \sqrt{\langle A^2\rangle - \langle A\rangle^2} = \sqrt{\int A^2 W - \left(\int AW\right)^2}.$$

Here we have used the shorthand $A = W_{\hat{A}}(\boldsymbol{q}, \boldsymbol{p})$ for the Stratonovich–Weyl corresponding distribution. We thus have $\Delta A = \Delta \hat{A}$ and in our example

$$\Delta q \Delta p \geq \frac{\hbar}{2}.$$

The product of the uncertainty in position and momentum must be greater than $\frac{\hbar}{2}$. This sets a minimum 'volume' of phase space that any Wigner function can occupy (as it is true regardless of the specific phase space distribution of the state). This has nothing to do with measurement - it is a fundamental property that any quantum probability density (Wigner) function will have. It is clear that, as $\hbar \to 0$, this constraint vanishes and classical physics is returned. Note that there is nothing about the measurement axioms in here, due to the fact that the uncertainty principle constrains the quantum state itself. That said, there will be implications for what can be measured. For example, we see if the state is 'squeezed' to have a very well-defined Wigner function in the coordinate, in order to satisfy the minimum uncertainty volume in phase space, it must become extended in momentum. This has nothing specifically to do with measurement disturbance. Nonetheless, if one does measure, e.g. position, then this degree of freedom becomes more well-defined. If the resolution of that measurement is enough for the above volume-of-phase-space argument to apply, then some deformation in the momentum degree of freedom must happen; the uncertainty relation holds just the same, *even if a measurement is never made!*

For the more general case, combining the phase space idea, $\{f, g\} = \lim_{\hbar \to 0} \{\{f, g\}\}$, and the Dirac quantisation scheme, $[\hat{A}, \hat{B}] = i\hbar \{\hat{A}, \hat{B}\}$, we arrive at the Heisenberg uncertainty relation expressed completely in phase space as

$$\Delta A \, \Delta B \geq \frac{\hbar}{2} |\langle \{\{A, B\}\} \rangle|.$$

Uncertainty can therefore be seen as arising as a consequence of deforming phase space structure by changing the Poisson bracket. When $\hbar \to 0$, so does the uncertainty. Away from this limit, quantisation of phase space using Moyal brackets leads to a minimum volume of phase space that any distribution representing a state can take. See Figure 4.2 for three examples.

4.3.6 Generalising Wigner Functions to Spins

One of the key aims of this text is to clarify the similarities as well as the differences between quantum and classical physics. As such, we feel the need to present a topic that is usually not discussed much outside the research literature: the treatment of spin in phase space. Firmly rooted in ideas of classical physics, it is perhaps unsurprising that phase space methods run into some difficulty when dealing with specifically quantum phenomena such as spin. Much progress has been made in this area, yet some questions currently remain unanswered. The results we present here are centred around using the Stratonovich–Weyl correspondence to determine phase space structure and can be used to reverse engineer a Moyal bracket. One may rightly ask if a more elegant solution can be found. We start this section by outlining a difficulty to which there is yet to be found a resolution (we do this to explain why we later take the approach that we do). Specifically, it would best link classical and quantum physics if one could consider a classical Lagrangian for spin, derive the

Figure 4.2 Examples showing how the phase space of the Wigner function changes so that the volume satisfies Heisenberg's uncertainty relation. Black is 0 and the lighter shades indicate positive values. (a) Wigner function of the vacuum state $|0\rangle$. This state, like all coherent states, minimises Heisenberg's uncertainty relation in position and momentum. (b) 'Squeezing' the position degree of freedom results in the Wigner function becoming extended in momentum. (c) 'Squeezing' position and momentum degrees of freedom results in the Wigner function becoming extended off-axis.

conjugate momenta, and use Poisson and Moyal brackets appropriately to study its classical and quantum physics properties. Considering the point particle as a sphere of 'fixed' radius and setting $dr = d\Theta = 0$, the Lagrangian is just the kinetic term

$$\mathscr{L} = J \cos \Theta \, \dot{\Phi}, \tag{4.82}$$

where we would assume $J = 1/2$ for spin-half particles. With $q = \Phi$ we have $p = \frac{\partial \mathscr{L}}{\partial \dot{q}} = J \cos \Theta$. This allows the construction of a Moyal bracket (we could do spin-one by generalising the above argument to the 4-sphere and so on). Unfortunately, such an approach seems to not work. Neither has an equivalent approach been engineered from the successful Stratonovich–Weyl correspondence body of work. That is, it has not been shown whether it is possible or impossible to take the kernel, construct a differential form of the Moyal bracket, identify coordinates and conjugate momenta, and then find the classical Lagrangian corresponding to the quantum idea of spin. Perhaps the difficulty lies in the fact that a proper understanding of spin only comes from relativistic quantum mechanics. We do not discuss this topic, but note that there have been few reports of progress in relativistic phase space quantum mechanics and spin in the research literature. Nevertheless, the framework described below does enable us to model spin (and other discrete systems) as a phase space Wigner function interpretable as a continuous quantum probability distribution over phase space – providing similar intuition as before (discrete versions of Wigner functions do exist, but we do not discuss those here).

Starting from the Stratonovich–Weyl correspondence, we will use the idea of displaced parity to generalise the Wigner function to spin systems. Here we will consider spin-half only, but the idea can be generalised to other spins, angular momentum (which we cover later), finite dimensional systems and arbitrary composites of these, and continuous variable systems. We start by observing that the displacement operator can be seen as an operator that translates the vacuum state of the system into another state with the same phase space distribution, but now centred at the point of displacement. However, we do not have a vacuum state for spin systems. The $\hat{\sigma}_z$ basis comprises only two states, $\{|+1\rangle, |-1\rangle\}$ (the same eigenstates as $\left|\pm\frac{1}{2}\right\rangle$ of $\hat{S}_z$). It turns out that either of these can have vacuum like properties; let us arbitrarily choose to discuss this with respect to $|-1\rangle$. It turns out that, just like with the vacuum state and coherent states, any other spin-state can be made from this state by using the displacement operator for spin:

$$\hat{U}(\theta, \phi, \Phi) = \exp(i\,\hat{\sigma}_z\phi)\exp(i\,\hat{\sigma}_y\theta)\exp(i\,\hat{\sigma}_z\Phi). \tag{4.83}$$

Here the angles, and the parameterisation of phase space, are in terms of Euler angles. We find that there is a parity which satisfies Stratonovich–Weyl conditions, namely

$$\hat{\Pi} = \frac{1}{2}\left[\hat{1} + \sqrt{3}\hat{\sigma}_z\right], \tag{4.84}$$

so that

$$\hat{\Pi}(\theta, \phi) = \hat{U}(\theta, \phi, \Phi)\hat{\Pi}\hat{U}^\dagger(\theta, \phi, \Phi), \tag{4.85}$$

for Euler angles θ, ϕ (note that as $\hat{\Pi}$ is diagonal in the $\hat{\sigma}_z$ basis, the third angle Φ_i cancels and plays no part in the Wigner function). The real power of this method shows when kernels for different systems are combined, enabling phase space representation of composite systems such as atoms and molecules. For a thorough introduction to the approach, see [73, 74, 86]. For examples on composite systems, see [14, 70–72].

Because we personally find this topic deeply interesting, we will give one concrete example of a Wigner function for a hybrid system. We will use the state

$$|\psi\rangle = \frac{1}{\sqrt{2}}\left(|\alpha = -2.5\rangle\left|m = \frac{1}{2}\right\rangle + |\alpha = 2.5\rangle\left|m = -\frac{1}{2}\right\rangle\right). \tag{4.86}$$

This is an entangled extension of the idea of a Schrödinger cat state shown in Figure 4.1c,d. As a composite of a $(\hat{q}, \hat{p})$ system and a spin system, the kernel needed to generate the Wigner function will simply be $\hat{\Pi}(q, p, \theta, \phi) = \hat{\Pi}(q, p) \otimes \hat{\Pi}(\theta, \phi)$ and the Wigner function the expectation values of this operator. Obviously, we cannot plot this full four-dimensional function. What we can do is to take slices or integrate out degrees of freedom to get a sense of the full phase space description and plot those in a careful way to get a sense of what the state looks like. We show one such example visualisation of the Wigner function for this state on the cover of the book. In this figure, red is positive and blue is negative. Before we describe the picture, please take a moment to note the similarities and differences of the cover image with Figure 4.1c,d[6] We made this plot by scanning 'pixel'-by-'pixel' across the (q, p) phase space. At each pixel we plotted a sphere which is the (θ, ϕ) dependence of the Wigner function at that (q, p) position. To help interpret this figure, note that for the Wigner function of a single spin the maximum positive value indicates the direction in which the spin is pointing. We displaced each sphere vertically by $W(q, p) \stackrel{\text{def}}{=} \int d\theta d\phi\, W(q, p, \theta, \phi)$ (multiplied by 1.2 for aesthetic reasons). This way we can retain both information on the (q, p) and the (θ, ϕ) degrees of freedom. The interference terms between Figure 4.1c,d and the cover image differ very much in their character. We find that this is a nice feature of the visualisation, as the fact that there is no amplitude in the interference terms in the cover image and that there is a rotation of the spin components is a visual indication of entanglement. We can understand this as being because the entanglement correlations are purely quantum mechanical and exist outside of the space of just (q, p) or just (θ, ϕ). While it is beyond the scope of this book, that the rich nature of highly correlated hybrid quantum states can be well visualised and understood using phase space methods like this, is something we find fascinating. As the displaced parity operator can be considered an observable, we tend to think of these visualisations as a form of quantum state spectroscopy.

A final note on spins: the reason we may struggle to find a nice connection to classical phase space is that the displacement operator is written in terms of Euler angles. Thinking in terms of quaternions, this is the natural connection to rotation in classical physics. Indeed, Euler angles may be used to make classical canonical coordinates and conjugate momenta and are particularly useful in, e.g., the study of spinning tops. Unfortunately, we are not aware of any successful attempts to quantise these in operator quantum mechanics (although research does continue in this direction). The fundamental difficulty may be that quantum spin has no classical analogue and, as already noted, is not fully understandable outside relativistic quantum mechanics.

4.3.7 Axioms of Quantum Mechanics Revisited

Let us once more revisit the axioms of quantum mechanics but this time phrase them in terms of phase space. One can see that some of these axioms are much more natural formulated this way, but some are not.

A1$_W$ That the complete information on a system's state is contained within a quasi-probability distribution termed the Wigner function. This distribution shares the same properties as a classical probability density function, with the exception that we do not restrict it to being non-negative.

---

6 Figures displaying Wigner functions of a Schrödinger cat state, such as that shown in Figure 4.1c, are considered by some as iconic of quantum superposition. At least one book [33] has chosen such a figure as its cover image. The reason we believe the image on the cover of this book is fitting, is that it takes the visualisation of quantum coherent superpositions further. Not only does the full system Wigner function display interference effects due to the macroscopically distinct superposition of coherent states but it also displays with some clarity the correlations due to the states' entanglement. To us, this is a figure of beauty as well as physical meaning.

A2$_W$ Measurable quantities are distributions, just as in classical statistical physics. They map onto their operator counterparts by the Stratonovich–Weyl correspondence. For those operators that have no classical counterpart, the corresponding phase space distribution is also given by the Stratonovich–Weyl correspondence.

A3$_W$ Where it exists, the Poisson bracket is replaced by a Moyal bracket. For systems with no classical analogue, there is an alternative Moyal bracket derivable from the Stratonovich–Weyl correspondence.

A4$_W$ Dynamics of any distribution are given by:

$$\frac{d\hat{A}_H}{dt} = \{\{\hat{A}_H, \hat{H}_H\}\} + \frac{\partial \hat{A}_H}{\partial t}.$$

The second condition in A3$_W$ is far from satisfactory, as it implies that the phase space formulation of quantum mechanics remains incomplete and subordinate to the others.

It is not clear that measurement axioms are needed in this statistical interpretation of quantum mechanics. For comparative purposes, below is a mapping of the Copenhagen interpretation of measurement into phase space:

M1$_W$ When an observable is measured, the only possible outcome of that measurement is a stargenvalue of the observable (that is, the solution to $A \star W = \lambda W = W \star A$).

M2$_W$ the probability of measuring a given stargenvalue is the expectation value of the current state of the system with the projection operator onto the subspace spanned by all the stargenfunctions of that stargenvalue.

M3$_W$ On measurement, the state of the system is projected into a normalised eigenstate of the system, associated with the measured stargenvalue.

These lack elegance and, as the measurement axioms are the most controversial of all those of quantum mechanics, it would be natural to seek to exclude them from the theory. We will return to that discussion in Chapter 10. There is no need in this framework for any additional axioms of choosing a representation. There are, however, alternative phase space representations and rules for mapping between them. That topic is beyond the scope of this text. One representation that is particularly noteworthy is that of the Weyl distribution. This has a kernel that is simply the displacement operator and can be interpreted as a Fourier transformation of the Wigner function.

4.4   The Path Integral Formulation of Quantum Mechanics

From our review of classical mechanics, we recall that Hamiltonian mechanics is derived from the more fundamental Lagrangian formulation, which in turn is derived directly from the deeper principle of least action (stationary-action). Recall that this states that the path followed by particles is one where the action

$$\mathscr{S} = \int_{t_i}^{t_f} dt \, \mathscr{L}(q_i(t), \dot{q}_i(t), t) \tag{4.87}$$

is at a turning point. The action is defined as a functional of the primitive quantities of coordinates and their velocities. It is often the case that new theories come about from old theories by reducing the restrictions or axioms of that theory (such as by the dropping of additional velocities in special relativity).

It is therefore natural to ask if it is possible to relax the principle of least action to recover a theory consistent with both classical and quantum phenomena. The success of phase space quantum mechanics in retaining the conceptual framework of classical Hamiltonian physics would hint that it may be so. To do this, we would want to generalise the principle of least action so that the Moyal, not the Poisson, bracket arises as a natural consequence of the arguments that led to Hamiltonian mechanics.

One could interpret the phase space picture of quantum mechanics as a wave theory. The probabilistic interpretation is the standard, but not only, one (matter waves is another, for example). To get an idea as to why the path integral formulation of quantum mechanics emerges in the way it does, let us consider another wave theory – electromagnetism. We can then think along the lines of the Huygens–Fresnel principle and see where this leads us. Loosely speaking, this states that every point on a wavefront can itself be considered a source for propagating the wave. In electromagnetism, we are used to thinking of the solution to the wave equation as electromagnetic fields with complex components. Specifically, let us consider one wave propagating in free space for some quantity f. The wave equation is

$$\nabla^2 E = \frac{1}{c^2} \frac{\partial^2 E}{\partial t^2},$$

which will have solutions of the form $E = A \exp(i\,\boldsymbol{k} \cdot \boldsymbol{q}) \exp(-i\,\omega t)$. If we take the solution at one point in space, q_0, and time, t_0 as an initial reference 'state', we can consider another solution at a different space, q_1, and time, t_1, as being related to the reference by a position phase and a time phase. In other words, one is a phasor in space and the other a phasor in time. We can think of these as little space and time 'clocks' (on the unit circle in the complex plane) whizzing round as we make these shifts. At one instant in time, we can apply the Huygens–Fresnel and superimpose all the contributions from each part of the wavefront (all have the same amplitude) by integrating this field over the path. Here the idea then is that over each defined path *with the same start and end points a and b*, the phasor clock hands move round as we traverse the path, and we add up all the hands together to get a total contribution from that path. Unlike the usual action, this results in a complex value. We can also ascribe an action based on a Lagrangian that is simply the distance along the path, Γ, from its starting position a that is $\mathscr{L} = 2\alpha x$, say, (for some constant *alpha* that depends on the wave-vector and frequency). So the action would be

$$S = \int_\Gamma \mathrm{d}x\,\mathscr{L} = \alpha x^2.$$

The phase picked up on the path, Γ, is then

$$\Phi_\Gamma = \exp(i\,\beta S_\Gamma) = \exp\left(i\,\alpha\beta x^2\right)$$

(for some other constant β that depends on the wave-vector and frequency). Now, applying the Huygens–Fresnel, we add up all the contributions from each path so that the total amplitude at the final point of the path,

$$\Phi(a, b) = \sum_{\text{all } \Gamma} \Phi_\Gamma = \sum_{\text{all } \Gamma} \exp\left[i\,\beta S_\Gamma\right] = \sum_{\text{all } \Gamma} \exp\left[i\,\beta \int_\Gamma \mathcal{L}\right].$$

When we do this, we see that the paths obeying Fermat's Principle of least time make a substantial contribution. This argument can be used to reproduce phenomena such as diffraction and interference (indeed, all wave theories can be written in terms of path integrals – but to do it properly involves some advanced mathematics that we do not cover here). It is worth taking the time to find some videos of Richard Feynman explaining this himself (using 'Feynman Vega science' in a good search engine should be sufficient). The path integral formulation of quantum mechanics takes a directly analogous approach and states that the complex (probability) amplitude between two points in phase space a and b is simply:

$$K(a, b) = \sum_{\text{all } \Gamma} \exp\left[\frac{i}{\hbar}(\text{constant}) \int_\Gamma dt\,\mathcal{L}(\boldsymbol{q}, \dot{\boldsymbol{q}}, t)\right]. \tag{4.88}$$

With some work, this can be shown to be consistent with the other formulations of quantum physics. We can consider the path integral formulation either as a sum over probabilities weighted by action or as the consequence of matter having a wave instead of particle nature. In this way we see that the path integral formulation is consistent with both Copenhagen and realist interpretations of quantum mechanics (and can be reasoned to be consistent with any other too). In the probabilistic analogy, we have abandoned 'least action' for 'all actions', and in the wave view we have abandoned a particle view of the system's state – interestingly, both have the same outcome. In this way, we find that all the complications of commutators or Moyal brackets arise from trying to represent this rather simple 'sum over all paths' idea in a Hamiltonian framework. Unfortunately, the framework in which quantum mechanics is most easily understood (Feynman path-integrals) is also the one where it is most difficult to perform calculations. For this reason, most quantum mechanics texts at this level present almost all of their discussion in the other frameworks, and ours is no exception here. See [28] for an excellent introductory text dedicated to the path integral formulation.[7]

4.5 Closing Remarks

Classical mechanics can take very different forms (Lagrangian, Hamiltonian, Newtonian, Koopman–von Neumann). In this chapter, we have discussed some of the ways in which quantum mechanics may also take apparently different forms. In classical mechanics, the base axiom that underpins all formulations is the principle of least action. In contrast, the main assumption of quantum mechanics is that we change this principle to one that sums over all actions. Each action principle is a global view of possible trajectories between well-defined space–time events. Just as in classical mechanics, we see that in quantum physics,

---

7 Note that one might be tempted to conclude that the Feynman path integral formulation of quantum mechanics is able to deal with situations where the Lagrangian is at most linear in the velocity and the Hamiltonian formulations are not. Dirac showed it is possible to extend the commutation relation of conventional quantum mechanics to accommodate such systems. This topic is advanced and beyond our current scope.

Hamiltonian mechanics is a geometric view of these action principles (the Poisson and Moyal Brackets).

We can also derive value from different formulations of quantum mechanics to better understand the philosophical foundations of the subject. For example, if we wished to assert the correspondence principle, each formulation allows a different way to make the argument for recovering classical physics from quantum theory. In the phase space view, the non-local geometric deformation is the thing that shrinks at the classical limit. That is, as the dynamics become large with respect to an area of phase space defined by Planck's constant, the non-local phase space contributions are more likely to cancel out. We might alternatively argue that the geometric view of all paths being followed (sum of actions as probabilities) reduces to the Poisson bracket being replaced by the Moyal bracket.

While each formulation may look very different from the others, they are essentially equivalent theories. The connection between operator and phase space quantum mechanics can be made by recognising that the star product is a function version of matrix multiplication, so the Moyal bracket is equivalent algebraically to the commutator. Conversely, the global view of geometric quantisation with the embedded Weyl ordering of the star product/Moyal brackets is the same as the sum over all paths of the path integral formulation. Selecting the one we chose to use at any one time is a matter of judgement based on what we are trying to achieve.

5

Vectors and Angular Momentum

"Our new idea is simple: to build a physics valid for all coordinate systems."

Albert Einstein

5.1 Introduction

In Section 3.4.1.2 we introduced quantum mechanics in three-dimensional Cartesian space. We chose, as an example, $(\hat{q}_x, \hat{q}_y, \hat{q}_z)$ as a complete set of commuting observables (CSCO). This meant that we could talk about the representation of states in terms of the basis vectors $|q_x, q_y, q_z\rangle$, or $|\boldsymbol{q}\rangle$ for short. We then introduced the wave-function as $\psi(\boldsymbol{q}) = \langle \boldsymbol{q}|\psi\rangle$, and we used this to motivate vector position and momentum operators as the Cartesian products $\hat{\boldsymbol{q}} = (\hat{q}_x, \hat{q}_y, \hat{q}_z)$ and $\hat{\boldsymbol{p}} = (\hat{p}_x, \hat{p}_y, \hat{p}_z)$. We spoke without too much concern of writing a vector operator in terms of basis vectors as, e.g. $\hat{\boldsymbol{q}} = \hat{q}_x\mathbf{i} + \hat{q}_y\mathbf{j} + \hat{q}_z\mathbf{k}$. We did this despite the fact that the axioms of operators only specify closure under multiplication by a scalar (recall that we stated $|\hat{X}\alpha u\rangle = \alpha|\hat{X}u\rangle$ in Section 1.3.1). In doing this, we have made some assumptions that work for Cartesian coordinates but do not hold true in general. We start this chapter with a discussion that explores this subject and seeks to understand what it means to define vector operators.

The subject of defining vectors is closely related to the study of angular momentum. To understand why this is the case, let us look more carefully at what we mean by scalar and vector quantities. Core to the following argument is the principle of relativity. Specifically, we will apply the constraint that a physical scalar or vector quantity must not be dependent on the orientation of the reference frame in which it is represented (while components of a vector may change for a given rotation of a coordinate system, the vector itself must remain pointing in the same direction in space). Recall that angular momentum is defined as $\boldsymbol{L} = \boldsymbol{r} \times \boldsymbol{p}$ and that it has the important symmetries $\{L_x, L_y\} = L_z$ (and cyclic permutations). We also recall that Poisson brackets tell us how things change in classical mechanics. We are used to thinking about $\{\cdot, \mathscr{H}\}$ telling us how a quantity change over time increments. In the case of angular momentum, $\{\cdot, L_k\}$ determines how quantities change under small rotations around each axis k. A true Euclidean scalar α must therefore satisfy $\{\alpha, L_k\} = 0$ for $k \in x, y, z$ as α should not change on rotation around any axis. Hence, we see that $q_x + q_y + q_z$ is not a scalar, but $\boldsymbol{q}^2$, the kinetic energy, and L^2 are. So we see that

Quantum Mechanics: From Analytical Mechanics to Quantum Mechanics, Simulation, Foundations, and Engineering, First Edition. Mark Julian Everitt, Kieran Niels Bjergstrom and Stephen Neil Alexander Duffus.
© 2024 John Wiley & Sons Ltd. Published 2024 by John Wiley & Sons Ltd.
Companion website: www.wiley.com/go/everitt/quantum

to be (Galilean relativistically) invariant under the rotation of a reference frame, a scalar quantity is not simply a quantity without any direction – it must also satisfy the symmetry property $\{\alpha, L_k\} = 0$. Similarly, a Euclidean spatial vector, say v, should have its own (Galilean relativistic) symmetry properties. The first is that the kth component should be invariant under rotations about the kth axis. This is expressed by the constraint that $\{v_k, L_k\} = 0$ for $k \in x, y, z$. Choosing rotations around the x axis as an example, we see that the z component will be continuously rotated into the y direction and the y component into the $-z$ direction. Mathematically, this is expressed as $\{v_z, L_x\} = v_y$ and $\{v_y, L_x\} = -v_z$ (and all cyclic permutations of these expressions). A spatial vector is therefore not just any quantity with magnitude and direction; it should also satisfy these relationships for it to transform in a way that is physically meaningful with regard to Galilean relativity.[1]

While we are able to define $\hat{L} = \hat{q} \times \hat{p}$, it turns out that the Poisson bracket/commutation relationships not only contain most of the important physics, their study also leads us to understand how to generalise the theory of angular momentum to arbitrary spin systems. This is because the vector algebraic structure, we discussed above, generalises the idea of rotation in three dimensions to any system where the rotation is important. For this reason, we once again turn to the Dirac scheme of setting $[\cdot, \cdot] = i\hbar\{\cdot, \cdot\}$ (noting that there are no operator ordering ambiguity issues here). The components of an orbital angular momentum operator satisfy $[\hat{L}_x, \hat{L}_y] = i\hbar\{\hat{L}_x, \hat{L}_y\} = i\hbar\hat{L}_z$, and the generalisation to an arbitrary angular momentum is a set of three operators that satisfy $[\hat{J}_1, \hat{J}_2] = i\hbar\hat{J}_3$ (and cyclic permutations).[2] Once angular momentum is defined in this way, we can in turn define vector operators, $\hat{V}$, using the same approach as the classical one. That is, we demand that their components transform under rotation like the components of a vector $[\hat{V}_i, \hat{L}_j] = i\hbar\{\hat{V}_i, \hat{L}_j\}$, specifically: $[\hat{V}_x, \hat{L}_x] = 0$, $[\hat{V}_z, \hat{L}_x] = i\hbar\hat{V}_y$, and $[\hat{V}_y, \hat{L}_x] = -i\hbar\hat{V}_z$ and all cyclic permutations.

As a consequence of the elegance and utility of the above approach, there is a historical imperative that means most modern treatments of angular momentum and vectors are phrased only in terms of the above argument, as commutation relations in terms of their Cartesian components. As the outcome of this analysis is sufficient to understand all the important physics, any discussion of alternative coordinate systems is usually avoided. One of the central themes of this books it to make clear not only the differences but also the similarities between quantum and classical physics. Given the frequent application of other coordinate systems in the treatment of classical rotating objects, the exclusive adoption of Cartesian coordinates is not what we find in the classical mechanics literature. One might rightly ask if it is possible to define vector quantities in terms of basis vectors in an analogous way to that used in classical mechanics. As one of the core principles of relativity is that no coordinate system should be considered any 'better' than any others, it is natural to want some assurance that quantum mechanics is not coordinate system dependent. We have, in previous chapters, asserted that the Dirac quantisation scheme $[\cdot, \cdot] = i\hbar\{\cdot, \cdot\}$ works for at least canonical position and momenta. With Groenewold's theorem exposing

---

1 It may be of interest to note that in this definition vectors are defined in terms of angular momentum, which is itself a pseudovector and not a vector (in the language of differential geometry). As such, this definition of vectors in terms of their Poisson brackets or commutators with angular momentum does not recognise the difference between the two – we need to also bring in the notion of parity to resolve this difference.

2 Which in turn leads to $\hat{D}(\hat{n}, \phi) = \exp\left(-i\phi\boldsymbol{n} \cdot \hat{\boldsymbol{L}}/\hbar\right)$ being the generator of a rotation ϕ around an axis $\boldsymbol{n}$.

surprising issues for quantum physics, one might even be concerned that there are also some hidden surprises with non-Cartesian coordinate systems. Specifically, you may ask if it is possible to define canonical position and momentum in non-Cartesian coordinates, and vector operators in those systems. There is a lot to be learned by the discussion from a fundamental point of view. Once we understand some of the complexities of that approach, we will move our discussion back onto the development of the standard treatment of quantum mechanics. We will conclude the chapter by exploring specific examples of spin and orbital angular momentum of central importance to understanding phenomena such as the quantum description of atoms.

5.2 On Curvilinear Coordinates (Using Spherical Coordinates as an Example)

5.2.1 The Coordinate Representation: A Statement of the Problem

Recall from Section 3.4.1 on page 78 that, in deriving the Cartesian position representation, we start with the Dirac scheme of $[\hat{q}_i, \hat{p}_j] = i\hbar \{\hat{q}_i, \hat{p}_j\} = i\hbar \delta_{ij}$ for $i, j \in \{x, y, z\}$. We then apply the postulates of measurement to argue that $\hat{q}_i |q_i\rangle = q_i |q\rangle$, where q_i is any real number (as that is what we presume we can measure). We also impose that the set of position operators form a CSCO so that $\langle q | q' \rangle = \delta(q - q')$. We then show that the action of the position operator in this representation is that of multiplication by the position eigenvalue: $\langle q |\hat{q}| \psi \rangle = q\psi(q)$ (where $\psi(q) = \langle q | \psi \rangle$). Using the Displacement operator $\hat{D}(Q) = \exp\left[-\frac{i}{\hbar} Q\hat{p}\right]$ together with McCoy's theorem $([A, B] = \alpha \implies [\hat{A}, f(\hat{B})] = \alpha f'(\hat{B}))$, we have:

$$\hat{p}\psi(q) = -i\hbar \frac{\partial \psi(q)}{\partial q}.$$

We might assume that following this exact same process but using, for example, spherical polar coordinates would lead to the coordinate representation for that system. Let us develop that argument and see where it fails, as there is a useful lesson to be learned here.

Recall that spherical polar coordinates are defined as:

$$q_x(r, \theta, \phi) = r\sin\theta \cos\phi, \quad r(q_x, q_y, q_z) = \sqrt{q_x^2 + q_y^2 + q_z^2}$$

$$q_y(r, \theta, \phi) = r\sin\theta \sin\phi, \quad \theta(q_x, q_y, q_z) = \arccos\frac{q_z}{r} \tag{5.1}$$

$$q_z(r, \theta, \phi) = r\cos\theta, \quad \phi(q_x, q_y, q_z) = \arctan\frac{q_y}{q_x},$$

where $r \in [0, \infty)$, $\theta \in [0, \pi]$, and $\phi \in [0, 2\pi)$. Here, $\arctan x/y$ is the standard shorthand for the function $\arctan(x, y)$. Note that these functions are the inverses of each other, e.g. $r(q_x(r, \theta, \phi), q_y(r, \theta, \phi), q_z(r, \theta, \phi)) = r$. For systems where the potential energy is not velocity dependent, we can then derive the classical canonical momenta from the kinetic energy:

$$\mathscr{T} = \frac{1}{2}m\dot{q}^2 = \frac{1}{2}m(\dot{r}\tilde{e}_r + r\dot{\theta}\tilde{e}_\theta + r\dot{\phi}\sin\theta\tilde{e}_\phi)^2$$

$$= \frac{1}{2}m(\dot{r}^2 + r^2\dot{\theta}^2 + r^2\dot{\phi}^2\sin^2\theta) \tag{5.2}$$

in the usual way to obtain:

$$p_r = \frac{\partial \mathcal{L}}{\partial \dot{r}} = m\dot{r}$$

$$p_\theta = \frac{\partial \mathcal{L}}{\partial \dot{\theta}} = mr^2 \dot{\theta}$$

$$p_\phi = \frac{\partial \mathcal{L}}{\partial \dot{\phi}} = mr^2 \sin^2\theta \, \dot{\phi}.$$

Note that the tilde over the basis vectors in Eq. (5.2) is used to indicate that these are the normalised unit vectors. As these momenta are canonically conjugate to the corresponding coordinate, we know that

$$\{q_i, q_j\} = 0, \{p_i, p_j\} = 0, \{q_i, p_j\} = \delta_{ij} \text{ where } i, j \in \{r, \theta, \phi\}. \tag{5.3}$$

Following the approach used for Cartesian coordinates would lead us to postulate that we can choose as a CSCO operator equivalents of the spherical coordinates $\hat{q}_i$, where $i \in \{r, \theta, \phi\}$. Our quantisation scheme, $[\hat{X}, \hat{Y}] = i\hbar \{\hat{X}, \hat{Y}\}$, implies

$$[\hat{q}_i, \hat{q}_j] = 0, [\hat{p}_i, \hat{p}_j] = 0, [\hat{q}_i, \hat{p}_j] = \delta_{ij}, \text{ where } i, j \in \{r, \theta, \phi\}, \tag{5.4}$$

just as with $\{x, y, x\}$. The key difference comes when we impose our measurement ansatz of restricting the domain of the coordinate observables to also apply to eigenvalues of their corresponding quantum mechanical operators so that, for $\hat{q}_i |q_i\rangle = q_i |q_i\rangle$, we have $q_r \in [0, \infty)$, $q_\theta \in [0, \pi]$, and $q_\phi \in [0, 2\pi)$. A direct, but incorrect, repetition of our displacement operator argument at this point leads us to the wrong result, that the conjugate momenta look just like their Cartesian equivalents $-i\hbar \frac{\partial}{\partial q_i}$. This approach fails for a number of reasons.

Exercise 5.1 Before reading further, review Section 3.4.1 on page 78 and try to find where the argument used might be inapplicable to spherical coordinates (do not worry if you cannot – part of the point of this exercise is to demonstrate that this is not easy). ∎

The first issue we discuss is that there are hidden, but incorrect, assumptions in using the same argument about Cartesian coordinates for spherical ones unaltered. The restriction of the domains of the eigenvalues of the observables to $\{r, \theta, \phi\}$ has non-trivial implications. Recall that the structure of the mathematical space in which quantum systems are represented comprises not only the vectors themselves, but also a specified inner product. The same is true for the coordinate space, where the definition of the inner product must take into account the domain and volume element of the coordinate space over which it is defined.[3] As properties, such as the Hermiticity of operators, are defined with respect to the inner product, this is an important consideration – the operators that are Hermitian under one inner product may not be under another. In the specific example of spherical coordinate representation space, the inner product is

$$\langle \psi | \psi' \rangle = \int d\omega \, \psi^*(r, \theta, \phi) \psi'(r, \theta, \phi), \tag{5.5}$$

---

3 Or, more precisely, the Haar measure which is inherited from the underlying metric of the space.

where $d\omega$ is the volume element $r^2 \sin\theta \, dr \, d\theta \, d\phi$ and the integral is over the entire domain of $\{r, \theta, \phi\}$ (and not $\mathbb{R}^3$).

Exercise 5.2   With respect to the above inner product show that the operators $-i\,\hbar\frac{\partial}{\partial q_i}$ are not Hermitian for $i \in \{r, \theta, \phi\}$. ∎

Furthermore, for our current argument of constructing a coordinate representation, the form of the inner product is important, as it changes the orthonormality condition on which the definition of the basis depends. Change the inner product, and the vectors that form an orthonormal basis must also change with it. This means that our assumption $\langle q | q' \rangle = \delta(q - q')$ is incorrect, as it does not appropriately account for this inner product. We will return to consider this point again shortly.

Before doing so, we also indicate another issue that it is worth being mindful of when dealing with operators. In this example, the issue lies in assuming we can make use of expressions such as McCoy's theorem, $[\hat{q}, f(\hat{p})] = i\,\hbar f'(\hat{p})$. The problem is that this expression assumes that $f(\hat{p})$ has a power series expansion such as $\exp(i\lambda\hat{p}) = \sum (i\lambda)^n \hat{p}^n / n!$. As such, McCoy's theorem is only guaranteed to hold for *bounded operators*. For unbounded operators, alternative definitions of the exponential may be needed, and McCoy's theorem may not hold (but that is a topic for a 'functional analysis' book). Whether or not an operator is bounded depends very much on its domain, and not just its form. This turns out to be an issue for momentum in curvilinear coordinates. With operators in general there may even be the further concern that, while the operators themselves may be well-defined, their commutator may not be (if two operators are defined over different domains, what should the domain of the commutator be?). In this example, application of McCoy's theorem leads us to the wrong result because it is not applicable.

One lesson we can draw from the above discussion is that even arguments that seem to move logically from one point to the next can contain many hidden assumptions in each step. While operator algebra for finite dimensional systems is relatively straightforward, we need great care when applying the same logic to unbounded systems.

5.2.2   Canonical Quantisation in Spherical Coordinates

This section can be safely skipped, as a more heuristic approach to determining the spherical coordinate representation is given in the next section. This discussion is included because we feel it is important to show that the canonical quantisation scheme is not limited to Cartesian coordinates if appropriate care is taken. Here we summarise a first principle argument by DeWitt [15] to address the derivation of the coordinate representation in non-Cartesian systems (the findings are actually restricted to those coordinate systems that are point transformations of Cartesian coordinates). The full argument requires an understanding of differential geometry that we do not wish to require of the reader. As such, we only outline the main argument.

As always, we start with the canonical commutation relation $[\hat{q}_i, \hat{p}_j] = i\,\hbar\,\{\hat{q}_i, \hat{p}_j\} = i\,\hbar\delta_{ij}$ for $i, j \in \{r, \theta, \phi\}$ with $[\hat{q}_i, \hat{q}_j] = [\hat{p}_i, \hat{p}_j] = 0$ (note that DeWitt took care to differentiate between contravariant position and covariant momentum $\left[\hat{q}^i, \hat{p}_j\right] = i\,\hbar\delta_j^i$ but we will not concern ourselves with this level of detail).

We next take the usual step of choosing the CSCO as $\{\hat{q}_r, \hat{q}_\theta, \hat{q}_\phi\}$ where

$$\hat{q}_r |r, \theta, \phi\rangle = r |r, \theta, \phi\rangle \tag{5.6}$$

$$\hat{q}_\theta |r, \theta, \phi\rangle = \theta |r, \theta, \phi\rangle \tag{5.7}$$

$$\hat{q}_\phi |r, \theta, \phi\rangle = \phi |r, \theta, \phi\rangle, \tag{5.8}$$

and the eigenvalues are restricted, just as with the inner product, to the same domain of spherical coordinates (as these are the values that can be measured).

The crucial observation is that orthonormalisation of the basis vectors is not expressed in terms of the usual Dirac delta function, but rather its generalisation. Specifically, we require that

$$\langle r, \theta, \phi | r', \theta', \phi' \rangle = \delta(r, \theta, \phi | r', \theta', \phi'), \tag{5.9}$$

where $\delta(r, \theta, \phi | r', \theta', \phi')$ is the generalisation of the Dirac delta function defined by

$$\delta(r, \theta, \phi | r', \theta', \phi') = 0 \text{ if } r, \theta, \phi \neq r', \theta', \phi' \tag{5.10}$$

$$\int d\omega\, f(r, \theta, \phi)\delta(r, \theta, \phi | r', \theta', \phi') = f(r', \theta', \phi'). \tag{5.11}$$

Here $d\omega = r^2 \sin\theta\, dr\, d\theta\, d\phi$, and the integral is over the entire domain of $\{r, \theta, \phi\}$. The matrix elements of the momentum operator in this basis can then be determined from the canonical commutation relations (in De Witt's work i $\hbar\delta_j^i\delta(q, q') = (q^i - q'^i)\langle q|\hat{p}_j|q'\rangle$ in a more general case). This leads to a set of differential equations that, when solved for the specific case of spherical coordinates, are:

$$\hat{p}_r = -i\,\hbar\frac{1}{r}\frac{\partial}{\partial r}r \tag{5.12}$$

$$\hat{p}_\theta = -i\,\hbar\left[\frac{\partial}{\partial\theta} + \frac{1}{2}\cot\theta\right] = -i\,\hbar\frac{1}{\sqrt{\sin\theta}}\frac{\partial}{\partial\theta}\sqrt{\sin\theta} \tag{5.13}$$

$$\hat{p}_\phi = -i\,\hbar\frac{\partial}{\partial\phi}. \tag{5.14}$$

One might notice that there are issues at the poles, but these are expected for spherical coordinates and their consideration in an unnecessary sophistication for our discussion.

We note before proceeding further that $\hat{p}_\phi = \hat{L}_z$. We will seek to show in the next section (i) that the same results can be obtained directly from the Cartesian operators and the appropriate use of a change of spatial basis and (ii) that quantities such as $\hat{\mathbf{p}}, \hat{p}^2$, and $\hat{L}^2$ are arrived at in agreement with the equivalent Cartesian counterparts.

5.2.3 Spherical Coordinates, Vectors, and Momenta

We opened this chapter by noting that the axioms of operators only specify closure under multiplication by a scalar. This means that an expression such as $\hat{\mathbf{q}} = \hat{q}_x\mathbf{i} + \hat{q}_y\mathbf{j} + \hat{q}_z\mathbf{k}$ is not well-defined. We introduce here a resolution that is non-standard, which we believe causes no problems for Cartesian coordinates, and which enables us to have a conversation about vector operators and coordinate systems. For a comprehensive standard treatment of vector operators in quantum mechanics, see complement D_X in [12]. Our approach will allow us to define vectors in a way that is closer to the usual classical treatment of vectors defined in terms of a coordinate system. This avoids restricting the conversation to the Cartesian component form that we discussed in the introduction.

In classical mechanics we can define the basis vectors according to

$$e_x = \frac{\partial q}{\partial q_x}, \ e_y = \frac{\partial q}{\partial q_y} \text{ and } e_z = \frac{\partial q}{\partial q_z}.$$

When writing q as the Cartesian product (q_x, q_y, q_z), we see these equate to $(1, 0, 0)$, $(0, 1, 0)$, and $(0, 0, 1)$, respectively. They can thus be seen to also correspond to the column vectors $\mathbf{i}, \mathbf{j}$, and $\mathbf{k}$ in that representation of Euclidean space. We will take a non-standard way to think about basis vectors for operators. In the spirit of full disclosure there may be issues with this approach that we are unaware of but it has the benefit of casting the basis vectors as Hermitian operators and therefore on the mathematical same footing as other operators. The reason for arguing this way will become clear shortly. We will define

$$\hat{e}_x = \frac{\partial \hat{q}}{\partial \hat{q}_x}, \ \hat{e}_y = \frac{\partial \hat{q}}{\partial \hat{q}_y} \text{ and } \hat{e}_z = \frac{\partial \hat{q}}{\partial \hat{q}_z}.$$

Here the hats over the e's denote operator status and not that they are normalised. This means, for example, that from the Cartesian product $(\hat{q}_x, \hat{q}_y, \hat{q}_z)$ we can equate $\hat{e}_x = (\hat{1}, \hat{0}, \hat{0})$, $\hat{e}_y = (\hat{0}, \hat{1}, \hat{0})$, and $\hat{e}_z = (\hat{0}, \hat{0}, \hat{1})$, where $\hat{1}$ is the identity and $\hat{0}$ the null operator. The simple form of these vectors and the often-used shorthand of 1 for the identity and 0 for the null operator may make this seem like an unnecessary complication (it may even be so). The Cartesian components of any vector will be $\hat{a}_i = \hat{e}_i \cdot \hat{a} = \hat{a} \cdot \hat{e}_i$ for $i \in \{x, y, z\}$. We can then define the dot product of vector operators as $\hat{a} \cdot \hat{b} = \hat{a}_x \hat{b}_x + \hat{a}_y \hat{b}_y + \hat{a}_z \hat{b}_z$ (where operator ordering is respected). Finally, let us note that $\hat{e}_i \cdot \hat{e}_j = \delta_{ij} \hat{1}$ generalises the notion of orthonormality to vector operators. We will later see that we will want to consider commutators of operators with basis vectors; hence, it makes a degree of sense to do this. For Cartesian coordinates everything commutes, and defining objects such as $\hat{q} = \hat{q}_x \hat{e}_x + \hat{q}_y \hat{e}_y + \hat{q}_z \hat{e}_z$ is reasonable (issues such as the product of two Hermitian operators not necessarily being Hermitian do not arise in this case as $\hat{1}$ and $\hat{0}$ commute with every other operator). It is worth noting that $\langle x, y, z | \hat{e}_x | \psi \rangle = e_x \psi(x, y, z)$, $\langle x, y, z | \hat{e}_y | \psi \rangle = e_y \psi(x, y, z)$, and $\langle x, y, z | \hat{e}_z | \psi \rangle = e_z \psi(x, y, z)$.

Just as in differential geometry, we can also seek to define basis vectors for spherical coordinates in the same way. Classically we have:

$$e_r = \frac{\partial q}{\partial q_r}, \ e_\theta = \frac{\partial q}{\partial q_\theta} \text{ and } e_\phi = \frac{\partial q}{\partial q_\phi}.$$

Exercise 5.3   Verify that this approach does indeed yield the correct basis vectors for spherical coordinates. Note that these basis vectors will *not* be normalised unit vectors. ∎

The operator version is then

$$\hat{e}_r = \frac{\partial \hat{q}}{\partial \hat{q}_r}, \ \hat{e}_\theta = \frac{\partial \hat{q}}{\partial \hat{q}_\theta}, \text{ and } \hat{e}_\phi = \frac{\partial \hat{q}}{\partial \hat{q}_\phi}.$$

Note that we can recover the position representation of each operator so that

$$\langle r, \theta, \phi | \hat{e}_r | \psi \rangle = e_r, \langle r, \theta, \phi | \hat{e}_\theta | \psi \rangle = e_\theta \text{ and } \langle r, \theta, \phi | \hat{e}_\phi | \psi \rangle = e_\phi.$$

Studying the conversion between Cartesian and spherical coordinates, Eq. (5.1), and noting that all of these quantities commute with one another, we see that

$$\hat{\boldsymbol{e}}_r = \sin \hat{q}_\theta \cos \hat{q}_\phi \, \hat{\boldsymbol{e}}_x + \sin \hat{q}_\theta \sin \hat{q}_\phi \, \hat{\boldsymbol{e}}_y + \cos \hat{q}_\theta \, \hat{\boldsymbol{e}}_z, \tag{5.15}$$

$$\hat{\boldsymbol{e}}_\theta = \hat{q}_r \left(\cos \hat{q}_\theta \cos \hat{q}_\phi \, \hat{\boldsymbol{e}}_x + \cos \hat{q}_\theta \sin \hat{q}_\phi \, \hat{\boldsymbol{e}}_y - \sin \hat{q}_\theta \, \hat{\boldsymbol{e}}_z \right), \tag{5.16}$$

$$\hat{\boldsymbol{e}}_\phi = \hat{q}_r \sin \hat{q}_\theta \left(- \sin \hat{q}_\phi \, \hat{\boldsymbol{e}}_x + \cos \hat{q}_\phi \, \hat{\boldsymbol{e}}_y \right). \tag{5.17}$$

From these expressions we can see one complication that is going to arise in spherical coordinates: that the momenta are not going to commute with the basis vectors. This is not a fundamental issue, as something similar happens classically, but it will make doing calculations tricky. Note that we can make use of $\hat{\boldsymbol{e}}_i \cdot \hat{\boldsymbol{e}}_j = \delta_{ij} \hat{\mathbb{1}}$ or $i, j \in \{x, y, z\}$ to do calculations with the radial basis vector operators.

In classical mechanics, we find that we can express the canonical momentum as $p_r = \boldsymbol{e}_r \cdot \boldsymbol{p}$, $p_\theta = \boldsymbol{e}_\theta \cdot \boldsymbol{p}$, and $p_\phi = \boldsymbol{e}_\phi \cdot \boldsymbol{p}$ in direct analogy to the Cartesian case. Importantly, we once more note that to obtain the canonical momenta we make use of the basis vectors defined according to $\partial \hat{\boldsymbol{q}} / \partial q_i$ and not the normalised/unit basis vectors. As $\left(\hat{A} \hat{B} \right)^\dagger = \hat{B}^\dagger \hat{A}^\dagger$, we can see that the product of two *non-commuting* Hermitian operators is not usually Hermitian. Applying the above scheme from classical mechanics directly to the quantum equivalent will not work. Following [49], we can make an Hermitian operator through the Weyl symmetrisation:

$$\hat{p}_r = \frac{1}{2} \left[\hat{\boldsymbol{e}}_r \cdot \hat{\boldsymbol{p}} + (\hat{\boldsymbol{e}}_r \cdot \hat{\boldsymbol{p}})^\dagger \right] = \frac{1}{2} \left[\hat{\boldsymbol{e}}_r \cdot \hat{\boldsymbol{p}} + \hat{\boldsymbol{p}} \cdot \hat{\boldsymbol{e}}_r \right] \tag{5.18}$$

$$\hat{p}_\theta = \frac{1}{2} \left[\hat{\boldsymbol{e}}_\theta \cdot \hat{\boldsymbol{p}} + (\hat{\boldsymbol{e}}_\theta \cdot \hat{\boldsymbol{p}})^\dagger \right] = \frac{1}{2} \left[\hat{\boldsymbol{e}}_\theta \cdot \hat{\boldsymbol{p}} + \hat{\boldsymbol{p}} \cdot \hat{\boldsymbol{e}}_\theta \right] \tag{5.19}$$

$$\hat{p}_\phi = \frac{1}{2} \left[\hat{\boldsymbol{e}}_\phi \cdot \hat{\boldsymbol{p}} + (\hat{\boldsymbol{e}}_\phi \cdot \hat{\boldsymbol{p}})^\dagger \right] = \frac{1}{2} \left[\hat{\boldsymbol{e}}_\phi \cdot \hat{\boldsymbol{p}} + \hat{\boldsymbol{p}} \cdot \hat{\boldsymbol{e}}_\phi \right]. \tag{5.20}$$

We then should check that applying symmetrisation is in agreement with our core quantisation hypothesis and confirm that $[\cdot, \cdot] = i \hbar \{\cdot, \cdot\}$ is satisfied. We can do this in longhand using the above definition of the basis vector operators; or we can take note of the fact that we already know the Cartesian position representation of the three-dimensional momentum operator. In its coordinate independent form, this is $\hat{\boldsymbol{p}} = -i \hbar \nabla$. So we can compute the spherical coordinate representation from the above Weyl symmetrisation scheme and the spherical polar form of ∇. This approach is less elegant than the purely algebraic approach, but much more easy to follow. Taking the radial momentum first, we find that

$$\begin{aligned}
\hat{p}_r \psi(r) &= \frac{1}{2} \left[\hat{\boldsymbol{e}}_r \cdot \hat{\boldsymbol{p}} + \hat{\boldsymbol{p}} \cdot \hat{\boldsymbol{e}}_r \right] \psi(r) \\
&= \frac{1}{2} \left[\boldsymbol{e}_r \cdot \hat{\boldsymbol{p}} \psi(r) + \hat{\boldsymbol{p}} \cdot \boldsymbol{e}_r \psi(r) \right] \\
&= -\frac{i \hbar}{2} \left[\frac{\partial}{\partial r} \psi(r) + \nabla \cdot \boldsymbol{e}_r \psi(r) \right] \\
&= -\frac{i \hbar}{2} \left[2 \frac{\partial}{\partial r} \psi(r) + \psi(r) \nabla \cdot \boldsymbol{e}_r \right] \\
&= -\frac{i \hbar}{2} \left[2 \frac{\partial}{\partial r} \psi(r) + \psi(r) \frac{1}{r^2} \frac{\partial}{\partial r} r^2 \right] \\
&= -\frac{i \hbar}{r} \frac{\partial}{\partial r} r \psi(r).
\end{aligned} \tag{5.21}$$

Next, we find that the inclination momentum is

$$\hat{p}_\theta \psi(\theta) = \frac{1}{2} \left[\hat{e}_\theta \cdot \hat{p} + \hat{p} \cdot \hat{e}_\theta \right] \psi(\theta)$$

$$= \frac{r}{2} \left[e_\theta \cdot \hat{p} \psi(\theta) + \hat{p} \cdot e_\theta \psi(\theta) \right]$$

$$= -\frac{i\,\hbar r}{2} \left[\frac{1}{r} \frac{\partial}{\partial \theta} \psi(\theta) + \nabla \cdot e_\theta \psi(\theta) \right]$$

$$= -\frac{i\,\hbar r}{2} \left[\frac{2}{r} \frac{\partial}{\partial \theta} \psi(\theta) + \psi(\theta) \nabla \cdot e_\theta \right]$$

$$= -\frac{i\,\hbar r}{2} \left[\frac{2}{r} \frac{\partial}{\partial \theta} \psi(\theta) + \psi(\theta) \frac{1}{r \sin \theta} \frac{\partial}{\partial \theta} \sin \theta \right]$$

$$= -i\,\hbar \left[\frac{\partial}{\partial \theta} + \frac{1}{2} \cot \theta \right] \psi(\theta)$$

$$= -i\,\hbar \frac{1}{\sqrt{\sin \theta}} \frac{\partial}{\partial \theta} \sqrt{\sin \theta} \psi(\theta) \tag{5.22}$$

and finally the azimuthal momentum

$$\hat{p}_\phi \psi(\phi) = \frac{1}{2} \left[\hat{e}_\phi \cdot \hat{p} + \hat{p} \cdot \hat{e}_\phi \right] \psi(r)$$

$$= \frac{r \sin \theta}{2} \left[e_\phi \cdot \hat{p} \psi(\phi) + \hat{p} \cdot e_\phi \psi(\phi) \right]$$

$$= -\frac{i\,\hbar r \sin \theta}{2} \left[\frac{1}{r \sin \theta} \frac{\partial}{\partial \phi} \psi(\phi) + \nabla \cdot e_\phi \psi(\phi) \right]$$

$$= -\frac{i\,\hbar r \sin \theta}{2} \left[\frac{2}{r \sin \theta} \frac{\partial}{\partial \phi} \psi(\phi) + \psi(\phi) \nabla \cdot e_\phi \right]$$

$$= -\frac{i\,\hbar r \sin \theta}{2} \left[\frac{2}{r \sin \theta} \frac{\partial}{\partial \phi} \psi(\phi) + \frac{1}{r \sin \theta} \frac{\partial e_\phi \cdot e_\phi}{\partial \phi} \right]$$

$$= -i\,\hbar \frac{\partial}{\partial \phi} \psi(\phi). \tag{5.23}$$

To summarise:

$$\hat{p}_r = -i\,\hbar \frac{1}{r} \frac{\partial}{\partial r} r \tag{5.24}$$

$$\hat{p}_\theta = -i\,\hbar \left[\frac{\partial}{\partial \theta} + \frac{1}{2} \cot \theta \right] = -i\,\hbar \frac{1}{\sqrt{\sin \theta}} \frac{\partial}{\partial \theta} \sqrt{\sin \theta} \tag{5.25}$$

$$\hat{p}_\phi = -i\,\hbar \frac{\partial}{\partial \phi} \tag{5.26}$$

in exact agreement with the more canonical approach of DeWitt. Now that we have derived the spherical components of momentum, let us expand our discussion to check their validity.

Unlike their Cartesian counterparts, the momenta in spherical polar coordinates do not in general commute with basis vectors.

Exercise 5.4   To make this point clear, show that

$$\left[\hat{p}_r, \hat{e}_\theta \right] = -2i\,\hbar \hat{e}_\theta.$$

Now convince yourself that the *only* pair of basis vectors and momentum operators that commute are $\hat{e}_r$ and $\hat{p}_r$. ∎

The non-trivial relationship between the momentum operators and the basis vectors has substantial implications. The product of any of the basis vectors $\{\hat{\boldsymbol{e}}_r, \hat{\boldsymbol{e}}_\theta, \hat{\boldsymbol{e}}_\phi\}$ or the unit vectors $\boldsymbol{e}_i$ with $\hat{p}_i$ cannot be Hermitian if their commutator is non-zero (which is zero for only the case of the radial basis vector with the radial momentum). The Weyl symmetrisation process will therefore need to be employed wherever such issues arise (and each time we need to check for consistency with the Cartesian equivalent to ensure that it is a valid method to apply). In doing symmetrisation calculations, the below result becomes rather useful:

$$\frac{1}{2}(\hat{A}\hat{B} + \hat{B}\hat{A}) = \frac{1}{2}(\hat{A}\hat{B} + \hat{B}\hat{A} + \hat{A}\hat{B} - \hat{A}\hat{B}) \tag{5.27}$$

$$= \hat{A}\hat{B} + \frac{1}{2}[\hat{B}, \hat{A}]. \tag{5.28}$$

To confirm that this scheme actually works, let us apply it to check that we get the correct form for $\hat{\boldsymbol{p}}$. Using the above commutator trick, we find:

$$\hat{\boldsymbol{p}} = \sum_{i \in \{r,\theta,\phi\}} \frac{1}{2}(\hat{p}_i \hat{\boldsymbol{e}}_i + \hat{\boldsymbol{e}}_i \hat{p}_i) \tag{5.29}$$

$$\sim \boldsymbol{e}_r \hat{p}_r + \frac{1}{r}\left(\boldsymbol{e}_\theta \hat{p}_\theta + \frac{i\hbar}{2}\boldsymbol{e}_r + \frac{1}{\sin\theta}\boldsymbol{e}_\phi \hat{p}_\phi + \frac{i\hbar}{2}[\boldsymbol{e}_\theta \cot\theta + \boldsymbol{e}_r]\right) \tag{5.30}$$

$$= \boldsymbol{e}_r\left[\hat{p}_r + \frac{i\hbar}{r}\right] + \frac{\boldsymbol{e}_\theta}{r^2}\left[\hat{p}_\theta + \frac{i\hbar}{2}\cot\theta\right] + \frac{\boldsymbol{e}_\phi}{r^2\sin^2\theta}\hat{p}_\phi. \tag{5.31}$$

By substituting the explicit expressions for each momentum and simplifying, we find:

$$\hat{\boldsymbol{p}} = -i\hbar\left(\boldsymbol{e}_r\left[\frac{1}{r}\frac{\partial}{\partial r}r - \frac{1}{r}\right] + \hbar\frac{\boldsymbol{e}_\theta}{r^2}\frac{\partial}{\partial\theta} + \frac{\boldsymbol{e}_\phi}{r^2\sin^2\theta}\frac{\partial}{\partial\phi}\right)$$

$$= -i\hbar\left(\boldsymbol{e}_r\frac{\partial}{\partial r} + \hbar\frac{\boldsymbol{e}_\theta}{r^2}\frac{\partial}{\partial\theta} + \frac{\boldsymbol{e}_\phi}{r^2\sin^2\theta}\frac{\partial}{\partial\phi}\right)$$

$$= -i\hbar\nabla \tag{5.32}$$

as required. Note that the factors of r and $r\sin\theta$ of the θ and ϕ component may look different to what you expect due to the fact that the basis vectors are not normalised.

The requirement to ensure Weyl ordering through repeated symmetrisation in the above argument would benefit from some justification. This can be found in phase space methods, which provide an alternative and more straightforward way to describe quantum systems in generalised coordinates. This is because the quantum mechanics are moved into the algebra of phase space by replacing function multiplication with the star product. The Moyal product ensures that Weyl ordering is respected and that a consistent theory is ensured. Both Poisson and Moyal brackets work just as well for any canonical pairs of position and momentum, and the above arguments just reflect the operator counterpart of phase space quantum physics. This is a fascinating, but mathematically detailed, topic and we refer the interested reader to [6].

The discussion presented in this section is important since it means that quantum mechanics could be a general theory that is not coordinate system dependent. So while from a 'foundations' perspective the above discussion is reassuring, in practising the discipline we need a less awkward way to solve problems and make predictions. Our discussion has been far from comprehensive and we have not touched on other coordinate systems. DeWitt's work implies that we can do quantum mechanics in any coordinate system that is a point transformation of Cartesian coordinates (such as cylindrical polar or skew systems).[4] Even if, e.g. spherical coordinates are a valid choice for representing quantum systems, it is clear that, even for simple systems, the resulting analysis will contain some quite complex differential equations to solve.

There is an alternative CSCO for systems with spherical symmetries, where the algebraic methods make the analysis much easier. We will discuss this approach to angular momentum, and its generalisation, in the next section.

5.3 The Theory of Orbital and General Angular Momentum

5.3.1 From Classical to Quantum Angular Momentum

The idea of angular momentum starts by asking what the momentum for a rotation around a given axis is. Let us pick the z axis and call ϕ the rotation about that axis (as is common to both spherical and cylindrical coordinates). This angular momentum is defined as $L_z \stackrel{\text{def}}{=} \partial \mathscr{L} / \partial \dot{\phi}$ where the potential is assumed to not be dependent on $\dot{\phi}$ and we find

$$L_z = p_\phi = q_x p_y - q_y p_x. \tag{5.33}$$

By symmetry we see that we can also define analogues around the other Cartesian axes $L_x = q_y p_z - q_z p_y$, $L_y = q_z p_x - q_x p_z$. We can combine these quantities into a single angular momentum vector,

$$\boldsymbol{L} = \boldsymbol{q} \times \boldsymbol{p} = \begin{vmatrix} \mathbf{i} & \mathbf{j} & \mathbf{k} \\ q_x & q_y & q_z \\ p_x & p_y & p_z \end{vmatrix}.$$

Exercise 5.5 Show $\{L_x, L_y\} = L_z$ and that this also holds for cyclic permutations. Note that you should not need to do any calculus if you use the Poisson brackets of the coordinates and momenta. ∎

Exercise 5.6 Noting that $L^2 = L_x^2 + L_y^2 + L_z^2$, show that $\{L^2, L_x\} = \{L^2, L_y\} = \{L^2, L_z\} = 0$. ∎

In spherical coordinates this is

$$\boldsymbol{L} = p_\theta \tilde{\boldsymbol{e}}_\phi - \frac{1}{\sin \theta} p_\phi \tilde{\boldsymbol{e}}_\theta = \frac{p_\theta}{r} \boldsymbol{e}_\phi - \frac{p_\phi}{r \sin^2 \theta} \boldsymbol{e}_\theta.$$

We now use this to write the kinetic energy as

$$\mathscr{T} = \frac{1}{2m} \left(p_r^2 + \frac{1}{r^2} p_\theta^2 + \frac{1}{r^2 \sin^2 \theta} p_\phi^2 \right) = \frac{1}{2m} \left(p_r^2 + \frac{L^2}{r^2} \right).$$

---

4 There are other systems such as Euler angles that may be used as the basis for canonical coordinates and deriving conjugate momenta in classical systems such as the spinning top. It is not clear to us that the approaches described above could be successfully applied to Euler angles (phase space would be the natural place to start such an exploration but we are not aware of any work in this area). At least for the moment, the true independence of quantum mechanics of coordinate system remains an open question to us.

This form makes it particularly clear that $\{\mathcal{T}, L^2\} = 0$. We can then deduce that that for any Hamiltonian which is rationally symmetric (such as for a particle in a central potential) conservation of angular momentum will hold.

The conventional textbook approach to considering angular momentum in quantum mechanics is to construct the operators from the Cartesian perspective

$$\hat{L} = \hat{q} \times \hat{p} = \begin{vmatrix} \mathbf{i} & \mathbf{j} & \mathbf{k} \\ \hat{q}_x & \hat{q}_y & \hat{q}_z \\ \hat{p}_x & \hat{p}_y & \hat{p}_z \end{vmatrix},$$

which has components $\hat{L}_x = \hat{q}_y \hat{p}_z - \hat{q}_z \hat{p}_y$, $\hat{L}_y = \hat{q}_z \hat{p}_x - \hat{q}_x \hat{p}_z$, and $\hat{L}_z = \hat{q}_x \hat{p}_y - \hat{q}_y \hat{p}_x$, where we have the usual canonical commutation relations $[\hat{q}_i, \hat{p}_j] = i\hbar \delta_{ij}$ where $i, j \in \{x, y, z\}$.

Exercise 5.7 Noting that $\hat{L}^2 = \hat{L}_x^2 + \hat{L}_y^2 + \hat{L}_z^2$, show that $[\hat{L}^2, \hat{L}_x] = [\hat{L}^2, \hat{L}_y] = [\hat{L}^2, \hat{L}_z] = 0$. ∎

Exercise 5.8 In classical spherical coordinates

$$\hat{L} = \frac{\hat{p}_\theta}{r} \mathbf{e}_\phi - \frac{\hat{p}_\phi}{r \sin^2 \theta} \mathbf{e}_\theta.$$

Using Weyl symmetrisation, we note that we expect, as $\hat{p}_\theta$ and $\mathbf{e}_\phi$ commute, the quantum equivalent to be

$$\hat{L} = \frac{\hat{p}_\theta}{r} \mathbf{e}_\phi - \frac{1}{r \sin^2 \theta} \frac{1}{2} \{\hat{p}_\phi, \mathbf{e}_\theta\}_+. \tag{5.34}$$

Here we use the subscript + to distinguish an anticommutator from a Poisson bracket. Show that this is equivalent to $-i\hbar \mathbf{r} \times \nabla$ as expected. Note that care needs to be taken to use the correct length on the basis vectors. ∎

Exercise 5.9 Taking care with respect to the momentum not commuting with basis vectors, use Eq. (5.34) to show that

$$\hat{L}^2 = \hat{L} \cdot \hat{L} = -\hbar^2 \left[\frac{\partial^2}{\partial \theta^2} + \cot \theta \frac{\partial}{\partial \theta} + \frac{1}{\sin^2 \theta} \frac{\partial^2}{\partial \varphi^2} \right]. \tag{5.35}$$

Again, the correct length of the basis vectors needs to be accounted for. ∎

As an alternative to the Cartesian or spherical polar basis here is a third option that turns out to be even more useful.

As $\hat{L}^2$ commutes with each Cartesian component of angular momentum, we could use any pair of these operators to specify a basis (forming part of a CSCO). That both of these operators would also commute with Hamiltonians for spherically symmetric problems adds to the appeal of this approach. As a further observation, since we can easily show $\hat{L}_z = \hat{p}_\phi$ for spherical and cylindrical polar coordinates, we note that $\{L^2, L_z\}$ is *a set of operators that are common to multiple coordinate systems* (as such our consequent findings should be of wide applicability).

We will also make use of our findings from Exercise 5.13 on page 17 and find ladder operators for $\hat{L}_z$. We will use the standard notation $\hat{L}_\pm$ for these so that

$$[\hat{L}_z, \hat{L}_\pm] = \pm \hbar \hat{L}_\pm.$$

We therefore know that, if $\hat{L}_z$ has eigenvalues m with corresponding eigenvectors $|m\rangle$ (i.e. $\hat{L}_z |m\rangle = \hbar m |m\rangle$), then $\hat{L}_\pm |n\rangle$ is also an eigenstate of $\hat{n}$ with eigenvalue $\hbar(m \pm 1)$.

Exercise 5.10   Using only $[\hat{L}_x, \hat{L}_y] = i\,\hbar\hat{L}_z$, show that

$$\hat{L}_\pm = \hat{L}_x \pm i\hat{L}_y \tag{5.36}$$

satisfy the requirement to be ladder operators: $[\hat{L}_z, \hat{L}_\pm] = \pm\hbar\hat{L}_\pm$.   ∎

Exercise 5.11   Show

$$\hat{L}_\pm = \hbar \exp(\pm i\,\phi) \left[\pm\frac{\partial}{\partial\theta} + i\cot\theta\frac{\partial}{\partial\phi}\right]. \tag{5.37}$$

∎

5.3.2   General Properties of Angular Momentum

It turns out that the algebraic properties we have outlined above can, on their own, tell us a great deal. For this reason, we now present some of the general theory of angular momentum. To indicate that our discussion is generic, we will use $\hat{J}_i$ to represent any angular momentum (orbital, intrinsic, and even composite). We impose that there must be three components of angular momentum that satisfy $[\hat{J}_x, \hat{J}_y] = i\,\hbar\hat{J}_z$ and cyclic permutations, and each must be a Hermitian operator. Defining $\hat{J}^2 = \hat{J}_x^2 + \hat{J}_y^2 + \hat{J}_z^2$, we find that $[\hat{J}^2, \hat{J}_i] = 0$. We choose the set $\{\hat{J}^2, \hat{J}_z\}$ to form part of a CSCO which will help us define a basis and thus a representation (since $\hat{J}^2$ commutes with $\hat{J}_z$ they share a common set of eigenvectors). We will see that this set of observables is not yet complete, and so more observables must be introduced to completely define the state. We choose the pair $\hat{J}^2$ and $\hat{J}_z$ as these are the standard choices and because in the specific case of orbital angular momentum we equate $\hat{J}_z$ to $\hat{L}_z = \hat{p}_\phi$.

We use the above framework to show that the eigenvalues of $\hat{J}^2$ can only be integer or half-integer, denoted j. For one given eigenvalue of $\hat{J}^2$ we find that (except for the zero eigenvalue) that there is a set of common eigenvectors of $\hat{J}_z$ that range between $-j$ and j in whole number steps. As the conversation is quite general at this stage, we will take advantage of this to discuss the addition of angular momentum before returning to specific examples.

Our first observation is that the eigenvalues of $\hat{J}^2$ must be positive. This can be seen by observing that each component is Hermitian and if, for any state $|\psi\rangle$, we define $\hat{J}_i |\psi\rangle = |\phi_i\rangle$ and so

$$\left\langle \psi \left| \hat{J}^2 \right| \psi \right\rangle = \sum_{i \in \{x,y,z\}} \left\langle \psi \left| \hat{J}_i^2 \right| \psi \right\rangle = \sum_{i \in \{x,y,z\}} \left(\langle \psi | \hat{J}_i \rangle \left(\hat{J}_i | \psi \rangle \right) \right)$$

$$= \sum_{i \in \{x,y,z\}} \langle \phi_i | \phi_i \rangle \geq 0. \tag{5.38}$$

If $|\psi\rangle$ is an eigenstate of $\hat{J}^2$ then $\hat{J}^2 |\psi\rangle = \lambda |\psi\rangle$ where λ is a real number as $\hat{J}^2$ is, by definition, Hermitian. So $\left\langle \psi \left| \hat{J}^2 \right| \psi \right\rangle = \lambda \langle \psi | \psi \rangle$. As $\left\langle \psi \left| \hat{J}^2 \right| \psi \right\rangle$ and $\langle \psi | \psi \rangle$ are both greater than or equal to zero, $\lambda \geq 0$. Noting that angular momentum is quantised in units of $\hbar$, the eigenvalue of $\hat{J}^2$ must be proportional to $\hbar^2$, and with the benefit of hindsight we will write the eigenvalue equation for $\hat{J}^2$ as

$$\hat{J}^2 \left| j^{(i)} \right\rangle = j(j+1)\hbar^2 \left| j^{(i)} \right\rangle \tag{5.39}$$

where $j \geq 0$ and the index (i) has been used to denote any degeneracy in the system.

Just as we did in Section 3.4.2 we will write the eigenvalue equation for $\hat{J}_z$ as

$$\hat{J}_z \left| m^{(i)} \right\rangle = m\hbar \left| m^{(i)} \right\rangle. \tag{5.40}$$

As they commute, we can find eigenvectors common to both as part of a basis:

$$\hat{J}^2 \left| j, m, \ldots \right\rangle = j(j+1)\hbar^2 \left| j, m, \ldots \right\rangle$$
$$\hat{J}_z \left| j, m, \ldots \right\rangle = m\hbar \left| j, m, \ldots \right\rangle. \tag{5.41}$$

Other observables, depending on the problem at hand, will usually be needed to completely specify a basis. In some cases, such as the spin-less particle in a central potential, this might be the eigenstates of the Hamiltonian itself, and in this example the specific operators $\hat{H}, \hat{L}^2, \hat{L}_z$ may form a CSCO.

An important and immediate application is to consider the action of the kinetic energy in this basis as

$$\hat{T} \left| l, m, \ldots \right\rangle = \frac{1}{2m} \left(\hat{p}_r^2 + \frac{\hat{L}^2}{r^2} \right) \left| l, m, \ldots \right\rangle$$
$$= \frac{1}{2m} \left(\hat{p}_r^2 + \frac{l(l+1)\hbar^2}{r^2} \right) \left| l, m, \ldots \right\rangle.$$

Here we have used l instead of j as it is the standard notation for orbital angular momentum. For classes of problems with a high degree of spherical symmetry this basis greatly simplifies analysis.

As the commutator algebra we used to show that $\hat{L}_\pm$ are ladder operators also holds for generalised angular momentum, we can immediately define

$$\hat{J}_\pm = \hat{J}_x \pm i\hat{J}_y \tag{5.42}$$

and note that the following commutation relations all hold:

$$\left[\hat{J}_z, \hat{J}_\pm \right] = \pm\hbar\hat{J}_\pm \tag{5.43}$$

$$\left[\hat{J}_+, \hat{J}_- \right] = 2\hbar\hat{J}_z \tag{5.44}$$

$$\left[\hat{J}^2, \hat{J}_\pm \right] = 0. \tag{5.45}$$

Exercise 5.12 Check that the above commutation relations are correct. ∎

In the following analysis, we will find expressions for $\hat{J}_+\hat{J}_-$ and $\hat{J}_-\hat{J}_+$ of utility. It will be important that these quantities can be written entirely in terms of $\hat{J}^2$ and $\hat{J}_z$ so their actions on the eigenstates $|j, m, ...\rangle$ can be determined. First, we note that

$$\hat{J}_+\hat{J}_- = (\hat{J}_x + i\hat{J}_y)(\hat{J}_x - i\hat{J}_y)$$
$$= \hat{J}_x^2 + \hat{J}_y^2 - i\hat{J}_x\hat{J}_y + i\hat{J}_y\hat{J}_x$$
$$= \hat{J}_x^2 + \hat{J}_y^2 + i\left[\hat{J}_y, \hat{J}_x\right]$$
$$= \hat{J}_x^2 + \hat{J}_y^2 + \hbar\hat{J}_z. \tag{5.46}$$

Following similar reasoning, we find $\hat{J}_-\hat{J}_+ = \hat{J}_x^2 + \hat{J}_y^2 - \hbar\hat{J}_z$. Noting that $\hat{J}^2 = \hat{J}_x^2 + \hat{J}_y^2 + \hat{J}_z^2$, we can write these products in terms of $\hat{J}^2$ and $\hat{J}_z$ only:

$$\hat{J}_+\hat{J}_- = \hat{J}^2 - \hat{J}_z^2 + \hbar\hat{J}_z$$
$$\hat{J}_-\hat{J}_+ = \hat{J}^2 - \hat{J}_z^2 - \hbar\hat{J}_z. \tag{5.47}$$

Adding these products yields an equation for $\hat{J}^2$ in terms of $\hat{J}_+, \hat{J}_-$ and $\hat{J}_z$ only:

$$\hat{J}_+\hat{J}_- + \hat{J}_-\hat{J}_+ = 2\hat{J}^2 - 2\hat{J}_z^2$$
$$\Rightarrow \hat{J}^2 = \frac{1}{2}(\hat{J}_+\hat{J}_- + \hat{J}_-\hat{J}_+) + \hat{J}_z^2. \tag{5.48}$$

These findings will be of great use as we now progress to explore what we may determine about the eigenvalues of $\hat{J}^2$ and $\hat{J}_z$.

5.3.3 Eigenvalues and Eigenvectors of Angular Momentum

Our goal is now to find the spectrum of $\hat{J}^2$ and $\hat{J}_z$, that is to say the permitted values that the eigenvalues j and m may take in the equations

$$\hat{J}^2 |j, m, ...\rangle = j(j + 1)\hbar^2 |j, m, ...\rangle$$
$$\hat{J}_z |j, m, ...\rangle = m\hbar |j, m, ...\rangle. \tag{5.49}$$

Throughout this section we will make the following definition:

$$|\psi\rangle = \hat{J}_+ |j, m, ...\rangle$$
$$|\phi\rangle = \hat{J}_- |j, m, ...\rangle \tag{5.50}$$

and we will make use of the facts that $\langle\psi|\psi\rangle \geq 0$ and $\langle\phi|\phi\rangle \geq 0$. Noting that $\hat{J}_\pm^\dagger = \hat{J}_\mp$, we have

$$\langle\psi|\psi\rangle = \langle j, m, ...|\hat{J}_-\hat{J}_+ |j, m, ...\rangle \geq 0$$
$$\langle\phi|\phi\rangle = \langle j, m, ...|\hat{J}_+\hat{J}_- |j, m, ...\rangle \geq 0. \tag{5.51}$$

We now use our expressions for $\hat{J}_+\hat{J}_-$ and $\hat{J}_-\hat{J}_+$ in Eq. (5.47) and substitute them into Eq. (5.51) to yield

$$\langle\psi|\psi\rangle = \langle j, m, ...|\hat{J}^2 - \hat{J}_z^2 - \hbar\hat{J}_z |j, m, ...\rangle \geq 0$$
$$\langle\phi|\phi\rangle = \langle j, m, ...|\hat{J}^2 - \hat{J}_z^2 + \hbar\hat{J}_z |j, m, ...\rangle \geq 0. \tag{5.52}$$

Now we can use the eigenvalue equations (Eq. (5.49)) to simplify this equation to

$$\langle \psi | \psi \rangle = \langle j, m, \ldots | \hbar^2 \left[j(j+1) - m(m+1) \right] | j, m, \ldots \rangle \geq 0$$

$$\langle \phi | \phi \rangle = \langle j, m, \ldots | \hbar^2 \left[j(j+1) - m(m-1) \right] | j, m, \ldots \rangle \geq 0. \tag{5.53}$$

Finally we note that we can assume that the basis is orthonormal, so $\langle j, m, \ldots | j, m, \ldots \rangle = 1$ which means we can conclude that:

$$\langle \psi | \psi \rangle = (j - m)(j + 1 + m) \geq 0$$

$$\langle \phi | \phi \rangle = (j + 1 - m)(j + m) \geq 0. \tag{5.54}$$

These inequalities will be satisfied if

$$-(j+1) \leq m \leq j$$

$$-j \leq m \leq j+1, \tag{5.55}$$

---

which means the magnetisation quantum number m must satisfy the inequality

$$-j \leq m \leq j. \tag{5.56}$$

Let us work from the bottom up and see what we can say regarding the case that m achieves its lower bound $-j$. We once again consider the inner product and substitute $m = -j$

$$\langle \phi | \phi \rangle = \hbar^2 \left[j(j+1) - m(m-1) \right] \tag{5.57}$$

$$= \hbar^2 \left[j(j+1) - (-j)((-j) - 1) \right] = 0. \tag{5.58}$$

The only vector that has the length zero is the null vector.

Therefore, we can conclude that when $m = -j$,

$$\hat{J}_- | j, -j, \ldots \rangle = | \theta \rangle \tag{5.59}$$

where $| \theta \rangle$ is the zero vector.

As $\hat{J}_-$ was constructed (i) so that $\hat{J}_-$ commutes with $\hat{J}^2$ and (ii) as a lowering ladder operator, we also know that for $m > -j$, $\hat{J}_- | j, -j, \ldots \rangle$ must also (i) be an eigenvector of $\hat{J}^2$ with the same eigenvalue (that is $\hat{J}^2 | \phi \rangle = j(j+1)\hbar^2 | \phi \rangle$); and (ii) be an eigenvector of $\hat{J}_z$ with eigenvalue $(m-1)\hbar$ (that is $\hat{J}_z | \phi \rangle = (m-1)\hbar | \phi \rangle$).

We can duplicate the above argument, but take a top-down perspective for the case that m achieves its upper bound j.

Exercise 5.13 Repeat the argument above but start from $m = j$ and use $\hat{J}_+$ instead of $\hat{J}_-$ to reach the following conclusion (in the box below): ∎

Therefore, we can conclude that when $m = j$,

$$\hat{J}_+ | j, -j, \ldots \rangle = | \theta \rangle, \tag{5.60}$$

where $| \theta \rangle$ is the zero vector.

As $\hat{J}_+$ was constructed (i) so that $\hat{J}_+$ commutes with $\hat{J}^2$ and (ii) as a raising ladder operator, we also know that for $m < j$ that $\hat{J}_+ |k, j, -j\rangle$ must also (i) be an eigenvector of $\hat{J}^2$ with the same eigenvalue (that is $\hat{J}^2 |\psi\rangle = j(j+1)\hbar^2 |\psi\rangle$); and (ii) be an eigenvector of $\hat{J}_z$ with eigenvalue $(m-1)\hbar$ (that is $\hat{J}_z |\psi\rangle = (m+1)\hbar |\psi\rangle$).

The above findings, when combined, can be used to show (see [12] for an expanded discussion):

(a) The only allowed values of j are of the form $\frac{1}{2}N$ where N is any integer greater than or equal to zero.

(b) For any given j the values of m are $-j, -j+1 \ldots j-1, j$. Hence m is an integral if j is an integral and m is a half-integral if j is a half-integral. For example, if $j = \frac{3}{2}$ then m can only, and will, take the values $m = -\frac{3}{2}, -\frac{1}{2}, \frac{1}{2}, \frac{3}{2}$. Likewise, if $j = 1$, then $m = -1, 0, 1$.

Let us now proceed to give what is termed the standard representation of angular momentum systems. We start by recalling from Eq. (5.53) that

$$||\hat{J}_\pm |j, m, \ldots\rangle||^2 = \hbar^2 \left[j(j+1) - m(m \pm 1) \right]$$

as well as

$$\hat{J}_z \hat{J}_\pm |j, m, \ldots\rangle = (m \pm 1)\hbar |j, m, \ldots\rangle$$

to conclude that the action of $\hat{J}_\pm$ on the state $|j, m, \ldots\rangle$ either promotes or reduces the state to $|j, m, \ldots \pm 1\rangle$ with a scale factor of

$$||\hat{J}_\pm |j, m, \ldots\rangle|| = \hbar \sqrt{j(j+1) - m(m \pm 1)}$$

and so, if we choose to set the scale factor for the new eigenvectors of $\hat{J}_3$ as

$$\hat{J}_\pm |j, m, \ldots\rangle = \hbar \sqrt{j(j+1) - m(m \pm 1)} |j, m, \ldots \pm 1\rangle, \tag{5.61}$$

we guarantee that the set of vectors generated by these ladder operators from any reference state will be orthogonal. In this basis, the matrix elements of $\hat{J}_z$ are given by:

$$\langle k', j', m' | \hat{J}_z | j, m, \ldots \rangle = m\hbar \delta_{k'k} \delta_{j'j} \delta_{m'm}. \tag{5.62}$$

The matrix elements for $\hat{J}_+$ and $\hat{J}_-$ are given by:

$$\langle k', j', m' | \hat{J}_\pm | j, m, \ldots \rangle = \hbar \sqrt{j(j+1) - m(m \pm 1)} \delta_{k'k} \delta_{j'j} \delta_{m'm \pm 1}. \tag{5.63}$$

5.3.4 Worked Examples of Matrix Construction

We will now make use of the procedures above to show the general expression for some matrix acting on the subspace of fixed j (we will drop the use of '...' as the contribution to the matrix elements from those labels will be Kronecker deltas). The matrix elements of some operator $\hat{B}$ acting in this subspace are given by $\langle j, m | \hat{B} | j', m' \rangle$, which forms the matrix:

$$
\begin{pmatrix}
\langle j,j | \hat{B} | j,j \rangle & \langle j,j | \hat{B} | j,j-1 \rangle & \cdots & \langle j,j | \hat{B} | j,-j \rangle \\
\langle j,j-1 | \hat{B} | j,j \rangle & \langle j,j-1 | \hat{B} | j,j-1 \rangle & \cdots & \langle j,j-1 | \hat{B} | j,-j \rangle \\
\vdots & \vdots & \ddots & \vdots \\
\langle j,-j | \hat{B} | j,j \rangle & \langle j,-j | \hat{B} | j,j-1 \rangle & \cdots & \langle j,-j | \hat{B} | j,-j \rangle
\end{pmatrix}
\tag{5.64}
$$

As a first example, consider the subspace for $j = 1$ where the only permitted values of m are given by $m = 1, 0, -1$. The matrix form of $\hat{J}_z$ is

$$
(\hat{J}_z)^{(j=1)} =
\begin{pmatrix}
\langle 1,1 | \hat{J}_z | 1,1 \rangle & \langle 1,1 | \hat{J}_z | 1,0 \rangle & \langle 1,1 | \hat{J}_z | 1,-1 \rangle \\
\langle 1,0 | \hat{J}_z | 1,1 \rangle & \langle 1,0 | \hat{J}_z | 1,0 \rangle & \langle 1,0 | \hat{J}_z | 1,-1 \rangle \\
\langle 1,-1 | \hat{J}_z | 1,1 \rangle & \langle 1,-1 | \hat{J}_z | 1,0 \rangle & \langle 1,-1 | \hat{J}_z | 1,-1 \rangle
\end{pmatrix}
\tag{5.65}
$$

$$
= \hbar
\begin{pmatrix}
1 & 0 & 0 \\
0 & 0 & 0 \\
0 & 0 & -1
\end{pmatrix}.
$$

Using the standard representations for $\hat{J}_+$, we find

$$
(\hat{J}_+)^{(j=1)} =
\begin{pmatrix}
0 & \hbar\sqrt{1(1+1) - 0(0+0)} & 0 \\
0 & 0 & \hbar\sqrt{1(1+1) + 1(-1+1)} \\
0 & 0 & 0
\end{pmatrix}
\tag{5.66}
$$

$$
= \hbar
\begin{pmatrix}
0 & \sqrt{2} & 0 \\
0 & 0 & \sqrt{2} \\
0 & 0 & 0
\end{pmatrix}
$$

and similarly

$$
(\hat{J}_-)^{(j=1)} = \hbar
\begin{pmatrix}
0 & 0 & 0 \\
\sqrt{2} & 0 & 0 \\
0 & \sqrt{2} & 0
\end{pmatrix}.
\tag{5.67}
$$

Recalling that

$$
\hat{J}_+ = \hat{J}_x + i\hat{J}_y
$$

$$
\hat{J}_- = \hat{J}_x - i\hat{J}_y
$$

we can construct the matrices $(\hat{J}_x^{(j=1)})$ and $(\hat{J}_y^{(j=1)})$:

$$
(\hat{J}_x)^{(j=1)} = \frac{\hbar}{\sqrt{2}}
\begin{pmatrix}
0 & 1 & 0 \\
1 & 0 & 1 \\
0 & 1 & 0
\end{pmatrix}
\text{ and } (\hat{J}_y)^{(j=1)} = \frac{\hbar}{\sqrt{2}}
\begin{pmatrix}
0 & -i & 0 \\
i & 0 & -i \\
0 & i & 0
\end{pmatrix}
\tag{5.68}
$$

Exercise 5.14 Use the same processes as above to find the matrices for $\hat{J}_x^{(j=1/2)}, \hat{J}_y^{(j=1/2)}, \hat{J}_z^{(j=1/2)}$. Confirm your findings are in agreement with $\hat{S}_x$, $\hat{S}_y$, and $\hat{S}_z$ in our discussion on spin in Section 3.4.2 on page 85. ∎

5.3.5 Orbital Angular Momentum Basis

By direct application of the general theory, we know that $\left\{ \hat{L}^2, \hat{L}_z, \hat{A} \right\}$ can form a basis for three-dimensional space. We will not yet define the $\hat{A}$ observable, and only at this stage require that it lifts the degeneracy of the basis. We will label its eigenvalues a. We also know that we can write

$$\hat{L}^2 |j, m, a\rangle = l(l+1)\hbar^2 |j, m, a\rangle$$

$$\hat{L}_z |j, m, a\rangle = m\hbar |j, m, a\rangle,$$

(5.69)

where we have again used l instead of j for indexing the eigenvalues of $\hat{L}^2$ as this is standard throughout the literature. We have will next use the fact that $\{\hat{q}_r, \hat{q}_\theta, \hat{q}_\phi\}$ form a CSCO to represent the basis vectors explicitly.

We know that $\hat{L}_z = \hat{p}_\phi$, so we can immediately write the last equation in its coordinate representation in the simple form

$$-i\hbar \frac{\partial}{\partial \phi} \langle r, \theta, \phi | j, m, a\rangle = m\hbar \langle r, \theta, \phi | j, m, a\rangle.$$

(5.70)

Similarly, we can use the solution to Exercise 5.9, Eq. (5.35), to write

$$-\hbar^2 \left[\frac{\partial^2}{\partial \theta^2} + \cot\theta \frac{\partial}{\partial \theta} + \frac{1}{\sin^2\theta} \frac{\partial^2}{\partial \varphi^2} \right] \langle r, \theta, \phi | j, m, a\rangle = l(l+1)\hbar^2 \langle r, \theta, \phi | j, m, a\rangle.$$

(5.71)

Clearly neither of these differential equations have any radial dependence, so we can write the position representation of these basis vectors, using separation of variables, as

$$\langle r, \theta, \phi | j, m, a\rangle = R^a(r)Y_l^m(\theta, \phi),$$

(5.72)

where you may notice that the $Y_l^m(\theta, \phi)$ use the standard notation of spherical harmonics (the superscripts a and m are indices, not powers). There is a good reason for this, since this is exactly what they are. The function $R(r)$ can be removed from the analysis for the moment. Looking at the first equation, again we have

$$\frac{\partial}{\partial \phi} Y_l^m(\theta, \phi) = i\, m Y_l^m(\theta, \phi).$$

(5.73)

We can again use separation of variables to note that this has solutions of the form

$$Y_l^m(\theta, \phi) = N_{lm} P_l^m(\theta) \exp(i\, m\phi),$$

(5.74)

where N_{lm} is a normalisation constant and P_l^m turns out to be related to the associated Legendre polynomial. We will not cover the details of the full derivation, as that can be found in existing texts such as [12]. Instead, we will list the salient features of the argument and summarise its findings. It is worth noting that the combination of algebraic (representation-free commutation relationships that lead to the general theory of angular momentum) and analytic (calculus in a specific representation) methods is very powerful.

The first observation is that $Y_l^m(\theta, \phi) = Y_l^m(\theta, \phi + 2\pi)$ and, as the only dependence on ϕ is $\exp(i\, m\phi)$, this can only happen if m is an integer. We also know that the values of m are $-l, -l+1 \ldots l-1, l$. This means that l must also only take on integer values.

From Exercise 5.37, Eq. (5.37), we see that applying $\hat{L}_+$ in its coordinate form to $Y_l^l(\theta, \phi)$ yields the null state

$$\hat{L}_+ Y_l^l(\theta, \phi) = \hbar \exp(i\,\phi) \left[\frac{\partial}{\partial\theta} + i\cot\frac{\partial}{\partial\phi} \right] Y_l^l(\theta, \phi) = 0.$$

Solving that equation yields

$$Y_l^l(\theta, \phi) = N_{ll} \sin^l\theta \exp(i\,l\phi).$$

Now repeated application of $\hat{L}_- = \hbar \exp(-i\,\phi) \left[-\frac{\partial}{\partial\theta} + i\cot\frac{\partial}{\partial\phi} \right]$ (Exercise 5.37, Eq. (5.37)) provides all the values of $Y_l^m(\theta, \phi)$ (with normalisation calculated for each basis).

Exercise 5.15 Apply $\hat{L}_\pm$ to $R^a(r)Y_l^m(\theta, \phi)$ and show that $R^a(r)$ cannot depend on m. ∎

From the above exercise, we can conclude that the basis vectors of well-behaved functions in three-dimensional space, where $\left\{ \hat{L}^2, \hat{L}_z, \hat{A} \right\}$ is a CSCO, can be written as

$$\langle r, \theta, \phi | j, m, a \rangle = R_l^a(r)Y_l^m(\theta, \phi), \tag{5.75}$$

where the specific form of $R_l^a(r)$ depends on the choice of $\hat{A}$. In Figure 5.1 we show one example basis state but encourage the reader to explore the many excellent online resources for this topic that can be found using a quick internet search.

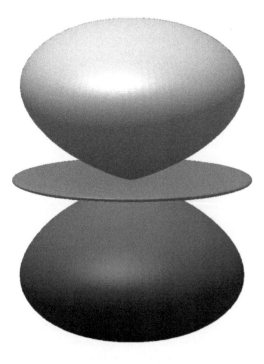

Figure 5.1 An example basis state showing the magnitude of $Y_l^m(\theta, \phi)$ by radius for $l = 2$ and $m = 0$. Note that there are many tables of spherical harmonics on-line, including some containing colourisation by the complex phase. If you have any experience with chemistry, you will likely have already drawn the correct conclusion that these states are closely related to atomic orbitals – a subject we will later return to.

5.4 Addition of Angular Momentum

5.4.1 The General Theory

In classical mechanics, the total angular momentum of a system of particles is the sum of the angular momenta of each particle. For a system of particles the total angular momentum may be an integral of the motion but the individual angular momenta are not. We now consider the quantum case of the angular momentum of a system of N 'particles' (by particle we mean a subsystem for which an angular momentum can be defined). The spaces of each particle are denoted in parentheses, so that $\hat{J}_z^{(2)}$ is the z component of angular momentum in the second space and $\hat{J}_{(4)}^2$ is the square of the angular momentum in the fourth space. To reduce the burden of notation, we will use the same symbol for the operator within each space and its extension to the full tensor product space. The commutation relation of each component of angular momentum is given by

$$\left[\hat{J}_i^{(a)}, \hat{J}_j^{(b)}\right] = i\,\hbar\varepsilon_{ijk}\delta_{ab}\hat{J}_k^{(a)}. \tag{5.76}$$

Here, ε_{ijk} is the Levi–Civita antisymmetric symbol that describes the permutations of $i, j, k \in \{1, 2, 3\}$ and δ_{ab} denotes the commutation of operators in different subspaces. We will make use of a shorthand for the state

$$|j_1, m_1, \dots\rangle_1 \otimes |j_2, m_2, \dots\rangle_2 \otimes \dots = |j_1 j_2, \dots; m_1 m_2 \dots; \dots\rangle. \tag{5.77}$$

The operators $\left\{\hat{J}_{(i)}^2, \hat{J}_z^{(i)}\right\}$ can be used to form a CSCO. This is very useful for some problems (especially a set of two-level systems, where this would be a CSCO). For qubits (quantum bits) this is the computational basis and it is also a standard choice for studying spin chains. The actions of $\hat{J}_z^{(a)}$ and $\hat{J}_{(a)}^2$ on the ket $|j_1 j_2, \dots; m_1 m_2 \dots; \dots\rangle$ is, as before, given by:

$$\hat{J}_{(i)}^2 |j_1 j_2 \dots; m_1 m_2 \dots; \dots\rangle = j_i(j_i + 1)\hbar^2 |j_1 j_2 \dots; m_1 m_2 \dots; \dots\rangle$$

$$\hat{J}_z^{(i)} |j_1 j_2 \dots; m_1 m_2 \dots; \dots\rangle = m_i \hbar |j_1 j_2 \dots; m_1 m_2 \dots; \dots\rangle. \tag{5.78}$$

Just as with classical mechanics there are a number of systems where the total angular momentum may commute with the system Hamiltonian, but the $\hat{J}_{(i)}^2$ do not. For those systems, an alternative basis is of greater utility. The general argument for constructing this basis is as follows: We first define the total angular momentum $\hat{J}$ as the Cartesian product

$$\hat{\mathbf{J}} = \left(\sum_i \hat{J}_x^{(i)}, \sum_i \hat{J}_y^{(i)}, \sum_i \hat{J}_z^{(i)}\right). \tag{5.79}$$

We then check all the commutation relations between components of the angular momentum starting with

$$[\hat{J}_x, \hat{J}_y] = \left[\sum_i \hat{J}_x^{(i)}, \sum_k \hat{J}_y^{(k)}\right] = \sum_{i,k} \left[\hat{J}_x^{(i)}, \hat{J}_y^{(k)}\right] = \sum_i i\,\hbar\hat{J}_z^{(i)} = i\,\hbar\hat{J}_z.$$

Cyclic permutations of the above argument yield

$$[\hat{J}_i, \hat{J}_j] = i\,\hbar\varepsilon_{ijk}\hat{J}_k. \tag{5.80}$$

By demonstrating that the sum of angular momenta $\hat{J}$ satisfies these commutation relations, we have confirmed that $\hat{J}$ is itself an angular momentum in the general sense. As such, we know that it must satisfy the following eigenequations:

$$\hat{J}^2 |j, m, \ldots\rangle = j(j+1)\hbar^2 |j, m, \ldots\rangle$$
$$\hat{J}_z |j, m, \ldots\rangle = m\hbar |j, m, \ldots\rangle, \tag{5.81}$$

as these equations hold for any angular momentum. This gives us an alternative basis of the eigenvectors common to $\hat{J}^2$ and $\hat{J}_z$, $|j, m, \ldots\rangle$, to that of the separable basis common to each component angular momentum $|j_1 j_2 \ldots; m_1 m_2 \ldots; \ldots\rangle$. The natural next question is to ask how these two bases are connected, or in other words, how do we convert from one basis to the other? An immediate follow-up question is: how do we better define $|j, m, \ldots\rangle$? These vectors are currently only constrained by two quantum numbers and it is natural to ask if we can find eigenvectors common to components $\hat{J}^2_{(i)}$ and $\hat{J}^{(i)}_z$ that we can use to better define this basis. A full analysis of this problem gets very messy, but like with many complex problems, we can make progress by studying the total angular momentum from its component parts. Specifically, we can decompose the total angular momentum into a hierarchy of pairwise angular momenta

$$\hat{J} = \underbrace{\left(\hat{J}^{(1)}_x, \hat{J}^{(1)}_y, \hat{J}^{(1)}_z\right) + \left(\hat{J}^{(2)}_x, \hat{J}^{(2)}_y, \hat{J}^{(2)}_z\right)}_{\text{Total angular momentum for spaces 1 and 2}} + \left(\hat{J}^{(3)}_x, \hat{J}^{(3)}_y, \hat{J}^{(3)}_z\right) + \ldots \ . \tag{5.82}$$

Total angular momentum for spaces (1 and 2) and 3

We can build the labelling of our new basis by decomposing the total angular momentum in this hierarchical way (the specific ordering of the pairs does not matter so long as once it has been chosen, the labels are fixed). The final labelling will look something like this:

$$|j_1, \ldots, j_N; j_{12}, ; j_{123}; \ldots; jm\rangle \, .$$

For this reason, it is sufficient to consider the addition of only two angular momenta at a time. A basis defined by all the angular momenta can then be found by recursively applying the pairwise analysis until the full basis over N particles has been defined.

5.4.2 Two-particle Systems

To determine the basis defined from the total angular momenta of two particles, we start by noting that

$$\hat{J}^2 = (\hat{J}_1 + \hat{J}_2) \cdot (\hat{J}_1 + \hat{J}_2)$$
$$= \hat{J}^2_{(1)} + 2(\hat{J}^{(1)}_x \hat{J}^{(2)}_x + \hat{J}^{(1)}_y \hat{J}^{(2)}_y + \hat{J}^{(1)}_z \hat{J}^{(2)}_z) + \hat{J}^2_{(2)}. \tag{5.83}$$

The commutation relation between $\hat{J}^2$ and $\hat{J}^{(i)}_z$ is given by

$$\left[\hat{J}^2, \hat{J}^{(1)}_z\right] = 2i\,\hbar(\hat{J}^{(1)}_x \hat{J}^{(2)}_y - \hat{J}^{(2)}_x \hat{J}^{(1)}_y) \neq 0. \tag{5.84}$$

This shows that $\hat{J}^2$ and $\hat{J}_z^{(i)}$ cannot form a CSCO and therefore we cannot use the z-components of the angular momenta and the m_i to help us define the basis of eigenvectors common to $\hat{J}^2$ and $\hat{J}_z$ ($\hat{J}^2$ and $\hat{J}_z^{(i)}$ do not commute, are not compatible observables, and do not share eigenstates).

Exercise 5.16 Calculate the commutator shown in Eq. (5.84) and show that $\hat{J}^2$ and $\hat{J}_3^{(a)}$ do not form a set of commuting observables. ∎

It is clear from Eq. (5.83) that

$$\left[\hat{J}^2, \hat{J}_{(i)}^2\right] = 0.$$

It is also clear, as $\hat{J}_{(i)}^2$ commutes with each element of $\hat{J}_z$, that

$$\left[\hat{J}_z, \hat{J}_{(i)}^2\right] = 0.$$

We therefore see that $\hat{J}_{(i)}^2$ is compatible with both $\hat{J}^2$ and $\hat{J}_z$ and must share a common set of eigenvectors. Thus, we can conclude that the following set of operators can be used to help form a complete set of commuting observables and define a basis:

$$\hat{J}_{(1)}^2 |j_1 j_2; jm; \ldots\rangle = j_1(j_1 + 1)\hbar^2 |j_1 j_2; jm; \ldots\rangle \tag{5.85}$$

$$\hat{J}_{(2)}^2 |j_1 j_2; jm; \ldots\rangle = j_2(j_2 + 1)\hbar^2 |j_1 j_2; jm; \ldots\rangle \tag{5.86}$$

$$\hat{J}^2 |j_1 j_2; jm; \ldots\rangle = j(j + 1)\hbar^2 |j_1 j_2; jm; \ldots\rangle \tag{5.87}$$

$$\hat{J}_z |j_1 j_2; jm; \ldots\rangle = m\hbar |j_1 j_2; jm; \ldots\rangle. \tag{5.88}$$

Note that this $\{|j_1 j_2; jm\rangle\}$ basis exists in the tensor product space and cannot simply be written in terms of separable basis state such as $\{|j_1 j_2; m_1 m_2\rangle = |j_1 m_1\rangle_1 \otimes |j_2 m_2\rangle_2\}$. However, by inserting an appropriate resolution of the identity we can write:

$$|j_1 j_2; jm\rangle = \sum_{m_1 m_2} |j_1 j_2; m_1 m_2\rangle \langle j_1 j_2; m_1 m_2| j_1 j_2; jm\rangle$$

$$= \sum_{m_1 m_2} |j_1 j_2; m_1 m_2\rangle \langle j_1 j_2; m_1 m_2| j_1 j_2; jm\rangle. \tag{5.89}$$

The constants $\langle j_1 j_2; m_1 m_2| j_1 j_2; jm\rangle$ are termed *Clebsch–Gordan coefficients* – a detailed study of these can be found in many other textbooks. Many useful resources can easily be found online or in computational physics literature (such as [2]). These include tables of the coefficients and open source code, in a number of programming languages, for their calculation. When using Clebsch–Gordan coefficients, problems will either be for so small an angular momentum that an existing table of values will suffice for the task at hand, or so complex that you will want a computer program to perform the calculation for you. We do note some important properties of Clebsch–Gordan coefficients. Firstly, they vanish *unless*

$$m = m_1 + m_2 \text{ and } |j_1 - j_2| \le j \le j_1 + j_2 \tag{5.90}$$

and secondly, there is the additional restriction that m can take on only the following values:

$$-(j_1 + j_2), -(j_1 + j_2) + 1, \ldots, (j_1 + j_2) - 1, (j_1 + j_2).$$

To better understand the application of these restrictions, let us consider a system of two angular momenta, $j_1 = 1$ and $j_2 = 2$. The permitted values of the combined angular momentum are given by

$$j = 1, 2, 3. \tag{5.91}$$

Let us now consider the permitted values of m for these different values of j:

For $j = 1$: $m = 1, 0, -1$

For $j = 2$: $m = 2, 1, 0, -1, -2$ $\tag{5.92}$

For $j = 3$: $m = 3, 2, 1, 0, -1, -2, -3$.

All permitted values of m can be found by considering the individual values, j_1 and j_2, separately. When $j_1 = 1$, m_1 can only take on the values $m_1 = 1, 0, -1$. Likewise, when $j_2 = 2$, $m_2 = 2, 1, 0, -1, -2$. The combined values of these constituents can be found easily using the following table:

$m_2 \setminus m_1$	-1	0	1
-2	-3	-2	-1
-1	-2	-1	0
0	-1	0	1
1	0	1	2
2	1	2	3

Exercise 5.17   Consider two angular momenta $j_1 = 3/2$, $j_2 = 3$. What are the permitted values of the total angular momentum j and total magnetic quantum number m?   ∎

5.4.3 Example: Addition of Spins

One of the most common systems is that of a spin $j = \frac{1}{2}$. All two-level systems can be mapped onto this basis. Let us consider adding two spin half systems together, i.e. $j_1 = j_2 = \frac{1}{2}$. In this case, $\left\{ \hat{J}^2_{(1)}, \hat{J}^2_{(2)}, \hat{J}^{(1)}_z, \hat{J}^{(2)}_z \right\}$ is a CSCO, and the following defines a basis:

$$\hat{J}^2_{(i)} \left| j_1 = \frac{1}{2} j_2 = \frac{1}{2}; m_1 m_2 \right\rangle = \frac{3}{4} \hbar^2 \left| j_1 = \frac{1}{2} j_2 = \frac{1}{2}; m_1 m_2 \right\rangle$$

$$\hat{J}^{(i)}_z \left| j_1 = \frac{1}{2} j_2 = \frac{1}{2}; m_1 m_2 \right\rangle = m_i \hbar \left| j_1 = \frac{1}{2} j_2 = \frac{1}{2}; m_1 m_2 \right\rangle \tag{5.93}$$

where $m_i = \pm\frac{1}{2}$. This notion is somewhat cumbersome and as $j_1 = \frac{1}{2}$ and $j_2 = \frac{1}{2}$ throughout this example we will omit their labels from the basis and use the shorthand

$$\left| j_1 = \frac{1}{2} j_2 = \frac{1}{2}; m_1 m_2 \right\rangle = |m_1 m_2\rangle$$

for brevity and clarity. As such, we see that the above eigenequations take the form

$$\hat{J}^2_{(i)} |m_1, m_2\rangle = \frac{3}{4} \hbar^2 |m_1, m_2\rangle$$

$$\hat{J}^{(i)}_z |m_1, m_2\rangle = m_i \hbar |m_1, m_2\rangle \tag{5.94}$$

and the basis is the set of vectors

$$\left\{ \left| \frac{1}{2},\frac{1}{2} \right\rangle, \left| -\frac{1}{2},\frac{1}{2} \right\rangle, \left| \frac{1}{2},-\frac{1}{2} \right\rangle, \left| -\frac{1}{2},-\frac{1}{2} \right\rangle \right\},$$

where the ordering of m_1 and m_2 means we do not need to write these labels explicitly in each ket.

In the total angular momentum basis, we seek the common eigenvectors of

$$\hat{J}^2_{(1)} |j_1 j_2; jm\rangle = \frac{3}{4}\hbar^2 |j_1 j_2; jm\rangle$$

$$\hat{J}^2_{(2)} |j_1 j_2; jm\rangle = \frac{3}{4}\hbar^2 |j_1 j_2; jm\rangle$$

$$\hat{J}^2 |j_1 j_2; jm\rangle = j(j+1)\hbar^2 |j_1 j_2; jm\rangle$$

$$\hat{J}_z |j_1 j_2; jm\rangle = m\hbar |j_1 j_2; jm\rangle.$$

Again, $j_1 = j_2 = \frac{1}{2}$. To keep it easy to distinguish these eigenstates from $|m_1, m_2\rangle$, we will shorten our notation of these basis vectors but we will retain the labels j and m.

As a consequence of the general theory of angular momentum and the properties of Clebsch–Gordan coefficients, we know that

$$|j_1 - j_2| \le j \le j_1 + j_2,$$

so that $j = 0, 1$. For $j = 0$ it is clear that $m = 0$ is the only allowed value and for $j = 1$ we know that m can only take the values 0, 1, and -1. This means that without doing any further analysis we already know that these basis state will be

$$\{ |j = 0, m = 0\rangle, |j = 1, m = -1\rangle, |j = 1, m = 0\rangle, |j = 1, m = 1\rangle \}.$$

As there is only one state with $j = 0$, it is referred to as the singlet state and the basis vectors with $j = 1$ are collectively known as the triplet states.

We have several choices about how to go about writing these eigenstates in terms of the ones we know in the $|m_1, m - 2\rangle$ basis. The first way would be to use the matrix representation in the $|m1, m2\rangle$ basis and turn the handle, finding eigenvectors; another to note that $|\frac{1}{2},\frac{1}{2}\rangle$ (or $|-\frac{1}{2},-\frac{1}{2}\rangle$) is $|j = 1, m = 1\rangle$ (or $|j = 1, m = -1\rangle$) and apply $\hat{J}_-$ (or $\hat{J}_+$) to find the other triplet states. We can then use orthogonality to construct the singlet state. Or, in such a simple case as this, we can just use logic to argue that these states will be

$$|j = 1, m = -1\rangle = \left| \frac{1}{2},\frac{1}{2} \right\rangle$$

$$|j = 1, m = 0\rangle = \frac{1}{\sqrt{2}} \left(\left| \frac{1}{2},-\frac{1}{2} \right\rangle + \left| -\frac{1}{2},\frac{1}{2} \right\rangle \right)$$

$$|j = 1, m = 1\rangle = \left| \frac{1}{2},\frac{1}{2} \right\rangle$$

$$|j = 0, m = 0\rangle = \frac{1}{\sqrt{2}} \left(\left| \frac{1}{2},-\frac{1}{2} \right\rangle - \left| -\frac{1}{2},\frac{1}{2} \right\rangle \right).$$

Thus a system of two spin-$1/2$ particles is only permitted to exist in states that are: an anti-symmetric singlet state when the total spin $j = 0$, and a superposition of symmetric triplet with total spin $j = 1$ state $|1, 0\rangle$. Both states with $m = 0$ are entangled and only live in the total tensor product space.

Exercise 5.18 Check that the above states do indeed satisfy the requirements for this basis (i.e. are eigenstates of $\left\{ \hat{J}^2_{(1)}, \hat{J}^2_{(2)}, \hat{J}^2, \hat{J}_z \right\}$ with the correct eigenvalues) ∎

6

Some Analytic and Semi-analytic Methods

"If you want to accomplish something in the world, idealism is not enough – you need to choose a method that works to achieve the goal."

Richard Stallman

6.1 Introduction

In this chapter, we introduce a number of analytic and semi-analytic techniques that are useful in studying specific quantum systems. We are limiting our discussion to those topics that we deem essential for any textbook on quantum mechanics, and to some methods that we have used ourselves with great utility, that are not usually described in the existing textbook literature. In all cases, we believe that the descriptions are sufficiently detailed to make the arguments behind the methods clear. We have, however, stopped short of providing full derivations because (i) this material is available elsewhere; and (ii) we do not wish to distract from the key conceptual points with excessive mathematical detail. We have chosen not to cover techniques, such as scattering, that we do not heavily use ourselves. While clearly important, it is our view that someone wishing to apply these methods should study texts from authors with substantial experience using them. This is because each technique contains limitations in its applicability and use that can only be properly informed by real-world experience.

In terms of essential techniques, we cover the interaction picture, perturbation theory, and the variational method. All of those are of great use in finding the approximate behaviour of quantum systems whose mathematical analysis is not otherwise possible. In terms of methods not well represented in the rest of the literature, we consider two semi-analytic techniques using as their core idea dynamic basis. These can be used to support the development of sophisticated numerical methods. Next, we consider Floquet's method for the analysis of time periodic systems, as this is not present in many textbooks on the subject of quantum mechanics. Two-state systems, e.g. electron spin, orientations of the ammonia molecule, or in simplifications of complex systems such as dihydrogen, are often encountered in quantum mechanics. For this reason, we dedicate Section 6.7 of this chapter to some general analysis of two-state systems.

Quantum Mechanics: From Analytical Mechanics to Quantum Mechanics, Simulation, Foundations, and Engineering, First Edition. Mark Julian Everitt, Kieran Niels Bjergstrom and Stephen Neil Alexander Duffus.
© 2024 John Wiley & Sons Ltd. Published 2024 by John Wiley & Sons Ltd.
Companion website: www.wiley.com/go/everitt/quantum

6.2 Problems of the Form $\hat{H}_0 + \hat{W}$

The idea of splitting a problem into different parts is a standard tactic in physics problem-solving. We may reduce the problem to one of a previously solved problem and an unsolved part. We may also simplify out of a problem those bits that we are not currently interested in. In quantum mechanics we often come across problems where this divide can be found in systems where the Hamiltonian can be written as

$$\hat{H}_{\text{total}} = \hat{H}_0 + \hat{W}. \tag{6.1}$$

In problems such as these we may either know everything about H_0 or simply not be interested in its properties. In this section, we consider several approaches to problems of this type.

6.2.1 The Interaction Picture

> **Prerequisite Material:** §4.2.6: The Evolution Operator

The interaction picture is often presented as an alternative to the Schrödinger and Heisenberg pictures. Mathematically it is in some ways, but physically it does not have the same fundamental status. The interaction picture can be used to analyse problems of the form

$$\hat{H}_{\text{total}} = \hat{H}_0(t) + \hat{W}(t), \tag{6.2}$$

where one knows the evolution operator, $\hat{U}_0(t, t_0)$, for $\hat{H}_0(t)$. If we define $\hat{W}_I(t) = \hat{U}_0^{\dagger}(t, t_0)$ $\hat{W}(t)\hat{U}_0(t, t_0)$ and $|\psi_I(t)\rangle = \hat{U}_0^{\dagger}(t, t_0)|\psi(t)\rangle$, we can show that

$$i\hbar \frac{\partial}{\partial t}|\psi_I(t)\rangle = \hat{W}(t)|\psi_I(t)\rangle, \tag{6.3}$$

which is the Schrödinger equation for $\hat{W}$ on its own.

We can understand this representation by its effect on a given set of basis vectors $\{|n\rangle\}$. The matrix representation of an operator $\hat{A}$ will be $A_{nm} = \langle n|\hat{A}|m\rangle$. As the evolution operator is unitary, we know that $\hat{U}_0^{\dagger}(t, t_0)\hat{U}_0(t, t_0) = \hat{U}_0(t, t_0)\hat{U}_0^{\dagger}(t, t_0) = \hat{I}$ and so these matrix elements can be written as

$$\begin{aligned} A_{nm} &= \langle n|\hat{I}\hat{A}\hat{I}|m\rangle \\ &= \langle n|\hat{U}_0(t, t_0)\hat{U}_0^{\dagger}(t, t_0)\hat{A}\hat{U}_0(t, t_0)\hat{U}_0^{\dagger}(t, t_0)|m\rangle \\ &= \langle n|\hat{U}_0(t, t_0)\hat{A}_I\hat{U}_0^{\dagger}(t, t_0)|m\rangle, \end{aligned} \tag{6.4}$$

where we have defined $\hat{A}_I = \hat{U}_0^{\dagger}(t, t_0)\hat{A}\hat{U}_0(t, t_0)$ as the interaction picture version of $\hat{A}$. We can then see that in the interaction picture, the basis vectors $\hat{U}_0^{\dagger}(t, t_0)|m\rangle$ are the ones that would evolve backward in time if the Hamiltonian were simply $\hat{H}_0$. The basis vectors, which are stationary in the Schrödinger picture, are dynamic in the interaction picture in a way that eliminates the motion associated with the reference Hamiltonian H_0.

The interaction picture takes out the evolution from the $\hat{H}_0$ contribution to the system Hamiltonian so that we can study the contribution to the dynamics from $\hat{W}$ alone.

The Schrödinger picture state $|\psi(t)\rangle$ is recovered from the interaction picture by noting that

$$|\psi(t)\rangle = \hat{U}_0(t, t_0) |\psi_I(t)\rangle. \tag{6.5}$$

We will find this method of great utility when we later discuss open quantum systems.

6.2.2 Time-independent Perturbation Theory

When studying real physical systems, there are very few occasions where the Schrödinger equation has a nice analytic solution. More often than not, we have to resort to computational methods in order to numerically analyse the system-of-interest. A serious drawback of numerical techniques is that it can be hard to understand why the system behaves the way it does. Some insight can be gained using approximation methods such a perturbation theory.

Time-independent perturbation theory addresses the problem of solving the time-independent Schrödinger equation (TISE) (i.e. finding the eigenvalues and eigenvectors) of a Hamiltonian we can write in the form

$$\hat{H}(\lambda) = \hat{H}_0 + \lambda\hat{W}, \tag{6.6}$$

where the eigenvalues and states of $\hat{H}_0 = \hat{H}(0)$ are known, and where $\lambda\hat{W}$ is considered a small perturbation to $\hat{H}_0$. We quantify the perturbative nature of $\lambda\hat{W}$ by imposing the requirement that no matrix element of $\hat{W}$ is significantly greater than those of $\hat{H}$ and $\lambda \ll 1$. The eigenvalues/states of $\hat{H}_0$ are known, and we will assume they also form a discrete orthonormal basis.

Assuming a discrete set of eigenvalues, we can write

$$\hat{H}_0 |n^{(a)}\rangle = E_n |n^{(a)}\rangle, \tag{6.7}$$

where the index (a) distinguishes degenerate eigenstates $\hat{H}_0$.

The idea of perturbation theory starts by noticing that the eigenvalues and vectors of $\hat{H}$ and $\hat{H}_0$ will be the same for $\lambda = 0$. As λ deviates from zero, the eigenvectors and values of $\hat{H}$ will change continuously from those of $\hat{H}_0$ (we are assuming that the eigenvectors and eigenvalues of $\hat{H}(\lambda)$ are well-behaved). For this reason, we can write the TISE for the full Hamiltonian as

$$\hat{H}(\lambda) |n(\lambda)\rangle = E_n(\lambda) |n(\lambda)\rangle. \tag{6.8}$$

Notice that there is no index indicating degeneracy. This is because the perturbation may (or may not) lift the degeneracy of the eigenvalues of $\hat{H}_0$. For convenience, we will assume that none of the eigenvalues of $\hat{H}_0$ are degenerate (a discussion of the degenerate case is already well represented in the literature and, as the discussion of such cases is more of a mathematical than physics nature, we will not cover it here).

With this assumption in mind we can write the unperturbed TISE as

$$\hat{H}(0) |n(0)\rangle = E_n(0) |n(0)\rangle$$

and we want to know the solutions to

$$\left(\hat{H}(0) + \lambda \hat{W}\right) |n(\lambda)\rangle = E_n(\lambda) |n(\lambda)\rangle.$$

Now we use the assumption that $\hat{E}_n(\lambda)$ and $|n(\lambda)\rangle$ are well-behaved functions of λ, so we can use a Taylor series:

$$E_n(\lambda) = \sum_k \frac{\lambda^k}{k!} \frac{d^k E_n(\lambda)}{d\lambda^k}\bigg|_{\lambda=0} \tag{6.9}$$

$$|n(\lambda)\rangle = \sum_k \frac{\lambda^k}{k!} \frac{d^k |n(\lambda)\rangle}{d\lambda^k}\bigg|_{\lambda=0}. \tag{6.10}$$

We now note that, as the derivatives of the eigenvalues will be just numbers and the derivatives of the eigenstates will remain vectors, we can define the kth order correction to the eigenenergies and vectors according to

$$E_n^{(k)} = \frac{1}{k!} \frac{d^k E_n(\lambda)}{d\lambda^k}\bigg|_{\lambda=0} \tag{6.11}$$

$$|n_k\rangle = \frac{1}{n!} \frac{d^n |i(\lambda)\rangle}{d\lambda^n}\bigg|_{\lambda=0}. \tag{6.12}$$

Therefore, we can write the power series expansion as

$$E_n(\lambda) = \sum_k \lambda^k E_n^{(k)} \text{ and } |n(\lambda)\rangle = \sum_k \lambda^k |n_k\rangle. \tag{6.13}$$

Note that the first terms $E_n^{(0)}$ and $|n_0\rangle$ in these expansions are the unperturbed n^{th} eigenvalues and eigenvectors of $\hat{H}_0$.

We can now determine each of these terms directly by substituting these power series into the Schrödinger equation $\hat{H} |n(\lambda)\rangle = E_n(\lambda) |n(\lambda)\rangle$ to obtain

$$\left(\hat{H}_0 + \lambda \hat{W}\right) \sum_{k=0}^{\infty} \lambda^k |n_k\rangle = \sum_{l=0}^{\infty} \lambda^l E_n^{(l)} \sum_{k=0}^{\infty} \lambda^k |n_k\rangle. \tag{6.14}$$

Rearranging gives

$$\sum_{k=0}^{\infty} \left(\lambda^k \hat{H}_0 + \lambda^{k+1} \hat{W}\right) |n_k\rangle = \sum_{l,l=0}^{\infty} \lambda^{k+l} E_n^{(l)} |n_k\rangle. \tag{6.15}$$

Now we equate terms with the same power of λ to obtain a hierarchical set of equations

$$\hat{H}_0 |n(0)\rangle = E_n(0) |n(0)\rangle \tag{6.16}$$

$$\hat{W} |n(0)\rangle + \hat{H}_0 |n_1\rangle = E_n(0) |n_1\rangle + E_n^{(1)} |n(0)\rangle \tag{6.17}$$

$$\hat{W} |n_1\rangle + \hat{H}_0 |n_2\rangle = E_n^{(1)} |n_1\rangle + E_n(0) |n_2\rangle + E_i^{(2)} |n(0)\rangle \tag{6.18}$$

$$\vdots$$

to solve, where the first line is the unperturbed problem, whose solution we already know. Note that while $|n_0\rangle = |n(0)\rangle$ we will use each notation depending on context – if we wish to emphasise the dependence on λ, we use $|n(0)\rangle$, and if we want to emphasise its place in the power series, we use $|n_0\rangle$.

At this stage we have made no approximations, and if the full set of equations could be solved, we would find the exact values of $E_n(\lambda)$ and $|n(\lambda)\rangle$. For many problems, the series expansion is well-behaved, but for many others (such as for unbounded potentials) it is not. The approximate nature of perturbation theory comes into play when we choose to truncate these expressions and limit the order to which we estimate $|n(\lambda)\rangle$ and $E_n(\lambda)$. When we compute these approximations ourselves, the calculations can be rather involved; hence, human analysis is usually limited to the first few orders. With modern automated algebraic and numerical methods, we turn to computation for higher-order analysis. The perturbative expansion above, Eq. (6.17), and pre-multiplying on the left by $\langle n(0)|$ gives us:

$$\left\langle n(0) \left| \hat{W} \right| n(0) \right\rangle + \left\langle n(0) \left| \hat{H}_0 \right| n_1 \right\rangle = E_n(0) \langle n|n_1\rangle + \overbrace{E_n^{(1)}}^{1} \langle n(0)|n(0)\rangle$$

$$\left\langle n(0) \left| \hat{W} \right| n(0) \right\rangle + \cancel{E_n \langle n(0)|n_1\rangle} = \cancel{E_n \langle n(0)|n_1\rangle} + E_n^{(1)}$$

$$\left\langle n(0) \left| \hat{W} \right| n(0) \right\rangle = E_n^{(1)} \tag{6.19}$$

and, recalling the power series expansion of $E_i(\lambda)$, yields the intuitive result that the first-order correction to the energy comes from how much $\hat{W} |n(0)\rangle$ remains in the direction of $|n(0)\rangle$

$$E_n(\lambda) = E_n + \lambda \left\langle n(0) \left| \hat{W} \right| n(0) \right\rangle + \mathcal{O}(\lambda^2). \tag{6.20}$$

Before proceeding to evaluate the first-order correction to the energy eigenstates, let us note that any eigenvector, including $|n(\lambda)\rangle$, can only be resolved up to some arbitrary phase factor. We will now make a set of constraints based on this observation. These seem ubiquitous in the literature but we are unsure as to their original source (we follow an argument constructed from several sources including [12, 77]). We will require the eigenvectors $|n(0)\rangle$ of $\hat{H}_0$ to be orthonormal and so $\lim_{\lambda \to 0} \langle n(0)|n(\lambda)\rangle = 1$. We will further require the inner product $\langle n(0)|n(\lambda)\rangle = \langle n(0)|n_0\rangle + \lambda \langle n(0)|n_1\rangle + \mathcal{O}(\lambda^2)$ to be real, which implies that $\langle n(0)|n_1\rangle$ must also be real. Noting that $\langle n(0)|n_0\rangle = \langle n_0|n_0\rangle = 1$, we can write

$$\langle n(\lambda)|n(\lambda)\rangle = \left[\langle n_0| + \lambda \langle n_1| + \mathcal{O}(\lambda^2)\right] \left[|n_0\rangle + \lambda |n_1\rangle + \mathcal{O}(\lambda^2)\right]$$

$$= 1 + \lambda \left[\langle n_1|n_0\rangle + \langle n_0|n_1\rangle\right] + \mathcal{O}(\lambda^2). \tag{6.21}$$

The only way for $\langle n(\lambda)|n(\lambda)\rangle$ to be normalised together with the constraint that $\langle n(0)|n_1\rangle$ must be real, is for the first-order correction to the eigenstate to be orthogonal to the unperturbed system $\langle n(0)|n_1\rangle = 0$ (this makes intuitive sense as the biggest correction to the direction of a vector will be to add to it an orthogonal vector).

That our argument has led to the first-order correction to the eigenstate being orthogonal to the unperturbed eigenstates of $\hat{H}_0$, provides the clue as to how to proceed in determining the first-order correction to the eigenstates of the system. The vector $|n_1\rangle$ must live in the space of vectors orthogonal to $|n(0)\rangle$. We can use the fact that the projector for the state $|n(0)\rangle$ is

$$\hat{\mathbb{P}}_{|n(0)\rangle} = |n(0)\rangle \langle n(0)|, \tag{6.22}$$

and the resolution of the identity in terms of the basis $\{|n(0)\rangle\}$, to define the complementary projector

$$\hat{\mathbb{P}}_{|n(0)\rangle}^{\perp} = \hat{\mathbb{I}} - \hat{\mathbb{P}}_{|n(0)\rangle} = \sum_{k \neq n} |k(0)\rangle \langle k(0)| \tag{6.23}$$

is that operator which projects any state into an orthogonal subspace of $|n(0)\rangle$. Note that the identity must be the sum of a projector and its complement $\hat{\mathbb{I}} = \hat{\mathbb{P}}_{|n(0)\rangle} + \hat{\mathbb{P}}_{|n(0)\rangle}^{\perp}$ (this concept can be generalised to arbitrary partitions of the vector space in any basis and is a useful trick to know).

One of two standard mathematical tricks is to multiply by one, another is to add zero to a problem in a way that leads to a useful simplification. In this case we start with $\hat{W}$ and pre-multiply by $\hat{\mathbb{I}}$

$$
\begin{aligned}
\hat{\mathbb{I}}\hat{W}|n_0\rangle &= \left(\hat{\mathbb{P}}_{|n_0\rangle} + \hat{\mathbb{P}}_{|n_0\rangle}^{\perp}\right)\hat{W}|n_0\rangle \\
&= \hat{\mathbb{P}}_{|n_0\rangle}\hat{W}|n_0\rangle + \hat{\mathbb{P}}_{|n_0\rangle}^{\perp}\hat{W}|n_0 i\rangle \\
&= |n_0\rangle \underbrace{\left\langle n_0 \left| \hat{W} \right| n_0 \right\rangle}_{E_n^{(1)}} + \hat{\mathbb{P}}_{|n_0\rangle}^{\perp}\hat{W}|n_0\rangle \\
&= E_n^{(1)}|n_0\rangle + \hat{\mathbb{P}}_{|n_0\rangle}^{\perp}\hat{W}|n_0\rangle .
\end{aligned}
$$

Substituting this into the first-order perturbation expansion in λ Eq. (6.17), we get:

$$\hat{W}|n(0)\rangle + \hat{H}_0|n_1\rangle = E_n(0)|n_1\rangle + E_n^{(1)}|n(0)\rangle$$

$$\cancel{E_n^{(1)}|n(0)\rangle} + \hat{\mathbb{P}}_{|n(0)\rangle}^{\perp}\hat{W}|n(0)\rangle + \hat{H}_0|n_1\rangle = E_n(0)|n_1\rangle + \cancel{E_n^{(1)}|n(0)\rangle}$$

$$\hat{\mathbb{P}}_{|n(0)\rangle}^{\perp}\hat{W}|n(0)\rangle = \left(E_n(0) - \hat{H}_0\right)|n_1\rangle .$$

Now we will make use of the fact that the unperturbed eigenstates of $\hat{H}_0$ are orthonormal, and premultiply by another eigenstate $\langle m(0)|$, where $m \neq n$, to obtain

$$\left\langle m \left| E_n(0) - \hat{H}_0 \right| n_1 \right\rangle = \left\langle m(0) \left| \hat{\mathbb{P}}_{|n(0)\rangle}^{\perp} \hat{W} \right| n(0) \right\rangle . \tag{6.24}$$

This can be simplified to find that the mth component of the first-order correction is

$$\langle m(0)|n_1\rangle = \frac{\left\langle m(0) \left| \hat{W} \right| n(0) \right\rangle}{E_n(0) - E_m(0)}, \tag{6.25}$$

and we finally obtain the first-order corrected eigenstates as

$$|n(\lambda)\rangle = |n(0)\rangle + \lambda \sum_{m \neq n} \frac{\left\langle m(0) \left| \hat{W} \right| n(0) \right\rangle}{E_n(0) - E_m(0)} |m(0)\rangle + \mathcal{O}(\lambda^2). \tag{6.26}$$

The process now continues iteratively with the first-order corrections to the energy and the state, allowing the calculation of the second-order correction to the energy and show that

$$E_n(\lambda) = E_n(0) + \lambda \left\langle n(0) \left| \hat{W} \right| n(0) \right\rangle + \lambda^2 \sum_{m \neq n} \frac{\left| \left\langle m(0) \left| \hat{W} \right| n(0) \right\rangle \right|^2}{E_n(0) - E_m(0)} + \mathcal{O}(\lambda^3). \tag{6.27}$$

As one can see, these sums are already getting rather cumbersome and rarely taken any further by hand. In some cases, the eigenvectors of $\hat{H}_0$ are degenerate. As the development of this theory is mostly mathematical linear algebra rather than physics, we refer the reader to the existing printed or on-line literature for details. We will, however, note that the usual physical effect of a perturbation in such cases is to lift the degeneracy of the system. A very important example is the Linear Stark effect (see [77] for example).

6.2.3 Time-dependent Perturbation Theory

As before, we consider a Hamiltonian of the form:

$$\hat{H} = \hat{H}_0 + \lambda \hat{W}(t) \tag{6.28}$$

where, again, the eigenvalues/states of $\hat{H}_0 = \hat{H}(0)$ are known and, once more, no matrix element of $\hat{W}$ is significantly greater than that of $\hat{H}$, and $\lambda \ll 1 \in \mathbb{R}$. The only difference now is that the perturbation $\lambda \hat{W}(t)$ is time-dependent, and we will be interested in finding approximate solutions to the time-dependent Schrödinger equation (TDSE).

As before, the eigenstates of $\hat{H}_0 |n(0)\rangle = E_n(0) |n(0)\rangle$ can be assumed to form a complete orthonormal, time-independent, basis. We can therefore write any state, including the initial condition, in the form:

$$|\psi(t)\rangle = \sum_n c_n(t) |n(0)\rangle . \tag{6.29}$$

We now substitute this power series expansion into the Schrödinger equation

$$\hat{H} |\psi(t)\rangle = i\hbar \frac{d}{dt} |\psi(t)\rangle$$

to yield

$$\sum_n c_n(t) \left(E_n |n(0)\rangle + \lambda \hat{W}(t) |n(0)\rangle \right) = i\hbar \sum_n \frac{dc_n(t)}{dt} |n(0)\rangle .$$

We will use a trick we used in time-independent perturbation theory and premultiply by $\langle m(0)|$ to yield

$$\langle m(0)| \sum_n c_n(t) \left(E_n(0) |n(0)\rangle + \lambda \hat{W}(t) |n(0)\rangle \right) = \langle m(0)| i\hbar \sum_n \frac{dc_n(t)}{dt} |n(0)\rangle \tag{6.30}$$

which, making use of the fact that $\langle m(0)|n(0)\rangle = \delta_{nm}$, simplifies to

$$\frac{d}{dt} c_m(t) = -\frac{i}{\hbar} c_m(t) E_m(0) - \frac{i\lambda}{\hbar} \sum_n c_n(t) \left\langle m(0) \left| \hat{W}(t) \right| n(0) \right\rangle . \tag{6.31}$$

As with the first stages of time-independent perturbation theory, our analysis is exact until this point. Solving the above first-order differential equation would determine $|\psi(t)\rangle$ completely. Importantly, note that it is only through the matrix element $\left\langle n \left| \hat{W}(t) \right| k \right\rangle$ that different eigenstates are coupled together, so if $\lambda = 0$ (or $\left\langle n \left| \hat{W}(t) \right| k \right\rangle = 0$) then

$$c_n(t) = c_n(0) \exp \left(-i \frac{E_n}{\hbar} t \right) ,$$

exactly as we would expect.

Motivated by the last expression, let us define (for any λ) a new set of coefficients, $b_n(t)$, according to

$$c_n(t) = b_n(t) \exp\left(-i\frac{E_n}{\hbar}t\right) \tag{6.32}$$

which yields a component-wide Schrödinger-type equation that is equivalent to that of the interaction picture

$$\frac{d}{dt}b_n(t) = -\frac{i\lambda}{\hbar}\sum_k b_k(t) \exp\left(i\frac{E_n - E_k}{\hbar}t\right)\left\langle n\left|\hat{W}(t)\right|k\right\rangle$$

$$= -\frac{i\lambda}{\hbar}\sum_k b_k(t) \exp(i\,\omega_{nk}t)\left\langle n\left|\hat{W}(t)\right|k\right\rangle, \tag{6.33}$$

where $\omega_{nk} = \frac{E_n - E_k}{\hbar}$ are known as the Bohr frequencies. This analysis is still exact, and, since we can see that the dynamics associated with $\hat{H}_0$ have been effectively removed by moving from c_n to b_n coefficients, justifies our assertion that this is a component-wise version of the interaction picture. Indeed, we could have started this analysis in the interaction picture, as stating that the eigenvalues and eigenvectors of $\hat{H}_0$ are known also means that we know the evolution operator through its eigen decomposition.

In terms of numerical analysis, the above tricks can be very useful. For perturbation theory and finding analytic expressions to help us understand the effect of adding the perturbation $\lambda\hat{W}$ we will use the same trick as in time-independent perturbation theory and write $b_n(t)$ as a power series expansion

$$b_n(t) = b_n^{(0)}(t) + \lambda b_n^{(1)}(t) + \lambda^2 b_n^{(2)}(t) + \ldots + \lambda^a b_n^{(a)}(t) + \ldots. \tag{6.34}$$

Substituting into the general equation for $b_n(t)$ yields:

$$\frac{d}{dt}\sum_a \lambda^a b_n^{(a)}(t) = -\frac{i}{\hbar}\sum_k\sum_a \lambda^{a+1} b_k^{(a)}(t) \exp(i\,\omega_{nk}t)\left\langle n(0)\left|\hat{W}(t)\right|k(0)\right\rangle$$

and, just as we did in the time-independent case, equating powers of λ gives:

$$\frac{d}{dt}b_n^{(a)}(t) = -\frac{i}{\hbar}\sum_k \exp(i\,\omega_{nk}t)\left\langle n(0)\left|\hat{W}(t)\right|k(0)\right\rangle b_k^{(a-1)}(t). \tag{6.35}$$

Now the zeroth-order coefficient stays the same, independent of the time-dependent perturbation, and so we can determine two facts from this condition. The first is that they cannot change over time, so

$$\frac{d}{dt}b_n^{(0)}(t) = 0. \tag{6.36}$$

In addition, the expansion must hold for any perturbation and so the coefficients must also be the initial condition. This means that for $t = 0$ all the higher-order corrections are 0. We bring all the above arguments together and conclude that to the first order in λ the perturbative correction is

$$b_n^{(1)}(t) = -\frac{i}{\hbar}\int dt \, \exp(i\,\omega_{nm}t)\left\langle n(0)\left|\hat{W}(t)\right|m(0)\right\rangle. \tag{6.37}$$

6.3 The Variational Method

It is possible to show that the mean value $\langle \hat{H} \rangle$ of the Hamiltonian is stationary in the neighbourhood of its discrete eigenvalues. As such, we should be able to apply a variational method to calculate all such eigenstates of the Hamiltonian. In practice, the method is inherently error-prone, and a proper exposition is beyond the scope of this text. We will limit our discussion to a case that is much more robust: the problem of finding the ground state of the TISE

$$\hat{H} |\psi_n\rangle = E_n |\psi_n\rangle.$$

Any random initial guess $|\psi\rangle$ can be written in terms of the (unknown) eigenvectors of $\hat{H}$

$$|\psi\rangle = \sum_k c_n |\psi_n\rangle.$$

The mean value of the Hamiltonian with for the state $|\psi\rangle$ is

$$\langle \hat{H} \rangle = \frac{\langle \psi | \hat{H} | \psi \rangle}{\langle \psi | \psi \rangle} = \frac{\langle \psi | \hat{I} \hat{H} \hat{I} | \psi \rangle}{\langle \psi | \psi \rangle} = \sum_{k,k'} \frac{\langle \psi | | k' \rangle \langle k' | \hat{H} | k \rangle \langle k | | \psi \rangle}{\langle \psi | \psi \rangle}$$

$$= \sum_{k,k'} E_k \frac{\langle \psi | k' \rangle \overbrace{\langle k' | k \rangle}^{\delta_{k'k}} \langle k | \psi \rangle}{\langle \psi | \psi \rangle} = \sum_k E_k \frac{\langle \psi | k \rangle \langle k | \psi \rangle}{\langle \psi | \psi \rangle}$$

$$\geq \sum_k E_0 \frac{\langle \psi | k \rangle \langle k | \psi \rangle}{\langle \psi | \psi \rangle} = E_0 \sum_k \frac{\langle \psi | k \rangle \langle k | \psi \rangle}{\langle \psi | \psi \rangle} = E_0 \frac{\langle \psi | \psi \rangle}{\langle \psi | \psi \rangle} = E_0. \tag{6.38}$$

This expression is a rather involved way of stating the intuitive fact that the ground state and its associated eigenvalue are the pair that minimise the mean value of the Hamiltonian

$$\langle \hat{H} \rangle = \frac{\langle \psi | \hat{H} | \psi \rangle}{\langle \psi | \psi \rangle} \geq E_0. \tag{6.39}$$

The challenge now is to minimise this expression. In some cases, we can choose a sensible first guess (called an *ansatz*) that is parameterised by some set of continuous variables and seek to minimise $\langle \hat{H} \rangle$ by varying these parameters. More often than not, we have to turn to computational methods. One can, for example, test a random sample of initial states and use some algorithm to search for local minima. If there are very many such minima, such algorithms can be challenging to use successfully. Often in complex problems such as those found in quantum chemistry, alternative methods such as Hartree–Fock are used, but these are a subject for specialised study in their own right and beyond our scope.

6.4 Instantaneous Energy Eigenbasis

When a system has no analytic solution and therefore has to be solved computationally, the choice of a basis with which to find the solutions of the time-dependent Schrödinger equation is of great importance. An inefficient choice of basis increases computational time dramatically. We can make a contribution to the numerical analysis by using some analytic methods in advance. We consider the general problem of solving the TDSE

$$\hat{H}(t)\,|\psi(t)\rangle = i\,\hbar \frac{\partial}{\partial t}\,|\psi(t)\rangle\,, \tag{6.40}$$

where we know analytically or numerically the instantaneous energy eigenvalues and vectors:

$$\hat{H}(t)\,|\kappa(t)\rangle = E_\lambda(t)\,|\kappa(t)\rangle\,. \tag{6.41}$$

As the Hamiltonian is Hermitian, we know that its eigenvalues are real. Additionally, we assume that the eigenvectors of this operator will (or can be made to) form a complete orthonormal basis set. We therefore know we can write $|\psi(t)\rangle$ as a linear combination of the states $|\kappa(t)\rangle$

$$|\psi(t)\rangle = \sum_\kappa c_\kappa(t)\,|\kappa(t)\rangle\,, \tag{6.42}$$

where the $c_\kappa(0)$ forms the known initial condition. By direct substitution into the Schrödinger equation:

$$\sum_\kappa \hat{H}(t)c_\kappa(t)\,|\kappa(t)\rangle = i\,\hbar \sum_\kappa \frac{\partial}{\partial t}\,\{c_\kappa(t)\,|\kappa(t)\rangle\}\,. \tag{6.43}$$

Now we use the fact that we are using eigenstates of the Hamiltonian as basis state, and we project on the left by another eigenstate $\langle \lambda(t)|$ to obtain

$$c_\lambda E_\lambda(t) = i\,\hbar \left\{ \frac{\partial}{\partial t}c_\lambda(t) + \sum_\kappa c_\kappa(t) \left\langle \lambda(t) \left| \frac{\partial}{\partial t} \right| \kappa(t) \right\rangle \right\}\,. \tag{6.44}$$

Rearranging, we see that

$$\frac{\partial}{\partial t}c_\lambda(t) = -\frac{i}{\hbar}c_\lambda E_\lambda(t) - \sum_\kappa c_\kappa(t) \left\langle \lambda(t) \left| \frac{\partial}{\partial t} \right| \kappa(t) \right\rangle\,, \tag{6.45}$$

which is a set of simple first-order differential equations, where the complexity is entirely encoded in the matrix element $\left\langle \lambda(t) \left| \frac{\partial}{\partial t} \right| \kappa(t) \right\rangle$. Just as we did in time-dependent perturbation theory, we can introduce a $c_n(t) = b_n(t)\exp\left(-i\frac{E_n}{\hbar}t\right)$ to simplify the equation further:

$$\frac{\partial}{\partial t}b_\lambda(t) = -\sum_\kappa b_\kappa(t)\exp\left(-i\,\omega_{\kappa\lambda}t\right)\left\langle \lambda(t) \left| \frac{\partial}{\partial t} \right| \kappa(t) \right\rangle\,, \tag{6.46}$$

where $\omega_{nk} = \frac{E_n - E_k}{\hbar}$.

To understand how $\left\langle \lambda(t) \left| \frac{\partial}{\partial t} \right| \kappa(t) \right\rangle$ may be efficiently determined, let us assume that each energy eigenstate can be written in terms of some time-independent basis

$$|\kappa(t)\rangle = \sum_n v_{\kappa,n}\,|n\rangle\,, \tag{6.47}$$

so that

$$\left\langle \lambda(t) \left| \frac{\partial}{\partial t} \right| \kappa(t) \right\rangle = \sum_m \sum_n v^*_{\lambda,m} \left(\frac{\partial}{\partial t} v_{\kappa,n} \right) \langle m|n \rangle$$

$$= \sum_n v_{\lambda,m} \frac{\partial v_{\kappa,m}}{\partial t}.$$

It is usual that the time-dependence can be expressed in terms of a function of time, say $q(t)$ (usually this would be a drive term), so that we can use the chain rule to write

$$\left\langle \lambda(t) \left| \frac{\partial}{\partial t} \right| \kappa(t) \right\rangle = \frac{\partial q(t)}{\partial t} \sum_n v_{\lambda,n} \frac{\partial v_{\kappa,n}}{\partial q(t)}. \tag{6.48}$$

The coefficients

$$V_{\lambda\kappa}(q) = \sum_n v_{\lambda,n} \frac{\partial v_{\kappa,n}}{\partial q(t)}$$

can all be pre-computed, sometimes very efficiently (especially if the external drive is a periodic or in some other way well bounded function).

$$\frac{\partial}{\partial t} b_\lambda(t) = -\frac{\partial q(t)}{\partial t} \sum_\kappa b_\kappa(t) \exp\left(-i\,\omega_{\kappa\lambda}t\right) V_{\lambda\kappa}(q(t)). \tag{6.49}$$

The solution to the problem of solving the TDSE is now reduced to finding the coefficients b_κ. This is just a set of first-order coupled differential equations.

6.5 Moving Basis

This trick for semi-analytical solving of the time-dependent Schrödinger equation is useful for supporting numerical methods and can be used in conjunction with the method described in Section 6.4. The idea is to use a dynamic basis so that a given basis remains an efficient one in which to represent the problem. This may be expressed in any sensible way, such as through the use of a displacement operator to centre a chosen basis near the 'middle' of the solution state vector.

Just as with the interaction picture, we assume that there is a some unitary operator, $\hat{U}(t)$, that makes the required transformation (evolution and displacement operators are two suitable examples). Noting that because of its unitary nature ($\hat{U}^\dagger(t)\hat{U}(t) = \hat{U}(t)\hat{U}^\dagger(t) = \hat{\mathbb{1}}$), the Schrödinger equation can be written as

$$\hat{H}(t)|\psi(t)\rangle = i\,\hbar \frac{\partial}{\partial t} |\psi(t)\rangle \tag{6.50}$$

$$\implies \hat{U}^\dagger(t)\hat{H}(t)\hat{U}(t)\hat{U}^\dagger(t) |\psi(t)\rangle = i\,\hbar\hat{U}^\dagger(t)\frac{\partial}{\partial t} |\psi(t)\rangle. \tag{6.51}$$

Now note that from the product rule:

$$\frac{\partial}{\partial t} \hat{U}^\dagger(t) |\psi(t)\rangle = \hat{U}^\dagger(t)\frac{\partial}{\partial t} |\psi(t)\rangle + \left(\frac{\partial}{\partial t}\hat{U}^\dagger(t)\right) |\psi(t)\rangle, \tag{6.52}$$

so that Eq. (6.51) can be written as

$$\left[\hat{U}^\dagger(t)\hat{H}(t)\hat{U}(t) - i\hbar U^\dagger(t)\frac{\partial}{\partial t}U(t)\right]\hat{U}^\dagger(t)\,|\psi(t)\rangle = i\hbar\frac{\partial}{\partial t}\hat{U}^\dagger(t)\,|\psi(t)\rangle. \tag{6.53}$$

We then define states $|\phi(t)\rangle = \hat{U}^\dagger(t)\,|\psi(t)\rangle$ and an equivalent (equivalent in the way that the expectation values of observables are unchanged) Hamiltonian by

$$\hat{H}(t) = \hat{U}^\dagger(t)\hat{H}(t)\hat{U}(t) - i\hbar\hat{U}^\dagger(t)\frac{\partial}{\partial t}\hat{U}(t) \tag{6.54}$$

with the Schrödinger equation being

$$\hat{H}(t)\,|\phi(t)\rangle = i\hbar\frac{\partial}{\partial t}\,|\phi(t)\rangle. \tag{6.55}$$

Let us see show this may be of use by considering the specific example where the Hamiltonian is of the form $\hat{H}(\hat{q} + q(t))$, as might happen with a time-dependent drive. Then the translation operator

$$U(t) = \exp\left(-i\frac{q(t)Q}{\hbar}\right) \tag{6.56}$$

can in some cases simplify calculations.

6.6 Time periodic Systems and Floquet Theory

Prerequisite Material: §4.2.6: The Evolution Operator

There are many periodic systems whose quantum mechanics is of great interest. Spatial periodicity of crystals leads to the development of the so-called Bloch theory, of which many excellent treatments exist elsewhere in the literature (see, for example, [12]). Bloch's theory makes use of the fact that the Hamiltonian of a system that has a periodic potential is invariant under some spatial translation (translation by the lattice spacing in the direction of the periodicity of the lattice in that direction). The conclusion from this is that the expectation values of observables and the probability density function must also respect this symmetry. So up to some complex phase, the wave function must be periodic. For large crystals this assumption works well, but as the crystals get smaller work needs to be done to understand the edge effects.

In many quantum systems, we deal with Hamiltonians that have a time-periodic potential. Any system exposed to a monochromatic source will usually be approximated by just such a Hamiltonian (dealing with the more physical situation of turning on and off that source in a physically meaningful way is rarely covered in the literature). While there has been recent activity in so-called time crystals for systems with such potentials on a practical tool level, the analysis of such systems leads to finding solutions to the TDSE, whose expectation values are also periodic in time with the same period as the potential. This allows us to efficiently compute the time-averaged expectation values of any observable. The analysis we use is due to Floquet and like many methods can be applied outside quantum mechanics to systems with time-periodic potentials[1].

We assume that the potential satisfies $\hat{V}(t) = \hat{V}(t + \tau)$, and we will use $\mathscr{T} = \{y(t) \mid y(t + \tau) = y(t) \forall t \in \mathbb{R}\}$ to denote the space of square integrable functions that are periodic in time with period τ.

We start by noting that we can define an extended inner product space over that used for normal quantum mechanics by integrating out all the time dependence of the usual inner product i.e. $\int_{-\infty}^{\infty} \langle \phi | \psi \rangle$. If the time dependence of the system is periodic, we can write this in the form

$$\langle \phi(t) \mid \psi(t) \rangle_{\mathscr{T}} = \frac{1}{\tau} \int_0^\tau dt \, \langle \phi(t) \mid \psi(t) \rangle, \tag{6.57}$$

which is simply the time-average of the product of two periodic functions (with the same period) over one period.

Recall that the evolution operator is an operator solution of Schrödinger's equation, i.e.

$$\hat{H}(t) \, \hat{U}(t, t_0) = i\,\hbar \frac{\partial}{\partial t} \hat{U}(t, t_0),$$

where $\hat{U}(t, t_0) \mid \psi(t_0) \rangle = \mid \psi(t) \rangle$, which leads us back to the functional form of Schrödinger's equation

$$\hat{H}(t) \mid \psi(t) \rangle = i\,\hbar \frac{\partial}{\partial t} \mid \psi(t) \rangle. \tag{6.58}$$

As the potential is periodic, we know that the Hamiltonian is also periodic in time with the same period τ, $\hat{H}(t + \tau) = \hat{H}(t)$. In this case, the evolution operator has the following interesting and useful properties:

- $\hat{U}(t + \tau, \tau) = \hat{U}(t, 0)$
- $\hat{U}(t + \tau, 0) = \hat{U}(t, 0) \, \hat{U}(\tau, 0)$
- The eigenvectors of the evolution operator are periodic in time, with the same period as the Hamiltonian.

We will adopt $\mid u_n \rangle$ to denote the eigenvectors of the evolution operator, and μ_n to denote its eigenvalues. As the evolution operator is a unitary operator, its eigenvalues must lie on the unit circle in the complex plane.

Let us assume that we have solutions to the time-dependent Schrödinger equation $\mid \psi_n(t) \rangle$. These states can be written in the so-called Floquet state solution form,

$$\mid \psi_n(t) \rangle = \exp\left(-\frac{i\,\varepsilon_n t}{\hbar}\right) \mid \varphi_n(t) \rangle, \tag{6.59}$$

where we impose that ε_n are real numbers. If $\mid \varphi_n(t + \tau) \rangle = \mid \varphi_n(t) \rangle$, the states $\mid \varphi_n(t + \tau) \rangle$ are termed Floquet modes and have much in common with the Bloch modes for systems where

---

1 The argument presented here is re-worked from the summary of Floquet theory in Everitt's DPhil thesis, which used [38] as a key reference.

the potential is spatially periodic. By substituting Eq. (6.59) into Schrödinger's equation (6.58), this can be rearranged to obtain

$$\left\{ \hat{H}(t) - i\hbar \frac{\partial}{\partial t} \right\} |\varphi_n(t)\rangle = \varepsilon_n |\varphi_n(t)\rangle. \tag{6.60}$$

As an eigenvalue equation, this looks identical in form to the TISE but for the extended inner product space we defined in the discussion around Eq. (6.57). We can emphasise this similarity by defining

$$\hat{\mathfrak{H}} = \left\{ \hat{H}(q,t) - i\hbar \frac{\partial}{\partial t} \right\}, \tag{6.61}$$

so that

$$\hat{\mathfrak{H}} |\varphi_n\rangle = \varepsilon_n |\varphi_n\rangle.$$

Because of this analogous form to the TISE, the ε_n are termed quasi-energies. This become meaningful when the $|\varphi_n\rangle$ are Floquet modes.

The eigenvalues and vectors of the evolution operator and these quasi-energies and Floquet modes are closely related. To make the connection, let us again consider TDSE for the evolution operator

$$\hat{H}\hat{U} = i\hbar \frac{\partial}{\partial t}\hat{U}.$$

We know that the eigenvectors $|u_n(t)\rangle$ of the evolution operator states are periodic in time and, if they solve the TDSE, would satisfy the condition of being Floquet modes. Let us therefore see if they are, and postmultiply both sides of the preceding equation by $|u_n(t)\rangle$ to obtain

$$\hat{H}\hat{U} |u_n(t)\rangle = i\hbar \frac{\partial}{\partial t}\hat{U} |u_n(t)\rangle$$

$$\hat{H}\mu_n |u_n(t)\rangle = i\hbar \frac{\partial}{\partial t}\mu_n(t) |u_n(t)\rangle$$

$$\mu_n \hat{H}(t) |u_n(t)\rangle = i\hbar \left[\mu_n(t) \frac{\partial}{\partial t} |u_n(t)\rangle + |u_n(t)\rangle \frac{\partial}{\partial t}\mu_n(t) \right]$$

$$\mu_n(t)\hat{\mathfrak{H}} |u_n(t)\rangle = i\hbar \left[|u_n(t)\rangle \frac{\partial}{\partial t}\mu_n(t) \right]$$

$$\hat{\mathfrak{H}} |u_n(t)\rangle = i\hbar \left[\frac{1}{\mu_n(t)} \frac{\partial}{\partial t}\mu_n(t) \right] |u_n(t)\rangle. \tag{6.62}$$

By comparison with Eq. (6.60), we see that we can define quasi-energies by solving the differential equation

$$i\hbar \frac{1}{\mu_n(t)} \frac{\partial}{\partial t}\mu_n(t) = \varepsilon_n, \tag{6.63}$$

which has solutions of the expected form

$$\mu_n(t) = \exp\left(-\frac{i\varepsilon_n}{\hbar}t \right), \tag{6.64}$$

as $\mu_n(0) = 1$ since $\hat{U}(0)$ is the identity operator. Note that it is the eigenvalues $\mu_n(t)$ of the evolution operator that are known, and the quasi-energies are simply the phase defined by the above equations. Recall that the time evolution of a stationary state of a time-independent Hamiltonian is given by $\exp\left(-\frac{\mathrm{i}\,E_n}{\hbar}t\right)$, where E_n are the eigenenergies of the Hamiltonian. Noting this makes the analogy between quasi-energies and the eigenenergies of the TISE clearer.

The way that we find the $|u_n(t)\rangle$ and $\mu_n(t)$ is through the Monodromy operator (occasionally also termed the Floquet operator), which is defined as the operator that propagates any initial state over one period,

$$\hat{\mathfrak{F}} \overset{\mathrm{def}}{=} \hat{U}(\tau, 0). \tag{6.65}$$

Knowing that the eigenvectors of the evolution operator $\hat{U}$ are periodic, then if we take the eigenvector $|u_n(\tau)\rangle$ of $\hat{\mathfrak{F}}$ we know that this is also the same as $|u_n(0)\rangle$. Hence, if we use $|u_n(\tau)\rangle$ as initial conditions for the Schrödinger equation, its solutions will be the Floquet modes of the system, and the eigenvalues of each $|u_n(\tau)\rangle$ allow us to calculate the quasi-energy ε_n. As a note for computational methods, one can numerically find the evolution operator at any time by evolving unit vectors in any given basis to that time. This works, as each unit vector evolves into the corresponding column of the evolution operator in that representation (for the Monodromy operator, this needs to be done over one time-period). Eigenvalues and vectors can then be found numerically. Noting that $\mu(0) = 1$, we can then solve for the Floquet solutions by once numerically solving the TDSE with these initial conditions. The algorithm thus integrates the Schrödinger equation over one choice period twice and involves solving one eigenproblem. This can be substantially more computationally efficient than taking time averages over very many periods of other states.

The $|u_n(t)\rangle$ are not the only Floquet modes of a system. In fact, for any given Floquet mode $|\varphi_n(t)\rangle$ we can construct infinitely more using

$$\left|\varphi_n^{(N)}(t)\right\rangle = \exp\left(-\mathrm{i}\,N\Omega t\right)|\varphi_n(t)\rangle,$$

where the index N is some integer and $\Omega = 2\pi/\tau$ is the frequency associated with the period of the Hamiltonian. By substitution into Eq. (6.60) we obtain

$$\hat{\mathfrak{H}}\left|\varphi_n^{(N)}(t)\right\rangle = \varepsilon_n\left|\varphi_n^{(N)}(t)\right\rangle$$

$$\left\{\hat{H}(t) - \mathrm{i}\,\hbar\frac{\partial}{\partial t}\right\}\exp\left(-\mathrm{i}\,N\Omega t\right)|\varphi_n(t)\rangle = \varepsilon_n\exp\left(-\mathrm{i}\,N\Omega t\right)|\varphi_n(t)\rangle$$

$$\hat{\mathfrak{H}}\,|\varphi_n(t)\rangle - N\hbar\Omega\,|\varphi_n(t)\rangle = \varepsilon_n\,|\varphi_n(t)\rangle$$

$$\hat{\mathfrak{H}}\,|\varphi_n(t)\rangle = (\varepsilon_n + N\hbar\Omega)\,|\varphi_n(t)\rangle. \tag{6.66}$$

We notice that each Floquet mode, $|\varphi_n(t)\rangle$, solves Eq. (6.60) but with quasi-energies shifted by $N\hbar\Omega$. Hence, in the same way as we define bands in momentum space (with period $\hbar k$) in Bloch theory, we can map all the quasi-energies (eigenvalues of $\hat{\mathfrak{H}}$) into an interval of length $\hbar\Omega$. This has led people to recently use terminology such as time crystals for such systems. In conventional quantum mechanics, time is not an observable and some caution should be used when comparing Bloch theory with Floquet theory, as there are some differences.

There is a special connection between the time-averaged energy expectation value and the Floquet energies. To show this, we start by recalling from Eq. (6.59) that the actual solutions to the TDSE are $|\psi_n(t)\rangle = \exp\left(-\frac{\mathrm{i}\,\varepsilon_n t}{\hbar}\right)|\varphi_n(t)\rangle$. For any such given solution, the time-averaged expectation value of any operator $\hat{A}$, $\overline{\langle A \rangle}$, can be found from the extended inner product, Eq. (6.57), and are

$$
\begin{aligned}
\overline{\langle A \rangle} &= \left\langle \psi_n(t) \left| \hat{A} \right| \psi_n(t) \right\rangle_{\mathcal{T}} \\
&= \left\langle e^{-\frac{i\,\varepsilon_n t}{\hbar}} \varphi_n(t) \left| \hat{A} \right| e^{-\frac{i\,\varepsilon_n t}{\hbar}} \varphi_n(t) \right\rangle_{\mathcal{T}} \\
&= \left\langle \varphi_n(t) \left| \hat{A} \right| \varphi_n(t) \right\rangle_{\mathcal{T}}.
\end{aligned}
\tag{6.67}
$$

This has simplified nicely due to the fact that time, because it is not an operator, commutes with $\hat{A}$. Next, observe that Eq. (6.60) may be expressed in the form

$$
\hat{H}(q, t) \, |\varphi_n(t)\rangle = \varepsilon_n \, |\varphi_n(t)\rangle + i\hbar \frac{\partial}{\partial t} |\varphi_n(t)\rangle.
$$

By substitution into Eq. (6.67) we find

$$
\begin{aligned}
\overline{\langle \hat{H} \rangle}_n &= \left\langle \varphi_n(t) \left| \hat{H} \right| \varphi_n(t) \right\rangle_{\mathcal{T}} \\
&= \varepsilon_n + i\hbar \left\langle \varphi_n \left| \frac{\partial}{\partial t} \varphi_n \right\rangle_{\mathcal{T}}.
\end{aligned}
\tag{6.68}
$$

Due to the periodic nature of the Floquet modes $|\varphi_n(t)\rangle$, we can expand them as a Fourier series of the form

$$
|\varphi_n(t)\rangle = \sum_k |c_k\rangle \exp(-i\, k\Omega t),
$$

which leads to

$$
\overline{\langle \hat{H} \rangle}_n = \varepsilon_n + \sum_k \hbar \Omega k \, \langle c_k | c_k \rangle.
$$

The time-averaged energy expectation value of the Hamiltonian can be understood as the quasi-energy plus a weighted distribution of the energy associated with each of the Floquet modes associated with that solution.

6.7 Two-level Systems

There are inherently two-state systems, such as electron spin. There are also many systems which can be described by two states-of-interest that are, to a good approximation, orthogonal. For example, there is the ammonia microwave amplification by stimulation emission of radiation (MASER) where the two such states correspond to the orientations of the nitrogen atom with regard to the plane containing the hydrogen atoms (see [27] for example). As we shall see in the next chapter, we can identify such states that are very useful for understanding the dihydrogen molecule. In each case, we understand that there are very complex state vectors and system Hamiltonian which describe the full state of the system, but we are only interested in two possible states, $|\psi_1\rangle$ and $|\psi_2\rangle$, that are approximately orthogonal and so satisfy $\langle \psi_1 | \psi_2 \rangle \approx \delta_{ij}$. For this reason, we will take $\{ |\psi_a\rangle, \{\psi_b\} \}$ as an approximate basis and write

$$|\psi_a\rangle \longrightarrow \begin{pmatrix} \langle \psi_1 | \psi_1 \rangle \\ \langle \psi_2 | \psi_1 \rangle \end{pmatrix} \approx \begin{pmatrix} 1 \\ 0 \end{pmatrix} \text{ and } |\psi_2\rangle \longrightarrow \begin{pmatrix} \langle \psi_1 | \psi_2 \rangle \\ \langle \psi_2 | \psi_2 \rangle \end{pmatrix} \approx \begin{pmatrix} 0 \\ 1 \end{pmatrix}. \tag{6.69}$$

Any state $|\psi\rangle$ can therefore be approximately written as

$$|\psi\rangle \longrightarrow \begin{pmatrix} \langle \psi_1 | \psi \rangle \\ \langle \psi_2 | \psi \rangle \end{pmatrix} \stackrel{\text{def}}{=} \begin{pmatrix} \psi_1 \\ \psi_2 \end{pmatrix}, \tag{6.70}$$

where we have defined the numbers $\psi_i = \langle \psi_i | \psi \rangle$. The Hamiltonian can be approximated by

$$H = \begin{pmatrix} \langle \psi_1 | \hat{H} | \psi_1 \rangle & \langle \psi_1 | \hat{H} | \psi_2 \rangle \\ \langle \psi_2 | \hat{H} | \psi_1 \rangle & \langle \psi_2 | \hat{H} | \psi_2 \rangle \end{pmatrix} \stackrel{\text{def}}{=} \begin{pmatrix} H_{11} & H_{12} \\ H_{12}^* & H_{22} \end{pmatrix}.$$

It is also often the case that the Hamiltonian is time-independent. In this case the evolution operator is

$$\hat{U}(t) = \exp\left\{ -\frac{i\hat{H}t}{\hbar} \right\},$$

which here has the rather simple expansion

$$\hat{U}(t) = \exp\left\{ -\frac{iE_1 t}{\hbar} \right\} |e_1\rangle \langle e_1| + \exp\left\{ -\frac{iE_2 t}{\hbar} \right\} |e_2\rangle \langle e_2|,$$

where E_i and $|e_i\rangle$ are the eigenvalues and eigenstates of the matrix H.

As an aside, let us note two other methods that can be of great usefulness for matrix methods. The first is Sylvester's formula which allows us to write an analytical function of a matrix (A) in terms of its eigen-decomposition

$$f(A) = \sum_{i=1}^{k} f(\lambda_i) A_i, \tag{6.71}$$

where λ_i are the eigenvalues of the matrix (A) and

$$A_i \equiv \prod_{j=1, j\neq i}^{k} \frac{1}{\lambda_i - \lambda_j} (A - \lambda_j \hat{\mathbb{1}}) \tag{6.72}$$

are the Frobenius covariants of A. The second, which we later discuss in the computational methods chapter, is an expression that is very useful, specifically when working with two-state systems, and using a scaled unit vector dot product with a vector of Pauli operators:

$$f(\boldsymbol{a} \cdot \boldsymbol{\sigma}) = \frac{f(a) + f(-a)}{2} \hat{\mathbb{1}} + \frac{f(a) - f(-a)}{2} (\boldsymbol{n} \cdot \boldsymbol{\sigma}), \tag{6.73}$$

particularly the special case

$$\exp i(\boldsymbol{a} \cdot \boldsymbol{\sigma}) = \cos a\, \hat{\mathbb{1}} + i \sin a(\boldsymbol{n} \cdot \boldsymbol{\sigma}). \tag{6.74}$$

In this section, we will apply the evolution operator in the eigenbasis of the Hamiltonian to understand the dynamics of several forms of time-independent Hamiltonians. The only constraint, we will impose, is that H_{12} is real, as this is the most common case considered (extension to the general case is not hard and left to the reader).

6.7.1 Non-degenerate Uncoupled System

This very simple example considers two distinct states that are completely independent of each other. We include this example simply for comparison with the following analysis of coupled systems. In this case, we assume $H_{11} = \alpha$, $H_{22} = \gamma$, and $H_{12} = 0$. So

$$(\hat{H}) = \begin{pmatrix} \alpha & 0 \\ 0 & \gamma \end{pmatrix}.$$

Unsurprisingly, the eigenvectors for this matrix are

$$|e_1\rangle = |\psi_1\rangle$$
$$|e_2\rangle = |\psi_2\rangle$$

with eigenvalues $E_1 = \alpha$ and $E_2 = \gamma$, respectively. The evolution operator is then

$$(\hat{U}(t)) = \exp\left\{-\frac{i\,\alpha t}{\hbar}\right\} \begin{pmatrix} 1 & 0 \\ 0 & 0 \end{pmatrix} + \exp\left\{-\frac{i\,\gamma t}{\hbar}\right\} \begin{pmatrix} 0 & 0 \\ 0 & 1 \end{pmatrix}$$

$$(\hat{U}(t)) = \begin{pmatrix} \exp\left\{-\frac{i\,\alpha t}{\hbar}\right\} & 0 \\ 0 & \exp\left\{-\frac{i\,\gamma t}{\hbar}\right\} \end{pmatrix}. \tag{6.75}$$

The action of the evolution operator on some arbitrary initial state $|\psi(0)\rangle = a\,|\psi_1\rangle + b\,|\psi_2\rangle$ gives

$$|\psi(t)\rangle = \begin{pmatrix} a\exp\left\{-\frac{i\,\alpha t}{\hbar}\right\} \\ b\exp\left\{-\frac{i\,\gamma t}{\hbar}\right\} \end{pmatrix}$$

and so each element of the initial state evolves independently with a phase determined by its eigenvalue

$$|\psi(t)\rangle = a\exp\left\{-\frac{i\,\alpha t}{\hbar}\right\}|\psi_1\rangle + b\exp\left\{-\frac{i\,\gamma t}{\hbar}\right\}|\psi_2\rangle,$$

which shows that the probability of obtaining $|\psi_1\rangle$ or $|\psi_2\rangle$ remains $|a|^2$ and $|b|^2$, respectively, for all time as expected (for completeness this is shown in Figure 6.1).

6.7.2 Non-degenerate Coupled System

We can generalise the previous approach for all cases of diagonal and off diagonal elements. We will consider a Hamiltonian whose matrix can be written

$$(\hat{H}) = \begin{pmatrix} \alpha & \beta \\ \delta & \gamma \end{pmatrix} \tag{6.76}$$

Figure 6.1 Time evolution plot of the probabilities associated to measuring state ψ_1 (solid) and ψ_2 (dashed).

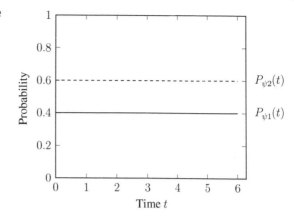

where, as $\hat{H}$ is Hermitian, $\beta = \delta^*$. The eigenvalues will be

$$E_1 = \alpha + \gamma - \Delta \tag{6.77}$$

$$E_2 = \alpha + \gamma + \Delta, \tag{6.78}$$

where $\Delta = \sqrt{(\alpha - \gamma)^2 + 4|\beta|^2}$. The eigenvectors are the normalised versions of

$$e_1 = \begin{pmatrix} (E_2 - \gamma)/c \\ 1 \end{pmatrix} \text{ and } e_2 = \begin{pmatrix} (E_1 - \gamma)/c \\ 1 \end{pmatrix}. \tag{6.79}$$

As in the previous examples, we can then write the evolution operator as a projector formed of these eigenvalues and eigenvectors.

However, as it is a simple example, we will use this as an opportunity to demonstrate the more general approach that Sylvester's formula, Eq. (6.71), makes available to us. To keep the analysis that follows as general as possible, we have not made use of the assumption that $\beta = \delta^*$. The eigenvalues are then

$$E_1 = \frac{\alpha + \gamma}{2} + \sqrt{\frac{(\alpha - \gamma)^2}{4} + \beta\delta} = C + D$$

$$E_2 = \frac{\alpha + \gamma}{2} - \sqrt{\frac{(\alpha - \gamma)^2}{4} + \beta\delta} = C - D.$$

We next calculate the Frobenius covariants, A_i, from Eq. (6.72). As there are only two eigenvalues, their product is not needed. If the matrices were larger, the product would come into effect here and one would need to cycle through all eigenvalues $\lambda_{j \neq i}$. The Frobenius covariants are given by:

$$A_1 = \frac{1}{E_1 - E_2}(A - E_2 \hat{\mathbb{I}})$$

$$= \frac{1}{2D}\begin{pmatrix} \alpha - (C - D) & \beta \\ \delta & \gamma - (C - D) \end{pmatrix}$$

$$A_2 = \frac{1}{E_2 - E_1}(A - E_1 \hat{\mathbb{I}})$$

$$= -\frac{1}{2D}\begin{pmatrix} \alpha - (C + D) & \beta \\ \delta & \gamma - (C + D) \end{pmatrix}.$$

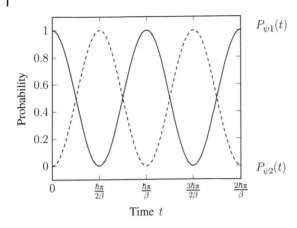

Probability

Time t

$P_{\psi 1}(t)$

$P_{\psi 2}(t)$

Figure 6.2 Time evolution plot of the probabilities associated to measuring state ψ_1 (solid) and ψ_2 (dashed).

The evolution operator can therefore be written as

$$\hat{U}(t) = \frac{1}{2D} \left[\exp\left(-\frac{i}{\hbar}(C+D)t\right) \begin{pmatrix} \alpha - (C-D) & \beta \\ \delta & \gamma - (C-D) \end{pmatrix} \right.$$
$$\left. - \exp\left(-\frac{i}{\hbar}(C-D)t\right) \begin{pmatrix} \alpha - (C+D) & \beta \\ \delta & \gamma - (C+D) \end{pmatrix} \right].$$

We will make use of this expression in the next chapter for the Jaynes–Cummings model and the Stern–Gerlach experiment. Its action on some, model-independent, arbitrary state is

$$|\psi(0)\rangle = a\,|\psi_1\rangle + b\,|\psi_2\rangle,$$

so that we can write

$$|\psi(t)\rangle = \hat{U}(t)\,|\psi(0)\rangle = a(t)\,|\psi_1\rangle + b(t)\,|\psi_2\rangle,$$

which yields

$$a(t) = \frac{1}{2D}\left(\exp\left(-\frac{i}{\hbar}(C+D)t\right)((\alpha - (C-D))a + \beta b)\right.$$
$$\left. - \exp\left(-\frac{i}{\hbar}(C-D)t\right)(\alpha - (C+D)) + \beta b)\right)$$
$$b(t) = \frac{1}{2D}\left(\exp\left(-\frac{i}{\hbar}(C+D)t\right)(\delta a + (\gamma - (C-D))b)\right.$$
$$\left. - \exp\left(-\frac{i}{\hbar}(C-D)t\right)(\delta a + (\gamma - (C+D))b)\right).$$

Here $a(t)$ and $b(t)$ are the time-dependent coefficients associated with each state $|\psi_1\rangle$ and $|\psi_2\rangle$, respectively. The significance of this result is that at time $t = 0$, the probabilities of obtaining $|\psi_1\rangle$ or $|\psi_2\rangle$ are no longer independent; this is why we call $\beta = \delta^*$ the coupling term. An example of the dynamics is shown in Figure 6.2.

6.7.3 Degenerate Coupled System

An important example occurs when the states $|\psi_1\rangle$ and $|\psi_2\rangle$ are degenerate (have the same energy expectation values), and where the Hamiltonian provides a non-zero coupling between the two states. Note that we take $|\psi_1\rangle$ and $|\psi_2\rangle$ to be orthogonal eigenstates of the uncoupled Hamiltonian (which may imply that each state contains a lot of hidden information if we are using, e.g. atomic eigenstates as each basis vector). The Hamiltonian can therefore be written

$$(\hat{H}) = \begin{pmatrix} \alpha & \beta \\ \beta & \alpha \end{pmatrix},$$

where, for simplicity, we are assuming β is real. The eigenvectors for this matrix are

$$|e_1\rangle = \frac{1}{\sqrt{2}}(|\psi_1\rangle + |\psi_2\rangle)$$

$$|e_2\rangle = \frac{1}{\sqrt{2}}(|\psi_1\rangle - |\psi_2\rangle)$$

with eigenvalues $E_1 = \alpha + \beta$ and $E_2 = \alpha - \beta$. For this type of system, we see that (i) the coupling 'lifts' the degeneracy of the eigenstates of the uncoupled systems; (ii) the symmetric and antisymmetric superpositions of the basis state are the energy states of the Hamiltonian; and (iii) depending on the sign of α and β, either the symmetric or antisymmetric superposition of states will have the lowest value and represent the system's ground state. Our construction of the evolution operator yields

$$(\hat{U}(t)) = \exp\left\{-\frac{i(\alpha + \beta)t}{\hbar}\right\} \begin{pmatrix} 1 & 1 \\ 1 & 1 \end{pmatrix} + \exp\left\{-\frac{i(\alpha - \beta)t}{\hbar}\right\} \begin{pmatrix} 1 & -1 \\ -1 & 1 \end{pmatrix},$$

which, using Euler's formula, can be rewritten as

$$(\hat{U}(t)) = \exp\left\{-\frac{i\alpha t}{\hbar}\right\} \begin{pmatrix} \cos\left(\frac{\beta t}{\hbar}\right) & -i\sin\left(\frac{\beta t}{\hbar}\right) \\ -i\sin\left(\frac{\beta t}{\hbar}\right) & \cos\left(\frac{\beta t}{\hbar}\right) \end{pmatrix}. \tag{6.80}$$

If we assume the system starts in one of the possible basis state, $|\psi(0)\rangle = |\psi_1\rangle$ for example, applying the evolution operator to give the states' dynamics gives

$$|\psi(t)\rangle = \hat{U}(t)|\psi_1\rangle$$

$$|\psi(t)\rangle = \exp\left\{-\frac{i\alpha t}{\hbar}\right\} \begin{pmatrix} \cos\left(\frac{\beta t}{\hbar}\right) & -i\sin\left(\frac{\beta t}{\hbar}\right) \\ -i\sin\left(\frac{\beta t}{\hbar}\right) & \cos\left(\frac{\beta t}{\hbar}\right) \end{pmatrix} \begin{pmatrix} 1 \\ 0 \end{pmatrix}$$

$$= \exp\left\{-\frac{i\alpha t}{\hbar}\right\} \begin{pmatrix} \cos\left(\frac{\beta t}{\hbar}\right) \\ -i\sin\left(\frac{\beta t}{\hbar}\right) \end{pmatrix}$$

and so the system oscillates back and forth between the two basis state according to

$$|\psi(t)\rangle = \exp\left\{-\frac{i\alpha t}{\hbar}\right\} \left(\cos\left(\frac{\beta t}{\hbar}\right)|\psi_1\rangle - i\sin\left(\frac{\beta t}{\hbar}\right)|\psi_2\rangle\right).$$

When we discuss systems such as dihydrogen in the next chapter, we will understand that one of the most important observations here is not the system's dynamics but rather that the energy eigenvectors are symmetric and antisymmetric superpositions of the basis state – can you guess why?

7

Applications and Examples

"Science is beautiful when it makes simple explanations of phenomena or connections between different observations."

Stephen Hawking

7.1 Introduction

In this chapter, we will take the tools that we have learned from previous chapters, and begin to apply them to some popular models within quantum mechanics, with the primary objective to find properties such as the energy eigenvalues, energy eigenstates, and the time evolution of the system. We will begin with the position representation, as that is the historical starting point of Schrödinger wave mechanics. We next consider the simple harmonic oscillator, which is a fundamental model due to the fact that most systems resemble a harmonic oscillator to some order of approximation. The analysis for the harmonic oscillator will be presented algebraically, and we will once more demonstrate the great utility of ladder operators in understanding and representing quantum systems (see Exercise 1.13 on page 17). We have previously seen these arising from the commutation relation of the position operator (page 78) and the displacement operator, as well as in our study of angular momentum (page 137). In the example of the simple harmonic oscillator, we use a ladder operator for the Hamiltonian, allowing us to determine the energy eigenvalues and vectors algebraically rather than by solving a differential wave equation. This is not only very elegant, but also proves to be a useful tool in representing quantum systems using matrix mechanics. The next system we study is hydrogen, and we do this to the level where we can see that quantum mechanics explains the physics behind spectroscopic notation. We next look at the dihydrogen$^+$ ion as an example of the power of approximating complex quantum systems in terms of a few physical basis state, as we obtain a qualitative understanding of atomic bonding from the analysis. The last two applications we look at are for composite systems. The first is the Jaynes–Cummings model, which is a very simplified representation of light–matter interaction that also demonstrates the Heisenberg matrix mechanics well. There is a very rich and growing body of research literature on this simple model. It is our intention that this section should introduce sufficient content to serve as a platform for engaging with that work and to illustrate that there is active research in topics that are quite

Quantum Mechanics: From Analytical Mechanics to Quantum Mechanics, Simulation, Foundations, and Engineering,
First Edition. Mark Julian Everitt, Kieran Niels Bjergstrom and Stephen Neil Alexander Duffus.
© 2024 John Wiley & Sons Ltd. Published 2024 by John Wiley & Sons Ltd.
Companion website: www.wiley.com/go/everitt/quantum

accessible to those new to the field. Finally, we consider the Stern–Gerlach experiment with more rigour than in Section 2.6.2. We use an approach which combines position and spin representations into one, known as the spinor representation; the Stern–Gerlach experiment provides a simple example of its utility.

7.2 Position Representation Examples of Particles and Potentials

In this section, we will consider modelling a particle subject to potential barriers of different kinds in the position representation. We will examine four different potentials that are simple in form and do not depend on time. We note that a first course in quantum mechanics is often delivered entirely in the position (and maybe momentum) representation. For this reason, many textbooks already exist which contain many more examples than we have space to provide here. Our examples are chosen because they give good insight into the nature of quantum mechanical systems without becoming too cumbersome in their analysis.

In all cases we can write the Hamiltonian for a particle with mass m, momentum $\hat{p}$, and position $\hat{q}$ as:

$$\hat{H} = \frac{\hat{p}^2}{2m} + V(\hat{q}) = -\frac{\hbar^2}{2m}\frac{\partial^2}{\partial q^2} + V(q). \tag{7.1}$$

As there is no time-dependence in the Hamiltonian, we will seek to solve the time-independent Schrödinger equation (TISE) that takes the specific form:

$$\left[-\frac{\hbar^2}{2m}\frac{\partial^2}{\partial q^2} + V(q) \right] \psi(q) = E\psi(q), \tag{7.2}$$

where the dynamics of each energy eigenstate is

$$\psi(q, t) = \exp\left(-\frac{iE}{\hbar}t\right)\psi(q).$$

Recall that expectation values of these states do not evolve with time; this is why they are referred to as stationary states. All superpositions of these states are also solutions to the Schrödinger equation and, unlike the stationary energy eigenstates, those may have properties whose expectation values are not stationary.

7.2.1 The Free Particle

In the absence of any potential, the TISE is

$$-\frac{\hbar^2}{2m}\frac{d^2}{dq^2}\psi(q) = E\psi(q), \tag{7.3}$$

and its solutions are plane waves of the form

$$\psi(q) = A\exp(ikq) \text{ where } E = \frac{\hbar^2 k^2}{2m}.$$

This is an interesting set of states as they are not normalisable and so do not quite fit into the theoretical framework that we have constructed. The resolution to this difficulty is to assert that no particle is truly free and that there is some constraint on the domain. The issue with this is that it is in effect like changing the potential and, as we shall later see, particles bound by a potential have discrete energy eigenvalues and not a continuum of these solutions. Aside from these concerns, as the solutions to the Schrödinger equation are plane waves, this example allows us some insight into why we might expect particles to diffract in interference experiments such as Young's slits.

7.2.2 Infinitely Deep Potential Well

The next model we will consider is the contrasting example of a particle confined to an infinite potential well in one dimension. Here, two 'walls' of infinite potential energy constrain an otherwise free particle to the region between them (Figure 7.1). The potential is

$$V(q) - \begin{cases} \infty, & q < 0 \\ 0, & 0 < q < a. \\ \infty, & q > a \end{cases} \tag{7.4}$$

Since the particle is constrained so that it can only exist between the walls of the well, we only need to find the energy eigenvalues and eigenstates of the particle in this region. Within the potential barrier the TISE is the same as it was for the free particle, that is,

$$-\frac{\hbar^2}{2m}\frac{d^2\psi(q)}{dq^2} = E\psi(q). \tag{7.5}$$

Unlike in the preceding section, we have boundary conditions that have to be satisfied. The general solution to this equation is in fact a combination of plane waves of opposite wave-number

$$\psi(q) = Ae^{ikq} + Be^{-ikq}, \tag{7.6}$$

where, again, $k = \sqrt{2mE}/\hbar$ is the wave number of the particle. Since the particle is bound, we expect a superposition of left and right travelling waves to give rise to standing wave solutions of the form

$$\psi(q) = C\cos(kq) + D\sin(kq), \tag{7.7}$$

Figure 7.1 Potential energy for a particle in an infinite well.

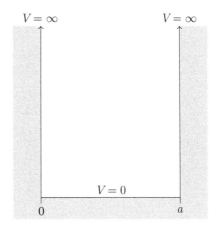

where $C = (A + iB)/2$ and $D = i(A - B)/2$ (according to Euler's formula). To find the constants C and D we use the boundary conditions of the infinite barrier height. We assume this means that the particle can only exist between the walls of the well, and nowhere else. This results in the condition that

$$\psi(0) = \psi(a) = 0. \tag{7.8}$$

Making use of this constraint on $\psi(q)$ at the $q = 0$ boundary gives us

$$\psi(0) = C\cos(0) + D\sin(0) = 0$$
$$\Rightarrow C = 0. \tag{7.9}$$

The wave function must therefore take the form

$$\psi(q) = D\sin(kq). \tag{7.10}$$

Our next step is to use the constraint at the other boundary:

$$\psi(a) = D\sin(ka) = 0. \tag{7.11}$$

For a non-trivial solution we need $D \neq 0$; thus, we can only conclude that $\sin(ka) = 0$ which means that k can only take on the following values

$$k = \frac{n\pi}{a}, \tag{7.12}$$

where n is a natural number. The solutions to the TISE are then

$$\psi(q) = D\sin\left(\frac{n\pi}{a}\right). \tag{7.13}$$

The fact that the wave number is $k = \frac{n\pi}{a} = \sqrt{2mE}/\hbar$ means that the energy eigenvalues can only be

$$E_n = \frac{\hbar^2 n^2 \pi^2}{2ma^2}. \tag{7.14}$$

We see that the energy spectrum of a bound particle is discrete. Recalling the measurement axiom of quantum mechanics, this result means that if we measure the energy of a particle in a potential well, we will only be able to measure one of these discrete eigenenergies. Moreover, immediately after the measurement the particle will be projected into the corresponding (non-degenerate) eigenstate. This result is an indication that quantum mechanics might well be able to explain the phenomenon behind emission and absorption spectra of atoms, where light of only very specific energy (or, equivalently, frequencies) is absorbed by atoms.

To determine the constant D, we use the normalisation condition

$$\langle \psi | \psi \rangle = 1 \tag{7.15}$$

which in position representation is

$$\int_{-\infty}^{\infty} \psi^*(q)\psi(q)dq = 1. \tag{7.16}$$

We can break this down into three regions:

$$\int_{-\infty}^{0} \psi^*(q)\psi(q)dq + \int_{0}^{a} \psi^*(q)\psi(q)dq + \int_{a}^{\infty} \psi^*(q)\psi(q)dq = 1. \tag{7.17}$$

Noting that the particle only exists in $0 < q < a$ means that $\psi(q)$ must be zero outside this region, so the first and last integrals are zero. Substitution of the wave-function in $0 < q < a$ into the above integral gives

$$\int_{0}^{a} D^2\sin^2\left(\frac{n\pi}{a}q\right)dq = 1. \tag{7.18}$$

Making use of the identity

$$\sin^2(\theta) = \frac{1}{2} - \frac{1}{2}\cos(2\theta) \tag{7.19}$$

we arrive at the result

$$D = \sqrt{\frac{2}{a}} \tag{7.20}$$

giving a final expression for the solution to the TISE:

$$\psi_n(q) = \sqrt{\frac{2}{a}}\sin\left(\frac{n\pi}{a}q\right), \tag{7.21}$$

where n denotes the nth energy eigenfunction of the system, with $n = 1$ corresponding to the ground state of the system.

Figure 7.2 shows the first three energy eigenfunctions for a particle in an infinite potential well displaced by the corresponding energy eigenvalues. We see that the energy levels become further and further apart as n increases. We also see that the number of nodes and anti-nodes increase with the energy level. The significance of this becomes apparent when we consider the probability of finding the particle in a particular region of the well at a fixed point in time. Since the probability density is given by the modulus-square of the wave function, we see that the particle is most likely to be found at an anti-node, and will not be found

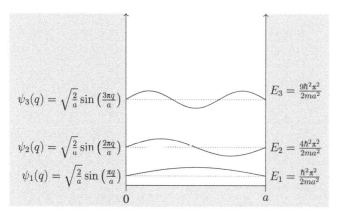

Figure 7.2 Energy eigenfunctions displaced vertically by their corresponding energy eigenvalues for a particle in an infinite well.

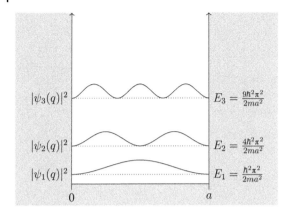

Figure 7.3 Probability densities for the first three energy eigenfunctions of the particle in an infinite well, displaced vertically by energy eigenvalue.

at a node. This is more clearly seen in the probability density functions $|\psi_n(q)|^2$ shown in Figure 7.3.

Exercise 7.1 Redo the analysis of this section but now define the well between $-a/2$ and $a/2$. Show that with this definition, even-numbered states have even symmetry and odd-numbered states have odd symmetry. ∎

7.2.3 The Finite Potential Barrier

The next standard example we present is that of a finite potential barrier. In Figure 7.4 we show a schematic of the potential, comprising a region where there is some finite potential, and a potential of zero everywhere else. We note that the particle is not bound to a finite domain; hence, we expect to find that it has a continuum of energy eigenvalues. The potential now takes the form

$$V(\hat{q}) = \begin{cases} 0, & q < 0 \\ V_0, & 0 < q < a, \\ 0, & a < q \end{cases} \tag{7.22}$$

where V_0 is a constant value. Solving the TISE for this potential is an exercise in solving one-dimensional partial differential equations in the three regions shown in Figure 7.4, and

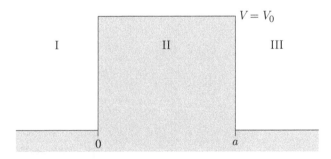

Figure 7.4 Schematic of a potential barrier of magnitude V_0 and width a with the regions needed to solve the time-independent Schrödinger equation labelled I, II, and III.

using matching conditions to find the solution over all space. In regions I and III, we see that the particle is a free particle biased at zero potential energy. The Schrödinger equation is that of our first example,

$$\hat{H}\psi(q) = -\frac{\hbar^2}{2m}\frac{d^2\psi(q)}{dq^2} = E\psi(q), \tag{7.23}$$

which has the general solution

$$\psi_I(q) = Ae^{ikq} + Be^{-ikq} \tag{7.24}$$

for region I and

$$\psi_{III}(q) = Fe^{ikq} + Ge^{-ikq} \tag{7.25}$$

for region III. In both cases $k = \sqrt{2mE}/\hbar$.

Note that since the particle is free without boundary we must assume that we have a linear combination of left- and right-travelling waves.

For region II we also have a free particle but with a non-zero potential energy, and the Schrödinger equation for this region is given by

$$\hat{H}\psi_{II}(q) = \left[-\frac{\hbar^2}{2m}\frac{d^2}{dq^2} + V_0\right]\psi_{II}(q) = E\psi_{II}(q), \tag{7.26}$$

which has the solution of the same form:

$$\psi_{II}(q) = Ce^{ik'q} + De^{-ik'q} \tag{7.27}$$

but with different wave-number $k' = \sqrt{2m(E - V_0)}/\hbar$. Our next step is to consider matching conditions. We are going to impose that the solutions to the Schrödinger equation are continuous so that they can be differentiated. The Schrödinger equation in this form is a wave equation, and as such we will borrow the language of waves from other topics, like electromagnetism.

If, as a specific example, we consider the case of an *incident wave from the left*, we would expect to see a chance of transmission over the barrier and a chance of reflection. We can say the same about the wave in region II, where transmission and reflection are also possible outcomes. In the third region, however, there is no boundary to reflect the particle, so we would only expect there to be a right-travelling particle in that region. Since there are no left-travelling waves in region III, we can constrain the solution in such a way that

$$G = 0. \tag{7.28}$$

We can also redefine our coefficients to better describe transmission and reflection with respect to the barrier. If we define the incident coefficient $A = 1$ as well as the reflection, r, and transmission, t, coefficients associated to the barrier $B = r$, $F = t$, we see that

$$\psi_I(q) = e^{ikq} + re^{-ikq} \tag{7.29}$$

$$\psi_{II}(q) = Ce^{ik'q} + De^{-ik'q} \tag{7.30}$$

$$\psi_{III}(q) = te^{ikq}. \tag{7.31}$$

It is our intention to find the remaining transmission and reflection amplitudes. To do this, we use the boundary conditions of our wave function, and eliminate the constants C and D. We note that we have imposed that the wave function must be smooth and continuous for all space. This means:

$$\psi_I(0) = \psi_{II}(0) \tag{7.32}$$

$$\psi_I'(0) = \psi_{II}'(0) \tag{7.33}$$

$$\psi_{II}(a) = \psi_{III}(a) \tag{7.34}$$

$$\psi_{II}'(a) = \psi_{III}'(a) \tag{7.35}$$

where once again $\psi'(q) = d\psi(q)/dq$ is the first-order derivative of the wave function with respect to position. Applying these conditions to each wave function, we get:

$$1 + r = C + D \tag{7.36}$$

$$ik(1 - r) = ik'(C - D) \tag{7.37}$$

$$Ce^{ik'a} + De^{-ik'a} = te^{ika} \tag{7.38}$$

$$ik'(Ce^{ik'a} - De^{-ik'a}) = ikte^{ika}. \tag{7.39}$$

Combining the first two equations yields expressions for C and D in terms of r:

$$C = \frac{1+r}{2} + \frac{k}{2k'}(1 - r) \tag{7.40}$$

$$D = \frac{1+r}{2} - \frac{k}{2k'}(1 - r). \tag{7.41}$$

Rearranging and equating the second two equations eliminates t:

$$Ce^{ik'a} + De^{-ik'a} = \frac{k'}{k}\left(Ce^{ik'a} - De^{-ik'a}\right). \tag{7.42}$$

Substitution of C and D into Eq. (7.42) and rearranging then gives an expression for the reflection coefficient r,

$$r = \frac{(k'^2 - k^2)\sin(k'a)}{2k'k\cos(k'a) - (k'^2 + k^2)i\sin(k'a)}. \tag{7.43}$$

This expression can subsequently be used to find the transmission coefficient

$$t = \frac{4k'ke^{-i(k-k')a}}{(k^2 + k'^2) - e^{2ik'a(k-k')^2}}. \tag{7.44}$$

When we take the modulus-square of these coefficients, we obtain the transmission and reflection probabilities. Doing so gives

$$T = |t|^2 = \frac{1}{1 + \frac{(k^2-k'^2)\sin^2(k'a)}{4k'^2k^2}} \tag{7.45}$$

$$R = |r|^2 = \frac{(k^2 - k'^2)\sin^2(k'a)}{4k^2k'^2\cos^2(k'a) + (k^2 + k'^2)\sin^2(k'a)}. \tag{7.46}$$

Note that the probabilities are conserved since $R + T = 1$.

What is of particular interest is the case when the particle's energy is lower than the potential barrier. We consider the transmission probability, T, once more while noting that $k = \sqrt{2mE}/\hbar$, and $k' = \sqrt{2m(E - V_0)}/\hbar$. This gives us the transmission probability in terms of the particle's energy and the barrier's height,

$$T = \frac{1}{1 + \frac{V_0^2 \sin^2(k'a)}{4E(E-V_0)}}.$$ (7.47)

Whilst a classical system would only see a particle 'bounce' off the potential barrier in this case, we see that due to the wave nature of the solution to the Schrödinger equation, there is a probability of the particle transmitting through the barrier, which is

$$T_{E<V_0} = \frac{1}{1 + \frac{V_0^2 \sinh^2(k'a)}{4E(V_0-E)}}.$$ (7.48)

This phenomenon is known as *tunnelling* and is seen in many systems that involve particles bound by a finite potential, such as alpha decay in radioactive elements and electron tunnelling through a $p - n$ junction in diodes.

7.2.4 Finite Potential Well

Our final one-dimensional position representation example looks at a particle in a finite potential well Figure 7.5. This is similar to our first example, but the walls of the well are finite in height and of infinite extent (the case of a barrier whose walls are finite in extent is a complication we avoid. Its analysis is left as an exercise to the keen reader). The potential energy for this example is

$$V(\hat{q}) = \begin{cases} V_0, \ q \leq -a/2 \\ 0, -a/2 < q < a/2, \\ V_0, \ q \geq a/2 \end{cases}$$

where, unlike in the first example, we have centred the well at the origin. We know from the potential barrier that when $E > V_0$, the solutions to the Schrödinger equation are plane waves whose wave numbers

$$k = \frac{\sqrt{2mE}}{\hbar}$$

$$k' = \frac{\sqrt{2m(E - V_0)}}{\hbar}$$

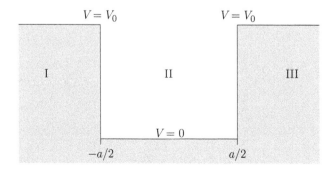

Figure 7.5 Particle in an infinite well.

are not constrained for the unbound particle, giving rise to a continuous spectrum of energy eigenvalues whose corresponding wave functions are comparable to the reflected and transmitted states seen in the potential barrier example (this is basically an upside-down barrier problem). For the case when $E < V_0$, we do have bound states, and so it is our goal to find the corresponding wave functions and permitted wave numbers. As with the potential barrier, we break the analysis up into three regions.

> Recall that the parity operator, $\hat{\Pi}\,|q\rangle = |-q\rangle$, has symmetric eigenstates with eigenvalues +1 and antisymmetric eigenstates with eigenvalues -1.

Note that, as the Hamiltonian is symmetric, it will commute with the parity operator. This means that there exist energy eigenfunctions which are also eigenstates of the parity operator (a powerful result for any system with such symmetry). For this reason, we will be able to assume that the energy eigenfunctions will be either of even or odd parity. We will use this fact in their construction. For regions I and III, when $E < V_0$, we expect solutions of an exponentially decaying form as seen in the potential barrier example:

$$\psi_I(q) = Ae^{-|k'|q} + Be^{|k'|q}$$
$$\psi_{III}(q) = Fe^{-|k'|q} + Ge^{|k'|q}.$$

In region II there is no potential and the wave functions will resemble those of the infinite potential well,

$$\psi_{II}(q) = C\cos(kq) + D\sin(kq).$$

Before we make use of the boundary conditions, we can find the values of some of the constants by considering the physicality of the solutions in regions I and III. For the limit $q \to -\infty$ for region I and $q \to \infty$ for region III the solutions are unbounded and non-physical if constants A and G are non-zero, leading to the conclusion that

$$A = G = 0$$

and so

$$\psi_I(q) = Be^{|k'|q}$$
$$\psi_{III}(q) = Fe^{-|k'|q}.$$

Within region II, recalling that the Hamiltonian commutes to the party operator, the solution will be either even or odd in parity. That is, in order to preserve this symmetry we are permitted two types of wave function within the well: even solutions

$$\psi_e(-q) = \psi_e(q)$$

and odd solutions

$$\psi_o(-q) = -\psi_o(q)$$

but not a combination of the two. Since this symmetry must prevail for all space, we come to the conclusion that

$$\psi_I(q) = Be^{|k'|q}$$

$$\psi_{II}(q) = C\cos(kq)$$
$$\psi_{III}(q) = Be^{-|k'|q}$$

for even solutions, and

$$\psi_{I}(q) = Be^{|k'|q}$$
$$\psi_{II}(q) = D\sin(kq)$$
$$\psi_{III}(q) = -Be^{-|k'|q}$$

for odd solutions. The types of solution can be seen in Figure 7.6, where we see exponential decay of the wave function outside the well, and standing wave solutions within it. Now making use of the boundary conditions of both odd and even solutions

$$\psi_{I}(-a/2) = \psi_{II}(-a/2)$$
$$\psi'_{I}(-a/2) = \psi'_{II}(-a/2)$$
$$\psi_{II}(a/2) = \psi_{III}(a/2)$$
$$\psi'_{II}(a/2) = \psi'_{III}(a/2).$$

Making use of these boundary conditions, we find that the wave numbers are constrained such that

$$|k'| = \frac{-k}{\tan(ka)}$$

for odd solutions, and

$$|k'| = k\tan(ka)$$

for even solutions. These conditions give a discrete spectrum for states bound to the finite potential well. Importantly, note that, despite the infinite extent of the finite height barrier, there is a finite chance of finding the particle outside the potential well, something which would not be allowed classically. If the barrier is also finite in extent there is an interesting quantum phenomenon, not seen in classical mechanics, the so-called quantum tunnelling effect. Specific details of how this manifests depends on symmetries and matching conditions between the inside and outside the well, and a systematic study are beyond the scope of this introductory text.

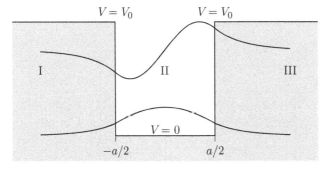

Figure 7.6 Examples of a ground and first excited state wave function for a particle in a finite potential well. The wave-functions have been displaced vertically by eigenvalue. Note the tunnelling into the potential barrier and the symmetry of each state.

7.3   The Harmonic Oscillator

One of the most important systems to study in quantum mechanics is the quantum harmonic oscillator (QHO). The QHO is a good approximation to many physical systems, such as the different modes of an electromagnetic field, or phonons within a lattice. Many more complex systems can also be approximated by the QHO, including superconducting quantum interference devices (but only for certain external magnetic flux bias), and even the helium atom (where the model is known as harmonium). Thus, the QHO is a model of particular importance, and for this reason it will be our starting point for this chapter. The Schrödinger equation for the QHO is exactly solvable, demonstrating once more the utility of ladder operators.

7.3.1   A Scheme for Creating Ladder Operators

Before we go into details of the harmonic oscillator, let us make the general observation that if any two Hermitian operators, $\hat{x}$ and $\hat{y}$, have a commutation relation of the form

$$[\hat{x}, \hat{y}] = i\beta, \tag{7.49}$$

where β does not need to be real – we have used the form $i\beta$ as it often is – then we can define a new operator

$$\hat{a} = \frac{1}{\sqrt{2}}(\hat{x} + i\hat{y}) \text{ so } \hat{a}^{\dagger} = \frac{1}{\sqrt{2}}(\hat{x} - i\hat{y}). \tag{7.50}$$

If we then additionally define $\hat{n} = \hat{a}^{\dagger}\hat{a}$, we find

$$[\hat{n}, \hat{a}] = -\beta\hat{a} \tag{7.51}$$

$$\left[\hat{n}, \hat{a}^{\dagger}\right] = \beta\hat{a}^{\dagger} \tag{7.52}$$

and so we see from Exercise 1.13 on page 17 that $\hat{a}$ and $\hat{a}^{\dagger}$ will be lowering and raising ladder operators for eigenstates of $\hat{n}$.

7.3.2   Ladder Operators and Some Eigenvalue Properties of the QHO

The Hamiltonian for the harmonic oscillator is given by

$$\hat{H} = \frac{\hat{p}^2}{2m} + \frac{1}{2}m\omega^2\hat{q}^2,$$

where m and ω are the constants attributed to the system's mass and natural angular frequency. Here $\hat{q}$ and $\hat{p}$ are canonical position and momentum operators that share the commutation relation

$$[\hat{q}, \hat{p}] = i\hbar. \tag{7.53}$$

We could work in the position representation and solve the resulting differential equation for the TISE. With the previous section in mind we see immediately from the commutation relation $[\hat{q},\hat{p}] = i\hbar$ that we can construct ladder operators for position and momentum and maybe this could be of use here. Specifically, if we equate

$$\hat{x} = \hat{Q} = \sqrt{\frac{m\omega}{\hbar}}\hat{q} \text{ and } \hat{y} = \hat{P} = \sqrt{\frac{1}{m\hbar\omega}}\hat{p}, \qquad (7.54)$$

then we have the simplification that $\beta = 1$, so $[\hat{x},\hat{y}] = [\hat{Q},\hat{P}] = i$. Note that we retain the $\hat{x}$ and $\hat{y}$ notations to relate this material to the previous section on ladder operators, but also introduce $\hat{Q}$ and $\hat{P}$ as dimensionless position and momentum to enable a more physical discussion later. We see that

$$\hat{a} = \frac{1}{\sqrt{2}}\left(\hat{Q} + i\hat{P}\right) \qquad (7.55)$$

$$\left[\hat{a},\hat{a}^{\dagger}\right] = \hat{\mathbb{1}} \qquad (7.56)$$

$$\left[\hat{n},\hat{a}\right] = -\hat{a} \qquad (7.57)$$

$$\left[\hat{n},\hat{a}^{\dagger}\right] = \hat{a}^{\dagger}. \qquad (7.58)$$

So, if $\hat{n}$ has an eigenvalue n with corresponding eigenvectors $|n\rangle$, then $\hat{a}\,|n\rangle$ (or $\hat{a}^{\dagger}\,|n\rangle$) is also an eigenstate of $\hat{n}$ with eigenvalue $n-1$ (or $n+1$).

Exercise 7.2 Using the definitions of $\hat{a}$ and $\hat{a}^{\dagger}$, show that

$$\frac{\hat{p}^2}{2m} + \frac{1}{2}m\omega^2\hat{q}^2 = \hbar\omega\left(\hat{a}^{\dagger}\hat{a} + \frac{1}{2}\right).$$

∎

From the above exercise, we see that we can factorise the Hamiltonian into the form

$$\hat{H} = \hbar\omega\left(\hat{a}^{\dagger}\hat{a} + \frac{1}{2}\right) = \hbar\omega\left(\hat{n} + \frac{1}{2}\right),$$

where $\hat{n} = \hat{a}^{\dagger}\hat{a}$. A first important observation that we find from writing the Hamiltonian in this factorised form is found by also noting that the action of an operator on a vector is simply another vector. So, for any state $|\psi\rangle$ we are able to define a new vector,

$$|\chi\rangle = \hat{a}\,|\psi\rangle$$

so that

$$\left\langle\psi|\hat{a}^{\dagger}\hat{a}|\psi\right\rangle = \langle\psi|\hat{n}|\psi\rangle = \langle\chi|\chi\rangle \geq 0.$$

This in turn implies

$$\left\langle\psi_0\left|\hat{H}\right|\psi_0\right\rangle = \left\langle\psi\left|\hbar\omega\left(\hat{a}^{\dagger}\hat{a} + \frac{1}{2}\right)\right|\psi_0\right\rangle$$

$$= \hbar\omega\underbrace{\left\langle\psi_0\left|\hat{a}^{\dagger}\hat{a}\right|\psi_0\right\rangle}_{\geq 0} + \frac{\hbar\omega}{2}\underbrace{\langle\psi_0|\psi_0\rangle}_{1}$$

$$\geq \frac{\hbar\omega}{2}.$$

Our first result of that is that the QHO has a zero-point energy, which for any state the energy expectation value must always match or exceed.

Now let us return to the consideration of $\hat{n}$ and the ladder operators. It is clear that $[\hat{H}, \hat{n}] = 0$. The fact that $\hat{H}$ and $\hat{n}$ commute means that they share eigenstates (and as these operators are Hermitian, we know that their eigenvectors can be made from an orthonormal basis). For this reason, we know that solving the TISE for the QHO reduces to the problem of solving the eigenvalue problem for $\hat{n}$. We have already seen that the expectation value for any state of $\hat{n}$ must be greater than or equal to zero. From the eigenvalue equation

$$\hat{n}\,|n\rangle = n\,|n\rangle$$

we see that $\langle n|\hat{n}|n\rangle = n$ and this means that $\hat{n} \geq 0$. We saw that ladder operators were useful in a similar situation when we discussed angular momentum, and so it is reasonable to expect the approach to be of use here. The first observation is that if n is a non-negative non-integer eigenvalue for eigenstate $|n\rangle$, we can repeatedly apply that $a\,|n\rangle$ is also an eigenstate of $\hat{n}$ but with the eigenvalue $n - 1$. In effect, we can do this so many times that eventually we would end up with a negative n (as the non-integer nature of the eigenvalue will skip 0). This contradicts our previous observation, that the eigenvalues n must be positive or zero. To satisfy both the general properties of ladder operators and the positivity of n, we see that (i) n must be a whole number; (ii) by repeated application of $\hat{a}$ to any non-zero $|n\rangle$ the lowest value of n is 0; and (iii) $\hat{a}\,|0\rangle = |\theta\rangle$, the zero or null vector (or else there are states with negative eigenvalues). As the eigenvalues of $\hat{n}$ are $0, 1, 2, \ldots$, it is known as the number operator. We now see that there is a non-zero eigenvector, the ground state $|0\rangle$, with eigenvalue zero $\hat{n}\,|0\rangle = 0\,|0\rangle$ (with $|0\rangle \neq |\theta\rangle$) and $\langle 0|\hat{n}|0\rangle = 0$ so $\left\langle 0\left|\hat{H}\right|0\right\rangle = \hbar\omega/2$ which means this is a zero-point energy that minimises the energy expectation $\left\langle \psi\left|\hat{H}\right|\psi\right\rangle$ for any state of a QHO.

Recall from Section 3.4.1.1 that the eigenstates of the position can be generated for the displacement operator according to

$$\hat{D}(q)\,|q = 0\rangle = |q\rangle, \tag{7.59}$$

where we have used the notation $|q = 0\rangle$ to distinguish the eigenstate of position with eigenvalue 0 from the QHO ground state $|0\rangle$. Now the position representation of the QHO ground state is

$$\psi_0(q) = \langle q|0\rangle = \left\langle q = 0\left|\hat{D}^\dagger(q)\right|0\right\rangle = \left\langle q = 0\left|\hat{D}(-q)\right|0\right\rangle. \tag{7.60}$$

As $|q = 0\rangle$ is not degenerate, and $\hat{D}(-q)$ just a linear operator, we can conclude that $|0\rangle$ is also not degenerate. Importantly, this means that all the eigenstates of the QHO, $|n\rangle$, are also non-degenerate (the mathematically minded might like to prove this by contradiction and using properties of the ladder operators alone). This means that any state, $|n\rangle$, may be generated from $|0\rangle$ by repeated application of $\hat{a}^\dagger$.

7.3.3 A Walk-through of Repeatedly Applying $\hat{a}^\dagger$ to $|0\rangle$

In this section, we will demonstrate the action of applying $\hat{a}^\dagger$ to generate the eigenstates of $\hat{n}$. The usual approach is to develop this argument in terms of the number operator, as this is mathematically nice. As $\hat{n}$ and the Hamiltonian commute and share eigenvectors, we us the slightly longer form of arguing in terms of the QHO Hamiltonian. This has the advantage that it better links the discussion to energy eigenvalues (and it gives a slightly different approach to the usual discussion found in many other texts on the subject).

To find the remaining energy eigenstates and eigenvalues for the harmonic oscillator we start by post multiplying the Hamiltonian by the creation operator

$$\hat{H}\hat{a}^{\dagger} = \hbar\omega \left(\hat{a}^{\dagger}\hat{a} + \frac{1}{2} \right)\hat{a}^{\dagger}$$

$$= \hbar\omega \left(\hat{a}^{\dagger}\hat{a}\hat{a}^{\dagger} + \frac{\hat{a}^{\dagger}}{2} \right)$$

$$= \hat{a}^{\dagger}\hbar\omega \left(\hat{a}\hat{a}^{\dagger} + \frac{1}{2} \right).$$

Using the commutation relation Eq. (7.56), we can write

$$\hat{a}\hat{a}^{\dagger} = \hat{\mathbb{1}} + \hat{a}^{\dagger}\hat{a}$$

which gives

$$\hat{H}\hat{a}^{\dagger} = \hat{a}^{\dagger} \left(\hat{H} + \hbar\omega \right).$$

If we act this product of operators onto the ground state of the harmonic oscillator we obtain

$$\hat{H}\hat{a}^{\dagger} |0\rangle = \hat{a}^{\dagger} \left(\hat{H} + \hbar\omega \right) |0\rangle$$

$$= \hat{a}^{\dagger} \left(\frac{\hbar\omega}{2} + \hbar\omega \right) |0\rangle$$

$$\Rightarrow \hbar\omega \left(\hat{a}^{\dagger}\hat{a} + \frac{1}{2} \right) \hat{a}^{\dagger} |0\rangle = \hbar\omega \left(1 + \frac{1}{2} \right) \hat{a}^{\dagger} |0\rangle$$

which shows that $\hat{a}^{\dagger} |0\rangle$ is an eigenstate of $\hat{H}$ with an eigenvalue of $3\hbar\omega/2$ and suggests that the action of $\hat{a}^{\dagger}\hat{a}$ on the state $\hat{a}^{\dagger} |0\rangle$ gives an eigenvalue of 1. We therefore conclude that

$$\hat{a}^{\dagger} |0\rangle = C |1\rangle, \tag{7.61}$$

where C is a normalisation constant. Assuming a normalised ground state, we can see that

$$\left\langle 0 | \hat{a}\hat{a}^{\dagger} | 0 \right\rangle = |C|^2 \overbrace{\langle 0|0\rangle}^{1} \Rightarrow |C| = 1. \tag{7.62}$$

As we are free to choose the phase of C, we will make it purely real. So, the action of $\hat{a}^{\dagger}$ on the ground state $|0\rangle$ gives the first excited state $|1\rangle$ exactly,

$$\hat{a}^{\dagger} |0\rangle = |1\rangle. \tag{7.63}$$

Now that we have the action of $\hat{H}\hat{a}^{\dagger}$ on the state $|1\rangle$, we repeat the process to find the second excited state $|2\rangle$ in terms of the ground state $|0\rangle$. We follow the same process as before, except this time we consider

$$\hat{H}\hat{a}^{\dagger} |1\rangle = \hat{a}^{\dagger} \left(\hat{H} + \hbar\omega \right) |1\rangle$$

$$= \hat{a}^{\dagger} \left(\frac{3\hbar\omega}{2} + \hbar\omega \right) |0\rangle$$

$$\Rightarrow \hbar\omega \left(\hat{a}^{\dagger}\hat{a} + \frac{1}{2} \right) \hat{a}^{\dagger} |1\rangle = \hbar\omega \left(2 + \frac{1}{2} \right) \hat{a}^{\dagger} |1\rangle,$$

which agrees with the fact that $\hat{a}^\dagger|1\rangle$ is an eigenstate of $\hat{a}^\dagger\hat{a}$ with eigenvalue 2 (as shown in the previous section). Again, we can write the state

$$\hat{a}^\dagger|1\rangle = C|2\rangle. \tag{7.64}$$

Normalising this state in a similar way to before gives

$$|2\rangle = \frac{1}{\sqrt{2}}|1\rangle = \frac{\hat{a}^{\dagger 2}}{\sqrt{2}\sqrt{1}}|0\rangle. \tag{7.65}$$

Exercise 7.3   Use the steps outlined from Eqs. (7.61) to (7.63) as well as relation Eq. (7.64), noting that $\hat{a}^\dagger\hat{a}|1\rangle = 1|1\rangle$, to show Eq. (7.65).   ∎

We can continue this process for all eigenstates $|n\rangle$ of $\hat{H}$ and find that we can express them in terms of the ground state $|0\rangle$ as

$$|n\rangle = \frac{\hat{a}^{\dagger n}}{\sqrt{n!}}|0\rangle \tag{7.66}$$

with the eigenvalues

$$E_n = \hbar\omega\left(n + \frac{1}{2}\right).$$

Finally we note that the TISE reads

$$\hat{H}|n\rangle = \hbar\omega\left(\hat{a}^\dagger\hat{a} + \frac{1}{2}\right)|n\rangle = \hbar\omega\left(n + \frac{1}{2}\right)|n\rangle.$$

> The energy eigenbasis, or number basis, of the harmonic oscillator is an example of the more general Fock basis, $\{|n\rangle\}$, which can be used to describe the number of particles, or quanta, in a particular ensemble whose states are related through ladder operators, such as the number of photons in a electromagnetic field. We shall see this later, in the Jaynes–Cummings model.

Exercise 7.4   Using the relation $C|n+1\rangle = \hat{a}^\dagger|n\rangle$, use proof by induction to show that Eq. (7.66) holds for all values of n.   ∎

7.3.4   The General Form of the Action of $\hat{a}$ and $\hat{a}^\dagger$ on $|n\rangle$

Generalising the argument above, the action of the number operator $\hat{n}$ on the state $\hat{a}|n\rangle$ can be found from

$$\hat{n}\hat{a}|n\rangle = (-\hat{a} + \hat{a}\hat{n})|n\rangle$$
$$= -\hat{a}|n\rangle + \hat{a}\hat{n}|n\rangle$$
$$= (n-1)\hat{a}|n\rangle.$$

We know that the number states are not degenerate, so there is no ambiguity in writing

$$\hat{a}\,|n\rangle = C\,|n-1\rangle.$$

Again, finding the inner product of $\langle n|\,\hat{a}^{\dagger}\hat{a}\,|n\rangle$ gives

$$\langle n|\,\hat{a}^{\dagger}\hat{a}\,|n\rangle = n\,\langle n|n\rangle = |C|^{2}\,\langle n-1|n-1\rangle$$

and so

$$|C| = \sqrt{n},$$

which lets us conclude that

$$\hat{a}\,|n\rangle = \sqrt{n}\,|n-1\rangle. \tag{7.67}$$

A similar analysis can be performed for the creation operator to obtain

$$\hat{a}^{\dagger}\,|n\rangle = \sqrt{n+1}\,|n\rangle.$$

A graphical representation of this ladder operator action is given in Figure 7.7.

Exercise 7.5 Follow the analysis used to obtain Eq. (7.67) to show the action of $\hat{a}^{\dagger}$ on the state $|n\rangle$. ∎

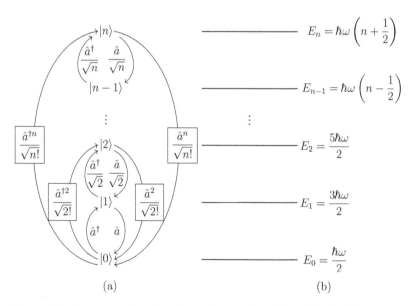

(a)   (b)

Figure 7.7 The energy eigenbasis for the harmonic oscillator (a) with their corresponding eigenvalues (b). Note how we can move through the basis from one state to another just by applying a series of scaled ladder operations.

7.3.5 Matrix Representation

For many systems, where Hamiltonians are easily expressible in terms of $\hat{n}$, $\hat{a}$, and $\hat{a}^{\dagger}$, the number basis is a particularly useful representation, especially for approximate computation. We can describe each state as an element in a column vector $\psi_n = \langle n|\psi \rangle$ and operators as their matrix elements $\left\langle n \left| \hat{A} \right| m \right\rangle$. For example, the basis vectors are

$$|0\rangle \sim \begin{pmatrix} 1 \\ 0 \\ 0 \\ 0 \\ \vdots \\ 0 \\ \vdots \end{pmatrix}, \quad |1\rangle \sim \begin{pmatrix} 0 \\ 1 \\ 0 \\ 0 \\ \vdots \\ 0 \\ \vdots \end{pmatrix}, \quad |n\rangle \sim \begin{pmatrix} 0 \\ 0 \\ 0 \\ 0 \\ \vdots \\ 1 \\ \vdots \end{pmatrix}.$$

The Hamiltonian takes the form of a diagonal (as we are in its eigenbasis) matrix with the elements

$$\langle n|\hat{H}|m\rangle = \left\langle n \left| \hbar\omega\left(\hat{n} + \frac{1}{2}\right) \right| m \right\rangle$$
$$= \hbar\omega\left(m + \frac{1}{2}\right)\langle n|m\rangle$$
$$= \hbar\omega\left(m + \frac{1}{2}\right)\delta_{nm},$$

or, writing this out in full,

$$(H) = \hbar\omega \begin{pmatrix} \frac{1}{2} & 0 & 0 & \cdots & 0 \\ 0 & \frac{3}{2} & 0 & \cdots & 0 \\ 0 & 0 & \frac{5}{2} & \cdots & 0 \\ 0 & 0 & 0 & \ddots & \vdots \\ 0 & 0 & 0 & \cdots & \left(n+\frac{1}{2}\right) \end{pmatrix}$$

we observe that the matrix representation for the harmonic oscillator presents a diagonal Hamiltonian.

Exercise 7.6 Show that the annihilation and creation operators can be represented by the matrices

$$(a) = \begin{pmatrix} 0 & \sqrt{1} & 0 & \cdots & 0 \\ 0 & 0 & \sqrt{2} & \cdots & 0 \\ 0 & 0 & 0 & \ddots & \vdots \\ 0 & 0 & 0 & \cdots & \sqrt{n} \\ 0 & 0 & 0 & \cdots & 0 \end{pmatrix}, \quad (a^{\dagger}) = \begin{pmatrix} 0 & 0 & 0 & \cdots & 0 \\ \sqrt{1} & 0 & 0 & \cdots & 0 \\ 0 & \sqrt{2} & 0 & \cdots & 0 \\ 0 & 0 & \ddots & \ddots & \vdots \\ 0 & 0 & \cdots & \sqrt{n} & 0 \end{pmatrix}$$

respectively. ∎

Either quasi-algebraically, or just using the above results with brute force, provides a rapid way to analyse many quantum systems computationally. We will discuss this and much more on that approach in the next chapter.

7.3.6 The Position Representation of Wave-functions

To finish solving the TISE, we need to determine the specific functional form of the number states. The usual approach is to start with

$$\hat{a}\,|0\rangle = |\theta\rangle$$

and look at the position representation

$$\frac{1}{2}\left\langle q\left|\sqrt{\frac{m\omega}{\hbar}}\hat{q} + \sqrt{\frac{1}{m\hbar\omega}}\hat{p}\right|0\right\rangle = \langle q|\theta\rangle,$$

which yields the following differential equation for the ground state

$$\frac{\partial}{\partial q}\psi_0(q) = -i\frac{m\omega}{\hbar}q\psi_0(q)$$

with the solution

$$\psi_0(q) \propto \exp\left(-\frac{1}{2}\frac{m\omega}{\hbar}q^2\right).$$

The constant of proportionality is found through normalisation, and we can apply the position representation of $\hat{a}^\dagger$ to generate all the other eigenfunctions, which we see are the normalised Hermite polynomials (details can be found in, e.g. [12]).

Here we will follow an argument presented in [75]. We do this because we have tried to be as algebraic as possible in developing the theoretical framework of the QHO. The argument we present is at least superficially algebraic,[1] demonstrates some useful algebraic techniques, and we believe does not appear in other textbooks. To begin our analysis we look at some important properties of non-commuting operators and first recall the Baker Campbell Hausdorff (BCH) formula

$$e^{\hat{A}}\hat{B}e^{-\hat{A}} = \hat{B} + [\hat{A}, \hat{B}] + \frac{1}{2!}\left[\hat{A}, [\hat{A}, \hat{B}]\right] + \frac{1}{3!}\left[\hat{A}, [\hat{A}, [\hat{A}, \hat{B}]]\right] + \cdots$$

$$= B + \sum_{n=1}^{\infty}\frac{1}{n!}[A, [A, , \dots, [A, B] \dots]]_n. \tag{7.68}$$

Exercise 7.7 If a function f has a power series expansion, show that

$$e^{\hat{A}}f(\hat{B})e^{-\hat{A}} = f\left(e^{\hat{A}}\hat{B}e^{-\hat{A}}\right). \tag{7.69}$$

■

From the BCH formula we can obtain the Hadamard Lemma,

$$e^{\hat{A}}e^{\hat{B}} = \exp\left(\hat{B} + [\hat{A}, \hat{B}] + \frac{1}{2!}\left[\hat{A}, [\hat{A}, \hat{B}]\right] + \cdots\right)e^{\hat{A}}.$$

If $[\hat{A}, \hat{B}]$ is just a number, then this expression reduces to

$$e^{\hat{A}}e^{\hat{B}} = e^{\hat{B}}e^{\hat{A}}e^{[\hat{A},\hat{B}]} \tag{7.70}$$

since both $\hat{A}$ and $\hat{B}$ will commute with $[\hat{A}, \hat{B}]$.

---

1 The method we follow does use some results of the exponential that could be argued to use some implicit results of calculus.

Exercise 7.8   If $\hat{A}$ and $\hat{B}$ will commute with $[\hat{A}, \hat{B}]$ show that Eq. (7.69) implies the so-called braiding relation:

$$e^{\hat{A}} f(\hat{B}) = f\left(\hat{B} + [\hat{A}, \hat{B}]\right) e^{\hat{A}}. \tag{7.71}$$

∎

With this mathematical background, we are now ready to embark on the derivation. We start with the displacement operator we saw in Section 3.4.1.1 and used on page 186:

$$\hat{D}(q) = \exp\left(-\frac{i\, q\hat{p}}{\hbar}\right).$$

In Section 3.4.1.1 we already mentioned that the displacement operator is a ladder operator for position eigenstates and a generator of translation, so that

$$\exp\left(-\frac{i\, q\hat{p}}{\hbar}\right) |q = 0\rangle = |q\rangle.$$

The energy eigenfunctions $\psi_n(q)$ for the harmonic oscillator are the position representation of each eigenstate

$$\psi_n(q) = \langle q|n\rangle.$$

We will now use the fact that $\langle q|$ can be written

$$\langle q| = \langle q = 0| \exp\left(\frac{i\, q\hat{p}}{\hbar}\right),$$

and any particular number state can be written as

$$|n\rangle = \frac{\hat{a}^{\dagger n}}{\sqrt{n!}} |n = 0\rangle.$$

Thus, the eigenfunctions are

$$\psi_n(q) = \langle q|n\rangle = \left\langle q = 0 \left| \exp\left(\frac{i\, q\hat{p}}{\hbar}\right) \frac{\hat{a}^{\dagger n}}{\sqrt{n!}} \right| n = 0 \right\rangle.$$

Writing $\hat{p}$ in terms of the ladder operators $\hat{a}$ and $\hat{a}^{\dagger}$ gives

$$\psi_n(q) = \langle q|n\rangle = \frac{1}{\sqrt{n!}} \left\langle q = 0 \left| \exp\left(\sqrt{\frac{m\omega}{2\hbar}}\, q\left(\hat{a} - \hat{a}^{\dagger}\right)\right) \hat{a}^{\dagger n} \right| n = 0 \right\rangle.$$

In order to reach the final form of the wave-functions for the harmonic oscillator we must make use of another formulation of the BCH formula. This formulation will enable the manipulation of the exponentiated ladder operators and make use of their action on the ground state of the harmonic oscillator. When the commutator of two operators is a number, the BCH formula can be written

$$e^{\hat{A}} e^{\hat{B}} = e^{\hat{A} + \hat{B} + \frac{1}{2}[\hat{A}, \hat{B}]}.$$

Applying this to $\psi_n(q)$ gives

$$\psi_n(q) = \frac{1}{\sqrt{n!}} \exp\left(-\frac{m\omega}{4\hbar}q^2\right)$$

$$\left\langle q = 0 \left| \exp\left(-\sqrt{\frac{m\omega}{2\hbar}}q\hat{a}^\dagger\right) \left[\exp\left(\sqrt{\frac{m\omega}{2\hbar}}q\hat{a}\right)\hat{a}^{\dagger n}\right] \right| n = 0 \right\rangle$$

and using the braiding relation Eq. (7.71) on the product in square brackets (so $\hat{A} = \sqrt{m\omega/2\hbar}q\hat{a}$ and $\hat{B} = \hat{a}^\dagger$), we obtain

$$\psi_n(q) = \frac{1}{\sqrt{n!}} \exp\left(-\frac{m\omega}{4\hbar}q^2\right) \left\langle q = 0 \left| \exp\left(-\sqrt{\frac{m\omega}{2\hbar}}q\hat{a}^\dagger\right) \right.\right.$$

$$\left.\left. \left[\left(\hat{a}^\dagger + \sqrt{\frac{m\omega}{2\hbar}}q\left[\hat{a},\hat{a}^\dagger\right]\right)^n \exp\left(\sqrt{\frac{m\omega}{2\hbar}}q\hat{a}\right)\right] \right| n = 0 \right\rangle.$$

Since

$$\left[\hat{a},\hat{a}^\dagger\right] = \hat{\mathbb{1}}$$

we have

$$\psi_n(q) = \frac{1}{\sqrt{n!}} \exp\left(-\frac{m\omega}{4\hbar}q^2\right) \left\langle q = 0 \left| \exp\left(-\sqrt{\frac{m\omega}{2\hbar}}q\hat{a}^\dagger\right) \right.\right.$$

$$\left.\left. \left[\left(\hat{a}^\dagger + \sqrt{\frac{m\omega}{2\hbar}}q\right)^n \exp\left(\sqrt{\frac{m\omega}{2\hbar}}q\hat{a}\right)\right] \right| n = 0 \right\rangle$$

and[2]

$$\exp\left(\sqrt{\frac{m\omega}{2\hbar}}q\hat{a}\right)|0\rangle = \hat{\mathbb{1}}|0\rangle$$

which means we can, rather surprisingly and *only for the vacuum state*, replace

$$\exp\left(\sqrt{\frac{m\omega}{2\hbar}}q\hat{a}\right)$$

with

$$\exp\left(-\sqrt{\frac{m\omega}{2\hbar}}q\hat{a}\right)$$

---

2 The action of $\exp\left(-\sqrt{\frac{m\omega}{2\hbar}}q\hat{a}\right)|n = 0\rangle$ returns the identity because its corresponding power series returns $\exp\left(-\sqrt{\frac{m\omega}{2\hbar}}q\hat{a}\right) = \hat{\mathbb{1}} - \sqrt{\frac{m\omega}{2\hbar}}q\hat{a} + O(\hat{a}^2)$. We have seen that the action of any power of $\hat{a}$ on the ground state $|n = 0\rangle$ returns the null vector, which means that the only term that remains from the expansion when acting on the ground state is the identity, and the actions of $\exp\left(\sqrt{\frac{m\omega}{2\hbar}}q\hat{a}\right)|0\rangle$ and $\exp\left(-\sqrt{\frac{m\omega}{2\hbar}}q\hat{a}\right)|0\rangle$ are equivalent.

to produce

$$\psi_n(q) = \frac{1}{\sqrt{n!}} \exp\left(-\frac{m\omega}{4\hbar}q^2\right) \left\langle q=0 \left| \exp\left(-\sqrt{\frac{m\omega}{2\hbar}}q\hat{a}^\dagger\right) \right.\right.$$

$$\left.\left. \left(\hat{a}^\dagger + \sqrt{\frac{m\omega}{2\hbar}}q\right)^n \exp\left(-\sqrt{\frac{m\omega}{2\hbar}}q\hat{a}\right) \right| n=0 \right\rangle.$$

We can use the fact that the term in the square brackets is of the same form as the rhs of the braiding relation Eq. (7.71) to switch the polynomial term with the exponential term to give

$$\psi_n(q) = \frac{1}{\sqrt{n!}} \exp\left(-\frac{m\omega}{4\hbar}q^2\right) \left\langle q=0 \left| \exp\left(-\sqrt{\frac{m\omega}{2\hbar}}q\hat{a}^\dagger\right) \exp\left(-\sqrt{\frac{m\omega}{2\hbar}}q\hat{a}\right) \right.\right.$$

$$\left.\left. \left(\hat{a}^\dagger + \sqrt{\frac{m\omega}{2\hbar}}q\right)^n \right| n=0 \right\rangle.$$

Now, using Eq. (7.70) and absorbing the phase factor outside the matrix element, we obtain

$$\psi_n(q) = \frac{1}{\sqrt{n!}} \exp\left(-\frac{m\omega}{2\hbar}q^2\right) \left\langle q=0 \left| \exp\left(-\sqrt{\frac{m\omega}{2\hbar}}q(\hat{a}+\hat{a}^\dagger)\right) \right.\right.$$

$$\left.\left. \left(\hat{a}^\dagger + \sqrt{\frac{2m\omega}{\hbar}}q\right)^n \right| n=0 \right\rangle.$$

We can rewrite the sum of the creation and annihilation operators in terms of the position operator

$$\hat{q} = \sqrt{\frac{\hbar}{2m\omega}}(\hat{a}+\hat{a}^\dagger)$$

which, as $|q=0\rangle$, is an eigenstate of $\hat{q}$ with eigenvalue zero, reduces the exponential inside the matrix element to $e^0 = 1$, so the wave-function simplifies to

$$\psi_n(q) = \frac{1}{\sqrt{n!}} \exp\left(-\frac{m\omega}{2\hbar}q^2\right) \left\langle q=0 \left| \left(\hat{a}^\dagger + \sqrt{\frac{2m\omega}{\hbar}}q\right)^n \right| n=0 \right\rangle.$$

Hence, we can immediately say that the ground state is

$$\psi_0(q) = \exp\left(-\frac{m\omega}{2\hbar}q^2\right) \langle q=0|n=0\rangle.$$

By finding the above expression, we have succeeded in our goal of deriving the ground state wave-function algebraically and without explicitly solving a differential equation. We have used some interesting and useful techniques to do so. Alas, to find the normalisation requires the calculus of integration as we need to solve

$$\int |\psi_0(q)|^2 = 1.$$

Performing this integral, or looking it up, or using a computational algebra package, we find

$$\langle q=0|n=0\rangle = \left(\frac{m\omega_0}{\pi\hbar}\right)^{\frac{1}{4}}.$$

Exercise 7.9 Use the normalisation condition for the ground state ψ_0 to verify that $\langle q = 0 | n = 0 \rangle = (m\omega/\pi\hbar)^{1/4}$. ∎

To connect our analysis with the usual approach, we introduce a function

$$H_n\left(\sqrt{\frac{m\omega_0}{\hbar}}q\right) = \frac{\sqrt{2^n}}{\langle q = 0 | n = 0 \rangle}\left\langle q = 0 \left| \left(\hat{a}^\dagger + \sqrt{\frac{2m\omega_0}{\hbar}}q\right)^n \right| n = 0 \right\rangle.$$

The inclusion of $\langle q = 0 | n = 0 \rangle$ in the denominator means that $H_0(q) = 1$. These are the Hermite polynomials $H_n(q)$, and the factor of $\sqrt{2^n}$ is included for historical reasons (it has to be cancelled out in the expression for $\psi_n(q)$). Writing our wave-functions in terms of the Hermite polynomials, then gives

$$\psi_n(q) = \frac{1}{\sqrt{n!}}\exp\left(-\frac{m\omega}{2\hbar}q^2\right)H_n\left(\sqrt{\frac{2m\omega}{\hbar}}q\right)\frac{\langle q = 0 | n = 0 \rangle}{\sqrt{2^n}}.$$

Simplifying this expression leads to our final expression for the eigenfunctions of the harmonic oscillator in terms of Hermite polynomials

$$\psi_n(q) = \frac{1}{\sqrt{n!2^n}}\left(\frac{m\omega}{\pi\hbar}\right)^{1/4}\exp\left(-\frac{m\omega}{2\hbar}q^2\right)H_n(q). \tag{7.72}$$

The final two exercises lay the framework for how to generate all of these functions from recursion relations and complete the analysis.

Exercise 7.10 Show that $H_1(x) = 2x$. ∎

Exercise 7.11 Show that $H_n(x) = 2xH_{n-1}(x) - 2(n-1)H_{n-2}(x)$. ∎

In Figure 7.8 we show the first six energy eigenfunctions (or eigenenergies) of the QHO. Note that the even eigenstates all have even parity and the odd ones odd parity. Also note that each state has a finite probability of being found outside its classically allowable positions for a particle with the same energy. Note that at every node, there is zero probability of finding a particle in that eigenstate in that position – just like with the particle-in-a-box example in the previous section.

7.3.7 Coherent States

7.3.7.1 Introduction
Coherent states is an important class of states that find many applications in quantum mechanics. We briefly discussed these states in Section 4.3.4 in connecting the Wigner phase space formulation of quantum mechanics to the usual operator formalisms. In this section, we continue that discussion and explore the deep connection between coherent states and the harmonic oscillator. There are two ways of defining coherent states, and we will later connect the two. One way is to consider them as displaced vacuum states. For the harmonic oscillator, the more historical, and in this case equivalent, way is as eigenstates of the annihilation operator

$$\hat{a}|\alpha\rangle = \alpha|\alpha\rangle. \tag{7.73}$$

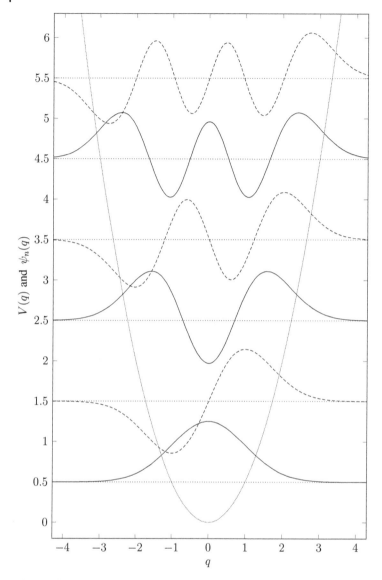

Figure 7.8 The dense dotted parabola shows the potential of the harmonic oscillator. Each eigenenergy is shown as a flat light dotted line. Then we show each energy eigenfunction, $\psi_n(q)$, displaced by its eigenenergy, the even number states are solid and the odd ones shown dashed for clarity.

We impose the normalisation $\langle \alpha | \alpha \rangle = 1$ and note that as $\hat{a}$ is not Hermitian, we do not expect eigenvectors with different eigenvalues to be orthogonal.

To help us understand the importance of these coherent states, let us first consider the classical dynamics which we find from the Hamiltonian

$$H = \frac{\hbar\omega}{2}\left(P^2 + Q^2\right). \tag{7.74}$$

Written in terms of classical dimensionless position and momentum that are analogues to Eq. (7.54),

$$Q = \sqrt{\frac{m\omega}{\hbar}} q \text{ and } P = \sqrt{\frac{1}{m\hbar\omega}} p. \tag{7.75}$$

We can then define the classical analogue of $\hat{a}$

$$\alpha = \frac{1}{\sqrt{2}}(Q + iP) \tag{7.76}$$

and see that

$$\frac{d\alpha(t)}{dt} = \{\alpha(t), H\} = -i\omega\alpha(t), \tag{7.77}$$

whose solution can be seen to describe circular motion in the complex phase space of α and is given by

$$\alpha(t) = \alpha_0 e^{-i\omega t} \tag{7.78}$$

where

$$\alpha_0 = \frac{1}{\sqrt{2}}(Q(0) + iP(0)).$$

Note that we can write $Q(t) = \sqrt{2}\Re[\alpha(t)]$ and $P(t) = \sqrt{2}\Im[\alpha(t)]$ in terms of this solution. Also note that the energy of the system is constant in time and equal to $\hbar\omega|\alpha_0|^2$.

In the quantum mechanical case we can apply Ehrenfest's theorem for the dynamics of $\langle\hat{a}\rangle$ so that

$$\frac{d}{dt}\langle\hat{a}\rangle(t) = -\frac{i}{\hbar}\left\langle[\hat{a}, \hat{H}]\right\rangle(t) = -i\omega\langle\hat{a}\rangle(t) \tag{7.79}$$

which implies that $\langle\hat{a}\rangle$ has the solution:

$$\langle\hat{a}\rangle(t) = \langle\hat{a}\rangle(0)e^{-i\omega t}. \tag{7.80}$$

Now, the state of the system is a coherent state

$$\langle\hat{a}\rangle = \langle\alpha|\hat{a}|\alpha\rangle = \alpha\langle\alpha|\alpha\rangle = \alpha. \tag{7.81}$$

By setting $|\psi(0)\rangle = |\alpha_0\rangle$ and by inspection and using Eq. (7.80), we see that

$$\alpha = \alpha_0 e^{-i\omega t}. \tag{7.82}$$

We see therefore that if the system starts in a coherent state, it evolves as coherent states whose eigenvalue dynamics take the same form as its classical counterpart Eq. (7.78). It is interesting to note that the energy expectation value

$$\langle\hat{H}\rangle = \hbar\omega\left\langle\alpha(t)\left|\hat{a}^\dagger\hat{a}\right|\alpha(t)\right\rangle + \frac{\hbar\omega}{2}$$

$$= \hbar\omega|\alpha_0|^2 + \frac{\hbar\omega}{2}. \tag{7.83}$$

This differs from the classical case by $\hbar\omega/2$. This difference becomes negligible in the limit, $|\alpha_0|^2 \gg 1$, which leads some to propose this as a classical correspondence limit for the harmonic oscillator. There are issues with this view but we will delay discussion of those until Chapter 10.

7.3.7.2 Coherent States in the Number Basis

As number states are very useful in problem-solving, especially in numerical analysis, we will now express $|\alpha\rangle$ using a series expansion of the basis state $\{|n\rangle\}$. That is, we will seek to determine $c_m(\alpha)$ in

$$|\alpha\rangle = \sum_m c_m(\alpha)|m\rangle. \tag{7.84}$$

By definition we know that $|\alpha\rangle$ is an eigenstate of $\hat{a}$. We will use this to determine the above unknown coefficients. We first apply $\hat{a}$ to give

$$\hat{a}|\alpha\rangle = \sum_m c_m(\alpha)\sqrt{m}|m-1\rangle. \tag{7.85}$$

Letting $m = n + 1$, we can equate the coefficients of $|n\rangle$ to yield

$$\hat{a}|\alpha\rangle = \sum_n c_{n+1}(\alpha)\sqrt{n+1}|n\rangle = \alpha|\alpha\rangle = \alpha\sum_n c_n(\alpha)|n\rangle$$

$$\Rightarrow c_{n+1}(\alpha)\sqrt{n+1} = \alpha c_n(\alpha). \tag{7.86}$$

Rearranging this equation to make $c_{n+1}(\alpha)$ the subject gives us the recurrence relation

$$c_{n+1}(\alpha) = \frac{\alpha}{\sqrt{n+1}}c_n(\alpha), \tag{7.87}$$

which allows us to find all coefficients $c_n(\alpha)$ in terms of $c_0(\alpha)$:

$$c_n(\alpha) = \frac{\alpha^n}{\sqrt{n!}}c_0(\alpha). \tag{7.88}$$

We now use normalisation of the state to determine the amplitude of $c_0(\alpha)$

$$1 = \sum_n |c_n(\alpha)|^2 = \sum_n \left|\frac{\alpha^n}{\sqrt{n!}}c_0(\alpha)\right|^2$$

$$= \sum_n \frac{|\alpha|^{2n}}{n!}|c_0(\alpha)|^2 \tag{7.89}$$

$$= |c_0(\alpha)|^2 \sum_n \frac{(|\alpha|^2)^n}{n!}$$

$$\Rightarrow 1 = |c_0(\alpha)|^2 e^{|\alpha|^2},$$

where the power series expansion $e^x = \sum_n x^n/n!$ has been used. Rearranging to make $c_0(\alpha)$ the subject, and noting that we are free to fix the phase of any eigenvector (so we choose $c_0(\alpha)$ to be real and positive), yields

$$c_0(\alpha) = e^{-\frac{|\alpha|^2}{2}}. \tag{7.90}$$

Substituting Eq. (7.90) into Eq. (7.84) leads to

$$|\alpha\rangle = e^{-\frac{|\alpha|^2}{2}} \sum_n \frac{\alpha^n}{\sqrt{n!}}|n\rangle. \tag{7.91}$$

Exercise 7.12 Show that all coherent states minimise the Heisenberg uncertainty relation, so that

$$(\Delta \hat{q})_\alpha (\Delta \hat{p})_\alpha = \frac{\hbar}{2}. \tag{7.92}$$

∎

7.3.7.3 Coherent States as Displaced Vacuum States

Recall that in Eq. (4.74) on page 115 we introduced the general form of the displacement operator

$$\hat{D}(\alpha) = e^{\alpha \hat{a}^\dagger - \alpha^* \hat{a}}. \tag{7.93}$$

We asserted that coherent states were displaced vacuum states, $\hat{D}(\alpha)|0\rangle = |\alpha\rangle$. We will now show that this is indeed the case. We can use a variant of the BCH formula that applies when two operators commute with their commutator

$$e^{\hat{A}} e^{\hat{B}} = e^{\hat{A} + \hat{B} + \frac{1}{2}[\hat{A},\hat{B}]}, \tag{7.94}$$

or equivalently,

$$e^{-\frac{1}{2}[\hat{A},\hat{B}]} e^{\hat{A}} e^{\hat{B}} = e^{\hat{A} + \hat{B}}. \tag{7.95}$$

Substituting $\hat{A} = \alpha \hat{a}^\dagger$ and $\hat{B} = \alpha^* \hat{a}$ into Eq. (7.95) gives

$$\hat{D}(\alpha) = e^{-\frac{|\alpha|^2}{2}} e^{\alpha \hat{a}^\dagger} e^{-\alpha^* \hat{a}}. \tag{7.96}$$

The action of the displacement operator on the ground state $|0\rangle$ yields

$$\hat{D}(\alpha)|0\rangle = e^{-\frac{|\alpha|^2}{2}} e^{\alpha \hat{a}^\dagger} e^{-\alpha^* \hat{a}} |0\rangle. \tag{7.97}$$

We take advantage of the fact that the annihilation operator acting on the vacuum state is the null vector, together with the power series expansion of the exponential, to note that

$$e^{-\alpha^* \hat{a}}|0\rangle = \left[1 - \alpha^* \hat{a} + \frac{\alpha^{*2}}{2!} \hat{a}^2 + \cdots \right] |0\rangle = |0\rangle. \tag{7.98}$$

Substituting Eq. (7.98) into Eq. (7.97) then gives

$$\hat{D}(\alpha)|0\rangle = e^{-\frac{|\alpha|^2}{2}} e^{\alpha \hat{a}^\dagger} |0\rangle. \tag{7.99}$$

Once again applying a series expansion, we see that

$$\hat{D}(\alpha)|0\rangle = e^{-\frac{|\alpha|^2}{2}} \sum_n \frac{(\alpha \hat{a}^\dagger)^n}{n!} |0\rangle$$

$$= e^{-\frac{|\alpha|^2}{2}} \sum_n \frac{\alpha^n}{\sqrt{n!}} \frac{\hat{a}^{\dagger n}}{\sqrt{n!}} |0\rangle \tag{7.100}$$

$$= e^{-\frac{|\alpha|^2}{2}} \sum_n \frac{\alpha^n}{\sqrt{n!}} |n\rangle.$$

The final result in Eq. (7.100) matches exactly that in Eq. (7.91) which allows us to conclude:

$$\hat{D}(\alpha)|0\rangle = |\alpha\rangle, \tag{7.101}$$

which shows that a coherent state $|\alpha\rangle$ may indeed be created from the ground state of the harmonic oscillator through a displacement $\hat{D}(\alpha)$.

7.3.7.4 The Position Representation

We can use the fact that coherent states are displaced vacuum states to express their position representation as

$$\psi_\alpha(q) = \langle q | \alpha \rangle = \left\langle q \left| \hat{D}(\alpha) \right| 0 \right\rangle. \tag{7.102}$$

Exercise 7.13 By writing $\hat{a}^\dagger$ and $\hat{a}$ in terms of $\hat{q}$ and $\hat{p}$, and rearranging, show that we can express $\alpha \hat{a}^\dagger - \alpha^* \hat{a}$ as shown:

$$\alpha \hat{a}^\dagger - \alpha^* \hat{a} = \sqrt{\frac{m\omega}{\hbar}} \left(\frac{\alpha - \alpha^*}{\sqrt{2}} \right) \hat{q} - \frac{i}{\sqrt{m\hbar\omega}} \left(\frac{\alpha + \alpha^*}{\sqrt{2}} \right) \hat{p}. \tag{7.103}$$

∎

We can use the BCH formula to write Eq. (7.93) as

$$\hat{D}(\alpha) = e^{\alpha \hat{a}^\dagger - \alpha \hat{a}}$$

$$= e^{\sqrt{\frac{m\omega}{\hbar}} \left(\frac{\alpha - \alpha^*}{\sqrt{2}} \right) \hat{q}} e^{-\frac{i}{\sqrt{m\hbar\omega}} \left(\frac{\alpha + \alpha^*}{\sqrt{2}} \right) \hat{p}} e^{\frac{\alpha^{*2} - \alpha^2}{4}}. \tag{7.104}$$

By substituting this result into our expression for the coherent state, Eq. (7.102), we get:

$$\psi_\alpha(q) = e^{\frac{\alpha^{*2} - \alpha^2}{4}} \left\langle q \left| e^{\sqrt{\frac{m\omega}{\hbar}} \left(\frac{\alpha - \alpha^*}{\sqrt{2}} \right) \hat{q}} e^{-\frac{i}{\sqrt{m\hbar\omega}} \left(\frac{\alpha + \alpha^*}{\sqrt{2}} \right) \hat{p}} \right| 0 \right\rangle$$

$$= e^{\frac{\alpha^{*2} - \alpha^2}{4}} e^{\sqrt{\frac{m\omega}{\hbar}} \left(\frac{\alpha - \alpha^*}{\sqrt{2}} \right) q} \left\langle q \left| e^{-\frac{i}{\sqrt{m\hbar\omega}} \left(\frac{\alpha + \alpha^*}{\sqrt{2}} \right) \hat{p}} \right| 0 \right\rangle$$

$$= e^{\frac{\alpha^{*2} - \alpha^2}{4}} e^{\sqrt{\frac{m\omega}{\hbar}} \left(\frac{\alpha - \alpha^*}{\sqrt{2}} \right) q} \left\langle q \left| e^{-\frac{i\lambda\hat{p}}{\hbar}} \right| 0 \right\rangle \tag{7.105}$$

$$= e^{\frac{\alpha^{*2} - \alpha^2}{4}} e^{\sqrt{\frac{m\omega}{\hbar}} \left(\frac{\alpha - \alpha^*}{\sqrt{2}} \right) q} \left\langle q \left| \hat{D}_q(\lambda) \right| 0 \right\rangle,$$

where we have defined $\lambda = \sqrt{\frac{\hbar}{2m\omega}} (\alpha + \alpha^*)$ within $\hat{D}_q(\lambda) = e^{-\frac{i\lambda\hat{p}}{\hbar}}$, being the displacement operator in position only. The action of the translation operator $\hat{D}_q(\lambda)$ on $\langle q |$ causes a shift in the vector by $-\lambda$, that is to say

$$\langle q | \hat{D}(\lambda) = \langle q - \lambda |. \tag{7.106}$$

Substituting this result into Eq. (7.105) yields

$$\psi_\alpha(q) = e^{\frac{\alpha^{*2} - \alpha^2}{4}} e^{\sqrt{\frac{m\omega}{\hbar}} \left(\frac{\alpha - \alpha^*}{\sqrt{2}} \right) q} \left\langle q - \sqrt{\frac{\hbar}{2m\omega}} \right| (\alpha + \alpha^*) | 0$$

$$= e^{\frac{\alpha^{*2} - \alpha^2}{4}} e^{\sqrt{\frac{m\omega}{\hbar}} \left(\frac{\alpha - \alpha^*}{\sqrt{2}} \right) q} \psi_0 \left(q - \sqrt{\frac{\hbar}{2m\omega}} (\alpha + \alpha^*) \right) \tag{7.107}$$

$$= e^{i\theta_\alpha} e^{i \langle \hat{p} \rangle_\alpha q / \hbar} \psi_0(q - \langle \hat{q} \rangle_\alpha),$$

where α and α^* have been written in terms of $\langle \hat{q} \rangle_\alpha$ and $\langle \hat{p} \rangle_\alpha$, while the global phase factor $e^{i\theta_\alpha} = e^{\frac{\alpha^{*2} - \alpha^2}{4}}$ has also been introduced. Equation (7.107) shows that the coherent state wave function $\psi_\alpha(q)$ can be obtained from the ground state wave function $\psi_0(q)$ of the harmonic oscillator by translating the function by $\langle \hat{q} \rangle_\alpha$ along the q plane and multiplying it by the oscillating exponential $e^{i \langle \hat{p} \rangle q / \hbar}$. Since the global phase factor has no physical significance, it is often omitted.

Finally, we can write the coherent state wave function using the explicit expression for the ground state in the position representation $\psi_0(q)$ to obtain:

$$
\begin{aligned}
\psi_\alpha(q) &= e^{i\theta_\alpha}\left(\frac{m\omega}{\pi\hbar}\right)^{\frac{1}{4}}\exp\left\{-\left[\frac{\beta(q-\langle\hat{q}\rangle_\alpha)}{\sqrt{2}}\right]^2+i\langle\hat{p}\rangle_\alpha\frac{q}{\hbar}\right\} \\
&= e^{i\theta_\alpha}\left(\frac{m\omega}{\pi\hbar}\right)^{\frac{1}{4}}\exp\left\{-\left[\frac{q-\langle\hat{q}\rangle_\alpha}{2\Delta\hat{q}_\alpha}\right]^2+i\langle\hat{p}\rangle_\alpha\frac{q}{\hbar}\right\},
\end{aligned}
\tag{7.108}
$$

where $\Delta\hat{q}_\alpha = 1/(\sqrt{2}\beta) = \sqrt{\hbar/2m\omega}$. The probability amplitude is then

$$
\begin{aligned}
|\psi_\alpha(q)|^2 &= e^{i\theta_\alpha}\left(\frac{m\omega}{\pi\hbar}\right)^{\frac{1}{4}}\exp\left\{-\frac{1}{4}\left[\frac{q-\langle\hat{q}\rangle_\alpha}{\Delta\hat{q}_\alpha}\right]^2+i\langle\hat{p}\rangle_\alpha\frac{q}{\hbar}\right\}\times \\
&\quad e^{-i\theta_\alpha}\left(\frac{m\omega}{\pi\hbar}\right)^{\frac{1}{4}}\exp\left\{-\frac{1}{4}\left[\frac{q-\langle\hat{q}\rangle_\alpha}{\Delta\hat{q}_\alpha}\right]^2-i\langle\hat{p}\rangle_\alpha\frac{q}{\hbar}\right\} \\
&= \sqrt{\frac{m\omega}{\pi\hbar}}\exp\left\{-\frac{1}{2}\left[\frac{q-\langle\hat{q}\rangle_\alpha}{\Delta\hat{q}_\alpha}\right]^2\right\}.
\end{aligned}
\tag{7.109}
$$

The above analysis shows that the position representation of $|\alpha\rangle$ is a Gaussian wave packet. This is to be expected, since coherent states are displaced vacuum states and the vacuum state is itself a Gaussian.

7.4 The Hydrogen Atom

7.4.1 Introduction

This subject provides an introduction to one of quantum mechanics' biggest successes: the quantum description of hydrogen, which begins to explain the absorption and emission spectra of atoms. Our approach also lays the foundation of the theory that explains the periodic table of elements and modern chemistry from a quantum perspective. From the point of view of learning quantum mechanics, the study of hydrogen provides some valuable lessons in choosing the right representation of a system in terms of a complete set of commuting observables (CSCO).

From experimental observation, we know that hydrogen is composed of a proton and an electron. The proton is much heavier than the electron and the attractive force is the electromagnetic one (there is a gravitational attraction too, but it is so much smaller than the electromagnetic that we can safely ignore it). As such, we might expect the classical analysis that simplifies studying particles in a central potential to be of use here. The Lagrangian for a system comprising only one proton and one electron with a coulomb interaction is

$$
\mathcal{L} = \frac{1}{2}m_p\dot{\mathbf{q}}_p^2 + \frac{1}{2}m_e\dot{\mathbf{q}}_e^2 - \mathcal{V}(\mathbf{q}_p - \mathbf{q}_e),
$$

where $\mathcal{V}$, the Coulomb potential, is

$$
\mathcal{V}(\mathbf{q}_p - \mathbf{q}_e) = \frac{-e^2}{4\pi\varepsilon_0}\frac{1}{|\mathbf{q}_p - \mathbf{q}_e|},
$$

and so the Hamiltonian is

$$\mathcal{H} = \frac{\mathbf{p}_p^2}{2m_p} + \frac{\mathbf{p}_e^2}{2m_e} + \mathscr{V}(\mathbf{q}_p - \mathbf{q}_e).$$

(7.110)

Because the proton is much heavier than the electron, we can apply the same sort of simplification used when studying gravitational systems like the earth and the moon. We note that the heavy particle is approximately immovable compared with the lighter one. We define a centre of mass coordinate,

$$\mathbf{q}_G = \frac{m_p \mathbf{q}_p + m_e \mathbf{q}_e}{m_e + m_p}.$$

As $m_p \gg m_e$ we see that this approximates $\mathbf{q}_p$. We also define a second 'relative' particle coordinate:

$$\mathbf{q}_R = \mathbf{q}_p - \mathbf{q}_e.$$

Since $\mathbf{q}_p$ does not change much due to the motion of the electron, $\mathbf{q}_R$ is a radial coordinate giving the approximate relative position of the electron (like the moon) compared with the origin, which is identified at the position of the proton (like the earth). These new coordinates greatly simplify the form of the potential to

$$\mathscr{V}(\mathbf{q}_p - \mathbf{q}_e) = \frac{-e^2}{4\pi\varepsilon_0} \frac{1}{|\mathbf{q}_R|}.$$

Exercise 7.14   (i) By re-writing the Lagrangian in terms of $\mathbf{q}_G$ and $\mathbf{q}_R$, show that the canonically conjugate momenta are

$$\mathbf{p}_G = \mathbf{p}_p + \mathbf{p}_e \text{ and } \mathbf{p}_R = \frac{m_e \mathbf{p}_p - m_p \mathbf{p}_e}{m_e + m_p},$$

and that the Hamiltonian can then be written in the form:

$$\mathcal{H} = \frac{\mathbf{p}_G^2}{2M} + \frac{\mathbf{p}_R^2}{2\mu} + \mathscr{V}(\mathbf{q}_R)$$

where

$$M = m_e + m_p \text{ and } \mu = \frac{m_e m_p}{m_e + m_p}.$$

Note that in keeping with the approximation for centre of mass and reduced particle: M is the total mass (and almost the same as that of the proton) and μ is referred to as the reduced mass (and almost the same as that of the electron).

(ii) Using Poisson brackets for all combinations of $\mathbf{q}_e, \mathbf{p}_e, \mathbf{q}_p,$ and $\mathbf{p}_p$ show that $\mathbf{q}_G, \mathbf{p}_G, \mathbf{q}_R,$ and $\mathbf{p}_R$ have an equivalent relationship as canonical pairs. (Note: this duplicates some of the effort above but will help understand some of the quantum methods to follow.) ∎

The solutions to the questions in the exercise above show that the Hamiltonian can be written in the form:

$$\mathcal{H} = \mathcal{H}_G + \mathcal{H}_R,$$

(7.111)

where

$$\mathscr{H}_G = \frac{p_G^2}{2M} \quad \text{and} \quad \mathscr{H}_R = \frac{p_R^2}{2\mu} + \mathscr{V}(\boldsymbol{q}_R)$$

are two independent systems, and

$$\{\mathscr{H}_G, \mathscr{H}\} = \{\mathscr{H}_R, \mathscr{H}\} = \{\mathscr{H}_G, \mathscr{H}_R\} = 0.$$

Here $\mathscr{H}_G$ describes a free particle and $\mathscr{H}_R$ a particle in a central potential, both of which are integrals of the motion, and each of which can be studied and solved independently, since their equations of motion are not linked.

Exercise 7.15   Using Hamilton's equations, verify that the equations of motion for $\boldsymbol{q}_G$ and $\boldsymbol{p}_G$ are indeed independent of those for $\boldsymbol{q}_R$, and $\boldsymbol{p}_R$.   ∎

The centre of mass is obviously just a free particle and will have correspondingly simple dynamics. The interesting subsystem is the *'particle in a central potential'*. As the Hamiltonian

$$\mathscr{H}_R = \frac{p_R^2}{2\mu} + \mathscr{V}(\boldsymbol{q}_R) \tag{7.112}$$

is dependent on only the square of the momentum and the modulus of the position, it displays spherical symmetry. For this reason, we choose to represent the relative particle subsystem in spherical polar coordinates defined according to

$$(\boldsymbol{q}_R)_x = r \sin\theta \, \cos\varphi, \tag{7.113}$$

$$(\boldsymbol{q}_R)_y = r \sin\theta \, \sin\varphi, \tag{7.114}$$

$$(\boldsymbol{q}_R)_z = r \cos\theta, \tag{7.115}$$

where $r \in [0, \infty)$, $\theta \in [0, \pi]$ and $\phi \in [0, 2\pi)$. The potential can then be written as

$$\mathscr{V}(\boldsymbol{q}_R) = \frac{-e^2}{4\pi\varepsilon_0} \frac{1}{|\boldsymbol{q}_R|} = \frac{-e^2}{4\pi\varepsilon_0} \frac{1}{r} = \mathscr{V}(r), \tag{7.116}$$

and the kinetic energy takes the form

$$\mathscr{T}_R = \frac{1}{2}\mu \dot{\boldsymbol{q}}_R^2 = \frac{1}{2}\mu(\dot{r}\boldsymbol{e}_r + r\dot{\theta}\boldsymbol{e}_\theta + r\dot{\phi}\sin\theta \boldsymbol{e}_\phi)^2 \tag{7.117}$$

$$= \frac{1}{2}\mu(\dot{r}^2 + r^2\dot{\theta}^2 + r^2\dot{\phi}^2 \sin^2\theta). \tag{7.118}$$

Thus the Lagrangian is

$$\mathscr{L} = \frac{1}{2}\mu(\dot{r}^2 + r^2\dot{\theta}^2 + r^2\dot{\phi}^2 \sin^2\theta) - \mathscr{V}(r) \tag{7.119}$$

and, just as for our discussion in Section 5.2, the canonically conjugate momenta are

$$p_r = \frac{\partial \mathscr{L}}{\partial \dot{r}} = \mu\dot{r} \tag{7.120}$$

$$p_\theta = \frac{\partial \mathscr{L}}{\partial \dot{\theta}} = \mu r^2\dot{\theta} \tag{7.121}$$

$$p_\phi = \frac{\partial \mathscr{L}}{\partial \dot{\phi}} = \mu r^2 \sin^2\theta \, \dot{\phi} \tag{7.122}$$

and

$$\mathcal{H} = \frac{p_r^2}{2\mu} + \frac{p_\theta^2}{2\mu r^2} + \frac{p_\phi^2}{2\mu r^2 \sin^2\theta} + \mathcal{V}(r). \tag{7.123}$$

Now we can use the fact that orbital angular momentum is

$$\boldsymbol{L} = \mu \boldsymbol{q}_R \times \dot{\boldsymbol{q}}_R = \mu r^2 \left(\dot{\theta} \boldsymbol{e}_\phi - \dot{\phi} \sin\theta \boldsymbol{e}_\theta \right)$$

to write the Hamiltonian as

$$\mathcal{H} = \frac{p_r^2}{2\mu} + \frac{L^2}{2\mu r^2} + \mathcal{V}(r). \tag{7.124}$$

This rather elegantly makes clear the contribution of radial and orbital angular momenta to the kinetic energy.

Exercise 7.16 Complete any intermediate steps in the derivation between Eqs. (7.119) and (7.124). ∎

7.4.2 Quantum Analysis

Prerequisite Material: Section 5.2: On Curvilinear Coordinates (Using Spherical Coordinates as an Example)

7.4.2.1 Choosing a Representation

In the classical argument above, we wrote the Hamiltonian in three different forms:

$$\mathcal{H}(\boldsymbol{q}_p, \boldsymbol{p}_p, \boldsymbol{q}_e, \boldsymbol{p}_e) = \frac{p_p^2}{2m_p} + \frac{p_e^2}{2m_e} + \mathcal{V}(\boldsymbol{q}_p - \boldsymbol{q}_e) \qquad \text{Case 1 (Eq. 7.110)}$$

$$\mathcal{H}(\boldsymbol{q}_G, \boldsymbol{p}_G, \boldsymbol{q}_R, \boldsymbol{p}_R) = \frac{p_G^2}{2M} + \frac{p_R^2}{2\mu} + \mathcal{V}(\boldsymbol{q}_R) \qquad \text{Case 2 (Eq. 7.111)}$$

$$\mathcal{H}(\boldsymbol{q}_G, \boldsymbol{p}_G, r, p_r, \boldsymbol{L}) = \frac{p_G^2}{2M} + \frac{p_r^2}{2\mu} + \frac{L^2}{2\mu r^2} + \mathcal{V}(r). \qquad \text{Case 3 (Eq. 7.124)}$$

There is no time dependence in the Hamiltonian, so we shall be interested in solving the TISE

$$\hat{H} |\psi\rangle = E |\psi\rangle.$$

The states we find from this analysis will be the states that hydrogen is predicated to be in directly following any projective measurement of the energy of an isolated hydrogen atom, and the eigenenergies are the values we are predicted to be able to measure. When quantising the hydrogen atom and seeking solutions to the TISE, any basis is theoretically suitable. This problem is, however, much more tractable if we choose one that respects the symmetries of the system. For this reason, we will take some care in making an appropriate choice of CSCO.

Exercise 7.17 Each form the Hamiltonian will suggest possible choices for a CSCO – can you guess at least two for each Hamiltonian? ∎

Note that a $1/r$ potential will have energies that correspond to both bound and unbound particles. Those with negative energies with respect to the Coulomb potential and contained within the well it generates, are the bound states (classical or quantum). In the argument that follows, we will focus on these bound states. As such, we are looking for states that are localised in position or radius, so when choosing a CSCO we will prefer position or radius operators over their momentum operator counterparts.

In **Case 1**, when we perform canonical quantisation, we have the usual commutation relations

$$\left[(\hat{\boldsymbol{q}}_a)_i, (\hat{\boldsymbol{p}}_b)_j\right] = i\,\hbar\,\left\{(\boldsymbol{q}_a)_i, (\boldsymbol{p}_b)_j\right\} = i\,\hbar\,\delta_{ij}\delta_{ab}, \tag{7.125}$$

where $i, j \in \{x, y, z\}$ and $a, b \in \{p, e\}$. In other words, components in orthogonal directions commute and all operators from different particles commute. If we choose the position representation, this implies choosing

$$\{\boldsymbol{q}_e, \boldsymbol{q}_p\} = \left\{(\hat{\boldsymbol{q}}_e)_x, (\hat{\boldsymbol{q}}_e)_y, (\hat{\boldsymbol{q}}_e)_z, (\hat{\boldsymbol{q}}_p)_x, (\hat{\boldsymbol{q}}_p)_y, (\hat{\boldsymbol{q}}_p)_z\right\}$$

as a CSCO. In this representation the Hamiltonian is

$$\hat{H} = -\frac{\hbar^2}{2m_p}\nabla_p^2 - \frac{\hbar^2}{2m_e}\nabla_e^2 - \frac{e^2}{4\pi\varepsilon_0}\frac{1}{|\boldsymbol{q}_p - \boldsymbol{q}_e|} \tag{7.126}$$

and the Schrödinger equation is:

$$\left[-\frac{\hbar^2}{2m_p}\nabla_p^2 - \frac{\hbar^2}{2m_e}\nabla_e^2 - \frac{e^2}{4\pi\varepsilon_0}\frac{1}{|\boldsymbol{q}_p - \boldsymbol{q}_e|}\right]\psi(q_{e_x}, q_{e_y}, q_{e_z}, q_{p_x}, q_{p_y}, q_{p_z})$$
$$= E\psi(q_{e_x}, q_{e_y}, q_{e_z}, q_{p_x}, q_{p_y}, q_{p_z}).$$

Expressing the Schrödinger equation in this basis may be useful for numerical simulation but is not at all suited to analytical methods.

In **Case 2** the situation improves. We again perform canonical quantisation and have the usual commutation relations

$$\left[(\hat{\boldsymbol{q}}_a)_i, (\hat{\boldsymbol{p}}_b)_j\right] = i\,\hbar\,\left\{(\boldsymbol{q}_a)_i, (\boldsymbol{p}_b)_j\right\} = i\,\hbar\,\delta_{ij}\delta_{ab}, \tag{7.127}$$

where $i, j \in \{x, y, z\}$ but now $a, b \in \{G, R\}$. Again, components in orthogonal directions commute and all operators from different 'particles' commute (as in the classical picture, the centre of mass and relative particles commute).

Above we have made use of the fact that we have already shown that $\boldsymbol{q}_G$ and $\boldsymbol{p}_G$ as well as $\boldsymbol{q}_R$ and $\boldsymbol{p}_R$ form canonically conjugate pairs, and imposed the commutation relation directly. We could also have done this the long way by inheriting the commutation relationship from the position and momentum operators from the electron and proton, applying an analogy to the classical argument. We start by defining a centre of mass position operator,

$$\hat{\boldsymbol{q}}_G = \frac{m_p\hat{\boldsymbol{q}}_p + m_e\hat{\boldsymbol{q}}_e}{m_e + m_p},$$

and a relative particle position operator,

$$\hat{\boldsymbol{q}}_R = \hat{\boldsymbol{q}}_p - \hat{\boldsymbol{q}}_e.$$

Inspired by a classical argument, we could also define:

$$\hat{\boldsymbol{p}}_G = \hat{\boldsymbol{p}}_p + \hat{\boldsymbol{p}}_e \text{ and } \hat{\boldsymbol{p}}_R = \frac{m_e \hat{\boldsymbol{p}}_p - m_p \hat{\boldsymbol{p}}_e}{m_e + m_p}.$$

We can then verify, using commutation relations, that

$$\left[(\hat{\boldsymbol{q}}_G)_j, (\hat{\boldsymbol{p}}_G)_k \right] = i\,\hbar \delta_{jk} \text{ and } \left[(\hat{\boldsymbol{q}}_R)_j, (\hat{\boldsymbol{p}}_R)_k \right] = i\,\hbar \delta_{jk},$$

so these two new sets of observables have the same operator algebra as position and momentum of the electron and proton. From the classical Poisson brackets we already know that these position and momenta form conjugate pairs, so this is not a surprising outcome.

If we use the position representation – but now using the centre of mass and relative particle variables – this is chosen as a CSCO:

$$\{\boldsymbol{q}_G, \boldsymbol{q}_R\} = \left\{ (\hat{\boldsymbol{q}}_G)_x, (\hat{\boldsymbol{q}}_G)_y, (\hat{\boldsymbol{q}}_G)_z, (\hat{\boldsymbol{q}}_R)_x, (\hat{\boldsymbol{q}}_R)_y, (\hat{\boldsymbol{q}}_R)_z \right\}.$$

In this representation the Hamiltonian is

$$\hat{H} = -\frac{\hbar^2}{2M} \nabla_G^2 - \frac{\hbar^2}{2\mu} \nabla_R^2 - \frac{e^2}{4\pi\varepsilon_0} \frac{1}{|\boldsymbol{q}_R|} \tag{7.128}$$

and the Schrödinger equation is:

$$\left[-\frac{\hbar^2}{2M} \nabla_G^2 - \frac{\hbar^2}{2\mu} \nabla_R^2 - \frac{e^2}{4\pi\varepsilon_0} \frac{1}{|\boldsymbol{q}_R|} \right] \psi(q_{Gx}, q_{Gy}, q_{Gz}, q_{Rx}, q_{Ry}, q_{Rz})$$

$$= E\psi(q_{Gx}, q_{Gy}, q_{Gz}, q_{Rx}, q_{Ry}, q_{Rz}).$$

We see that the situation has improved considerably, as there is now a separation of variables, and we can look for solutions in the form

$$\psi(q_{Gx}, q_{Gy}, q_{Gz}, q_{Rx}, q_{Ry}, q_{Rz}) = \psi_G(q_{Gx}, q_{Gy}, q_{Gz})\psi_R(q_{Rx}, q_{Ry}, q_{Rz}).$$

Importantly for both this case and the next one, $\hat{H}_G |\psi\rangle_G = E_G |\psi\rangle_G$ in the $|\boldsymbol{q}_G\rangle$ representation takes the form:

$$-\frac{\hbar^2}{2M} \nabla_G^2 \psi(\boldsymbol{q}_G) = E_G \psi(\boldsymbol{q}_G).$$

Just as in the classical case, the mechanics here are also the mechanics of a free particle. Specifically, this is the TISE for the centre of mass 'particle' and is exactly the same TISE as that of a free particle, which has as its solution plane wave eigenfunctions. *This result is the origin of the idea of the wavelike nature of propagating atoms, and why we expect them to diffract in experiments such as Young's slits!*

In terms of the relative particle, $\hat{H}_R |\phi\rangle_R = E_R |\phi\rangle_R$ in the $|\boldsymbol{q}_R\rangle$ representation takes the form:

$$\left(-\frac{\hbar^2}{2\mu} \nabla_R^2 + V(\boldsymbol{q}_R) \right) \psi(\boldsymbol{q}_R) = E_R \psi(\boldsymbol{q}_R).$$

This is the Schrödinger equation for the relative particle. However, in this form it is hard to solve. Most textbooks will convert this differential equation into spherical polar coordinates and then follow the logic of that calculus. Our approach, using case 3, will avoid the need to do that.

Before proceeding to the last case, let us look, as we did for the classical case, at the separation of variables. We can write $\hat{H} = \hat{H}_G + \hat{H}_R$ where

$$\hat{H}_G = \frac{\hat{\boldsymbol{p}}_G^2}{2M} \text{ and } \hat{H}_R = \frac{\hat{\boldsymbol{p}}_R^2}{2\mu} + V(\hat{\boldsymbol{q}}_R).$$

Just as the Poisson bracket commuted for the components of the Hamiltonian in the quantum case, we have

$$[\hat{H}, \hat{H}_G] = [\hat{H}, \hat{H}_R] = [\hat{H}_G, \hat{H}_R] = 0.$$

So, as with the classical case, the centre of mass and relative particle Hamiltonians are integrals of the motion and, moreover, they are also compatible observables. As all the (Hermitian) Hamiltonians commute, we can conclude that there exists a basis of eigenvectors common to all three Hamiltonians. Thus, for some common eigenvector $|\psi\rangle$, we have:

$$\hat{H}_G |\psi\rangle = E_G |\psi\rangle$$
$$\hat{H}_R |\psi\rangle = E_R |\psi\rangle$$
$$\implies \hat{H} |\psi\rangle = (E_G + E_R) |\psi\rangle,$$

which means that $\hat{H}_G$ and $\hat{H}_R$ are also candidate operators for forming a CSCO.

Let us finally consider **case 3**, where the Hamiltonian takes the form:

$$\hat{H}(\hat{\boldsymbol{q}}_G, \hat{\boldsymbol{p}}_G, \hat{r}, \hat{p}_r, \hat{L}^2) = \hat{H}_G(\hat{\boldsymbol{q}}_G, \hat{\boldsymbol{p}}_G) + \hat{H}_R(\hat{r}, \hat{p}_r, \hat{L}^2) = \frac{\hat{\boldsymbol{p}}_G^2}{2M} + \frac{\hat{p}_r^2}{2\mu} + \frac{\hat{L}^2}{2\mu\hat{r}^2} + \hat{V}(\hat{r}).$$

Note that division by an operator may not be well-defined, as the radius of convergence of the associated power series may not match the domain of the operator. That said, this argument leads to the same conclusion as mathematically precise ones. As the Hamiltonian can still be split in the same way as in case 2,

$$[\hat{H}, \hat{H}_G] = [\hat{H}, \hat{H}_R] = [\hat{H}_G, \hat{H}_R] = 0,$$

the previous observations about separation of variables hold and the discussion of $\hat{H}_G(\boldsymbol{q}_G, \boldsymbol{p}_G)$ does not change. Once more choosing $\boldsymbol{q}_G$ as part of our set of a CSCO yields the TISE:

$$-\frac{\hbar^2}{2M} \nabla_G^2 \psi(\boldsymbol{q}_G) = E_G \psi(\boldsymbol{q}_G),$$

and the nucleus is still considered a free particle with plane wave eigenfunctions of any energy. The difficult Hamiltonian remains

$$\hat{H}_R(\hat{r}, \hat{p}_r, \hat{L}^2) = \frac{\hat{p}_r^2}{2\mu} + \frac{\hat{L}^2}{2\mu\hat{r}^2} + \hat{V}(\hat{r}).$$

Now $\hat{L}^2$ commutes with itself and $\hat{L}_z$, so we also have

$$\left[\hat{H}_R, \hat{L}^2\right] = [\hat{H}_R, \hat{L}_z] = 0,$$

and so $\hat{H}_R$, $\hat{L}^2$, and $\hat{L}_z$ commute and share common eigenvectors. This means that $\hat{L}^2$ and $\hat{L}_z$ are also integrals-of-the-motion and therefore representative of conserved quantities. This suggests that instead of $\{\hat{q}_{Rx}, \hat{q}_{Ry}, \hat{q}_{Rz}\}$ or even $\{\hat{q}_r, \hat{L}^2, \hat{L}_z\}$, we may in fact use $\{\hat{H}_R, \hat{L}^2, \hat{L}_z\}$ as our CSCO. Recall the general properties for angular momentum,

$$\hat{L}^2 |k, j, m\rangle = l(l+1)\hbar^2 |k, j, m\rangle$$

$$\hat{L}_z |k, j, m\rangle = m\hbar |k, j, m\rangle,$$

where the eigenstates in this representation can be written in terms of their radial and angular parts as

$$\psi_{lkm}(r, \theta, \varphi) = R_{kl}(r)Y_l^m(\theta, \varphi), \tag{7.129}$$

where $Y_l^m(\theta, \varphi)$ are the spherical harmonics. Now the TISE is

$$\left(-\frac{\hbar^2}{2\mu}\frac{1}{r}\frac{\partial^2}{\partial r^2}r + \frac{1}{2\mu r^2}\hat{L}^2 + V(r)\right)\psi(r, \theta, \phi) = E_R\psi(r, \theta, \phi). \tag{7.130}$$

Substitution of the general form of the eigenfunctions yields

$$\left(-\frac{\hbar^2}{2\mu r}\frac{\partial^2}{\partial r^2}r + \frac{l(l+1)\hbar^2}{2\mu r^2} + V(r)\right)R_{kl}(r)Y_l^m(\theta, \varphi) = E_R R_{kl}(r)Y_l^m(\theta, \varphi)$$

$$\cancel{Y_l^m(\theta, \varphi)}\left(-\frac{\hbar^2}{2\mu r}\frac{\partial^2}{\partial r^2}r + \frac{l(l+1)\hbar^2}{2\mu r^2} + V(r)\right)R_{kl}(r) = E_R R_{kl}(r)\cancel{Y_l^m(\theta, \varphi)}$$

$$\left(-\frac{\hbar^2}{2\mu r}\frac{\partial^2}{\partial r^2}r + \frac{l(l+1)\hbar^2}{2\mu r^2} + V(r)\right)R_{kl}(r) = E_{kl}R_{kl}(r). \tag{7.131}$$

A simplification of this equation can be obtained if we define

$$R_{kl}(r) = \frac{1}{r}u_{kl}(r), \tag{7.132}$$

yielding the following one-dimensional eigenproblem

$$\left(-\frac{\hbar^2}{2\mu}\frac{\partial^2}{\partial r^2} + \frac{l(l+1)\hbar^2}{2\mu r^2} + V(r)\right)u_{kl}(r) = E_{kl}u_{kl}(r). \tag{7.133}$$

This is just like a one-dimensional Schrödinger equation with the effective potential:

$$V_{\text{eff}}(r) = \frac{l(l+1)\hbar^2}{2\mu r^2} + V(r) = \frac{l(l+1)\hbar^2}{2\mu r^2} - \frac{e^2}{4\pi\varepsilon_0 r}. \tag{7.134}$$

As this is not a book on differential equations, we shall not solve this equation directly, but we do note that it has the following important properties:

- The energy eigenvalues for negative energies take are

$$E_{kl} = -\frac{\mu e^4}{32\pi^2\varepsilon_0^2\hbar^2}\frac{1}{(k+l)^2}.$$

The magnitude of the coefficient gives the ionisation energy of hydrogen as it take this energy to excite an electron from the ground state into the unbound energy eigenstates.
- Regardless of the form of the potential, the fact that m does not appear in the radial term implies that there will always be a $(2l + 1)$-fold degeneracy corresponding to the different possible values of l. This is termed essential degeneracy.
- The additional degeneracy associated with being able to interchange values of k and l is termed accidental degeneracy and does not exist for all central potentials.

- For the coulomb potential $\hat{H}_r$, $\hat{L}^2$, and $\hat{L}_z$ do form a CSCO. The quantum numbers k, l, and m uniquely determine each state in this eigenbasis.
- Choosing the right set of CSSOs greatly reduced the complexity of the problem – why do you think this is?

When looking at the energy eigenvalues of the Hamiltonian, we see a discrete structure that is not dissimilar to the emission and absorption spectra of hydrogen. There are, however, some noticeable differences that need to be explained in order to be confident that quantum mechanics is the right theory for understanding this physical phenomenon. It turns out that the resolution of this is taking into account the intrinsic spin of the electron, which is what we will discuss next.

7.4.3 Fine Structure of Hydrogen: Spin–Orbit Coupling

Experimentally, we know that in addition to its mass and electrical charge, the electron possesses an intrinsic magnetic moment we call spin. Thinking classically in terms of the rest-frame of the electron, we see that any motion of the proton through this magnetic field, due to the proton's charge, results in a force between the electron and proton. This is known as spin–orbit coupling. Using apparatus such as the Stern–Gerlach experiment, we can determine that the spin of the electron is an angular momentum with $j = \frac{1}{2}$ (so the only possible value of m are $\pm\frac{1}{2}$). It turns out that the contribution to the Hamiltonian for the spin–orbit coupling for a particle in a central potential takes the form

$$\hat{W}_{SO} \propto \frac{\hat{L} \cdot \hat{S}}{r^3}, \tag{7.135}$$

where $\hat{L}$ is the usual angular momentum and $\hat{S} = \frac{\hbar}{2}(\hat{\sigma}_x \mathbf{i} + \hat{\sigma}_y \mathbf{j} + \hat{\sigma}_z \mathbf{k})$. A full analysis would find that this extra contribution to the Hamiltonian can be considered a perturbation of the Hamiltonian we considered in the previous section as

$$\mathcal{O}\left(\frac{\hat{W}_{SO}}{\hat{H}_0}\right) \approx \left[\frac{e^2}{4\pi\varepsilon_0\hbar c}\right]^2 = \alpha^2 = \frac{1}{137^2},$$

where α is the fine structure constant (see, for example, Ref. [12] for details).

Importantly, adding the spin–orbit coupling term to the Hamiltonian $\hat{H}_R$ stops the orbital angular momentum from being an integral of the motion. However, the total angular momentum $\hat{J} = \hat{L} + \hat{S}$ is. This means that we can bring to bear the theory of addition of angular momentum. The CSCO is now modified, and the new basis defined to be the simultaneous eigenvectors:

$$\hat{L}^2 |j_1 j_2; jm\rangle = l(l+1)\hbar^2 |j_1 j_2; jm\rangle \tag{7.136}$$

$$\hat{S}^2 |j_1 j_2; jm\rangle = s(s+1)\hbar^2 |j_1 j_2; jm\rangle \tag{7.137}$$

$$\hat{J}^2 |ls; jm\rangle = j(j+1)\hbar^2 |j_1 j_2; jm\rangle \tag{7.138}$$

$$\hat{J}_z |ls; jm\rangle = m\hbar |j_1 j_2; jm\rangle. \tag{7.139}$$

If we add to this the labels for the bound states, $n = k + l$, of $\hat{H}_R$, this forms a CSCO for the relative particle.

We can specify each basis state with an alternative notation:

$$n^{2s+1}l_j, \tag{7.140}$$

where l is the orbital-angular momentum eigenvalue that is denoted $s = 0, p = 1, d = 2, \ldots$. Hence, for hydrogen we have the states:

	$l = 0$	$l = 1$	$l = 2 \ldots$
$n = 1$	$1^2 s_{\frac{1}{2}}$		
$n = 2$	$2^2 s_{\frac{1}{2}}$	$2^2 p_{3/2}, 2^2 p_{1/2}$	
$n = 3, \ldots$	$3^2 s_{\frac{1}{2}}$	$3^2 p_{3/2}, 3^2 p_{1/2}$	$3^2 d_{3/2}, 3^2 d_{5/2}$

Those with a background in chemistry or spectroscopy will recognise this as spectroscopic notation, which predates quantum mechanics. A full analysis that constructs the energy eigenvalue distribution for hydrogen shows that these energies are in good agreement with experimental observation of the energy levels of hydrogen found from absorption and emission spectra (full agreement needs an even deeper analysis and is covered in specialist books on the subject). The fact that quantum mechanics is able to explain such phenomena as atomic spectra by replacing the Poisson bracket algebra of classical mechanics with the appropriate non-commutative algebra is truly phenomenal. To obtain a complete understanding of why atoms have the structure they do, requires application of the Pauli exclusion principle, which we briefly discuss on page 358.

7.5 The Dihydrogen Ion

Prerequisite Material: Section 6.7: Two-level Systems

In this section, we explore a greatly simplified model of the simplest of molecules – the dihydrogen ion, which comprises two protons and a single electron. Although simple, the model we will use still gives good insight into how atomic states combine to form molecular states and why certain states are found in nature while others are not Figure 7.9. Each particle in the dihydrogen ion possesses kinetic and potential energies, and there are electrostatic forces acting between each of the constituent particles. The Hamiltonian for the total system is therefore given by:

$$\hat{H} = \frac{\hat{p}_1^2}{2m_p} + \frac{\hat{p}_1^2}{2m_p} + \frac{\hat{p}_e^2}{2m_e} + V_{12}(\hat{q}_1 - \hat{q}_2) + V_{1e}(\hat{q}_1 - \hat{q}_e) + V_{2e}(\hat{q}_2 - \hat{q}_e), \tag{7.141}$$

where $\hat{p}_i$, $\hat{q}_i$, and m_i are, respectively, the momentum, position and mass of each proton ($i = 1, 2$) and electron ($i = e$), whilst V_{ij} denotes the electrostatic potential between each particle. The resulting Schrödinger equation is difficult to solve, and an alternative method is to consider a combination of two pictures. The idea is as follows: let us consider a state of the system that is separable, partitioning the system into a state vector for one of the protons and an electron in a tensor product with a state vector for the other proton. We might

Figure 7.9 Diagram showing the H_2^+ ion system comprising two protons and one electron with their respective kinetic and potential energies. Note that this is an entirely classical picture, and we should remember that quantum mechanically, the system is not composed of particles but of state vectors (or wave functions, Wigner functions, etc. but never point particles).

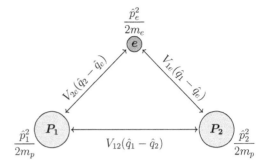

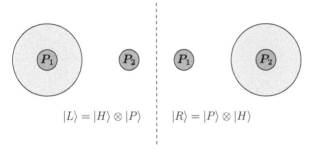

$$|L\rangle = |H\rangle \otimes |P\rangle \qquad |R\rangle = |P\rangle \otimes |H\rangle$$

Figure 7.10 Diagram showing the two possible pictures used to describe the dihydrogen ion. Note that in this convention we define two states, $|L\rangle$ and $|R\rangle$, which distinguish the two pictures and refer to the hydrogen atom being on the 'left' or the 'right'.

imagine for example that the state vector for one of the protons and the electron is an eigenstate of hydrogen, as this is also a system comprising a proton and an electron (remember this is the free particle wave function *and* the orbital functions). We might also consider the state of the proton to be the energy eigenstate of the proton Hamiltonian (i.e. just that of a free particle). As the protons are identical particles, and we do not know which proton is included in the 'hydrogen', we have two possible states that are orthogonal by construction, with the same energy expectation value: one has a 'hydrogen' on the 'left' with a proton on the 'right', the other the opposite, as shown in Figure 7.10. For finite separation, neither state will be an eigenstate of the full Hamiltonian for dihydrogen$^+$ (as there are coupling terms between every particle). If we think classically, we might imagine that – at infinite separation of the 'hydrogen' with the 'electron' – $|L\rangle$ and $|R\rangle$ are, at this limit, in eigenstates of the full Hamiltonian (as the interaction energy is negligible). As the eigenstates of each system include, e.g. plane wave components, this is not the best reasoning, but it may help understand the logic of the approach. In any case, we can conclude by the symmetry of the Hamiltonian that $\left\langle L\left|\hat{H}\right|L\right\rangle = \left\langle R\left|\hat{H}\right|R\right\rangle \stackrel{\text{def}}{=} E_0$ and, because $|L\rangle$ and $|R\rangle$ are not eigenstates, that $\left\langle L\left|\hat{H}\right|R\right\rangle = \left\langle R\left|\hat{H}\right|L\right\rangle \stackrel{\text{def}}{=} -A \neq 0$.

As mentioned in Section 6.7, we can use $\{|L\rangle, |R\rangle\}$ as a basis and write the Hamiltonian matrix as

$$(\hat{H}) = \begin{pmatrix} \langle L|\hat{H}|L\rangle & \langle L|\hat{H}|R\rangle \\ \langle R|\hat{H}|L\rangle & \langle R|\hat{H}|R\rangle \end{pmatrix} = \begin{pmatrix} E_0 & -A \\ -A & E_0 \end{pmatrix}. \tag{7.142}$$

Using the analysis from Section 6.7 we see that the energy eigenvalues for this system are

$$E_a = E_0 + A$$
$$E_b = E_0 - A. \tag{7.143}$$

The associated energy eigenvectors are

$$|\psi_a\rangle = \frac{1}{\sqrt{2}} \begin{pmatrix} 1 \\ -1 \end{pmatrix} = \frac{1}{\sqrt{2}}(|L\rangle - |R\rangle)$$
$$|\psi_b\rangle = \frac{1}{\sqrt{2}} \begin{pmatrix} 1 \\ 1 \end{pmatrix} = \frac{1}{\sqrt{2}}(|L\rangle + |R\rangle). \tag{7.144}$$

We can interpret the physical significance of these two eigenstates as follows: the state $|\psi_a\rangle$ with the higher energy represents the antisymmetric superposition and is the difference between the left and the right wave functions when working in the position representation. Because the difference in the wave functions leaves a void between the two states, we call this the *antibonded state* (hence the subscript a). The symmetric superposition $|\psi_b\rangle$ with lower energy describes the total wave function that we consider a *bonded state* (subscript b). In the position representation, and ignoring the plane wave components, thinking in terms of atomic orbitals, we see there will be a region between both protons where there is a high probability of finding the electron. An important observation is that the energy eigenvalue for the bonded state provides a lower energy state than that of the antibonded state; it is for this reason that one can argue that many atoms brought into close proximity to one another will share electrons and form bonds, since it is energetically more favourable to do so.

In Figure 7.11 we show a schematic of the different states considered in the above discussion before and after bonding. They are arranged vertically with their corresponding energy expectation value.

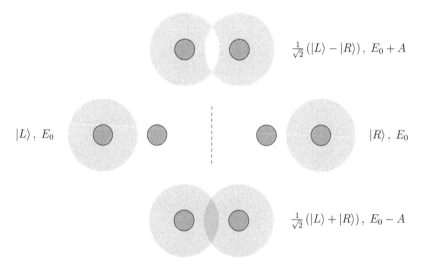

$$\frac{1}{\sqrt{2}}(|L\rangle - |R\rangle), \quad E_0 + A$$

$|L\rangle, \ E_0$

$|R\rangle, \ E_0$

$$\frac{1}{\sqrt{2}}(|L\rangle + |R\rangle), \quad E_0 - A$$

Figure 7.11 Diagram showing the bonded and antibonded states of the dihydrogen ion, with their corresponding energy eigenvalues. Note that the bonded state sits at a lower energy than the separate and antibonded states.

In our example of the dihydrogen ion, we used two identical particles interacting with one electron. The type of bond formed here is an exact sharing, known as a *covalent bond*. The covalent bond comes from the symmetry of the problem, that $\langle L|\hat{H}|L\rangle = \langle R|\hat{H}|R\rangle = E_0$, but covalent bonds can be formed by different atoms, provided the energies of each are similar. The ultimate factor in quantifying this energy gap between the 'left' atom and 'right' atom is the potential felt by the electron in each. For atoms with substantially larger potentials (or a larger electrostatic pull on the electron) the states still superpose, but the bonded electron will have a far greater chance of being found in the proximity of the larger potential. Examples of this type of bond are *polar bonds* and the *ionic bonds* seen when alkali metals bond with halogens. To evaluate such a problem, one would follow the framework mentioned above, but apply the non-degenerate, coupled approach discussed in Section 6.7.2.

7.6 The Jaynes–Cummings Model

In this section, we introduce a simple but a fully quantum mechanical treatment of atoms interacting with fields; this is important in areas such as quantum optics. We will use this analysis to uncover some novel features to quantum mechanical fields, namely Rabi oscillations and the phenomena of collapse and revival . The derivation of this model starts from the theoretic quantum field description of the set-up shown in Figure 7.12, and is therefore outside the scope of this book. The outline of the derivation is as follows: we begin by writing the Hamiltonian density for the electromagnetic field interacting with an atom. Following a quantisation procedure, we note that the electromagnetic field can be expressed as a set of QHOs that we term field modes (as each has its own natural frequency). Our assumption is that the frequency of the cavity is very well-defined, and that there is only significant coupling between one mode of the electromagnetic field and two of the energy levels in the atom. Using the assumption that the applied electromagnetic field mode is in near resonance with the atomic transition, we apply the so-called rotating wave approximation which ignores the high-frequency terms in the atom–field coupling term in the Hamiltonian, which greatly simplifies the analysis.

7.6.1 The Hamiltonian

As already mentioned, it can be shown that under certain conditions the behaviour of the electromagnetic field mode can be represented by a single QHO – one field mode, which

Figure 7.12 A schematic of the physical set-up that the Jaynes–Cumming model represents – it consists of an atom interacting with an electromagnetic field mode within a cavity.

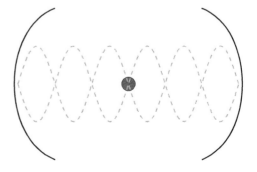

has the Hamiltonian

$$\hat{H}_f = \hbar \omega_f \hat{n}. \tag{7.145}$$

Note that it is a convention in quantum optics to often neglect the zero-point energy, as it only adds a phase factor to the dynamics of the state and its omission does not affect the experimentally determinable predictions of the model.

The atom with which the field mode interacts is considered to have only two levels of relevance: a ground state and an excited state. To distinguish between these two states, the Hamiltonian for the atom is given by

$$\hat{H}_a = \frac{\hbar \omega_a}{2} \hat{\sigma}_z, \tag{7.146}$$

where ω_a denotes the frequency of the atom. The total Hamiltonian, without interaction, is given by:

$$\hat{H} = \hat{H}_f \otimes \hat{\mathbb{1}}_a + \hat{\mathbb{1}}_f \otimes \hat{H}_a = \hbar \omega_f \hat{n} \otimes \hat{\mathbb{1}}_a + \hat{\mathbb{1}}_f \otimes \frac{\hbar \omega_a}{2} \hat{\sigma}_z. \tag{7.147}$$

Each Hamiltonian term has its own natural basis, the number basis $\{|n\rangle\}$ for the field, and the σ_z basis (where we use the standard labels for this model, $\{|g\rangle, |e\rangle\}$, where $|g\rangle \sim \left|m = -\frac{1}{2}\right\rangle$ and $|e\rangle \sim \left|m = \frac{1}{2}\right\rangle$) for the atom. Thus, the actions of these Hamiltonians on their respective eigenstates are:

$$\hat{H}_f |n\rangle = \hbar \omega_f n |n\rangle \tag{7.148}$$

$$\hat{H}_a |g\rangle = -\frac{\hbar \omega_a}{2} |g\rangle \tag{7.149}$$

$$\hat{H}_a |e\rangle = \frac{\hbar \omega_a}{2} |e\rangle, \tag{7.150}$$

which gives the two separate energy spectra given in Figure 7.13. When the frequencies of the field mode and the atom are equal to each other, the energy levels (apart from the ground state) become twofold degenerate. The interaction between the field and the atom is, as outlined in the introduction, given by the rotating wave approximated coupling

$$\hat{H}_I = \hbar \lambda \left(\hat{a} \hat{\sigma}_+ + \hat{a}^\dagger \hat{\sigma}_- \right). \tag{7.151}$$

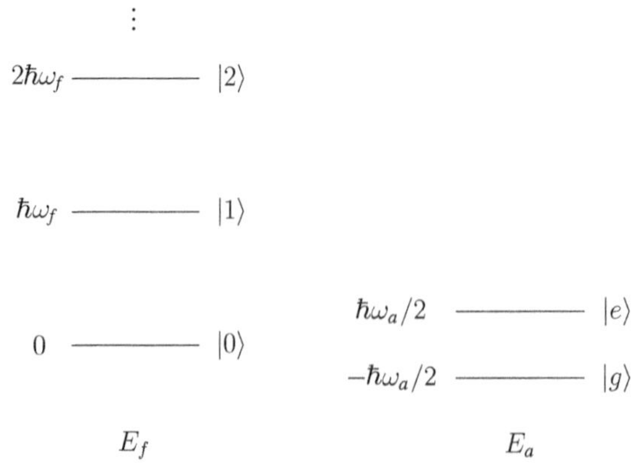

$\vdots$

$2\hbar\omega_f$ ———— $|2\rangle$

$\hbar\omega_f$ ———— $|1\rangle$

$\hbar\omega_a/2$ ———— $|e\rangle$

0 ———— $|0\rangle$

$-\hbar\omega_a/2$ ———— $|g\rangle$

E_f E_a

Figure 7.13 Diagram showing the energy level structures, E_f and E_a, of the field mode and the atom, respectively.

The total Hamiltonian of the system, including interaction, is then

$$\hat{H} = \hat{H}_f + \hat{H}_a + \hat{H}_I$$
$$= \hbar\omega_f \hat{n} + \frac{\hbar\omega_a}{2}\hat{\sigma}_z + \hbar\lambda\left(\hat{a}\hat{\sigma}_+ + \hat{a}^\dagger\hat{\sigma}_-\right). \tag{7.152}$$

7.6.2 The Eigenstates and Eigenvalues

We seek to solve the TISE and find the energy eigenstates and eigenvalues of Eq. (7.152). With the benefit of hindsight, we will start by examining the effect of the Hamiltonian on two eigenstates for the case $\lambda = 0$, $\{|n, e\rangle, |n+1, g\rangle\}$. We define the following as shorthand for these states:

$$|1_n\rangle = |n\rangle \otimes |e\rangle = |n, e\rangle$$
$$|2_n\rangle = |n+1\rangle \otimes |g\rangle = |n+1, g\rangle. \tag{7.153}$$

The action of the Hamiltonian on $|1_n\rangle$ is

$$\hat{H}|1_n\rangle = \left[\hbar\omega_f \hat{n} + \frac{\hbar\omega_a}{2}\hat{\sigma}_z + \hbar\lambda\left(\hat{a}\hat{\sigma}_+ + \hat{a}^\dagger\hat{\sigma}_-\right)\right]|n, e\rangle$$
$$= \left[\hbar\omega_f n + \frac{\hbar\omega_a}{2}\right]|n, e\rangle + \hbar\lambda\sqrt{n+1}\,|n+1, g\rangle$$
$$= \left[\hbar\omega_f n + \frac{\hbar\omega_a}{2}\right]|1_n\rangle + \hbar\lambda\sqrt{n+1}\,|2_n\rangle. \tag{7.154}$$

Likewise for $|2_n\rangle$, we see that

$$\hat{H}|2_n\rangle = \left[\hbar\omega_f(n+1) - \frac{\hbar\omega_a}{2}\right]|2_n\rangle + \hbar\lambda\sqrt{n+1}\,|1_n\rangle. \tag{7.155}$$

We see that both of the above expressions are closed in terms of $|1_n\rangle$ and $|2_n\rangle$ (that is, no other states are involved). Also note that these states form an orthonormal basis. By labelling the basis in this way, we make clear that the Hamiltonian can be arranged in a block-diagonal form, where each block is of the form:

$$(\hat{H})_n = \begin{pmatrix} \langle 1_n|\hat{H}|1_n\rangle & \langle 1_n|\hat{H}|2_n\rangle \\ \langle 2_n|\hat{H}|1_n\rangle & \langle 2_n|\hat{H}|2_n\rangle \end{pmatrix}$$
$$= \hbar\begin{pmatrix} \omega_f n + \dfrac{\omega_a}{2} & \lambda\sqrt{n+1} \\ \lambda\sqrt{n+1} & \omega_f(n+1) - \dfrac{\omega_a}{2} \end{pmatrix}. \tag{7.156}$$

The block-diagonal form of the Hamiltonian reduces the TISE eigenproblem to that of solving for the eigenvectors and the eigenvalues of this two-state Hamiltonian. From Section 6.7 we know that the eigenvalues of the Hamiltonian on the reduced basis are:

$$E_n^\pm = \hbar\left[\omega_f\left(n + \frac{1}{2}\right) \pm \frac{1}{2}\sqrt{(\omega_a - \omega_f)^2 + 4\lambda^2(n+1)}\right]$$
$$= \hbar\left[\omega_f\left(n + \frac{1}{2}\right) \pm \Omega_n\right], \tag{7.157}$$

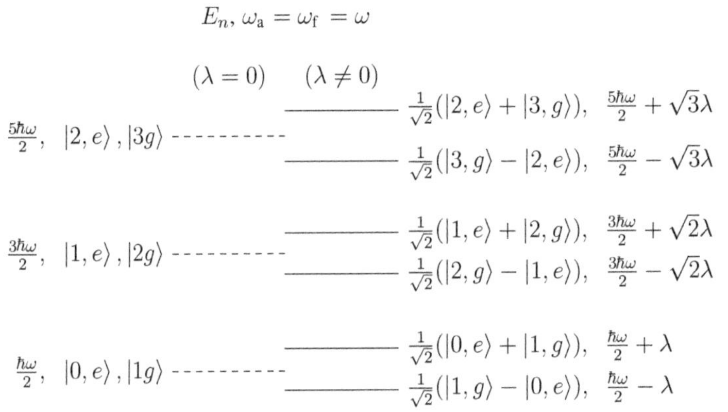

$$E_n, \; \omega_a = \omega_f = \omega$$

$$(\lambda = 0) \qquad (\lambda \neq 0)$$

$$\frac{1}{\sqrt{2}}(|2,e\rangle + |3,g\rangle), \quad \frac{5\hbar\omega}{2} + \sqrt{3}\lambda$$

$$\frac{5\hbar\omega}{2}, \; |2,e\rangle, |3g\rangle \; ---------$$

$$\frac{1}{\sqrt{2}}(|3,g\rangle - |2,e\rangle), \quad \frac{5\hbar\omega}{2} - \sqrt{3}\lambda$$

$$\frac{1}{\sqrt{2}}(|1,e\rangle + |2,g\rangle), \quad \frac{3\hbar\omega}{2} + \sqrt{2}\lambda$$

$$\frac{3\hbar\omega}{2}, \; |1,e\rangle, |2g\rangle \; ---------$$

$$\frac{1}{\sqrt{2}}(|2,g\rangle - |1,e\rangle), \quad \frac{3\hbar\omega}{2} - \sqrt{2}\lambda$$

$$\frac{1}{\sqrt{2}}(|0,e\rangle + |1,g\rangle), \quad \frac{\hbar\omega}{2} + \lambda$$

$$\frac{\hbar\omega}{2}, \; |0,e\rangle, |1g\rangle \; ---------$$

$$\frac{1}{\sqrt{2}}(|1,g\rangle - |0,e\rangle), \quad \frac{\hbar\omega}{2} - \lambda$$

Figure 7.14 Diagram showing the breaking of degeneracy when $\lambda \neq 0$. Note the new eigenstates and eigenvalues of the first two energy levels.

where $\Omega_n = \frac{1}{2}\sqrt{\left(\omega_a - \omega_f\right)^2 + 4\lambda^2 \, (n+1)}$ which, as we shall later see, determines the frequency of oscillation between the ground and excited states of the atom. For the resonant case where $\omega_a = \omega_f = \omega$ we see that $\hat{H}$ spans a degenerate subspace for $\lambda = 0$. For $\lambda \neq 0$ the degeneracy is broken, and the energy eigenvalues are given by:

$$E_n^{\pm} = \hbar \left[\omega \left(n + \frac{1}{2} \right) \pm \lambda \sqrt{n+1} \right] \tag{7.158}$$

with the eigenvectors:

$$|+_n\rangle = \frac{1}{\sqrt{2}} \left(|1_n\rangle + |2_n\rangle \right)$$
$$|-_n\rangle = \frac{1}{\sqrt{2}} \left(|2_n\rangle - |1_n\rangle \right). \tag{7.159}$$

The interaction between the atom and the field mode therefore lifts the degeneracy as depicted in Figure 7.14. This distribution of energy levels is known as the Jaynes–Cummings ladder.

7.6.3 Dynamics of the Atomic Inversion

Prerequisite Material: As above, we will continue to directly apply results found in Section 6.7: Two-level Systems.

We will now study the dynamics of the Jaynes–Cummings model for the $|e\rangle$ atomic state coupled to different example states of the field mode as follows: (i) a number state (ii) superpositions of a few number states (iii) and finally, a coherent state. Rather than looking at the probabilities associated with the evolving state, we will calculate how the atomic inversion, $\langle \sigma_z(t) \rangle$ (the average atomic state), varies with time. For our first example we consider the initial state

$$|\psi(0)\rangle = |1_n\rangle = |n, e\rangle. \tag{7.160}$$

From Section 6.7 we know that the evolution operator can be written in terms of energy eigenvalues and eigenstates, and is given by

$$\hat{U}_n(t) = e^{-i \, E_n^+ t/\hbar} |+_n\rangle \langle +_n| + e^{-i \, E_n^- t/\hbar} |-_n\rangle \langle -_n|. \tag{7.161}$$

Writing this in matrix form in the $\{|1_n\rangle, |2_n\rangle\}$ basis gives

$$U_n(t) = e^{-i\omega(n+1/2)t}\begin{pmatrix} \cos(\Omega_n t) & -i\sin(\Omega_n t) \\ -i\sin(\Omega_n t) & \cos(\Omega_n t) \end{pmatrix}. \tag{7.162}$$

The dynamics of the state then is:

$$|\psi(t)\rangle = \hat{U}(t)\,|\psi(0)\rangle \sim U_n(t)\begin{pmatrix}1\\0\end{pmatrix} = e^{-i\omega(n+1/2)t}\begin{pmatrix}\cos(\Omega_n t)\\-i\sin(\Omega_n t)\end{pmatrix}. \tag{7.163}$$

The atomic inversion is then

$$\langle\hat{\sigma}_z(t)\rangle = \langle\psi(t)|\hat{\sigma}_z|\psi(t)\rangle = \cos^2(\Omega_n t) - \sin^2(\Omega_n t)$$
$$= \cos(2\Omega_n t). \tag{7.164}$$

Figure 7.15 shows this dynamics, and we see the that state oscillates between ground and excited states. These should not be confused with the so-called Rabi oscillations that you may read about elsewhere. Those occur in response to a sinusoidal classical drive on a two-state system. That said, the term 'Rabi oscillation' does now appear to be used in some of the literature in the context of the Jaynes–Cummings model; we prefer to keep this

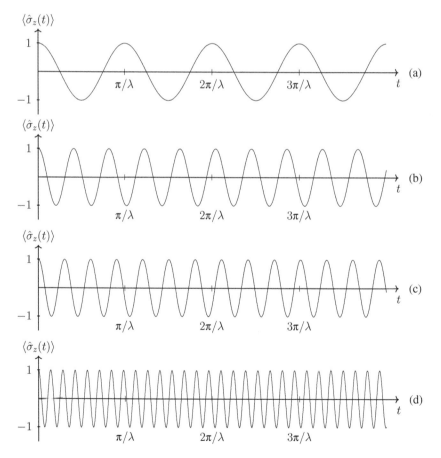

Figure 7.15 To help understand the later superposition examples, we show the effect of the choice of n in the initial state $|n, e\rangle$ for the resonant case $\omega_a = \omega_f$. We show the cases: Ω_0 (a), Ω_5 (b), Ω_{10} (c), and Ω_{50} (d).

distinct, as there are important differences between the two models [25]. We see that, as $\Omega_n = \frac{1}{2}\sqrt{(\omega_a - \omega_f)^2 + 4\lambda^2(n+1)}$, the higher the number state of the field mode, the higher the frequency of transition between the atomic states. In general, that field will not be in a pure number state $|n\rangle$ but some superposition

$$|\varphi_f\rangle = \sum_n c_n |n\rangle. \tag{7.165}$$

To simplify the algebra, we will continue to assume that the atom's initial state is $|e\rangle$. The initial state of the atom–field system will then be given by

$$|\psi_0\rangle = \sum_n c_n |n\rangle \otimes |e\rangle = \sum_n c_n |1_n\rangle \tag{7.166}$$

and, recalling that $|\pm_n\rangle = 1/\sqrt{2}(|1_n\rangle \pm |2_n\rangle)$, we can write this state in terms of energy eigenstates $|+\rangle, |-\rangle$ as

$$|\psi_0\rangle = \sum_n \frac{c_n}{\sqrt{2}} (|+_n\rangle - |-_n\rangle). \tag{7.167}$$

We do this so that we can once more evolve the state using the evolution operator, but this time we will use the fact that the state is expressed in terms of the energy eigenbasis to write

$$|\psi(t)\rangle = \hat{U}(t) |\psi_0\rangle$$

$$= \exp\left(-\frac{i\hat{H}t}{\hbar}\right) \sum_n \frac{c_n}{\sqrt{2}} (|+_n\rangle - |-_n\rangle) \tag{7.168}$$

$$= \sum_n \frac{c_n}{\sqrt{2}} \left[\exp\left(-\frac{i E_n^+ t}{\hbar}\right) |+_n\rangle - \exp\left(-\frac{i E_n^- t}{\hbar}\right) |-_n\rangle\right].$$

We introduce the following shorthand:

$$e_n^+ = \exp\left(-\frac{i E_+ t}{\hbar}\right) = e^{-i\omega t\left(n+\frac{1}{2}\right)} e^{-i\lambda\sqrt{n+1}t}$$

$$e_n^- = \exp\left(-\frac{i E_- t}{\hbar}\right) = e^{-i\omega t\left(n+\frac{1}{2}\right)} e^{i\lambda\sqrt{n+1}t}. \tag{7.169}$$

Then writing the eigenstates in terms of basis state once more, yields

$$|\psi(t)\rangle = \sum_n \frac{c_n}{2} \left[e^+ |1_n\rangle + e_n^+ |2_n\rangle + e_n^- |1_n\rangle - e_n^- |2_n\rangle\right]$$

$$= \sum_n c_n \left[\left(\frac{e_n^+ + e_n^-}{2}\right) |1_n\rangle + \left(\frac{e_n^+ - e_n^-}{2}\right) |2_n\rangle\right] \tag{7.170}$$

$$= \sum_n c_n e^{-i\omega t\left(n+\frac{1}{2}\right)} \left[\cos(\lambda\sqrt{n+1}t) |1_n\rangle - i\sin(\lambda\sqrt{n+1}t) |2_n\rangle\right].$$

This expression shows that the system dynamics comprises oscillations between the states $|1_n\rangle$ and $|2_n\rangle$ whose contribution is determined by the initial values of c_n. The associated atom inversion is

$$\langle \hat{\sigma}_z(t) \rangle = \sum_n |c_n|^2 \left[\cos^2(\lambda\sqrt{n+1}t) - \sin^2(\lambda\sqrt{n+1}t)\right]$$

$$= \sum_n |c_n|^2 \cos(2\lambda\sqrt{n+1}t). \tag{7.171}$$

Figure 7.16 The atomic inversion for an atom initially prepared in its excited state and coupled to a field mode that is a superposition of four number states: $\frac{1}{2}(|0\rangle + |5\rangle + |10\rangle + |50\rangle)$.

This expression is not a Fourier series but it does contain enough similar features for us to expect rich and complex behaviour in the atomic inversion for any non-trivial initial state. To illustrate this, let us consider an initial state based on the examples given in Figure 7.15:

$$\left|\varphi_f\right\rangle = \frac{1}{2}(|0\rangle + |5\rangle + |10\rangle + |50\rangle). \tag{7.172}$$

The resulting atomic inversion is shown in Figure 7.16. It illustrates that a simple superposition of four different field states can lead to an apparently complex dynamics in the state occupation of the atom. This four-state example, while instructive, is rather artificial in its construction. A much more common, but non-trivial, state of the field is the coherent state, since it is the idealisation of a laser source. We therefore now consider the case of an atom prepared in its excited state, coupled to a field prepared in a coherent state,

$$|\psi(0)\rangle = |\alpha\rangle \otimes |e\rangle, \tag{7.173}$$

where coherent states in the number basis are given by

$$|\alpha\rangle = \sum_n e^{-|\alpha|^2/2} \frac{\alpha^n}{\sqrt{n!}} |n\rangle. \tag{7.174}$$

Here $|\alpha|^2 = \langle n \rangle$ gives the average number of photons in the field mode. Performing the same analysis as in the previous superposition example and noting that

$$c_n = e^{-|\alpha|^2/2} \frac{\alpha^n}{\sqrt{n!}}, \tag{7.175}$$

we obtain the following expression for the atomic inversion:

$$\langle \hat{\sigma}_z(t) \rangle = \sum_n e^{-|\alpha|^2} \frac{|\alpha^2|^n}{n!} \cos(2\lambda\sqrt{n+1}t). \tag{7.176}$$

If we now use $|\alpha|^2 = \langle \hat{n} \rangle$, we see that the above expression reduces to

$$\langle \hat{\sigma}_z(t) \rangle = e^{-\langle \hat{n} \rangle} \sum_n \frac{\langle \hat{n} \rangle^n}{n!} \cos(2\lambda\sqrt{n+1}t). \tag{7.177}$$

We can find the approximate behaviour of the atomic inversion by evaluating this sum to some given level of truncation. In Figure 7.17 we show an example for the first 11 states, as this is easily calculable by hand in a simple plotting package. Much higher orders can be calculated easily in computing packages or programming languages (the topic of the next chapter). In this dynamics, we see a reduction and growth in the atomic inversion that is known as collapse and revival. In the true solution of the full sum, the oscillations in the

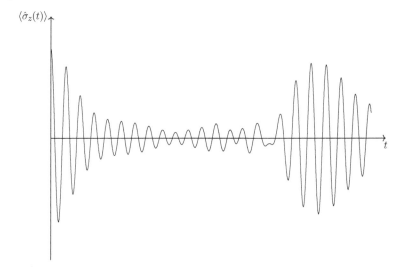

Figure 7.17 Diagram showing the approximate atomic inversion of the Jaynes–Cummings model, where the initial state is $\left|\alpha = \sqrt{10}, e\right\rangle$. The result is found by summing the first 11 terms in the power series Eq. (7.177). In the next chapter we will consider an alternative approach to the numerical analysis of this system.

collapse region almost completely die off. This can be seen more clearly in Figure 8.2 of the next chapter, where the atomic inversion resulting from an actual numerical solution is displayed. At time $t = 0$ the oscillations start in phase, but as the system evolves, these oscillations fall out of phase and destructively interfere, causing $\left\langle \hat{\sigma}_z(t) \right\rangle$ to fall to a minimum, hence the collapse. This is until the oscillations return to being in phase and constructively interfere, giving the revival part of the system's evolution. A full discussion of this model is beyond the scope of this text but we do note one important feature that makes the model interesting: in the middle of the collapse, the field becomes a Schrödinger cat state – making this system useful for the preparation of such states.

7.7 The Stern–Gerlach Experiment

Prerequisite Material: Section 2.6.2: The Stern–Gerlach Experiment, and Section 3.4.3: Spin and Position – The Spinor

When motivating the need for quantum mechanics in Section 2.6.2, we looked at the example of the Stern–Gerlach experiment. A diagram of the experiment is shown in Figure 7.18. This, to aid our physical analysis, is deliberately less of a schematic than the ones used in Section 2.6.2. The set-up comprises a source of atoms such as silver, often prepared in a furnace. The main apparatus is a collimator followed by a non-uniform magnetic field. The magnetic field is required to be non-uniform so that there is a gradient in the desired direction. The spatial degree of freedom of the atoms are then detected on

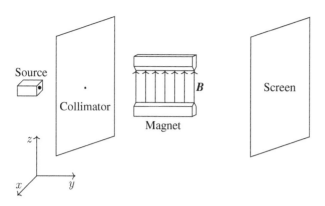

Figure 7.18 The set-up of the Stern–Gerlach experiment: a beam of Spin-1/2 particles is sent through a non-uniform magnetic field and their positions measured on an observation screen.

an observation screen at some distance from the magnetic field. Recall that, classically, the particles leaving the furnace would be expected to have a random magnetic moment associated with them. The action of the collimator and field one would expect to produce a Gaussian distribution, since the magnetic moment of each particle would interact with the field differently, and a random distribution of moments would produce a random distribution on the screen. The results of the Stern–Gerlach experiment are significant because this random distribution is not observed. Instead, a distribution of a number of 'blobs' corresponding to the m quantum number of the spin of the atoms is seen on the screen. The purpose of this section is to understand why this happens, and also to begin to understand the cause of the non-symmetric nature of real experimental results.

Assuming that we can start the atoms of as particles in free space emitted from the source, the Hamiltonian of the system will be of the form

$$\hat{H} = \frac{\hat{p}^2}{2m} + V(\hat{q}) - \boldsymbol{\mu} \cdot \boldsymbol{B}.$$

Here $\hat{p}$ is the momentum operator, m the mass of the particle, $V(q)$ includes the boundary conditions of the collimator and screen, $\boldsymbol{\mu}$ the magnetic moment of the particle, and $\boldsymbol{B}$ the magnetic field with which it interacts. We can approximate out the potential by assuming that the initial position state-vector is that which would result from plane waves incident on the collimator, and that the measurement screen is distant from the magnet. This approximation allows us to process an analytical solution, but a full analysis should have this level of detail. Ignoring $V(q)$ and writing the magnetic energy in terms of the spin of the particle gives

$$\hat{H} = \frac{\hat{p}^2}{2m} - \frac{g\mu_B}{\hbar} \hat{\boldsymbol{S}} \cdot \boldsymbol{B},$$

where $\hat{\boldsymbol{S}} = (\hat{S}_x, \hat{S}_y, \hat{S}_z)$, g is the gyromagnetic ratio (the ratio of the magnetic moment to the angular momentum), and $\mu_B = e\hbar/2m_e$ is the Bohr Magneton. To make this a specific example, we will assume the simplest non-trivial case of atoms whose total spin is one-half.

The first case, we will explore, will be that of the field in the z-direction being significant, such that fields in x and y directions are ignored. We make the approximation that

$B_x \approx 0$, $B_y \approx 0$, while the field in the z-direction is given by

$$B_z = a + bz,$$

where a and b are constants and the linear dependence on z gives the necessary gradient in the field. This is an unphysical approximation since the field components B_x and B_y in reality cannot be 0 if $B_z = a + bz$, as this would violate Gauss's law for magnetic fields, $\nabla \cdot \boldsymbol{B} = 0$. For now we will ignore this and explore the impact of the field on the spins in the z-direction as a benchmark example to understand the implications of setting $\nabla \cdot \boldsymbol{B} = 0$. The evolution operator will then be

$$\hat{U}(t) = \exp\left[-\frac{\mathrm{i}t}{\hbar}\left(\frac{\hat{\boldsymbol{p}}^2}{2m} - \frac{g\mu_B}{\hbar}\hat{S}_z(a+bz)\right)\right]$$

$$= \exp\left[-\frac{\mathrm{i}t}{\hbar}\frac{\hat{\boldsymbol{p}}^2}{2m}\right]\exp\left[\frac{g\mu_B\mathrm{i}t}{\hbar^2}\hat{S}_z(a+bz)\right]\exp\left[\frac{t^2}{2}\frac{g\mu_B}{2m\hbar}b\hat{S}_z\left[z,\hat{\boldsymbol{p}}^2\right]\right]$$

$$= \exp\left[-\frac{\mathrm{i}t}{\hbar}\frac{\hat{\boldsymbol{p}}^2}{2m}\right]\exp\left[\frac{g\mu_B\mathrm{i}t}{\hbar^2}\hat{S}_z(a+bz)\right]\exp\left[\frac{t^2}{2}\frac{g\mu_B}{m\hbar}b\hat{S}_z\hat{p}_z\right], \tag{7.178}$$

where we have used the Zassenhaus formula, Eq. (1.19) from page 22. So that if we start the system off in the state

$$|\Psi(0)\rangle = |\psi_+\rangle|+\rangle + |\psi_-\rangle|-\rangle \tag{7.179}$$

the dynamics is

$$\hat{U}(t)|\Psi(0)\rangle = \exp\left[-\frac{\mathrm{i}t}{\hbar}\frac{\hat{\boldsymbol{p}}^2}{2m}\right]\exp\left[\frac{g\mu_B\mathrm{i}t}{\hbar^2}(a+bz)\right]\exp\left[\frac{t^2}{2}\frac{g\mu_B}{m\hbar}b\hat{p}_z\right]|\psi_+\rangle|+\rangle$$

$$\exp\left[-\frac{\mathrm{i}t}{\hbar}\frac{\hat{\boldsymbol{p}}^2}{2m}\right]\exp\left[-\frac{g\mu_B\mathrm{i}t}{\hbar^2}(a+bz)\right]\exp\left[-\frac{t^2}{2}\frac{g\mu_B}{m\hbar}b\hat{p}_z\right]|\psi_-\rangle|-\rangle. \tag{7.180}$$

We see, that for this Hamiltonian, the initial ratio of spin superposition remains unaltered. So, for example, a state that is initially separable will always remain so. When it comes to actually modelling the experiment, we need to choose a representation. By inspection, we see that $\{\hat{\boldsymbol{q}}, \hat{S}_z\}$ is a sensible CSCO, which we recall leads to the spinor representation we discussed in Section 3.4.3. The Schrödinger equation in this representation takes the form

$$\left[\frac{1}{2m}\begin{pmatrix}\hat{\boldsymbol{p}}^2 & 0 \\ 0 & \hat{\boldsymbol{p}}^2\end{pmatrix} - \frac{g\mu_B}{2}\begin{pmatrix}B_z & 0 \\ 0 & -B_z\end{pmatrix}\right]\begin{pmatrix}\psi_+(\boldsymbol{q},t) \\ \psi_-(\boldsymbol{q},t)\end{pmatrix} = \mathrm{i}\hbar\frac{\partial}{\partial t}\begin{pmatrix}\psi_+(\boldsymbol{q},t) \\ \psi_-(\boldsymbol{q},t)\end{pmatrix}.$$

This can be written as two uncoupled differential equations

$$\left[\frac{-\hbar^2}{2m}\nabla^2 - \frac{g\mu_B}{2}(a+bz)\right]\psi_+(\boldsymbol{q},t) = \mathrm{i}\hbar\frac{\partial}{\partial t}\psi_+(\boldsymbol{q},t)$$

$$\left[\frac{-\hbar^2}{2m}\nabla^2 + \frac{g\mu_B}{2}(a+bz)\right]\psi_-(\boldsymbol{q},t) = \mathrm{i}\hbar\frac{\partial}{\partial t}\psi_-(\boldsymbol{q},t).$$

These equations can be solved, or the evolution operator, in this representation, applied to get the dynamics of any initial state. Because of the collimator, one assumes for the initial state a Gaussian beam times some spin superposition. The analysis gets somewhat involved for details of states and their dynamical properties. Please see [55, 66].

We will mention an interesting complication that arises for a divergenceless magnetic field. We consider the case when the x-component of the magnetic field is non-zero. The field we will use is given by

$$\mathbf{B} = -bx\mathbf{i} + (a + bz)\mathbf{k}.$$

Here we can see that B_z remains the same as before but we have introduced $B_x = -bx$ as a simple additional term that makes $\mathbf{B}$ divergenceless as required by Gauss's law of magnetism $\nabla \cdot \mathbf{B} = 0$. Now the evolution operator

$$\hat{U}(t) = \exp\left[-\frac{it}{\hbar}\left(\frac{\hat{p}^2}{2m} - \frac{g\mu_B}{\hbar}\hat{S}_z(a + bz) + \frac{g\mu_B}{\hbar}\hat{S}_x bx\right)\right] \tag{7.181}$$

does not have the nice Zassenhaus expansion as in the previous case due to the cyclic nature of the spin commutation relations (you may like to convince yourself of this by attempting to apply the expansion). Looking at the Schrödinger equation in the spinor representation, we also see that things are now more complex

$$\begin{pmatrix} \dfrac{-\hbar^2}{2m}\nabla^2 - \dfrac{g\mu_B}{2}(a + bz) & -\dfrac{g\mu_B}{2}bx \\[2mm] -\dfrac{g\mu_B}{2}bx & \dfrac{-\hbar^2}{2m}\nabla^2 + \dfrac{g\mu_B}{2}(a + bz) \end{pmatrix} \begin{pmatrix} \psi_+(\mathbf{q}, t) \\ \psi_-(\mathbf{q}, t) \end{pmatrix}$$

$$= i\hbar\frac{\partial}{\partial t}\begin{pmatrix} \psi_+(\mathbf{q}, t) \\ \psi_-(\mathbf{q}, t) \end{pmatrix}.$$

This results in two coupled differential equations for the components of the spinor:

$$\left[\frac{-\hbar^2}{2m}\nabla^2 - \frac{g\mu_B}{2}(a + bz)\right]\psi_+(\mathbf{q}, t) + -\frac{g\mu_B}{2}bx\psi_-(\mathbf{q}, t) = i\hbar\frac{\partial}{\partial t}\psi_+(\mathbf{q}, t)$$

$$-\frac{g\mu_B}{2}bx\psi_+(\mathbf{q}, t) + \left[\frac{-\hbar^2}{2m}\nabla^2 + \frac{g\mu_B}{2}(a + bz)\right]\psi_-(\mathbf{q}, t) = i\hbar\frac{\partial}{\partial t}\psi_-(\mathbf{q}, t).$$

Note that it is clear that the more physical Hamiltonian with a divergenceless magnetic field leads to dynamics that will entangle the spin and spatial degrees of freedom of any initial state. The Stern–Gerlach experiment is an example of how apparently simple quantum systems can rapidly become subtle, interesting, and hard to analyse. The picture becomes even more complex when one considers spins other than a half, and more realistic magnetic fields. Its study could easily expand to an entire book on just this one subject. It has been our intention to make this clear, and our analysis stops here.

8

Computational Simulation of Quantum Systems

"Computer science is no more about computers than astronomy is about telescopes, biology is about microscopes or chemistry is about beakers and test tubes. Science is not about tools. It is about how we use them, and what we find out when we do."

Edsger W. Dijkstra

8.1 Introduction

This chapter addresses two topics. The first is good practice, where our discussion focuses on principles of scientific programming design that should help you form your own considered approach to the subject. The second part attempts to illustrate by example many of the principles discussed in the first. We will look at the problem of numerically solving the Schrödinger equation for a simple system and develop a basic but powerful library[1] that provides the required capability to find the solution. Our discussion will use development of the source code to provide concrete guidance, tips, and tricks, for simulating quantum and other physical systems.

Our presentation of coding style and algorithm choice comes from practical experience, including: many years of scientific coding, having written hundreds of thousands of lines of code in various languages; teaching programming in a number of languages at undergraduate level; and supervising a number of PhDs whose research required advanced computational physics. The focus of this chapter will be on sharing an approach to computational physics with quantum systems as the example subject. We will look at a range of topics from high-level design choices, coding strategies, and language selection, through to specific numerical algorithms for solving frequently encountered problems.

---

1 Available at http://www.wiley.com/go/everitt/quantum or scan

Quantum Mechanics: From Analytical Mechanics to Quantum Mechanics, Simulation, Foundations, and Engineering, First Edition. Mark Julian Everitt, Kieran Niels Bjergstrom and Stephen Neil Alexander Duffus.
© 2024 John Wiley & Sons Ltd. Published 2024 by John Wiley & Sons Ltd.
Companion website: www.wiley.com/go/everitt/quantum

We will discuss some specific programming syntax, semantics, and design strategies (and their relevance to scientific coding) that are important to consider when selecting a language to use. We try to make the ideas we discuss as accessible as possible, and the code that is presented should be mostly self-explanatory. Even for those completely new to programming, our presentation should thus still have real value (and maybe our discussion will help you choose your first language). Furthermore, we have made the core library we develop and discuss for this chapter open-source, so you may download it, explore it, and experiment with it yourself.

In our experience, the only way to learn how to code is to tackle non-trivial problems yourself – working at it, making mistakes, and learning from them – allowing you to form programming strategies and a programming world view that works for you. If you need to code in a team it is a similar process, but one that needs some empathy of others' approaches and techniques, and processes that enable organised co-contribution. What is presented here is based on our experience and needs; these will differ from yours, so it is important to critically evaluate the advice given here for merit and relevance to your individual circumstances. There is no single right approach to programming – but there are some very wrong ones!

This chapter avoids taking a prescriptive form (*'here is a language and this is how to simulate quantum systems using it'*). There are already entire textbooks and internet resources devoted to that approach. But, there exist very few texts seriously addressing the philosophy of code design for scientific applications. We also avoid discussing simple code, such as that which might be generated by an artificial intelligence (AI). Recent advances in AI have led to substantial capability for simple coding tasks where there is existing code that can be used as a resource. For example, `write code in [language of choice] for finding the dynamics of a Duffing oscillator using Runge–Kutta method` might produce perfectly good code (even making good use of abstraction if well prompted). However, AI methods currently struggle with open-ended (i.e. research) problems: performing computational experiments to test physical theories; process and analyse data (i.e. where physical insight is needed); optimising code performance and high-level design; and developing new methods (again where deeper insight is needed). This chapter focuses on those conceptually high-level programming topics where AI methods have yet to develop real capability. Even though we will not go into detail in this chapter, we encourage you to investigate the potential for AI to be of use to you. For example, an AI may be able to, e.g. `write code in Swift for multiplying two matrices using Strassen's algorithm and generic types`. You might like to see (i) how well written, correct, robust, and optimised output from such a query is, and (ii) how you could use the output of an AI for a query like this in the library we develop here. A key lesson drawn from our experience is that mistakes in code design can generate a vast amount of unnecessary work in the long term. If we design good code from the outset, we can write better, clearer, and, most importantly, more trustworthy code. The latter is particularly important if the outcome of your code or simulation may cause harm. The use of quantum technology in fields such as sensing and metrology is growing; with emerging medical, defence, and other critical applications, coding errors could have serious ramifications.

It may be important that you understand what is needed for verification and validation of your code, its associated programme logic and behaviours under such circumstances.

You may not agree with some of the opinions that are put forward in this chapter – and as computing style can be somewhat subjective, the chance of you accepting all the points that are made is quite small. Where you do disagree, though, it is worth making sure you are clear in your own mind *why you think your approach is better for you*. Articulating to yourself why you code the way that you do is worthwhile, your style should be a conscious and reasoned choice. The very process of reflecting on our own coding strategies and understanding what sort of programmer we wish to be is, in and of itself, a valuable exercise.

8.2 General Points for Consideration

8.2.1 On Code Clarity and Performance

Here we present an example driver routine (a portion of code which solves a complete problem) for investigating the quantum dynamics of the Jaynes–Cummings model. We deliberately do this before introducing any syntax to make the points that (i) code should be readable, and (ii) starting a project by writing a driver routine can be a useful code design strategy. Note that throughout this chapter we have used formatting for the code suited for the printed page rather than a text editor or integrated development environment (IDE).

Recall from Section 7.6 that the Jaynes–Cummings model is an approximate description of an atom (subscripts a) coupled to a continuous field mode (subscripts f). The Hamiltonian is:

$$\hat{H}_{JC} = \hbar\omega_f\left(\hat{n}_f + \frac{1}{2}\right) + \frac{1}{2}\hbar\omega_a\hat{\sigma}_z + \hbar\lambda\left(\hat{\sigma}_+\hat{a} + \hat{\sigma}_-\hat{a}^\dagger\right). \tag{8.1}$$

Here the field mode is taken as a simple harmonic oscillator, the atom is a two-level system, and the coupling term models a simple exchange process between the two. If we choose the field mode as a reference, we can define a dimensionless time $\tau = \omega_f t$; the Schrödinger equation is

$$\frac{\partial}{\partial\tau}|\psi\rangle = -i\frac{\hat{H}_{JC}}{\hbar\omega_f}|\psi\rangle$$

and, as our focus here is on coding, we will further simplify this example by setting $\omega_f = \omega_a$ (we could also ignore the constant term, as it does not affect dynamics, but we choose not to as it is useful in supporting our discussion on coding). We thus obtain an effective Hamiltonian and Schrödinger equation of the form

$$\hat{H} = \left(\hat{n}_f + \frac{1}{2}\right) + \frac{1}{2}\hat{\sigma}_z + \gamma\left(\hat{\sigma}_+\hat{a} + \hat{\sigma}_-\hat{a}^\dagger\right), \quad \text{and} \quad \frac{\partial}{\partial\tau}|\psi\rangle = -i\hat{H}|\psi\rangle, \tag{8.2}$$

where the coupling constant $\gamma = \lambda/\omega_f$ will be set to the usual textbook example value of 1. It is this set-up that our example driver seeks to model and solve for the initial condition $|\psi(t=0)\rangle = |m = \frac{1}{2}\rangle_a \otimes |\alpha = \sqrt{15}\rangle_f$. Importantly, note that tensor product is implied, so $\hat{n}_f$ is a shorthand for $\hat{n}_f \otimes \hat{1}_a$.

Exercise 8.1 Please study the code in Listing 8.1 and translate it back into the mathematics from which it came. The important thing to note here is that you should be able to do this with knowledge of the physics but without any knowledge of this language. You will need to know that Runge–Kutta is an iterative method for solving systems of ordinary differential equations. ∎

Exercise 8.2 We will sometimes use mathematics over physics nomenclature as the application of our code may be to problems beyond the scope of physics. From the example driver routine, why, in this case, is it better to choose to term the type that accounts for the space in which vectors and operators live as `StateSpace` rather than the mathematical types of `VectorSpace`, `InnerProductSpace` or `HilbertSpace`? ∎

Exercise 8.3 What are the advantages and disadvantages of using `evolve(by: ...)` as a function name instead of `doRungeKuttaStep(by: ...)` in this example? ∎

8.2.1.1 On Comments

Note the few comments in the example code. This is intentional, and for very good reasons. Rob Pike in his *Notes on Programming in C* in 1989 summarised the issue with comments very nicely:

> A delicate matter, requiring taste and judgement. I tend to err on the side of eliminating comments, for several reasons. First, if the code is clear, and uses good type names and variable names, it should explain itself. Second, comments aren't checked by the compiler, so there is no guarantee they're right, especially after the code is modified. A misleading comment can be very confusing. Third, the issue of typography: comments clutter code.

A further observation by Kevlin Henney (`@KevlinHenney`, Twitter 12:47 pm, 20 Sep 2013) reinforces this

> A common fallacy is to assume authors of incomprehensible code will somehow be able to express themselves lucidly and clearly in comments.

and, as 'Bob' Martin points out in Clean Code (2008) [54]:

> Nothing can be quite so helpful as a well-placed comment. Nothing can clutter up a module more than frivolous dogmatic comments. Nothing can be quite so damaging as an old crufty comment that propagates lies and misinformation.

Comments should be used with caution to explain intent, convey warnings and information to other programmers, and to clarify something in the code where there is no better way to do so. Before adding a comment, ask yourself *'can I change my code design to make this comment redundant?'*. This could be as simple as changing the naming of a variable or method. Hopefully the examples in this chapter should demonstrate that comments are rarely needed.

```
 1  // Jaynes--Cummings Schrodinger Dynamics driver.
 2  /* For specifics on model, units and example/test parameter
       values and data see Fig 1 and associated text in
       doi:10.1103/PhysRevA.79.032328           */
 3  let couplingConstant = 1.0
 4  let fieldBasisSize = 35
 5
 6  let fieldSpace = StateSpace(dimension: fieldBasisSize)
 7  let fieldHamiltonian = fieldSpace.numberOperator
 8                       + 0.5 * fieldSpace.identityOperator
 9
10  let spinSpace = StateSpace(dimension: 2)
11  let spinHamiltonian = spinSpace.angularMomentumZ
12
13  let totalSpace = fieldSpace.tensorProduct(spinSpace)
14
15  let interactionTerm = couplingConstant * (
16      totalSpace.tensorProduct(
17          fieldSpace.creationOperator,
18          spinSpace.angularMonentumLoweringOperator
19          )
20      +
21      totalSpace.tensorProduct(
22          fieldSpace.annhiliationOperator,
23          spinSpace.angularMonentumRaisingOperator
24          )
25      )
26
27  let jaynesCummingsHamiltonian = totalSpace.sum(fieldHamiltonian,
       spinHamiltonian)
28                  + interactionTerm
29  let timeIncrement = 0.01
30  let endTime = 10.0
31
32  let spinState = spinSpace.makeVector(from: [1.0, 0.0])
33  let fieldState = fieldSpace.makeCoherentState(
34          alpha: Complex(modulus: sqrt(15.0),
35          argument: 0.0))
36
37  var jaynseCummingsSystem = QuantumSystem(
38    startTime: 0.0,
39    initialstate: totalSpace.tensorProduct(of: spinState,
40                                  with: fieldState),
41    hamiltonian: jaynesCummingsHamiltonian)
42
43  while jaynesCummingsTDSE.currentTime < endTime {
44    jaynesCummingsTDSE.doRungeKuttaStep(by: timeIncrement)
45  }
```

Listing 8.1: Example driver code for a simple simulation.

In our above example driver routine, there was only a title comment, including a reference to where the mathematics and physics can be found. Here is the first weakness of this code: units are specified in another document, so the code does not stand in isolation. In the worst case, the reference document could be lost, or updated in a way that no longer matches the code – even in the best case it is an added inconvenience to the reader. In this case a published source was 'used' which has the advantage that it mitigates the first two risks. It also has the advantage that a coherent presentation of the science will be given, and results will be available that a new user can use to validate the output of their own programme.

8.2.1.2 Clear Code

In the old days of programming, and until somewhat more recently, great care had to be taken when considering the efficiency of a given algorithm. Integer arithmetic is much faster than floating point arithmetic, floating point arithmetic at different precisions executes at different speeds, addition and subtraction are much less costly than multiplication and division. Unnecessary function calls may add noticeable run time, and just changing the order of some loop optimisations (or using function methods like map instead) could result in drastic performance increases. Modern compilers do much of this optimisation for us, and usually do a better job than we can[2].

We should therefore strive to write clear code that we can reason about, and that presents itself in a way that a human (and not just the compiler) can understand[3]. The core idea that we are presenting here can be expanded into a philosophy for coding, such as that of 'clean coding', for which entire textbooks exist.

8.2.2 Should I Use Third-party Libraries?

A number of numerical libraries for quantum physics already exist. Examples include: *'A C++ library using quantum trajectories to solve quantum master equations'* [78]; a *'A computational toolbox for quantum and atomic optics'* [85]; *'A Julia framework for simulating open quantum systems'* [45] and *'QuTiP: An open-source Python framework for the dynamics of open quantum systems'* [43]. The purpose of this discussion will not be 'which of these libraries is best?' as that depends very much on your own requirements, but rather: *are third-party resources right for me, and if we choose to use them,*

---

2 If clear code runs too slowly, we can utilise profiler tools in order to find out where the bottlenecks are and improve those specific bits of the code base.

3 In many languages it is possible to replace method calls such as `.times()` with characters (or sets of characters) such as `*` so that we can write `a*b` instead of `a.times(b)`. These are referred to as custom infix operators (prefix, postfix, and assignment operators are also possible). Sometimes this is a good idea and sometimes it is not. The benefit of doing this is more concise code that can look much more like mathematics results. There are some potential disadvantages though. One is that there may be multiple operations that could use the same symbol, thus introducing ambiguity. Another example is in UTF8-encoded languages that would allow for use of symbols such as $\otimes$ (UTF circled times) as infix operators so we could write A$\otimes$B directly in code for a tensor product. My problems with this approach are (i) it will make housekeeping the tensor product structure harder; (ii) it requires the reader to infer that this is what we have done (although unlikely to be an issues in this case); and (iii) there are some different but similar-looking UTF characters and it can be a pain to make sure one continues to use the correct one. As such I tend to use custom operators sparingly.

what do we need to consider? As libraries and languages do fall in and out of favour, such consideration will be important for some but not others.

The reduced time to write code that gets useful results and the convenience of being able to follow examples written by other people make third-party library use very appealing for many programmers.[4] Maybe better questions would be: *how do I know which libraries to use?* or *what are the reasons I might not want to use these effort-saving tools that get me results quickly?* While the answer to the first of these questions may seem obvious at first (pick the most popular tools in the language I know or want to learn) the answer to the second question may change your view on this.

An example of a subtle error that has been seen in some libraries is the inappropriate use of random numbers in modelling noise (which needs the specific calculus of Itô or Stratonovich to be properly accounted for). *Prima facie*, such code can look like it is behaving as it should, but in fact it is not. This leads us to our first evaluation question: *if I use this code, do I know enough to determine that it is working correctly?* If many people use the same code, there is a good chance that errors will be spotted and fixed by the community (at least the obvious ones). The flip side is that with many different contributors to the library, it is also more likely that an error or other unwanted behaviour will be introduced. If this is in a portion of the code that is not often used, then there is the risk that it will go unnoticed for some time.

Less extreme examples happen quite frequently when modifying complex projects. So the next question is, supposing I knew my code that relied on third-party libraries used to work, *how often do I need to check that everything still works?* The answer is every time I update any library I use and the numerous libraries on which it depends. The next question is *how many tests will I need to run?* We will cover this later in a more full discussion on testing code.

Another consideration is future proofing. Libraries do not have an indefinite lifetime; people, for one reason or another, stop working on projects. Even in actively maintained libraries, specific parts of the library may be retired or re-written. Large projects like the GNU Scientific Library for C and C++, or commercially available solutions, carry with them a greater security than smaller projects. So *if development of this library stops, how much of an issue will it be for me?*

If you understand what your specific requirements are, and the risks of using others' libraries and code, then you are well placed to make an informed decision about which approach is right for you.

8.2.3 Choice of Language

It is worth noting that languages come and go. There was a time when almost all scientists used FORTRAN, then many of us adopted C, then C++, for some Java, and C# became popular. At the time of writing, Python has become very popular and is now possibly the dominant language for scientific programming. However, there are new competitors that show great promise for scientific computation, such as Julia. There is also an emerging interest in programming paradigms such as functional programming in languages such as

---

4 This is particularly true when people are beginning their career in scientific coding and want to make a quick start. It is most definitely worth playing with some of the very good tools that exist to get a feel for what they can do (e.g. by following some of the many tutorials that can be found on-line).

lisp, Haskell, and Clojure (indeed some are using these languages, traditionally considered to be quite abstract, to teach physics [88]). To complicate matters even further, there are paradigm-hybrid languages such as Scala that combine object-oriented and functional programming in one language, or Apple's Swift that mixes several paradigms in one language.

Not everyone has a choice in the language that they use. Sometimes one needs to adopt the languages of a team or organisation of which they are a part (although it may be possible to instigate change or be asked to change languages in an established project). Other times, there may be external factors that force or narrow your choice. Even if you are completely free to choose, this may be a difficult decision to make and a careful matter of balancing your requirements, personal taste, and other constraints. Also be aware that your circumstances may change. Some environments such as Matlab are very powerful but not free. You may have access to such tools, e.g. as a student at a University, but lose this access on graduating. Do you want to invest effort into writing for a platform that may carry a future cost for continued use?

If you have the freedom to choose a language, play with some until you find one that you like (numerous good candidates are listed above). The factors I consider in making this choice, in order of importance, are:

- Will it allow me to focus on the physics? We need to be able to reason about code, and my priority is to focus on getting trustworthy simulation results. I also want to spend more time thinking about physics than coding. For this reason I always prefer languages that encourage expressive clear coding styles with all the information needed to reason about the code in one place (so no header files please, or funny built-in functions whose purpose or actions are unclear).
- Can I trust the results? In other words, how sure can we be that the code really does what we want it to do? Functional languages usually lead in this respect, but they require a style of programming that many find hard to learn. My preference is, at the very least, to use a strongly typed language.
- Do I enjoy coding in it? Life is short. In many cases, code can run in the background while we do other things. So long as we can get results in an acceptable time-frame, this is what matters (you can then think about other physics problems while the computer works on the last one you gave it). Try to make the time spent solving physics problems as efficient and enjoyable as possible. Try to avoid a language or environment where you find yourself 'babysitting' the code, where you are frequently doing maintenance.
- Is it powerful enough? Problems in quantum mechanics can quickly become so computationally demanding that they strain even large supercomputers. If these are the sort of problems that we are solving - can we easily take good advantage of the hardware available to us (while many languages have capability in this respect, a few like Scala have such considerations at their core)?
- How much work will continued use of a platform entail? As time goes on, language features and core code change (e.g. the substantial changes in the Python 2 to 3 transition, or in the early development stages of most languages). For some languages and libraries, installation and maintenance can be very time-consuming, and may necessitate changes to your own code base.

Ultimately, do not worry too much about language choice as a new programmer. We include this discussion because it is important you are mindful that the choice can make a real difference and there are those who will champion their favourite language regardless of its suitability. It is our intention that you are equipped to think critically about this question and use this in your decision-making process. Until you are familiar with a number of languages, it is hard to really appreciate what they offer, and to make an informed selection. There will, from time to time, be reasons to consider changing languages. New languages with new features are continually emerging and these can enable you to write better, more powerful, and easier to read code. As compilers get better and better, the need for low-level fast (but often hard to read) code is much reduced. I have been involved in and led porting very large scientific libraries a number of times. Every time it was worth it. In porting your projects to a new language, you get an opportunity to re-plan your code base and fix big design issues. If done correctly, you should end up with faster, better code that reduces development time and makes solving problems easier.

Then there are aspects to any language that can undermine the trustworthiness of the code at a fundamental level. Let us look at code posted by user Veedrac in response to a Reddit question[5]. It does not matter if you do not follow this code:

```
1  import ctypes
2  def deref(addr, typ):
3      return ctypes.cast(addr, ctypes.POINTER(typ))
4  deref(id(29), ctypes.c_int)[6] = 100
5  #>>>
6  29
7  #>>> 100
8  29 ** 0.5
9  #>>> 10.0
```

What is important here is releasing that this changes the number 29 into 100 – something you might not expect to be possible in a language until you are told that everything in Python is an object (we will later discuss in more detail that this can be done with objects as they are an example of something called a reference type). This is an extreme example of unwanted behaviour that may also not be detected by automated tests (and it is enough for me personally to avoid using languages where reference types are the only option, even if specific issues like this are later fixed).

We use Swift here, and there are a number of ways that this could be justified. It is a good first language – used in schools as part of Apple's *Everyone Can Code* initiative (which provide some of the best resources and tools available for learning to code, such as Swift Playgrounds). It is a modern, clean, expressive, and easy-to-read language. For me, a really important consideration is that there has been a lot of care taken to make the language safe (unlike, e.g. Python which is dynamically typed, uses late binding, indentation to determine structure, and a number of other potential sources of error). It is fast, powerful, open source, and cross-platform (macOS, Linux, and growing support for Windows). It is an industry standard – in demand and growing (currently in many of the top 10 lists of languages to learn), so learning Swift is useful even if you do not stay in science (which is generally true of programming).

---

5 http://www.reddit.com/r/Python/comments/2441cv/can_you_change_the_value_of_1/

There are many other languages which share many of these attributes, but the bottom line explaining why I use Swift is that, above all of these attributes, I enjoy coding in it.

8.3 Some Overarching Coding Principles

In this section, we build on the discussion above, and distil some language-independent principles that should govern our approach to coding.

8.3.1 Have Clear Objectives

Writing good code on any subject starts from understanding that subject very well. In many introductory texts on programming one encounters the example of a 'bank account' but often presented in a way that differs from real-world solutions. Such an approach immediately fails to recognise the subtleties of financial programming where regulatory compliance, reliability, security, speed, and convenience are all key considerations. This leads to false expectations of what is expected from programmers and their creations. In scientific coding we have our own priorities, concerns, and constraints. Having clear objectives helps to design code that respects our discipline. Our code should take account of core principles of physics such as rigour and reproducibility (replicability and repeatability). As we will expand on below, science is more than this, peer review is part of the process, so writing clear code that enables external review should also be part of what we do. As a last example, when we later discuss logging, note that this method can be used to ensure appropriate data recording. It is important that we try to maintain this physics context as clear objectives because it can be easy to neglect some of the above considerations as we get involved in coding projects, and our focus shifts to details of program logic rather than design.

8.3.2 Trust Your Code

The most important principle by far is the accuracy and reliability of our program. Getting this wrong at best means we don't get an accurate picture of the physics of our system and may arrive at incorrect conclusions. In the worst case, this could lead to actual harm (e.g. in medical physics or autonomous vehicle control). So our first principle must be to *write code that we can trust*. There are some key corollaries to this. The first is that the *code that we write must be understandable by ourselves and our peers*. One should be able to see what calculations have been performed, and where inaccuracies may arise. If you return to some code that you have written many months ago and cannot figure out how or why it does what it does, can you really trust it? That said, sometimes some numerical algorithms are just hard to follow. This leads to our next corollaries, that we: *document appropriately* and *test, test, test*. If a project lasts for any length of time, it will grow in complexity and we need to *manage the complexity of the code* and *avoid unintended consequences* by adding new code to a project. For this reason, it is a good idea to try to separate the complicated bits of an algorithm from the rest of your code so that they can be thoroughly tested and then used with confidence. In addition to this, there may be opportunity in code for users supplying inappropriate inputs (missing, erroneous or null) or executed code could produce erroneous output (such as not-a-number after a divide-by-zero). We gain more confidence in our code if we can handle such situations well – this approach is a form of defensive programming.

We can further establish our level of trust on code output by introducing a degree of logging associated with any run (and saved to an output file). There are several levels of information that we can log. The first is the debug-level used to diagnose problems, mainly in the development phase. The next is general information, which might for example confirm that the normalisation of a state vector has stayed within acceptable limits. The following level is a 'warning' to indicate something unexpected may have happened (an input state vector that is not normalised might trigger this, for example). Then we may have different levels of error that may or may not be correctable, such as might be encountered when integrating a system of differential equations results in divergent behaviour. Even if we cannot correct that behaviour at run time, we would in general want to capture what happened so that we can analyse and later rectify the situation.

Exercise 8.4   For the example code later in this chapter, consider what information should be included in a log.   ∎

8.3.3   Plan for the Future

Planning programming projects, especially in research or early technology contexts, can be very challenging. By its very nature, some research starts ill-defined. We want to write code that not only addresses the initial problem, but is also extensible to whole classes of future problems.[6] So we need to try to *code in a way that will allow substantive future design changes*. For this reason, it is important to think strategically about how we code and to keep potential future applications in mind. Implementing a quick fix to a problem in the short term may cause a lot of additional work in the long term. So *do not be afraid to play* and try out different ways to approach a problem so that you can get a feel for what may or may not be a good approach. Also, if you see a potential problem, *fix it early* before too much code depends on something fragile. Use of pseudocode (a natural language description of an algorithm) and flowcharts can help us plan specific algorithms. Other techniques such as 'user stories' or forming a requirements specification can help us think strategically about what we will need the entire code base to do. Even a small effort spent planning can lead to huge rewards later, so it is important not to neglect this stage in the excitement of getting any non-trivial project started.

---

6 For example, there is now an emerging field of applying machine learning to problems in the physical sciences, including quantum physics and chemistry. Even if the current problem we are interested in is so simple that we have no interest in machine learning, should we plan to make sure that we can use such techniques in the future? In addition, the emergence of new language features can change radically what the art-of-the-possible is. For example, when a language (such as Java) introduces a new feature (such as generics in v5 c.2004), it can be possible to simplify complex code. This reduction in complexity can help us make code more manageable and trustworthy, but if we have made some poor design decisions, we may be unable to do this without extensive work.

8.3.4 Test, Test, Test

It goes without saying that scientific code should be well tested. Due to the scaling problems in quantum physics, numerical simulations can take a very long time indeed – so we do want to make sure in advance that we can trust a program's output before committing to long run-times. The points of testing are twofold; the first is to *verify* that the code does what it should do (check 'truth' and accuracy of each bit of the code) and the second is to *validate* its fitness for purpose (e.g. is it usable for solving the problems that need to be solved or does the machine learning algorithm that is correctly implemented actually solve the problem?) Our focus here is on the first part - verifying that our code is 'correct' (validation is too large a topic to be covered here).

Typically, scientific programmers will test their code by reproducing analytical, text-book and published results of calculations which establish trust that the code works as expected. This can work rather well if one has the self-control to not write any code that is unnecessary to achieve that specific result. Most people, myself included, often cannot resist the temptation to add features to code as we write it. Let us say, for example, that we are trying to reproduce some result from a publication that just needs the matrix operations of addition and multiplication by a scalar (such as a simple initial value problem solver). We may be tempted to also code matrix subtraction and multiplication, as we know we will probably use them eventually. We could then successfully reproduce the published result without ever running those lines of code. This coding behaviour leaves odds and ends in the code base that have not been tested and leads to a code base that cannot be trusted.

An extreme approach to coding exists, which attempts to combat this behaviour, called test driven development [4] (TDD). Here, the 'mantra' is that one writes the test before one writes any code. Code is written, the test run, and the code modified if necessary, until the test is passed. The idea of TDD is to ensure all code works as intended by driving development through tests that must be passed before more code is written. The basic process here is to: (i) start by writing the test code that will verify what you want to do; (ii) run all tests [the one you just made and any preceding ones], the new test may fail but all the others should pass; (iii) if the new code fails the test, adjust it sufficiently to allow the new test to pass, implementing what you intended; (iv) check that all tests now pass; and (v) improve the code (e.g. for clarity and performance) and repeat the last two steps until satisfied. If each test is encapsulated and does not depend on other tests, it is termed a Unit Test and coding at this level of granularity is termed Unit Testing (Unit Testing is not about testing units of code). Along with other techniques such as pair programming, TDD can be useful in scientific programming contexts - not just to reduce the risk of making or introducing errors, but to also provide a framework where we can improve our code with confidence. Any methodology is only effective if it will actually be used, and very few scientific programmers embrace TDD. The problem is that TDD can be a slow and frustrating process, and can also make it hard to redesign code. Part of the reason for this is that in scientific coding the *implementation* of the code (e.g. matrix multiplication) is usually tested and not its *behaviour*. Here, by behaviour, we mean the ability the library provides us to write code that solves quantum physics problems and is understandable to any physicist with sufficient knowledge of quantum mechanics. To explain what we mean by testing behaviour, consider the process that was used to develop the library attached to this book.

The behaviour that we wanted is encapsulated by the use-case listed on page 231 for the Jaynes–Cummings model. Drafting this 'end-user' code is a way of specifying our requirements and begins to define some design choices. Adopting a TDD approach was done by (i) commenting out every single line of code in this 'driver' file and then recursively (ii) uncommenting one line of code and (iii) writing enough of the back-end library for that code to run successfully. At this stage, it is not essential to worry about performance, but only clarity and functionality. Functionality such as basic complex number, vector, and matrix arithmetic was not directly tested, as much indirect testing was achieved by reproducing the quantum dynamics of the Jaynes–Cummings model (some *ad hoc* unit tests were used in this phase, but these might later be removed since some were too implementation focused).

After this development phase, we have a reasonably mature library. Other (unit) tests were then added, so that if we later change algorithms in the library, automated testing can be done to verify that we have not broken the code (of course - the Jaynes–Cummings code can also be run as a validation check). In an ideal world these tests would cover all the functionality of the code but without being implementation specific (the way we do this is by making use of physics-specific concepts such as verifying that certain commutation relations are satisfied – simultaneously checking operator initialisation, complex number, and matrix operations without worrying about how these are done). With such tests in place, we can now improve our code, e.g. performance and use the tests to verify that we have not broken our code-base.

A key benefit of TDD is that it can help avoid scope creep. Once all the tests pass, work stops. Software applications frequently suffer from developers adding functionality that is not needed now but might be useful in the future. In a commercial context, this should be avoided to ensure work does not exceed deadlines or budgets. In a scientific context, we should spend some time ensuring the rigour of our code, and writing code that is not tested would undermine this. TDD is useful in preventing this, but only if one has the discipline to adhere to it.

As a last note, an often-overlooked part of testing is verifying the input of the user (another example of defensive programming). For anything important, we should take steps to ensure that what is being run is as expected and sensible,[7] as garbage in leads to garbage out. In quantum mechanics, we might want to check that a state vector has unit norm or that we only multiply operators that belong to the same state space. Our testing regime should consider what happens when users supply inputs that are not necessarily sensible and for code being used in an incorrect or inappropriate way. In a mature library, such code should fail gracefully and return a considered response to bad inputs.

Exercise 8.5 The code that has been supplied with this chapter has not been well unit tested and does not cope well with misuse. Consider how it could be unit tested and respond to bad user input. Remember that good tests check behaviour, not implementation. The unit test framework for Swift is XCTest. For example, something to consider might be: if a state vector is not normalised, should a programme continue to execute and provide a warning or should it exit? ∎

---

7 This may even take the form of a copy or printout of the program's parameters to be signed off in an on-line or physical lab book.

8.3.5 Object-oriented Design

Physics is about understanding the fundamental laws that govern matter, force, energy, and their behaviour and interactions. We may be concerned with understanding fundamental particles, composite systems, or some higher-level phenomenology. If we look at the models we put together, we can take a slightly alternative perspective. Fundamentally, physical systems comprise conceptual objects that are subject to, and coupled by, some underlying laws of interaction. We are used to expressing these mathematically. In software design, objects (structures or classes) are a mechanism to encapsulate properties and data. Methods belonging to those objects are used to enact changes of state and interactions between objects. There are therefore natural synergies between physical objects and software objects. We can use this connection to naturally represent physical systems and their interactions within a physics context. We will illustrate this approach by example when we start developing the library, and the first object we create will be that of a complex number (which is a mathematical object that encapsulates some properties [the real and imaginary parts or modulus and argument] and have methods which belong to them, such as conjugation, and of interaction such as arithmetic).

As in physics, we can generalise classes of objects under common headings. For example, many physical systems are particles, or they are directional (represented by vectors) or share properties such as electric charge. Beyond the encapsulation of properties and methods that represent a physical object, there are some other important aspects of object-oriented design that have natural physics analogies. For example, the idea of an electron as a specific kind of charged particle is one of inheritance. We can define a structure or class for a charged particle, and define an electron as a subclass or structure. The 'electron' inherits the properties and methods common to all charged particles (such as a position or velocity) and adds or overrides the specifics applying to the functionality of an electron (such as specific charge and mass). Another physics-like key concept in object-oriented design is that of an interface. Consider again the charged particle example. We could achieve much of the same outcome as our previous example if we could simply require that any 'charged particle' class or structure had certain properties and methods. Another example that illustrates the utility of deferring the implementation of a given object is the idea of a vector. Interfaces allow us to ensure that any 'vector' type will satisfy certain axiomatic properties, such as closure under addition or scalar multiplication. We will provide specific examples later in the chapter by defining vector interfaces using the Swift syntax called protocols. When we have an interface or a protocol, we can then write code that allows us to assume that any object that conforms to that protocol has those methods. For example, we could assume that all charged particles have a mass or all vectors have a scalar product. With inheritance, we have a way to replace an object with any sub-objects. Alternatively, with interfaces, we can write code that can use any example of any object that conforms to it. We term this interchangeability polymorphism.

A number of languages allow one object to inherit from multiple parent objects (examples include C++, Kotlin, Scala, Python, and R). For example, we may want to have a base class of `Particle` which has subclasses of `MassiveParticle` and `Charged-Particle`. Here, each particle may gain methods relevant to that type. For example, a massive particle may have a method for calculating the gravitational force it is subject to, and a charged particle the Lorentz force. A class that we use to define an `Electron` object that is a child of both `MassiveParticle` & `ChargedParticle`. This all seems very nice and a convenient way to design systems with physics-based separation of concerns. The issue with such an approach is what happens if `Particle` has, e.g. the requirement that there is a `totalForce()` function. It clearly cannot inherit the `totalForce()` from either `MassiveParticle` or `ChargedParticle`. If the user and compiler do not pick up on this, then undesirable behaviour will almost certainly emerge. This is an example of diamond structure that multiple inheritance enables and is illustrated in Figure 8.1. The issues that such design possibilities lead to can be mitigated by good practice, but still tend to result in code that is hard to reason about. This specific problem with multiple inheritance is so disliked by some that the structure in Figure 8.1 has even been referred to as the 'Deadly Diamond of Death' (*Java and C++ A critical comparison*, Robert C. Martin, 1997). Even outside of the 'Deadly Diamond', other coding problems can easily arise when using multiple inheritance.

In our previous example, consider what might have happened if `Particle` did not contain the requirement for a `totalForce()` function. We could have `Massive-Particle` with a `gravitationalForce()` function which is fine and without ambiguity. Let us imagine that for some reason the person who implemented `Charged-Particle` coded the force functions as `staticForce()` for $q\boldsymbol{E}$ and `dynamicForce()` for $q\boldsymbol{v} \times \boldsymbol{B}$. Now also imagine that the coder makes a error of naming and instead of defining a `lorentzForce()` function, they call it `totalForce()`. It is now the case that `Electron` will inherit `totalForce()` from `ChargedParticle`, which does not include the `gravitationalForce()` it inherits from `MassiveParticle`. Clearly this is not something that we would desire.

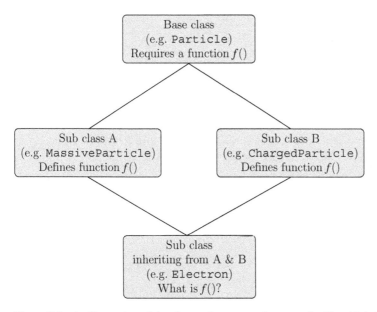

Figure 8.1 An illustration of the diamond structure that can arise if multiple inheritance patterns are used, and which leads to issues of complexity and ambiguity.

The 'Deadly Diamond of Death' and associated problems can be avoided by using languages that do not allow multiple inheritance such as Swift, Java, and some versions of FORTRAN and C#. Much of the capability that multiple inheritance provides can be retained by use of interfaces and protocols - but even these cannot protect against poor code design. With these warnings and caveats in mind, it is very much the case that good object-oriented design can be used to enable the meaning (physical or otherwise) and intent of our code to be clear.

Exercise 8.6 Can using interfaces or protocols instead of multiple inheritance prevent the problem that was described in our last example where `Electron` inherited an inappropriately named `totalForce()` function? ∎

8.3.6 Be a SOLID Scientific Programmer

SOLID is a contrived acronym referring to five principles of good programme design (especially with regard to an object/structure oriented approach). Each of these will be detailed below (with the appropriate part of the acronym in italic bold face). As a whole, they form a set of guiding principles and patterns laid down by Robert C Martin that are very helpful in architecting any software project. The ideas are very important, and in this section we seek to provide sufficient detail to enable an understanding of how these principles led to the design choices in the Quantum Library we will present later. Specifically, SOLID seeks to mitigate several key poor code designs that have the following symptoms: the programme is difficult to change even in simple ways (Rigidity); the programme breaks in many places whenever it is changed (fragility); you cannot reuse your code (immobility); making changes is slow and hard (viscosity). Note that these are principles only – it may make sense to not follow them from time to time, but if you do it is best that this is a conscious choice.

8.3.6.1 The *Single-responsibility* Principle
This is a *separation of concerns* principle – for us, an example might be that the code used to plot data should not be mixed with the code used to produce that data. If we want to change the format of the plot, we do not want to risk introducing errors into the process that was used to create the data set itself. A code that represents something within a coding container of some kind should have only one reason to change, specific to the function of that code (in Swift, this will be a structure or class – which we will define later). One way to achieve single responsibility is to use multiple short methods/functions in favour of longer methods. This will facilitate both re-usability and reliability of code. A good rule of thumb is that no single method should exceed one printed page. If it is longer than this, it is almost certainly attempting to do too much. Some numerical methods may be exceptions to this rule but even then, most can be broken down into smaller functions that perform specific sub-tasks. Such a breakdown can make it much easier to reason about complex code.

8.3.6.2 The *Open-closed* Principle
A module should be open for extension but closed for modification.

Robert C Martin

'Bob' said that the idea *originated from the work of Bertrand Meyer. It means simply this: We should write our modules so that they can be extended, without requiring them to be modified. In other words, we want to be able to change what the modules do, without changing the source code of the modules'*. In Swift, extensions allow us to do this with ease[8].

8.3.6.3   The *Liskov* Substitution Principle

If S is a subtype of T, then objects of type T may be replaced with objects of type S without breaking the program.

Barbara Liskov

While this may seem somewhat abstract, an exercise might help you get a concrete feel for the idea:

Exercise 8.7   Given that a column vector can be considered as a special kind of matrix, would deriving a vector class from a matrix class violate the Liskov substitution principle? ∎

Three hints for this question, that you may wish to postpone reading if you want to figure this out for yourself, follow. Hint 1: consider all the arithmetic operations you may do with matrices. Do all of these make sense for vectors? If the answer is no, then vectors should not be a subclass of matrix[9]. Hint 2: What about a function like R's `cbind()` that binds a new column onto an existing matrix? Hint 3: is the standard example that a square should not be a special case of a rectangle. This may seem counterintuitive, as in geometry a square is indeed the special case of a rectangle whose width is the same as its height. In coding, we can attach functions to the idea of a rectangle, such as one that changes its width. If square was a special case of a rectangle, it would also 'inherit' that functionality. If we take a square and change its width without changing its height, then it is no longer a square. The Liskov substitution principle dictates that a square should not be a special case of a rectangle, where this or any similar behaviour is possible.

8.3.6.4   The *Interface* Segregation Principle

Segregating interfaces is very much like the idea that mathematical axioms are separated into distinct items, such as closed under addition. In the same way that a group is (i) associative, (ii) has an identity element and (iii) an inverse element. Each can be associated with its own interface or protocol. An Abelian group is formed by also imposing that the group operation is commutative.

---

8 For example, it is possible to extend built-in types such as `Double`. We can even override + for arrays so that the operation it performs becomes element-wise addition rather than concatenation. That said, changing behaviour in this way can be a very dangerous thing to do if the scope of the modification is not limited; an example of the fact that just because a language allows you to do something, is not reason to do it. While we would not have modified the original code base, we would have modified code behaviour, which could lead to unintended consequences (e.g. since + may have been used in its original sense elsewhere).

9 For those already familiar with SOLID, Swift uses protocols (rather than interfaces in other languages) which aid design by contract and can be used to help ensure covariance and compliance with the Liskov substitution principle.

An interface (or in Swift, a protocol) is a specification of all functions and properties that an object must have in order to be of that kind. As with mathematical axioms, interfaces should be minimal so that only interfaces that are needed or used are implemented; or to put it another way, code should never be forced to depend on methods it does not use. One way to achieve this is to break interfaces/protocols into fragments and combine them so that one only implements the methods that are actually needed.

8.3.6.5 The *Dependency Inversion Principle*

This states: '*High-level modules should not import anything from low-level modules. Both should depend on abstractions (e.g., interfaces)*'. and '*Abstractions should not depend on details. Details (concrete implementations) should depend on abstractions*' [53].

To put this in the context of scientific coding, let us look at the simplest example of an initial value problem solver – the Euler step method. This takes a system of first-order differential equations

$$\frac{dy_i}{dt} = f_i(\boldsymbol{y}, t)$$

and by 'multiplying' through by dt gets an approximation to the gradient of $\boldsymbol{y} = (y_0, y_1, \ldots)$. So

$$dy_i(t) = f_i(\boldsymbol{y}, t)\, dt$$

and we can approximate the solution from t to $t + dt$ by

$$y_i(t + dt) \approx y_i(t) + f_i(\boldsymbol{y}, t)\, dt.$$

Applying this recursively, we can approximate the solution to the differential equation. If we study this last equation, we see that there are only two key features that are needed to integrate the system: that the type of $\boldsymbol{y}$ is closed under both addition and scalar multiplication.

The dependency inversion principle means that we should be able to implement an integrator based on these assumptions only (possibly aided by the Interface segregation principle). We will come back to this point once we have some code examples to work with for solving the Schrödinger equation in some discrete basis (such as Fock states of the harmonic oscillator). We later use the Euler method and interface abstraction via protocols to implement this principle and write code that works for any type that is closed under addition and multiplication by a scalar.

8.3.7 Be a Clean Coder

It is common to use variable names such as `time` or `frequency`. While this may seem OK, it is not clear, e.g. what frequency we are referring to or what system of units are being used. It is important to have a convention that enables you, and anyone else looking at your code, to be able to determine precisely what you are calculating at any particular time. Clean coding is an approach to writing code that expands on this principle. Your code is not clean if 'smells' bad, that is: it is difficult to change; can break in many places because of a change in one place; does not allow reuse of code and/or contains repetitious code; is complex where it could be simple; and is hard to understand or read (most well-written code does not need comments to explain what it does).

A lot of clean coding follows from common sense, but is also a matter of personal taste – see for example on-line videos by Bob Martin and Kevlin Heaney on certain free streaming services. In the following code examples, we have tried to demonstrate clean coding throughout. We have for the most part adopted the convention of: lower camel case for variables names and methods; upper camel case for structures and classes; and the use of appropriate verbs and nouns in method names to make the intent of written code transparent. We also tend to indent code so that it fits within a screen width or well on a printed page. We have made use of functions to avoid repetitious code, enabling re-use. If in places you think our code is a bit 'smelly' try to also think about how you might 'clean' it.

8.3.8 Continue to Master Your Craft

Keep up to date with new languages and features – then make use of them (a sentence so important it deserves its own subsection)!

8.4 A Small Generic Quantum Library

This section concludes our discussion on good coding practice and presents the development of a small quantum library for solving the time-dependent Schrödinger equation for a simple system. We will discuss how the library is built, making reference to developing functionality in-line with our example driver code, Listing 8.1.

Before getting to the specifics, we need to make clear that the library is by no means complete. It simply serves to illustrate the methodology and thinking behind the approach we advocate. The idea was to demonstrate what could be done with good practice but limited time. To that end we spent the equivalent of one person-week of time building the codebase from scratch. We believe that our approach demonstrates that an experienced, but by no means professionally trained, programmer, can develop their own powerful code-base in a reasonably short time frame.

Our approach will be one that follows the development of the code in the order it was written. This way, the presentation will mimic, as closely as is possible, the thought process that was followed. We do this as our principal aim is to share our process as a way of approaching scientific coding rather than teaching general programming (there are plenty of other resources for that). For this reason, we will introduce language syntax, grammar, and features as we go. Since much of the Swift programming language was learnt in order to write this library, this also is a fair reflection of the process. The art was in knowing what might be possible in a language, and then finding Swift's approach to solving such problems.

Our strategy leads to a rapid introduction to the language; if you find it too fast, there are lots of internet resources and on-line swift environments you can use to help support your study of this section[10].

8.4.1 Some Swift Basics

8.4.1.1 Comments
Comments are non-executable text that can be included in code:

```
1  // One line comments look like this
2
3  /* Multiline comments look like this
4     Just like in C */
```

We open with a comment that simply specifies the intent of solving for Jaynes–Cummings model dynamics and a reference to where the specifics of the physics can be found. These are the only comments needed in the driver. For completeness this is:

```
1  // Jaynes--Cummings Schrodinger Dynamics driver.
2  /* For specifics on model, units and example/test parameter
      values and data see Fig 1 and associated text in
      doi:10.1103/PhysRevA.79.032328          */
```

8.4.1.2 Primitive Types and Their Declaration
After the comments, the first lines of code are,

```
1  let couplingConstant = 1.0
2  let fieldBasisSize = 100
```

Swift provides a number of standard (primitive) types. The ones we will use the most are `Int` for integers, `Double` and `Float` for floating-point values, `Bool` for Boolean values, and `String` for text. Note that Swift will infer whole numbers as `Int`, those with a decimal point as `Double` and those enclosed in " " as strings. The syntax for making the type explicit is

```
1  let couplingConstant: Double = 1.0
```

Swift, like a number of other languages, uses constants and variables to associate a label with a value of a particular type. If we think about physical parameters, the notion of having a constant that cannot be changed once it is set is clearly very valuable. Constants are declared with the `let` keyword and variables with the `var` keyword. Note that we have chosen to use `let` in the above example, as we do not expect or want either of these values to change.

---

10 See http://www.swift.org for installation instructions and other useful resources. For simple programming tasks, while getting started, you might like to use on-line Swift editors and run-time environments. At the time of writing, good free editors for Swift code include XCode, Visual Studio Code, and Atom (and for those who like old-school, vi and Emacs are also possibilities). We will not go into details here, as specifics may change from release to release. Alternatively, if you have access to a Mac or an iPad, Swift playground is an interesting gamified way to begin coding Swift.

> Always declare using `let` unless you know you need a variable. This way the compiler can make sure you do not change, e.g. the speed of light by accident.

8.4.2 Complex Numbers and the 'Generic' Decision

8.4.2.1 Introducing Structure and Classes

The next line of code,

```
1   let fieldSpace = StateSpace(dimension: fieldBasisSize)
```

expresses the desire to be able to create spaces in which operators and vectors may live. We will need a number of tools to do this. We know that Hilbert spaces are defined over fields, including the complex numbers. The way we will model complex numbers will also provide a suitable introduction to some of the language features that we will use to set up `StateSpace`[11]. Hence, before we go further in building the code that directly fulfils the promise of `StateSpace`, we will first develop a capability to handle complex numbers.

There are several options for working with complex numbers, some of which are language dependent and some not. We could completely avoid defining a complex type or data structure, and implement complex arithmetic using appropriate functions. We could, for example, store complex vectors and matrices as real arrays by doubling up storage and making use of the fact that

$$Z = a1 + bI, \text{ where } 1 = \begin{pmatrix} 1 & 0 \\ 0 & 1 \end{pmatrix} \text{ and } I = \begin{pmatrix} 0 & 1 \\ -1 & 0 \end{pmatrix}$$

behaves in exactly the same way as complex numbers. Beyond the fact that this is not easily made efficient, such an approach leads to code that can be hard to read[12]. Like most modern programmers, we choose to introduce a complex type.

Swift offers a number of data types that can be used to collect relevant pieces of information, and to perform tasks on, or with, that information. They are **struct** (structures), **class**, and **actor**. These three have much in common, but for our present discussion we note that the main difference is that `struct` is a data-type and the other two are reference-types (actors are like classes but designed for asynchronous applications).

Exercise 8.8 There are other collection types such as **Array**, **Dictionary**, and **Set**. After finding out how they are defined in Swift, would these be useful in any way for dealing with complex numbers? ∎

As an example, let us consider `struct` and `class` and show some example code to see the difference in how data and reference types behave (the reason for the differences we

---

11 Before going into the details of complex numbers, we note that there is some close similarity to what we present here and that which independently emerged in the Swift `Numerics` project. In a real-world setting, we would probably adopt and extend that package (it makes use of advanced features such as `@inlinable` that are beyond the scope of this discussion). We include our `Complex` code here as it is useful in understanding some key coding ideas.

12 Although tricks like this can come in handy when using, but not wanting to re-write, certain algorithms designed for real types, such as eigen-solvers.

find is because of the way each are stored in memory). The code that follows generates a Complex structure:

```
1  struct Complex {
2      var real: Double
3      var imag: Double
4      init(real real_in: Double, imag imag_in: Double) {
5          real = real_in
6          imag = imag_in
7      }
8  }
```

the anatomy of this code is (i) the `struct` keyword that indicates that a structure is about to be defined followed by (ii) the name of that structure (Complex), next (iii) a '{' is used to indicate the beginning of the internal content of the structure (its partner '}' is at the end of the code) then (iv) two internal variables (termed **properties**) of type double are defined, representing the real and imaginary components of the complex number; finally (v) is a special method called the initialiser (abbreviated to `init`). When we make a structure, it is this method that does all the work. In this case, all it does is copy some provided numbers to the internal storage of the structure. A really nice feature is that functions can take optional argument labels for external use (here `real` and `imag` - not to be confused with the eponymous internal storage variables) as well as parameter names for use within the function body (here `real_in` and `imag_in`).

The following snippet indicates how this structure can be used:

```
1  var aNumber = Complex(real: 3.0, imag: 0.0)
2  let three = aNumber
3  aNumber.real = 0.0
```

Note that in the last line, we can access and change the structure's property using the `variable_name.property` syntax. At the end of execution aNumber = 0 and three = 3. In this example, struct is a data type and this code behaves as expected.

If, instead of using a struct, we were to make Complex a **class**, the definition would look like this:

```
1  class Complex {
2      var real: Double
3      var imag: Double
4      init(real real_in: Double, imag imag_in: Double) {
5          real = real_in
6          imag = imag_in
7      }
8  }
```

In spotting the difference we see that the only change in the code is changing the keyword `struct` to `class`. The behaviour, however, is now rather different. If we take exactly the same example usage code:

```
1  var aNumber = Complex(real: 3.0, imag: 0.0)
2  let three = aNumber
3  aNumber.real = 0.0
```

At the end of execution aNumber = 0 and three = 0. Not what we would expect or want, even though we used the keyword 'let' that specifies that three is immutable (the reference has not mutated!). This example should scare you very much.

As we would want complex numbers to behave like normal numbers, it makes sense for Complex to be a data-type and we will use the struct version in our code. There are times when a class would be the right choice and we will re-visit this topic later.

8.4.2.2 A First Look at Generics

In our definition of Complex we made both the real and imaginary properties double precision. For some physics simulations, this may exceed or not meet our precision requirements. It would therefore be beneficial to have the option to change precision in the Complex struct without changing the code.

Most modern typed languages have a capability to express that certain parts of the code may work with a variety of types. This is called generics, and it is very powerful; it is worth spending time learning to use generics early, as it affects the way you approach any coding problem. The new structure for Complex using a generic type instead of Double:

```
1  public struct Complex<T> {
2      public var real: T
3      public var imag: T
4  }
```

Here the notation <T> indicates that T is a type that is to be provided later, rendering the code 'generic'. For example Complex<Double> would be equivalent to the previous example and everywhere that the placeholder T appeared would be a Double. Note that the fact that real and imaginary are both forced to be of the same type adds a level of reliability to the code that one would not have with an untyped language.

This is nice, but it ignores the fact that complex numbers are always defined over some scalars (Complex<String> would be valid but meaningless code). It would make much more sense if we could constrain the types that Complex can use to those that are of some scalar type. Swift uses another idea called a **protocol** that enables us to express that a certain capability (function) is needed[13].

To make that discussion meaningful, we first need to introduce two other key language features – functions and then operator overloading.

8.4.2.3 Functions

Similar to mathematical functions, programming functions (sometimes termed procedures, subroutines, or methods) are self-contained units performing a specific task[14]. Just like sine and cosine, we give functions specific names, and it is best to make them as descriptive as possible. Function declarations look something like this in Swift:

```
1  func function_name (external_label internal_label: Type) -> Type {
2      // code goes here
3      return the_thing_to_return
4  }
```

---

13 Most other typed languages have similar features such as interfaces and abstract classes in C# and Java.
14 Some like to reserve the phrase *function* for units of code that return a value; we do not make that distinction in Swift.

for example, we can have

```
1  func times(_ this: Double, by that: Double) -> Double {
2      return this * that
3  }
4  print (times(3.0, by: 2.0))
```

Note that by using an underscore, the external label is not needed when calling the function. By using labels carefully, we can make the code read in a way that makes it self-explanatory (as in our previous `init` example for the `Complex` structure and class).

8.4.2.4 Operator Overloading

In maths, we are used to using operations such as '+' with numbers or '!' for factorial or '−' to negate. Many languages offer the same capability in terms of special functions termed **infix**, **postfix**, and **prefix** operators, respectively. In Swift, their declaration looks like:

```
1  public struct Complex {
2    public var real: Double
3    public var imag: Double
4    static func + (lhs: Complex, rhs: Complex) -> Complex  {
5        return Complex(real: lhs.real + rhs.real,
6                       imag: lhs.imag + rhs.imag)
7  }
```

Because the + character is already defined as an operator, this will work. The Swift notation seems less odd if you consider that we can define a function $f(a, b) = a + b$ but by replacing f with +, it reads $+(a, b) = a + b$ exactly like our code.

Exercise 8.9   Code the remaining arithmetic functions (you can start by cutting and pasting the above code into any on-line Swift environment). ∎

If we wish to use one or more character(s) as yet not designated as an operator, then we can add them using the **operator** command, e.g.

```
1            infix operator **
```

which is a common abbreviation for taking a power, $a^n = a$ ** n (as already discussed, be careful that you do not hide clarity of meaning if implementing non-standard operators).

8.4.2.5 Introducing Protocols and Extensions

A protocol is a bit like specifying mathematical axioms, insofar as it defines a blueprint of how something conforming to that protocol should behave. Like mathematical axioms, they also serve as a guarantee that things conforming to the protocol can do certain things and/or have certain properties. That is, the approach is akin to specifying properties of mathematical structures, such as groups, rings, and fields having closure under addition and multiplication[15]. As you might expect from the analogy, this is a very powerful language

---

15 There are some important differences since the approach of protocols is not the same as a proper axiomatic system. So the analogy should not be taken too literally.

feature. Mathematical axioms allow for developing entire disciplines by simply looking at the consequences of the set of chosen axioms (the theorems, corollaries, etc.). Protocols allow general code to be written for anything conforming to that protocol.

We can, for example, guarantee that a type can use the infix operator + by ensuring it conforms to an Addable protocol that we can define by

```
1  protocol Addable {
2     static func + (lhs: Self, rhs: Self) -> Self
3  }
```

Here **Self** is shorthand for any type that conforms to the protocol. For instance, as Double already has an appropriate +(lhs: Double, rhs: Double) -> Double defined, we can make it addable by simply writing

```
1  extension Double: Addable {}
```

We can make sure that the generic used of Complex is closed under addition by restricting the generic T by changing the first line of the declaration to

```
1  public struct Complex<T: Addable> {
```

Now that T is Addable, we can implement a generic version of + in Complex which in turn will make Complex closed under addition and therefore Addable. The full implementation of this is:

```
1  public struct Complex<T: Addable>: Addable {
2
3     public var real: T
4     public var imag: T
5
6     static func + (lhs: Self, rhs: Self) -> Self {
7        return Complex(real: lhs.real + rhs.real,
8                       imag: lhs.imag + rhs.imag)
9  }
```

The general nature of the above code is central to most of the following discussion[16].

Exercise 8.10 Complex numbers form a scalar field. Define protocols for Multipliable, Subtractable, Dividable. Then expand the code for Complex accordingly. Note that multiple protocols can be specified as a comma separated list, e.g. <T: Addable, Subtractable>. ∎

Long comma-separated lists of protocols can become unwieldy. There are two choices we can take to manage this. The first would be to extend Complex, adding one protocol at a time, for example:

---

16 Note that here we could have used Complex as the type instead of Self. The compiler would have inferred the same outcome. The use of Self has been adopted because it makes clear that it is the value's actual type at run-time. So the Self for Complex<Double> will be different from that of Complex<Float32>. Instances of each cannot be added together as + requires the same Self throughout.

```
1  extension Complex: Subtractable where T: Subtractable {
2    static func - (lhs: Self, rhs: Self) -> Self {
3      return Complex(real: lhs.real - rhs.real,
4                     imag: lhs.imag - rhs.imag)
5  }
```

Here the keyword **where** will constrain this extension to be made only when T actually conforms to the protocol Subtractable. This can be hugely useful when one needs a generic type to change behaviour depending on what specific concept types are later used. For instance, we will see this again when defining inner products of vectors that take different forms depending on whether the underlying scalar field is real or complex valued (one inner product requires a conjugate and the other does not).

The other approach to types that implement many protocols is based on combining multiple protocols into a single protocol. We know that 'complex numbers' is an example of a field. In physics, the elements of a field are more commonly referred to as scalars. So we can define a new protocol like this:

```
1  protocol Scalar: Addable, Multipliable,
2                   Subtractable, Dividable {}
```

Exercise 8.11 Does the above approach violate the Interface segregation principle? ∎

Actual scalars have a number of other properties that we might also want to add, such as the additive and multiplicative identities (0 and 1). We could add this functionality by including the protocols in the definition of Scalar:

```
1  protocol Has_getMultiplicativeIdentity {
2      static func getMultiplicativeIdentity() -> Self
3  }
4  extension Has_getMultiplicativeIdentity {
5      public static var one: Self
6          { return Self.getMultiplicativeIdentity() }
7  }
8
9  public protocol Has_getAddativeIdentity {
10     static func getAddativeIdentity() -> Self
11 }
12 extension Has_getAddativeIdentity {
13     public static var zero: Self
14         { return Self.getAddativeIdentity() }
15 }
```

Note that for one and zero we have used a **computed property**. We use the keyword static to make them **type properties** (this enables, e.g. Double.one to be valid code – as this is the representation of '1' that belongs to the type Double).

Exercise 8.12 Look up computed properties – why do you think protocols need to use the above approach and cannot just have stored properties? ∎

Exercise 8.13 Should `Scalar` conform to these protocols? Would doing so make the definition of `Scalar` more or less of a violation of the Interface segregation principle? ∎

Exercise 8.14 As it is not possible to enforce, e.g. commutativity of multiplication, what might be done to make sure that a given implementation of `Scalar` is good? Hint: what could you do with tests? ∎

The generic complex type could then have been defined according to

```
1  public struct Complex<T: Scalar>: Scalar {
2    // code to satisfy protocols ...
```

but better still, we can combine both approaches in a way that maximises the general nature of the code

```
1  public struct Complex<T> {
2      public var real: T
3      public var imag: T
4  }
5  extension Complex: Scalar where T: Scalar {
6    // code to satisfy protocols ...
```

8.4.2.6 An Aside on the Power of Extensions
A trivial example of the utility of Swift's extensions is

```
1  extension Multipliable {
2      func square() -> Self {
3          self * self
4      }
5  }
```

Now every type that conforms to `Multipliable` can be squared, and because this is a protocol that is part of `Scalar`, all such types can now be squared (for us this might be `Double`, `Float16`, and `Complex`). Finding the square is rather trivial, but more complex algorithms sometimes need more capability than one protocol may implement. Let us consider extending the idea of squaring something to a more useful function: taking an integer power of a number. To make a general power function, our code needs to be able to deal with the fact that $x^0 = 1$, but the notion of 1 is not guaranteed by `Scalar` (by the definition above).

We could ask the user to insert 1 as an argument. Then the below example would give a power function to any `Scalar`:

```
1  extension Mutipliable {
2  // en.wikipedia.org/wiki/Exponentiation_by_squaring
3  // #Basic_method 2021.10.10 11:01
4    func power(_ n: UInt, identity: Self) -> Self {
5      func exp_by_squaring (_ x: Self,_ n: UInt) -> Self {
6        if n == 0 {
7          return identity
8        } else if n == 1 {
```

```
9              return x
10           } else if n.isMultiple(of: 2) {
11             return exp_by_squaring(x * x,   n / 2)
12           } else {
13             return x * exp_by_squaring(x * x, (n - 1) / 2)
14           }
15        }
16      return exp_by_squaring(self, n)
17    }
18  }
```

Note that here we have defined a function within a function (you may want to reflect on when this is a good idea or not) and we have used the Swift built-in function isMultiple. But as, e.g. 3.0.power(4, identity: 1.0) is not only inelegant but also exposes us to the potential to make a mistake every time it is used, this is not so satisfactory. Instead, if we can ensure our type conforms to the protocol Has_getMultiplicative-Identity, then any such Scalar can be made to work with algorithms that need 1. With a few lines of code we can implement this type conformance and take advantage of it:

```
1  extension Mutipliable where
2            Self: Has_getMultiplicativeIdentity {
3    func power(_ n: UInt) -> Self {
4      self.power(n,
5              identity: Self.getMultiplicativeIdentity())
6    }
7  }
```

which makes use of the previous extension - so we have only gained functionality and can now write 3.0.power(4).

Exercise 8.15 Following on from Exercise 8.13, does the above example make you more or less inclined to add Has_getMultiplicativeIdentity and Has_get AddativeIdentity conformance to Scalar? ∎

Exercise 8.16 Note that we could have defined power using a custom infix operator like ** in Swift. Add this functionality. ∎

Exercise 8.17 If a type is dividable as well as multipliable, then it is also possible to define a power to a negative number. Extend, e.g. Scalar to have this capability. ∎

8.4.3 Adding Quantum Structure to the Code

Now we are in a position to return to our main driver routine and resume the exercise of examining it, a bit at a time, in our simulation of the process of constructing the library. Recall that the key idea is that we start from the sort of code that one wants to write to get results, and then write the back-end that makes this happen. We left off examining the driver at this line of code:

```
1  let fieldSpace = StateSpace(dimension: fieldBasisSize)
```

In quantum mechanics, we have two main primitive types: states and operators. However, both these mathematical objects always live within another structure, the state (Hilbert) space. The rationale for our approach is that we should not be allowed to do things like adding vectors from different state spaces, even if they are of the same dimension (it would then be easy to write code that would compile and run but be physically incorrect). We therefore desire a mechanism for allocating quantum things to the space in which they belong. The above line of code is the beginning of expressing the desire for such a capability. We next discuss a way to make this happen.

Exercise 8.18 In the following discussion on *Spaces* and then *Vectors and Operators*, continue to ask yourself if have we violated any of the Single-responsibility, Open-closed, Liskov substitution or Interface segregation or principles? If so, is it worth it? ∎

8.4.3.1 Spaces

If we analyse the requirement to specify a Hilbert (state) space, we see that this is built upon the more general (and reusable) concept of a vector space that can then be equipped with an inner product[17]. We will therefore begin by creating a `VectorSpace` type. But what should this type be? A single instance may be used a lot, associated with every vector and operator in that space[18]. Perhaps we would want the space to actually contain all the things that are conceptually inside it at some point[19], and if we do that, we would not want to make a clone if copying the space. This reasoning leads us to want `VectorSpace` to be a reference type (`class`) and *not* a data type (`struct`). Let us first give an implementation and then seek to understand it line-by-line:

```
1  public class VectorSpace<T: Scalar> {
2      public typealias ScalarFieldOfVectorSpace = T
3
4      public let dimension: Int
5      public let description: String
6      internal let identifier: Int
7      internal var Set: [VectorSpace]
8
9      public init(dimension: Int, label: String) {
10         self.dimension = dimension
11         self.label = label
12         self.identifier = space_counter
13         space_counter += 1
14         Set = []
15         self.Set.append(self)
16     }
17 }
```

---

17 In numerical modelling we cannot realise the 'completeness' of Hilbert spaces, so it would be more correct to talk about inner-product spaces. But as many who study physics do not study functional analysis in depth, we keep to the terminology that many will be familiar with.

18 Note that for `VectorSpace` we will not enable usage such as `oscillatorSpace.numberOperator` seen in the driver routine. This will be postponed until the introduction of `StateSpace`, as the number operator is a quantum mechanical concept.

19 Not implemented at the time of writing.

The first thing we do is to set the type of the field over which the vector space is defined. Using a **typealias** is a useful way to make self-documenting code, especially when used with generics. In this way it should be absolutely clear, without using any comments, that the scalar field over which the Hilbert space is defined is the generic T.

The next two properties are self-explanatory. It is worth noting that (i) using dimension implies we will always use a finite basis and (ii) I feel the need to force the adding of a description to any state space so that debugging and logging can be done with this information in place – it would be easy to extend the class to have an initialiser that provides an empty string as a label, but, since we wish to guarantee that each instance has a description, that is not something we will allow ourselves to do.

Then we have identifier, which is intended to uniquely identify each space. This will be needed as a mechanism for satisfying the Equitable protocol, and also for sorting spaces later, when we use this for housekeeping when constructing tensor product spaces[20]. If the space is a tensor product space, we want to keep track of its component spaces. As a single space can be viewed as a trivial tensor product, we store self in Set for now (we have to make the array before we put self in it[21]). We can now extend the above base class, e.g.:

```
1   public class StateSpace: VectorSpace<Complex<Double>>{
2       // some StateSpace specific code
3   }
```

and these new classes will gain the functionality we put into VectorSpace for free.

8.4.3.2 Vectors
Vector spaces are not of much use without vectors. In addition, the next line of the driver routine,

```
1   let fieldHamiltonian = fieldSpace.numberOperator
2                   + 0.5 * fieldSpace.identityOperator
```

makes clear that we are going to want to do algebra with some standard operators too. As our current concern is vectors, we will postpone discussion of operators such as fieldSpace.numberOperator for now.

Vectors have a number of properties that we will want to enforce; we shall do this through protocols. Luckily, we have already defined the protocols we need for a vector. Looking at the axioms of vectors, we know that they need to be closed under addition. We know that

---

20 We might want the counter of the number of spaces that we have created to be a static variable as it would then be shared among all the instances of VectorSpace. Unfortunately, we cannot do this because Swift does not allow static stored properties within generic types. As this class is defined over any scalar T, having a static space_counter property is not an option for us. To get round this we use a global counter

```
1   internal var space_counter = 0,
```

noting that we set the access to internal so that it cannot accidentally be changed from outside the package (this should potentially be a unsigned integer, UInt, as it will never be negative).
21 While we would ideally like to make this a let, this is not possible. Instead we have set the scope to internal to the package (we do not want to use private because access is useful for testing purposes, and as it is our library, we can be careful - we could always use a getter function instead if we wanted the code to be more safe [this decision depends on how much control there is over who writes which bits of the code, and how that effort is managed].

there needs to be an additive inverse; this means the vector must be subtractable, and it must be defined over a scalar field. So the only thing we need to add to the existing protocols is:

```
1  public protocol VectorType:
2          ClosedUnderScalarFieldMultiplication,
3          Addable,
4          Subtractable {
5      var space: VectorSpace<ScalarField> { get set }
6  }
```

where we ensure multiplication of our vector type by a scalar makes sense by also defining these base protocols[22]:

```
1  public protocol definedOverScalarField {
2    associatedtype ScalarField: Scalar
3  }
4  public protocol ClosedUnderScalarFieldMultiplication:
       definedOverScalarField {
5    static func * (left: Self, right: ScalarField) -> Self
6    static func * (left: ScalarField, right: Self) -> Self
7  }
```

Note that the coding order followed this thought process: a vector must live in a vector space; it must be closed under scalar field multiplication and closed under addition and subtraction. We therefore need some more protocols to implement the functionality of scalar field multiplication in a way that isolates the concerns of defining the scalar field, and then the closure relation (as something could be defined over a scalar field but not closed under scalar multiplication).

Unlike spaces, we mathematically think of vectors as more similar to numbers than to spaces. For instance, if we do $u = v$, then $v = 2v$, we do not expect u to change. This means that we want vectors to be a data type, and thus will use a `struct` and not a `class` for them. Our implementation, which we will analyse below, is:

```
1  public struct ColumnVector<T: Scalar>: VectorType {
2      public typealias ScalarField = T
3      public var space: VectorSpace<T>
4      var values: [ScalarField]
5      public init(in space: VectorSpace<T>) {
6          values = Array(repeating: ScalarField.zero,
7                         count: space.dimension)
8          self.space = space
9      }
10     init (values: [T], in space: VectorSpace<T>) {
11         self.values = values
12         self.space = space
13     }
```

---

22 A niggle with Swift here is that we cannot use `let` but are forced to use `var` because space is a computed property (if we used type alias instead of generics we could use normal properties and hence `let`, but then we would lose the flexibility of generics). We will just have to be careful not to change a space once it is set. Nevertheless, the advantages of protocol and generics-based programming make this a worthwhile risk.

```
14      public static func + (lhs: Self, rhs: Self) -> Self {
15          assert (lhs.space == rhs.space)
16          return Self(values: elementwiseBinaryOperation(
17                          thisArray: lhs.values,
18                          thatArray: rhs.values,
19                          operation: +),
20                      in: lhs.space)
21      }
22      //... rest of the code to satisfy protocols ...
```

As with `VectorSpace`, we start off with some housekeeping, followed by the values to be stored. We provide two initialisers for the anticipated most common usage and then implement the methods needed to satisfy the protocol. Note the way that the generic type is used to keep everything self-consistent. Also note that we use the `assert` syntax to ensure we only add vectors in the same state space.

Element-wise operations using common arithmetic operations are going to be an often-repeated pattern. We therefore delegate this to a utility function that lives outside the `struct`. This can be done in the following way[23]:

```
1  public func elementwiseBinaryOperation<T>(
2                      thisArray: [T],
3                      thatArray: [T],
4                      operation: (T,T)->T )
5  -> [T] {
6      assert(thisArray.count == thatArray.count)
7      return zip(thisArray,thatArray).map(operation)
8  }
```

The advantage of avoiding repetition across the code-base is that we can now change the logic (e.g. to a for loop, or for parallel execution) later. We can seek to increase the performance of our code by editing it *in only one place*[24].

Exercise 8.19 If you have not already done so, look up the standard functions `zip` and `map`, and check you understand how the above functions work. Try to replace these function calls with a `for` loop. Check the performance of each implementation. ∎

---

23 or, if you prefer, one could achieve a similar result by extending Array itself.

```
1  extension Array {
2      func elementwiseBinaryOperation(
3          _ that: [Element],
4          by: (Element,Element)->Element ) -> [Element] {
5          return zip(self,that).map(by)
6      }
7  }
```

Which do you prefer?

24 As compilers improve, it may be that we would even want to switch between algorithms depending on target platform and hardware constraints. This way of coding allows us to make those decision later without affecting the logical structure of the library. For example, in production code, e.g. matrix multiplication, we would want to make use of Strassen's algorithm for matrices where this gives an advantage. We might also want to track and optimise code for matrices with specific structure. For example we could have flags such as `isDiagonal` and `isHermitian` as part of a structure and execute different code depending on which flags are set.

Note that using a typed language means that we can ensure the vector space and the vector are defined over the same scalar field (it would not be possible to pass an incorrectly typed space into the constructor).

We can now add functionality to the vector and importantly, we can do this depending on what protocols are satisfied. For example, we can have a 'default' protocol for an inner product

```
1   extension ColumnVector   {
2       public func innerProduct(dualVector: Self) -> T {
3           assert (self.space == dualVector.space,
4                   "incompatible spaces" )
5           var sum = T.zero
6           for i in 0 ..< values.count {
7               sum = sum + values[i] * dualVector.values[i]
8           }
9           return sum
10      }
11  }
```

and an alternative for complex numbers, provided that we have made sure that the generic T which is already a Scalar also conforms to the Has_Conjugate protocol (in other words it is a complex number)[25]:

```
1   extension ColumnVector where T: Has_Conjugate {
2       public func innerProduct(dualVector: Self) -> T {
3           assert (self.space == dualVector.space,
4                   "incompatible spaces" )
5           var sum = T.zero
6           for i in 0 ..< values.count {
7               sum = sum +
8                   values[i] * dualVector.values[i].conjugate
9           }
10          return sum
11      }
12  }
```

The value of using templates can be seen if we consider extending ColumnVector to take the expectation value of an operator.

```
1   extension ColumnVector   {
2     public func expectationValue(of A: MatrixOperator<T>) -> T {
3       return (A * self).innerProduct(dualVector: self)
4     }
5   }
```

We can't implement this yet, as we have so far not defined operators (we will do that next). This example code is presented to make the point that the specific innerProduct that is used will depend on whether T conforms to Has_Conjugate or not. Also note that the high-level code does not depend on the low-level implementation.

---

25 Note that we are forcing the use of the label dualVector, since there is the potential for confusion. As an example, in the mathematics notation for an inner product we have the dual vector as the second argument (so $\langle u, \alpha v \rangle = \alpha^* \langle u, v \rangle$) but in Dirac notation it is the first (so $\langle \alpha \psi | \phi \rangle = \alpha^* \langle \psi | \phi \rangle$).

8.4.3.3 Operators

By definition, operators in a vector space need a vector to operate on, and both of these mathematical objects should belong to the same vector space. Unlike vectors, operators are also required to be closed under multiplication. This leads us to define a protocol for an operator type in this way:

```
1   public protocol OperatorType:
2       ClosedUnderScalarFieldMultiplication,
3       Addable,
4       Subtractable,
5       Multipliable
6   {
7       var space: VectorSpace<ScalarField> { get set }
8       associatedtype Vector: VectorType
9       static func * (lhs: Self, rhs: Vector) -> Vector
10  }
```

The implementation of a specific operator type will then follow a similar pattern to the one for vectors, so we won't go into details here (see the associated library for specifics). We simply provide enough of an outline to get a good idea of how the code is structured. The opening declarations are:

```
1   public struct MatrixOperator<T: Scalar>: OperatorType {
2       public typealias ScalarField = T
3       public var space: VectorSpace<T>
4       public var values: [ScalarField]
5
6       public init(in space: VectorSpace<T>) {
7           values = Array(repeating: ScalarField.zero,
8                              count: rows*columns)
9           self.space = space
10      }
11      // code to satisfy protocols follows...
```

We have chosen to store the matrix in a one-dimensional array as some operations, such as addition, can be done using the same functions as `ColumnVector` used as follows:

```
1   public static func + (lhs: Self, rhs: Self) -> Self {
2     assert (lhs.space == rhs.space)
3     return MatrixOperator<T>(
4             values: elementwiseBinaryOperation(
5                       thisArray: lhs.values,
6                       thatArray: rhs.values,
7                       operation: +),
8             in: lhs.space)
9   }
```

It is also the storage format assumed by a number of well-established numerical algorithms, so this choice will facilitate easier use of such resources.

As a note, Swift does provide the possibility to de-reference structures and classes according to the notation `A[i,j]` using a feature termed subscripts (we can do the same for vectors too but using only a single index). We can enable this in the main struct or as an extension with the following code:

```
1   /* See Matrix example at
2    docs.swift.org/swift-book/LanguageGuide/Subscripts.html
3   */
4   func indexIsValid(row: Int, column: Int) -> Bool {
5     return row >= 0 && row < space.dimension &&
6            column >= 0 && column < space.dimension
7   }
8   subscript(row: Int, column: Int) -> T {
9     get {
10        assert(indexIsValid(row: row, column: column),
11               "Index out of range")
12        return values[atIndex(row: row, column: column,
13                   nColumns: self.space.dimension)]
14    }
15    set {
16        assert(indexIsValid(row: row, column: column),
17               "Index out of range")
18        values[atIndex(row: row, column: column, nColumns:
19               space.dimension)] = newValue
20    }
21  }
```

where

```
1   func atIndex(row: Int, column: Int, nColumns: Int)
2   -> Int {
3       return (row * nColumns) + column
4   }
```

is a function available to the whole library as part of a set of utility functions[26].

8.4.4   Quantum Functionality

In this section, we see how extensions can be used to effectively add specific quantum functionality to the code base. Before we discuss the specifics, there are two questions on SOLID coding that are worth keeping in mind while we develop the discussion. There is a lot of detail in this section, but this is needed to support the narrative. On a first read, it may be best to focus on what the code seeks to achieve and its style (leaving details to be studied later if needed).

---

26 Note that we have used 'row major' order to allocate the two-dimensional array of a matrix into a one-dimensional array for storage. The use of separating out de-referencing to an `atIndex` function means that if we later want to change storage to 'column-major' format that would not be too hard (modern IDEs make refactoring easy). We can also implement a sparse matrix structure that satisfies the same protocols, which enables interchangeable use within the same syntax. While there are some interesting tricks in our implementation of `SparseMatrixOperator`, these are technical in nature and we leave the interested reader to study the library itself. For many tasks in quantum mechanics, sparse methods provide substantial speedup and are worth some effort to develop.

Exercise 8.20 We are making heavy use of the open-for-extension part of the Open-Closed principle. We now have enough substance to get a feel for the parts of the library that should be closed-for-modification. The `final` keyword can be used to help realise this second part of the principle. Look up its syntax and identify the parts of the code (including those already covered) that should be protected using `final`. ∎

Exercise 8.21 With reference to the Leskov substitution principle, does all the code in this section sub-type at appropriate levels of abstraction? ∎

The next couple of lines of code in the driver routine exemplify the usage of some standard operator arithmetic. Note how the use of instances of `StateSpace` makes the code easy to read and, when one is using an editor with auto-complete functionality, to write.

```
1  let fieldHamiltonian = fieldSpace.numberOperator
2                        + 0.5 * fieldSpace.identityOperator
3  let spinHamiltonian = spinSpace.angularMomentumZ
```

As the implementation is pretty standard coding, we will only give one example function and property here (you can find the other code in the supplementary material).

```
1  extension StateSpace {
2      public var numberOperator: MatrixOperator<T> {
3          return makeNumberOperator()
4      }
5
6      public func makeNumberOperator()
7          -> MatrixOperator<T>
8          {
9          var output = MatrixOperator(in: self)
10         for i in 0 ..< output.columns {
11             output[i, i] = T(i)
12         }
13         return output
14     }
15 }
```

A potentially non-obvious thing to note is that `numberOperator` is provided as a computed property here, rather than being made at initialisation. We made this choice, as we do not necessarily want to create every operator, even those we do not use, when we make a `StateSpace`.

Exercise 8.22 Should the `func` for number operator have been `private` and/or `final`? ∎

Exercise 8.23 Can you think of a way to adapt the `func` for number operator so that unnecessary computations are not made with multiple calls? Write a performance test to check if the compiler actually does this for you. Given that compiler behaviour can change over versions, what would be a good strategy to adopt here? (your answer will depend on perceived usage of `numberOperator`). ∎

The next line of the driver is the first example of how the original vision for the code should be open to change:

```
1  let totalSpace = fieldSpace.tensorProduct(spinSpace)
```

It turned out that, when coding the back-end for this code, this syntax led to a less-than-satisfactory implementation. The reason is that a tensor product is an idea that belongs to all vector spaces, so it should be a static method or an initialiser (`tensorProduct` does not conceptually belong to `fieldSpace`). We therefore changed our vision in order to take a more general approach:

```
1   extension VectorSpace {
2     convenience public init(
3                 tensorProductOf spaces: [VectorSpace],
4                 label: String) {
5      var tempDimension = 1
6      for space in spaces {
7        tempDimension *= space.dimension;
8      }
9      self.init(dimension: tempDimension,
10                 label: label + " (tensor product space)")
11     Set=spaces.sorted(by: {$0.identifier < $1.identifier})
12     for i in 0 ..< Set.count-1 {
13       assert(Set[i] != Set[i+1],
14              "Error: Two identical spaces in call")
15       tempDimension *= Set[i].dimension;
16     }
17  }
```

which, recalling that `StateSpace` is an extension of `VectorSpace`, led to a change in the driver routine

```
1  let totalSpace = StateSpace(
2          tensorProductOf: [fieldSpace, spinSpace],
3          label: "JC System")
```

which has the advantage that it makes clear that the returned tensor product space is also a `StateSpace`.

Exercise 8.24 In the above implementation an array was used to pass in the components spaces of the state space. We could have used variadic parameters instead. What would be the advantages and disadvantages of such an approach? ∎

Exercise 8.25 The above code is not built to handle tensor products of tensor product spaces. List two different ways to address this deficiency. Which you prefer and why? ∎

The next steps in the driver build the interaction term $\gamma \left(\hat{\sigma}_+ \hat{a} + \hat{\sigma}_- \hat{a}^\dagger \right)$

```
1  let interactionTerm = couplingConstant * (
2      totalSpace.tensorProduct(
3          fieldSpace.creationOperator,
4          spinSpace.angularMonentumLoweringOperator
5          )
6      +
7      totalSpace.tensorProduct(
8          fieldSpace.annhiliationOperator,
9          spinSpace.angularMonentumRaisingOperator
10         )
11     )
```

Coding general tensor products is a bit tricky. Recall our constraint of coding the backend library within a week. Coding a general tensor product for an arbitrary number of operators and easy usage is a task that was not achievable in this time-frame. The good news is that pairwise tensor (Kronecker) products of matrices and vectors is much easier and well-described in many sources. Here we have used the 'numbering from zero' definition of Wikipedia: $(A \otimes B)_{pr+v,qs+w} = a_{rs}b_{vw}$ (we have done this to illustrate some code commenting and notation suggestions). Note that when doing tensor (Kronecker) products of vectors, the vector is treated as a special case of a single column matrix. Hence, within VectorSpace we will have code for making tensor products of vectors and operators that will use the same underlying kronekerProduct function. For example:

```
1  public func tensorProduct(
2      of   A: MatrixOperator<ScalarFieldOfVectorSpace>,
3      with B: MatrixOperator<ScalarFieldOfVectorSpace>)
4      -> MatrixOperator<ScalarFieldOfVectorSpace>
5  {
6  // omitted code to check operators in the right spaces
7     let C = kronekerProduct(A: temp[0].values,
8                             rowsA: temp[0].rows,
9                             colsA: temp[0].columns,
10                            B: temp[1].values,
11                            rowsB: temp[1].rows,
12                            colsB: temp[1].columns)
13    return MatrixOperator(values: C, in: self)
14 }
```

and the Kronecker product function will be available to the whole library (i.e. it does not belong in VectorSpace) as it may be of use in other coding 'objects'.

```
1  // https://en.wikipedia.org/wiki/Kronecker_product
2  // [accessed: 12/01/2022 - see Definition]
3  public func kronekerProduct<T: Scalar>   (
4      A: [T], rowsA: Int, colsA :Int,
5      B: [T], rowsB: Int, colsB :Int)
6  -> [T]   {
7     assert(A.count == rowsA * colsA, "dimension of A bad")
8     assert(B.count == rowsB * colsB, "dimension of B bad")
```

```
 9      var C = Array(repeating: T.zero,
10                        count: rowsA*rowsB*colsA*colsB)
11
12      let colsC = colsA * colsB
13      let p = rowsB // so notation is same as wikipedia
14      let q = colsB
15
16      let index = { ( _ R: Int,_ C: Int,_ N: Int) -> Int in
17           return atIndex(row: R, column: C, nColumns: N) }
18
19      for r in 0 ..< rowsA {
20        for s in 0 ..< colsA {
21          let A_rs = A[index(r,s, colsA)]
22          for v in 0 ..< rowsB {
23            for w in 0 ..< colsB {
24              C[index(p * r + v, q * s + w, colsC)] =
25                               A_rs * B[index(v,w, colsB)]
26            }
27          }
28        }
29      }
30      return C
31  }
```

Note the usage, once again, of at Index to ensure that we are consistent in de-referencing the same way across the library. In this example we have used an anonymous function (closure) to, in our view, make the code clearer. We could have taken a different approach. Using the Subscript feature of Swift, we can de-reference anything of matrix form if we make it conform to a protocol such as:

```
1  public protocol matrixForm {
2       subscript(row: Int, col: Int) -> Int { get set }
3       var rows: Int { get }
4       var columns: Int { get }
5  }
```

which leads to a much more compact and elegant kronekerProduct function

```
 1  public func kronekerProduct<T: definedOverScalarField &
 2                                      matrixForm>
 3      (of A: T, with B: T, output C: inout T) {
 4      let p = B.rows // so notation is same as wikipedia otherwise
         redundant
 5      let q = B.columns
 6      for r in 0 ..< A.rows {
 7        for s in 0 ..< A.columns {
 8          for v in 0 ..< B.rows {
 9            for w in 0 ..< B.columns {
10              C[p * r + v, q * s + w] = A[r,s] * B[v,w];
11            }
12          }
```

```
13        }
14      }
15  }
```

Exercise 8.26   What are the advantages and disadvantages of each approach?   ∎

Exercise 8.27   In the above two listings of `kronekerProduct` there is at least one clear violation of one of the SOLID principles. Can you spot it and in which implementation is it most easily fixed. Hint: it has to do with scalars. Does this example violation change your answer to the previous Question?   ∎

In the next line of the driver, we sought to make the total Hamiltonian of the system:

```
1  let jaynesCummingsHamiltonian = totalSpace.sum(fieldHamiltonian,
       spinHamiltonian)
2                  + interactionTerm
```

Implementing this would involve writing a new function `sum` in `VectorSpace` requiring quite a lot of effort (have a go at writing it if you would like to verify this). If we look at our existing code base, we quickly realise that `sum` is not needed, as we recall that, e.g. $\hat{n}_f$ in the Hamiltonian is just a shorthand for $\hat{n}_f \otimes \hat{I}_a$ and + is already defined in the Vector space. This leads us to an alternative and better solution[27]. Our revised driver code is

```
1  let fieldHamiltonianExtension =
2          totalSpace.tensorProduct(
3              of: fieldHamiltonian,
4              with: atomSpace.identityOperator)
5  let atomHamiltonianExtension =
6          totalSpace.tensorProduct(
7              of: atomHamiltonian,
8              with: fieldSpace.identityOperator)
9
10 let jaynesCummingsHamiltonian =
11          fieldHamiltonianExtension +
12          atomHamiltonianExtension +
13          interactionTerm
```

If it occurred to you before this point that the above would be a better approach to the driver than the one originally proposed, then you were correct.
The next two lines of code are simple to understand

```
1  let timeIncrement = 0.01
2  let endTime = 10.0
```

but are they in the right place? The following two lines are also clear in intent

```
1  let spinState = spinSpace.makeVector(from: [1.0, 0.0])
2  let fieldState = fieldSpace.makeCoherentState(
3              alpha: Complex(modulus: sqrt(15.0),
4              argument: 0.0))
```

---

[27] as we have not had to write any unnecessary library code that would introduce extra scope for making errors.

but the second line needs a little bit of work to implement. Although the implementation is standard, we will discuss how it was achieved, as there is one trick worth noting and it provides another useful example of how to work with generic code. Here we wish to represent in code

$$|\alpha\rangle = e^{-\frac{|\alpha|^2}{2}} \sum_{n=0}^{\infty} \frac{\alpha^n}{\sqrt{n!}} |n\rangle.$$

In the Harmonic oscillator basis, this is

$$\langle k|\alpha\rangle = e^{-\frac{|\alpha|^2}{2}} \sum_{n=0}^{\infty} \frac{\alpha^n}{\sqrt{n!}} \langle k|n\rangle = e^{-\frac{|\alpha|^2}{2}} \sum_{n=0}^{\infty} \frac{\alpha^n}{\sqrt{n!}} \delta_{kn} = e^{-\frac{|\alpha|^2}{2}} \frac{\alpha^k}{\sqrt{k!}},$$

which we might be tempted to code as it is written mathematically. The problem for us is that factorial can exceed the maximum value of Int for quite low argument values[28]. We can avoid the issue by noting that

$$\frac{\alpha^k}{\sqrt{k!}} = \frac{\alpha^{k-1}}{\sqrt{(k-1)!}} \frac{\alpha}{\sqrt{k}},$$

which leads us to extend VectorSpace to include:

```
1  public func makeCoherentState(alpha: T)
2  -> ColumnVector<T> {
3      var output = ColumnVector(in: self)
4
5      let modulusAlphaSquared = alpha * alpha.conjugate
6      let prefactor = T.exp( -modulusAlphaSquared / T(2) )
7
8      output.values[0] = prefactor
9      for n in 1 ..< self.dimension {
10         output.values[n] = output.values[n-1] * alpha /
11                            T.sqrt( T(n) )
12     }
13     return output
14 }
```

where we have constrained T to conform to the protocols: Has_Conjugate, Has_Exp, Has_IntegerInitialiser, and Has_Sqrt.

Exercise 8.28   We have assumed the number state basis here. We have also used this and other standard bases for other operators. There is no indication of this assumption in the function name. Is an assumption of domain-specific knowledge of quantum physics enough to make makeCoherentState an acceptable addition to the vocabulary?   ∎

---

28 In languages like Python this would not be an issue. One of the advantages of making everything, including numbers, objects is that arbitrary precision arithmetic can be automatically invoked if needed. It is also worth noting that recomputing the factorial for every element would be inefficient coding.

8.4.5 Dynamics

Now we finally turn to the lines of code that solve the Schrödinger equation:

```
1  var jaynseCummingsSystem = QuantumSystem(
2      startTime:     0.0,
3      initialstate: totalSpace.tensorProduct
4                      (of: spinState, with: fieldState),
5      hamiltonian:  jaynesCummingsHamiltonian)
6
7  while jaynesCummingsTDSE.currentTime < endTime {
8      jaynesCummingsTDSE.doRungeKuttaStep(by: timeIncrement)
```

In a real-world situation, we would also want to process and output information such as $\langle \hat{\sigma}_z \rangle$ (in this example, only the computer knows what happened). Here we wish instead to focus on the design choices we take in forming the integrator, taking into account the discussion above (there are many freely available sources covering writing data to files).

While we wish to find a solution to the Schrödinger equation, solving initial value problems for systems of ordinary differential equations covers a vast range of applications. We therefore seek a solution with the widest possible level of applicability. That is, following our discussion of the dependency inversion principle on page 244, we wish to have a function that can be reused by any type where integration makes sense.

Exercise 8.29 Before reading further, review the SOLID programming content, especially the Dependency inversion principle. Write the most general Euler step routine you can that respects all these principles. The Euler step in vector component notation is $y_i(t + dt) \approx y_i(t) + f_i(\mathbf{y}, t)\, dt$. ∎

One of the advantages of a generic approach is that we should be able to develop a solution to hard problems by first solving simple problems where we know what the behaviour should be. In that vein, we may make life easier by testing any integrator we write, using the classical simple harmonic oscillator as our example. The equation of motion for the displacement x is:

$$\ddot{x} + 2\zeta \omega \dot{x} + \omega^2 x = \frac{F(t)}{m},$$

where ζ is the damping coefficient, ω the resonant frequency, m the mass, and $F(t)$ the driving force. If we define a vector $\mathbf{y}$ with components $y_0 = x$ and $y_1 = \dot{x}$, then

$$\frac{d\mathbf{y}}{dt} = \begin{pmatrix} \dot{y}_0 \\ \dot{y}_1 \end{pmatrix} = \begin{pmatrix} \dot{x} \\ \ddot{x} \end{pmatrix} = \begin{pmatrix} \dot{x} \\ \frac{F(t)}{m} - 2\zeta \omega \dot{x} - \omega^2 x \end{pmatrix} = \begin{pmatrix} y_1 \\ \frac{F(t)}{m} - 2\zeta \omega y_1 - \omega^2 y_0 \end{pmatrix}.$$

An example function that calculates this derivative is

```
1  func sho_derivative(y: [Double], t: Double) -> [Double]
2  {
3      let omega = 1.0, zeta = 1.0, m = 1.0
4      let F = cos(t) // for example
5      return [                              y[1] ,
6              F/m - 2.0*zeta*omega*y[1] - omega*omega*y[0] ]
7  }
```

and a conventional Euler step integrator might look like

```
1   func eulerStep(y: [Double], t: Double, dt: Double,
2                     _ derivs: ([Double],Double) -> [Double])
3   -> [Double] {
4       let dydt = derivs(y,t)
5       var yout = [Double]()
6       for i in 0 ..< y.count {
7           yout.append(y[i] + dydt[i]*dt)
8       }
9       return yout
10  }
```

Note that we are using less-than-ideal labels to help to format for the printed page. A specific solution could be implemented by something like

```
1   var state = [1.0, 0.0] // [position, velocity]
2   for i in 0 ..< 1000 {
3       let time = Double(i) * dt
4       state = eulerStep(t: time, h: dt, y: state,
5                           derivative_function: sho_derivative)
6   }
```

which is more than adequate for such a simple system. It is, however, not of much use for solving the Schrödinger equation over a vector space defined over Complex<T>.

Exercise 8.30   Swift provides a way of changing variables in-place (so they do not have to be created and destroyed in memory). This is done using the inout keyword in the argument list. Is it worth making yout an inout? What would you do to establish if an inout approach is faster than the above solution? Which approach is better?   ∎

In the above example, we made use of the vector component form of Euler step $y_i(t + dt) \approx y_i(t) + f_i(\mathbf{y}, t)\, dt$. If we instead consider the vector form $\mathbf{y}(t + dt) \approx \mathbf{y}(t) + \mathbf{f}(\mathbf{y}, t)\, dt$, looking at the right-hand side, we see that all we need to be able to do is to add vectors (an Addable type) and multiply a vector by the type of dt (the same type as the independent variable t). The most general doEulerStep we can think of is listed below. It may look very much like *extreme overkill* at the moment, but when we give examples of more complex integrators later, the approach will make much more sense.

```
1   public func doEulerStep <IndependentVariable,
2                               IntegrandType> (
3       t: IndependentVariable,
4       h: IndependentVariable,
5       y: IntegrandType,
6       derivative_function return_derivatives: (
7               IndependentVariable, IntegrandType)
8               -> IntegrandType,
9       add: (IntegrandType, IntegrandType)
10              -> IntegrandType,
11      times: (IndependentVariable, IntegrandType)
12              -> IntegrandType )
```

```
13   -> IntegrandType {
14       let dydx = return_derivatives(t,y)
15       return add(y, times(h, dydx))
16   }
```

It would be easy at this stage to discard the approach as overly complicated, but let us show its power by making all arrays of a 'sensible' type solvable using doEulerStep. Keep in mind that we only have to do this once, which is one of the reasons to bother with the approach. Furthermore, we will later see that we will not have to repeat the effort when we come to implement a better integration algorithm. The idea is to invest some considered effort upfront so that we can reap the rewards later. We start by defining some basic protocols:

```
1   public protocol OdeAddable {
2        static func odeAdd(lhs: Self, rhs: Self) -> Self
3   }
4   public protocol OdeScalarMutipliable {
5      associatedtype OdeScalar
6      static func odeMultiply(lhs: OdeScalar, rhs: Self)
7                    -> Self
8   }
9   public protocol OdeIntegrable:
10                     OdeAddable & OdeScalarMutipliable {}
```

Note that we have deliberately not reused Addable, as we will not want to change the existing behaviour of + in Array later (which is concatenation). We have also not reused ClosedUnderScalarMultiplation, as sometimes we may want to use different scalar fields for different purposes (such as complex numbers for a Vector space and floating point for time). With these protocols in place, we can now define an easier to use Euler step function that calls the base function

```
1   public func doEulerStep <T: OdeIntegrable> (
2        t: T.OdeScalar,
3        h: T.OdeScalar,
4        y: T,
5        derivative_function return_derivatives:
6                     (T.OdeScalar, T) -> T
7   ) -> T  {
8        return doEulerStep(t: t, h: h, y: y,
9                  derivative_function: return_derivatives,
10                 add: T.odeAdd, times: T.odeMultiply
11        )
12  }
```

One could argue that we could have started from this point. We chose not to, as we believe breaking the problem down makes it easier to reason about code. Finally, we make any suitable array OdeIntegrable by

```
1   extension Array: OdeAddable where Array.Element: Addable {
2        public static func odeAdd(lhs: Self, rhs: Self) -> Self {
3            return zip(lhs,rhs).map(+)
4        }
```

```
 5  }
 6  extension Array: OdeScalarMutipliable where Array.Element:
        Mutipliable {
 7      public static func odeMultiply(lhs: Array.Element, rhs:
        Array<Element>) -> Array<Element> {
 8          return rhs.map { $0 * lhs }
 9      }
10  }
11  extension Array: OdeIntegrable where
12                        Array.Element: Mutipliable & Addable {}
```

To enable use of our simple harmonic oscillator example, we would simply have to ensure that `Double` conforms to `Addable` and `Mutipliable`. We have, however, already done this in our library and do not need to repeat that process. With this in place, we recover our specific example usage

```
 1  var state = [1.0, 0.0]
 2  for i in 0 ..< 1000 {
 3      let time = Double(i) * dt
 4      state = doEulerStep(t: time, h: dt, y: state,
 5                  derivative_function: sho_derivative)
 6  }
```

Before looking at the Schrödinger equation, let us stick with the harmonic oscillator a little longer. Finding the dynamics in the previous example takes a procedural approach. This is not in-line with the physics view of an harmonic oscillator as an 'object' in its own right. An arguably more elegant approach would be to encapsulate the entire system, parameters, and physics, in a single entity. One such example would be:

```
 1  struct simpleHamonicOscillator {
 2    let omega, zeta: Double // Physical characteristics
 3    var time: Double        // dynamical properties
 4    var state: [Double]
 5    var position: Double { state[0] } // convenience code
 6    var velocity: Double { state[1] }
 7
 8    init(position: Double = 1.0,
 9         velocity: Double = 0.0,
10         omega: Double = 1.0,
11         zeta: Double = 0.0,
12         time: Double = 0.0) {
13      self.zeta = zeta    ; self.time = time
14      self.omega = omega  ; state = [position, velocity]
15    }
16
17    private func equationOfMotion(t: Double,
18                                  y: [Double])
19        -> [Double] {
20        return [ velocity ,
21            -2.0*zeta*omega*velocity - omega*omega*position]
22    }
23    public mutating func evolve(by dt: Double) {
```

```
24      state = doEulerSetp(t: time,
25                          h: dt,
26                          y: state,
27                          derivative_function:
28                                      equationOfMotion)
29      time+=dt
30    }
31  }
```

which tidily encapsulates a representation of an harmonic oscillator.

Exercise 8.31 Add a time-dependent drive to the above code example. Can you identify a way for the user to supply the drive function without having to edit the `struct` directly? Hint: Look up *'Escaping Closures'*. ∎

This simple harmonic oscillator structure can be 'driven' by a few lines of code to obtain the system's dynamics

```
1  var mySHO = simpleHamonicOscillator() // default values
2  for i in 0 ..< 1000 {
3      mySHO.evolve(by: dt)
4  }
```

When it comes to solving the Schrödinger equation, we can follow exactly the same approach. We first ensure that the `StateVector` conforms to `OdeIntegrable`. The first protocol is trivial as

```
1  extension ColumnVector: OdeAddable  {
2    public static func odeAdd(lhs: Self,
3                              rhs: Self) -> Self {
4      return lhs + rhs // already overloaded for this
5    }
6  }
```

is all it takes to satisfy the first protocol needed for `doEulerStep`. For the second protocol we will use a `typealias` to keep track of our type for any real number:

```
1  public typealias RealNumberType = Double
2  extension ColumnVector: OdeScalarMutipliable
3                          where T: Has_DoubleInitialiser {
4    public static func odeMultiply(lhs: RealNumberType,
5                                   rhs: Self) -> Self {
6      return lhs * rhs // already overloaded for this
7    }
8  }
```

hence we have

```
1  extension ColumnVector: OdeIntegrable where T:
      Has_DoubleInitialiser {}
```

and we are finally in a position where we can solve the Schrödinger equation:

```
1  public class QuantumSystem {
2      public var Ψ: StateVector
3      public var minus_i_H: Operator
4      public var time: ℝ
5
6      public init(initialstate: StateVector,
7                  hamiltonian: Operator) {
8          time = 0.0
9          Ψ = initialstate
10         minus_i_H = -C.I * hamiltonian
11     }
12     public func schrodingerEquation(time: ℝ,
13                  ψ: StateVector) -> StateVector {
14         return minus_i_H * ψ
15     }
16     public func evolve(by dt: ℝ) {
17         Ψ = doEulerStep(t: time, h: dt, y: Ψ,
18             derivative_function: schrodingerEquation)
19         time += dt
20     }
21 }
```

where we have indulged ourselves with UTF8 characters and also defined `typealias` as follows (do you like or dislike the use of UTF8 characters?):

```
1  public typealias ℝ = RealNumberType
2  public typealias C = Complex<ℝ>
3  public typealias StateVector = ColumnVector<C>
4  public typealias Operator = MatrixOperator<C>
```

Exercise 8.32   Modify the above `QuantumSystem` code to include a Hamiltonian with both static and time-dependent terms. ∎

Exercise 8.33   Draft the code needed to solve the Liouville–von Neumann equation $\frac{d\hat{\rho}}{dt} = -\frac{i}{\hbar}[\hat{H}, \hat{\rho}]$. ∎

As anyone with experience of Euler step will know, it is in practice an awful integrator because errors rapidly spiral out of control even for relatively simple systems and small step sizes. It is certainly not up to the task of integrating the Jaynes–Cummings model in any reasonable time. For this reason, we need to do better: it will prove sufficient to implement a fourth-order Runge–Kutta method. In the example below, our implementation is a based on example code in *Numerical Recipes in C* [67] but originally and substantially adapted to accommodate our general approach (*Numerical Recipes in C*, at time of printing, has a freely available on-line version). For those wishing to study numerical algorithms well suited to the physical sciences, sources such as this are an excellent starting point. The main reason for using this source is that by using similar notation, the burden of documenting our code and method is reduced. The routine takes the form

```
1  // based on rk4 in numerical.recipes/book/book.html
2  public func doRungeKuttaStep <IndependentVariable:
       Has_IntegerInitialiser & Addable & Dividable, IntegrandType>
       ( SEE EULER ) -> IntegrandType {
3      let dydx = derivs( t, y )
4      let hh = h / IndependentVariable(2)
5      let h6 = h / IndependentVariable(6)
6      let xh = t + hh
7      var yt = add( y, times( hh, dydx ) )
8      var dyt = derivs(xh,yt)
9      yt = add( y, times( hh, dyt ) )
10     var dym = derivs(xh, yt)
11     yt = add( y, times( h, dym ) )
12     dym = add( dym , dyt )
13     dyt = derivs( t + h, yt )
14     let sum1 = add( dydx , dyt )
15     let sum2 = add( dym , dym )
16     return add(y , times( h6, add( sum1, sum2 ) ) )
17 }
```

where we have omitted the function arguments as they are identical to those we used for doEulerStep. In the same way that we wrote a convenience wrapper for doEulerStep for types that are OdeIntegrable, we also write one for doRungeKuttaStep. The only differences are that T.OdeScalar now needs to conform to Addable, Dividable as well as Has_IntegerInitialiser. Finally, as Double already conforms to these protocols, QuantumSystem can be altered to use the Runge–Kutta method simply by changing doEulerStep to doRungeKuttaStep. Completing all these steps now puts us in a position to solve the Schrödinger equation as an initial value problem with sufficient accuracy to reproduce standard textbook and published results.

Exercise 8.34 Below is code generated by ChatGPT to implement a fourth-order Runge–Kutta method using the built-in protocols AdditiveArithmetic and one it assumed to, but which does not seem to, exist: ScalarMultiplication. Compare and contrast this with the one converted from numerical recipes. Is its algorithm correct and what needs to be done for it to compile? How easy would it be to adapt this to our above approach? Should we change our code to use AdditiveArithmetic? How hard would it be to make that change? ∎

```
1  func rk4<T: AdditiveArithmetic & ScalarMultiplication>
2      (y: T, dydt: (T) -> T, t: Double, h: Double) -> T
3  {
4    let k1 = h * dydt(y)
5    let k2 = h * dydt(y + k1/2)
6    let k3 = h * dydt(y + k2/2)
7    let k4 = h * dydt(y + k3)
8    let y_next = y + (k1 + 2 * k2 + 2 * k3 + k4) / 6
9    return y_next
10 }
```

Exercise 8.35 Sketch out an implementation of an adaptive step-size integration routine that is as generic as the above examples and can be equally well used in QuantumSystem. ∎

8.4.6 Plotting the Output

The Single-responsibility principle leads us to advocate separating concerns, and delegate plotting to a third-party program. We could write this ourselves if we wish, or seek existing solutions (which is the path we advocate). One of the most common formats that plotting packages can read is tab-delimited-text. Each line of the file represents a set of data points which are separated by the tab character. The independent variable is usually placed in the first column but need not be. The below code evolves the system, and then outputs the data needed for plotting:

```
1  while jaynseCummingsSystem.time < endTime {
2    jaynseCummingsSystem.doRungeKuttaStep(by: timeIncrement)
3    let inversion = jaynseCummingsSystem.Ψ.
4                      expectationValue(of: sigmaZextended)
5    print(''\(jaynseCummingsSystem.time)\t\(inversion.real)'')
6  }
```

When run from the command-line, file-redirection (./executableName > output .dat) provides a convenient way to get the data into a file. The plot of the output data shown in Figure 8.2 was produced in gnuplot. This is a free graphing utility for all major operating systems and even some less widely used ones. It is very powerful and has been used by us to produce many plots in our published works. It comes very highly recommended.

One of the most interesting behaviours of the Jaynes–Cummings model is its ability to generate Schrödinger states from coherent states. These cat states are best realised in the middle of the collapse region. We could not conclude this chapter without also showing one example of such a cat state, and this is shown as a Wigner function in Figure 8.2. This plot was generated using gnuplot but the code to produce the data added after the week's development time. We include that code as one of the appendices of this chapter.

8.5 Concluding Remarks

We achieve the goal of solving the Schrödinger equation for the Jaynes–Cummings model using domain-specific (i.e. quantum physics) language that meets the requirements implicitly provided by our original example driver routine in Listing 8.1. Our methodology was approximately one of test-driven development, as we implemented the driver a bit at a time (at each stage we would check that the code of our main driver routine did indeed perform correctly). The result of following this process is a library whose structure is well mapped out, and the proof of concept is in place. Being able to code like this makes it easy for us to take real physics research problems and rapidly code for their solution in a way that everyone working on the problem can clearly follow. More importantly, we have also guaranteed that the usage of the library meets our requirement of being able to write expressive,

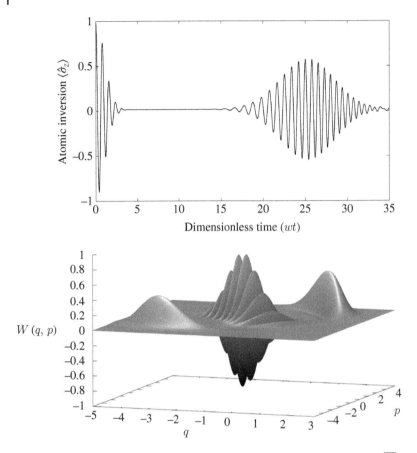

Figure 8.2 Top: Plot of data outputted by our code for the initial state $|\alpha = \sqrt{15}\rangle \otimes |m = \frac{1}{2}\rangle$ showing the expected collapse and revival behaviour of the Jaynes–Cummings model. Bottom: Wigner function of the field mode at $t = 10$ showing that the dynamics of the Jaynes–Cummings system creates a Schrödinger cat state.

self-documenting, code to solve quantum physics problems. While coding the library we also had the time to add basic optimisation using sparse matrix algebra and some better integration routines not documented here (please take a look at the code to learn more).

Substantial things that were not done in the time frame, which could to good effect be included in the library, include: (i) optimising matrix arithmetic for structure, size, serial, or parallel execution (the structure to easily enable this change has been written) (ii) adding an eigen-solver for Hermitian matrices. An eigen-solver was later added to the source code of the library accompanying the book in order to generate Figure 9.4. The approach taken was to use a wrapper to the LAPACK SGEEV library function contained in Swift's accelerate framework which, depending on implementation, may make this code platform depen-dent. Use on non-native platforms might require SGEEV to be called from a C library, or translated into Swift (We note that such code translation is something an AI might be able to do well in the, maybe near, future, but AI applications currently appear to struggle with producing highly optimised code) and (iii) adding some capability for visualisation of quantum states (e.g. by calculating Wigner functions, something we later did add to the code). Such features can be added to the library as and when required, and, by adhering to the Open-closed principle, we can ensure doing so is straightforward and does not affect existing functionality. This way every new feature may be added to satisfy a need, reducing bloat (which helps us reason our code), and facilitating testing (ensuring we do not break new or exiting functionality).

Our hope is that this chapter has laid out an approach to scientific coding that will not only help you write powerful expressive code but also make problem solving a more enjoyable, efficient, and rewarding process.

8.6 Appendix I – Some Useful Calculated Quantities

In some applications where, e.g. matrix exponentials are built in, it can be tempting to use general functions to compute approximations to these quantities. For some specific quantum examples, analytic methods can be used to advantage to both reduce numerical error and improve performance. This appendix contains a few explicit examples of such calculations that illustrate this approach.

8.6.1 Exponentials of the Annihilation and Creation Operators

We often need these in the number (Fock) basis, such as when calculating displacement operators, and sines and cosines of operators that can be expressed in terms of exponentials of the creation and/or annihilation operator. Let us start by considering $\exp \alpha \hat{a}$ where α is some complex number. Its matrix elements are

$$
\langle m \,|\exp \alpha \hat{a}|\, n \rangle = \left\langle m \left| \sum_{k=0}^{\infty} \frac{\alpha^k \hat{a}^k}{k!} \right| n \right\rangle
$$

$$
= \delta_{mn} + \sum_{k=1}^{\infty} \frac{\alpha^k}{k!} \left\langle m \left| \hat{a}^k \right| n \right\rangle
$$

$$
= \delta_{mn} + \sum_{k=1}^{\infty} \frac{\alpha^k}{k!} \left\langle m \left| \sqrt{\frac{n!}{(n-k)!}} \right| n - k \right\rangle
$$

$$
= \delta_{mn} + \sum_{k=1}^{\infty} \frac{\alpha^k}{k!} \sqrt{\frac{n!}{(n-k)!}} \langle m | n - k \rangle
$$

$$
= \delta_{mn} + \sum_{k=1}^{\infty} \frac{\alpha^k}{k!} \sqrt{\frac{n!}{(n-k)!}} \delta_{m,n-k}
$$

$$
= \begin{cases} \frac{\alpha^{(n-m)}}{(n-m)!} \sqrt{\frac{n!}{m!}} & \text{if } m < n, \\ 1 & \text{if } m = n, \\ 0 & \text{if } m > n. \end{cases} \tag{8.3}
$$

As we have previously pointed out, factorials can be a problem to calculate. Instead of computing each matrix element from the above expression, we can instead compute the matrix representing $\exp a\hat{a}$ by noting (i) that the lower diagonal of $\langle m | \exp a\hat{a} | n \rangle$ is zero, (ii) that the diagonal elements are always 1, and (iii) that

$$\langle m | \exp a\hat{a} | n+1 \rangle = \langle m | \exp a\hat{a} | n \rangle \frac{a\sqrt{n+1}}{n-m+1},$$

which leads to this function:

```
1  public func exponentialOfScaledAnnihilationOperator
2              (scaleFactor alpha: ScalarField)
3  -> MatrixOperator {
4    var output = MatrixOperator(in: self)
5    for m in 0 ..< dimension {
6      output[m,m] = ScalarField(real: 1.0)
7      for n in m + 1 ..< dimension {
8        output[m,n] = output[m,n-1] * alpha *
9          ScalarField(Double.sqrt(Double(n))/Double(n-m))
10     }
11   }
12   return output
13 }
```

Now for the counterpart, $\exp a\hat{a}^\dagger$, we have

$$\left\langle m \left| \exp a\hat{a}^\dagger \right| n \right\rangle = \left\langle m \left| \sum_{k=0}^{\infty} \frac{a^k (\hat{a}^\dagger)^k}{k!} \right| n \right\rangle$$

$$= \delta_{mn} + \sum_{k=1}^{\infty} \frac{a^k}{k!} \left\langle m \left| (\hat{a}^\dagger)^k \right| n \right\rangle$$

$$= \begin{cases} \frac{a^{(m-n)}}{(m-n)!} \sqrt{\frac{(n+1)!}{(m+1)!}} & \text{if } m > n, \\ 1 & \text{if } m = n, \\ 0 & \text{if } m < n, \end{cases} \tag{8.4}$$

which is just the transpose of the previous case (so while it may be a bit computationally inefficient, that is how we will calculate the matrix representation of this operator, as we are unlikely to call this function often and it reduces the amount of code that needs to be maintained).

8.6.2 The Pauli Vector: Euler Formula and Other Functions

For two by two matrices there are some further tricks we can make use of. For exponentials we can use the fact that for some vector $\boldsymbol{a}$ written in terms of its magnitude a and unit direction $\boldsymbol{n}$ (so $\boldsymbol{a} = a\boldsymbol{n}$) and for $\sigma = (\hat{\sigma}_x, \hat{\sigma}_y, \hat{\sigma}_z)$,

$$\exp i(\boldsymbol{a} \cdot \sigma) = \cos a\, \hat{\mathbb{1}} + i\, \sin a(\boldsymbol{n} \cdot \sigma). \tag{8.5}$$

This can be very useful, especially because there are many situations in computational problems where evaluating exponentiated Pauli matrices is needed. The following formula extends this result to any function:

$$f(\boldsymbol{a} \cdot \boldsymbol{\sigma}) = \frac{f(a) + f(-a)}{2}\hat{\mathbb{1}} + \frac{f(a) - f(-a)}{2}(\boldsymbol{n} \cdot \boldsymbol{\sigma}). \tag{8.6}$$

It is taken from problem 2.1 in [64] (so we will also leave it as an exercise here). From a computational perspective it makes evaluating many functions of Pauli operators very efficient.

8.6.3 Cosine of the Position Operator

Let us consider the operator

$$\cos(\alpha\hat{q} + \beta), \tag{8.7}$$

where α and β are real numbers. The best direct way to calculate this is to use the same approach as we did for the exponentials of the creation and annihilation operators. However, if we do not need to calculate this operator very often, there is a trick we can use that reduces the amount of code we need to write and therefore reduces the chance of us making an error. We start by noting that

$$\cos(\hat{A}) = \frac{1}{2}\left(\exp\left[i\hat{A}\right] + \exp\left[-i\hat{A}\right]\right). \tag{8.8}$$

All operators commute with themselves, so we know that an expression like this must work for any operator $\hat{A}$. Let us just consider the first term (evaluating the second term will be the same but with a sign charge):

$$\exp\left[i(\alpha\hat{q} + \beta)\right] = \exp\left[i\beta\right]\exp\left[i\alpha\hat{q}\right]$$
$$= \exp\left[i\beta\right]\exp\left[i\alpha\sqrt{\frac{\hbar}{2m\omega}}\left(\hat{a}^{\dagger} + \hat{a}\right)\right]$$
$$= \exp\left[i\beta\right]\exp\left[i\alpha'\left(\hat{a}^{\dagger} + \hat{a}\right)\right], \tag{8.9}$$

where $\alpha' = \alpha\sqrt{\frac{\hbar}{2m\omega}}$. Now, we know that as $\hat{a}^{\dagger}$ and $\hat{a}$ commute with their commutator, the Zassenhaus formula reduces to

$$\exp\left[\alpha''\left(\hat{a}^{\dagger} + \hat{a}\right)\right] = \exp\left[\alpha''\hat{a}^{\dagger}\right]\exp\left[\alpha''\hat{a}\right]\exp\left[-\frac{\alpha''^{2}}{2}\left[\hat{a}^{\dagger}, \hat{a}\right]\right]$$
$$= \exp\left[\frac{\alpha''^{2}}{2}\right]\exp\left[\alpha''\hat{a}^{\dagger}\right]\exp\left[\alpha''\hat{a}\right], \tag{8.10}$$

where $\alpha'' = i\alpha'$ and we have used $\left[\hat{a}, \hat{a}^{\dagger}\right] = 1$. This has reduced the problem to one that we have already solved, simplifying our coding problem – but is it the right thing to do?

Exercise 8.36 This approach introduces at least two types of numerical errors that would not be present if we had calculated the matrix elements of the cosine directly (as we did for the exponential). What are they and under what circumstances might they be an issue? ∎

Exercise 8.37 Implement code that calculates the cosine matrix elements directly. ∎

Exercise 8.38 Based on a critical comparison of the two methods, decide which approach you would use. ∎

8.7 Appendix II – Wigner Function Code

The code used to generate the Wigner function in Figure 8.2 used the original code base to evolve the system until $t = 10$ and then used the following code to generate the Wigner function:

```
1   let start_q = -5.0, end_q = 3.0
2   let start_p = -4.0, end_p = 5.0
3   let Nq = 800, Np = 800
4
5   let dq = (end_q - start_q)/Double(Nq)
6   let dp = (end_p - start_p)/Double(Np)
7
8   var outputText = ""
9
10  for qCount in 0 ..< Nq {
11    let q = start_q + Double(qCount) * dq
12    for pCount in 0 ..< Np {
13      let p = start_p + Double(pCount) * dp
14      let twoalpha = 2.0 * Complex(real: q, imag: p)
15      let abs2alphaSqrd = twoalpha.modulus *
16                         twoalpha.modulus
17      let pre_factor = exp( -0.5 * abs2alphaSqrd )
18  //journals.aps.org/pra/abstract/10.1103/PhysRevA.50.4488
19  // D(2a)Pi = D(a) Pi D^dag(a)
20      let displacementOperator =
21        Complex(real: pre_factor) *
22        fieldSpace.exponentialOfScaledCreationOperator
23        (scaleFactor: twoalpha) *
24        fieldSpace.exponentialOfScaledAnnihilationOperator
25        (scaleFactor: -(twoalpha.conjugate) )
26
27      let Pi = displacementOperator *
28                 fieldSpace.parityOperator
29
30      let PiExtended = totalSpace.tensorProduct
31        (of: Pi, with: spinSpace.identityOperator)
32
33      let W = PiExtended.expectationValue
34                 (of: jaynseCummingTDSE.Psi )
35      outputText += "\(q)\t\(p)\t\(W.real)\n"
36    }
37    outputText += "\n"
38  }
39  // code to output to file
```

Note that this is not a very efficient algorithm for computing the Wigner function - but it is one that works using the code from our library that was developed in one week (in order to fit the code on the page some sacrifice of good formatting was made).

9

Open Quantum Systems

"We have reversed the usual classical notion that the independent 'elementary parts' of the world are the fundamental reality, and that the various systems are merely particular contingent forms and arrangements of these parts. Rather, we say that inseparable quantum interconnectedness of the whole universe is the fundamental reality, and that relatively independent behaving parts are merely particular and contingent forms within this whole."

David Bohm

9.1 Introduction

9.1.1 Context

In this section, we will introduce the study of quantum mechanical systems subject to external environments. While the physics reasoning of open systems is mostly intuitive, an appropriate treatment of this topic requires some quite advanced mathematical methods. Despite the complexity, this topic is so important that we felt its inclusion was necessary. We have tried to follow Einstein's advice to *'Make everything as simple as possible, but not simpler'*. Our guide to the 'as simple as possible' requirement was to include sufficient detail to give enough clarity for the reader to understand where key steps come from and why they are made. With some work, it is intended that the reader will be able to follow the logic sufficiently to apply the methods here to derive master equations for any system they study. We also have included critical discussion of the method to enable an understanding of the scope of validity of those derived models due to the approximations that need to be made as part of the process.

9.1.2 Background

Up until this point, we have explored the behaviour of quantum systems that are isolated from any external influence beyond a time-dependent potential. Such objects that are isolated from their environment are termed closed systems. In this situation, the state of

Quantum Mechanics: From Analytical Mechanics to Quantum Mechanics, Simulation, Foundations, and Engineering,
First Edition. Mark Julian Everitt, Kieran Niels Bjergstrom and Stephen Neil Alexander Duffus.
© 2024 John Wiley & Sons Ltd. Published 2024 by John Wiley & Sons Ltd.
Companion website: www.wiley.com/go/everitt/quantum

the system can be represented as a vector and is termed a pure state. An object that is not isolated from its environment is termed an open system. Our only consideration of the effect of external factors was when we discussed the measurement postulates of quantum mechanics and the collapse of the wave-function. In reality, a system is never truly isolated and will always feel some external influence, as there is always some energy transfer between a system and its surroundings. Interacting quantum systems rapidly become inseparable, sharing quantum information and becoming interconnected in ways that classical systems do not.

Any quantum object or set of objects that we are interested in will be a subsystem of some larger system, and thus be open. In such cases, only the total system plus environment can be represented as a pure state. The state of the subsystem excluding the environment is not a valid idea. As we will see, we can approximate the information about the object by using density operators and tracing out the environmental degrees of freedom. This will yield a density matrix for the system-of-interest, but it will almost always be a mixed state (a statistical mixture of pure state density operators of the subsystem). Pure states are said to be coherent and the coherence or purity of a system is, broadly speaking, how unmixed it is (coherent states of the harmonic oscillator are also coherent pure states; the context should make clear the possible confusion of meaning). Decoherence is often used as an umbrella term for different dissipative channels within an open system, such as the dissipation of energy, and the dephasing of perturbed quantum states. When coherence is lost due to interaction with the environment, we term this effect environmental decoherence. In order to appreciate the physics of quantum objects, we need to have a theoretical framework capable of accommodating their interaction with their environments.

From a practical perspective, an understanding of open quantum systems will also be crucial for engineering quantum technologies. This is because the coherence of the quantum state is fragile in the presence of an environment, and the nature of these technologies requires the maintenance of some sufficient level of coherence. For example, quantum computers need to be able to entangle multiple sets of quantum bits (two-level quantum systems also known as qubits), and perform operations on them. It must do this in sufficiently short time so that the environmental decoherence (or imperfect operations) does not destroy the coherence needed to realise the computation. To produce quantum technologies, it will be necessary to prepare and maintain quantum states to a desired level of coherence, enabling them to perform some given task relating to sensing, communication, or computation. Much work is being undertaken to prolong the amount of time in which quantum states used in these technologies can remain in a coherent superposition, so that increasingly complex tasks can be performed. The grand challenge of quantum computing is maximising the number of qubits that can be entangled and coherently manipulated. There have been a number of attempts to prolong the coherence times of such technologies, including efforts to engineer suitable environments capable of maintaining and even enhancing the quantum coherence of the system-of-interest. In order to address these challenges, we need a strong understanding of open quantum systems and how a quantum system may behave when subjected to external influences. The main purpose of this chapter is to introduce this subject in sufficient detail for you to be able to understand the nature, magnitude, and subtleties of this challenge.

9.1.3 System Plus Bath Models

Our approach to understanding systems interacting with the environment will be to break the system plus environment into its component parts in terms of their energies. The environment can take different forms, such as a simple particle, an arrangement of particles, or a field. Whatever the form, it is simply the other part of the whole with which our system-of-interest interacts. Later in this chapter, we shall see that some environments can be modelled as a *reservoir* with infinite degrees of freedom. In the examples we consider, we will do this using an infinite set of harmonic oscillators (complex environments may need to be represented by multiple reservoirs with different properties). As this is an introduction to the subject, we will limit our discussion to consider only those reservoirs that remain at thermal equilibrium. Such reservoirs are known as *baths*. The total Hamiltonian can then be decomposed into the Hamiltonians of the system, the bath, and the interactions between them:

$$H = H_S + H_I + H_B, \tag{9.1}$$

where H_S is the Hamiltonian of the system, H_B is the Hamiltonian of the bath, and H_I is the interaction between the two.

The classical dynamics of the entire system, including the bath, can be described by the Liouville equation

$$\frac{\partial \rho}{\partial t} = \{H, \rho\},$$

where ρ is the probability density function. Quantum mechanically, the dynamics will be given by the Liouville–von Neumann equation,

$$\frac{d\hat{\rho}}{dt} = \frac{1}{i\hbar}[\hat{H}, \hat{\rho}], \tag{9.2}$$

where $\hat{\rho}$ is the density operator and $\hat{H}$ an Hamiltonian operator version of H. It is understood that the problem of solving for these total system plus environment models will be impossible. We will instead seek to understand the behaviour of marginal probability density functions of reduced density operators for the system, integrating out all environmental degrees of freedom.

By stochastic we mean any random process with a well-defined probability distribution; A master equation describes the probabilistic/statistical evolution of a system, and a trajectory or realisation is a single possible outcome of the process described by the master equation.

In this chapter, we will explore the process of deriving equations that approximate the dynamical evolution of the system's evolution in the presence of a bath. By analogy with Brownian motion, we might expect the effects of the environment on the system-of-interest to manifest themselves as stochastic processes. We could then ask what happens to the system statistically (to obtain a dynamical master equation) or to an individual realisation or trajectory. We will highlight the kinds of approximations that are typically made and discuss the utility and limits of applicability of such modelling.

9.2 Classical Brownian Motion

In order to continue our theme of highlighting similarities between quantum and classical physics, we will begin our technical discussion with open classical systems. We will consider a system subject to environmental effects that take the form of a random stochastic force (Brownian motion) as well as a dynamical friction force. As a concrete example, consider a simple pendulum that is subject to damping in the form of air resistance and other frictional forces. It is common practice to account for these frictional forces by introducing a term to the equation of motion that is dependent on the system's velocity. It seems sensible to assume that the frictional forces increase with the velocity of the pendulum. The equation of motion for such a pendulum is usually assumed to be given by

$$\frac{d^2\theta}{dt^2} + k\frac{d\theta}{dt} + \omega^2\theta = 0, \tag{9.3}$$

where $\theta(t)$ denotes the angular displacement from the pendulum's equilibrium point, k quantifies the damping within the system, and ω is the natural frequency of the pendulum. This model is considered appropriate because the net effect of the air resistance due to quasi-random interactions with surrounding air particles can on this large scale be averaged and quantified by the constant k. The same cannot be done if we reduce the mass of the pendulum to a level where collisions with particulates surrounding the pendulum are each significant. At this scale, the exact nature of the air resistance becomes important. Here, we must use more care when considering the impact of a system's environment on its dynamics.

Instead of a pendulum, let us now consider the simpler case of a free particle subject to random stochastic forces from its environment. A grain of pollen surrounded by vibrating water molecules is the classic example due to Brown. For this reason, the motion of a free particle interacting with its environment in this way is known as Classical Brownian Motion. Let us represent by $\xi(t)$ the stochastic forces felt by the particle subject to Brownian motion. The equation of motion is then

$$m\frac{d^2q}{dt^2} = -m\gamma\frac{dq}{dt} + \xi(t), \tag{9.4}$$

where q is the particle's position, m its mass, γ is the damping rate that quantifies the frictional force $-m\gamma\dot{q}$. Stochastic differential equations like this, containing deterministic and stochastic contributions, are referred to as *Langevin equations*. Their integration requires care and the use of Stratonovich or Itô calculus, as normal Riemann integration cannot deal with the random processes. The study of stochastics is advanced and beyond the scope of this book. Because of this, our discussion of Brownian motion will be superficial, presenting only the results we need for our later quantum analysis.

To help develop some intuition for Langevin equations, we will consider the case of a particle in a fluid. As there is no preference with regard to the direction in which the fluctuation term acts, our first observation is that $\langle\xi(t)\rangle = 0$. It is also intuitive that the fluctuation terms should be related in some way to the damping rate of the fluid, as the fluctuations will be more significant in less viscous fluids. To make this connection, we start by noting without proof that the damping rate in this example is

$$\gamma = \frac{6\pi\eta r_0}{m}, \tag{9.5}$$

where r_0 is the radius of the particle, while η is the viscosity of the fluid within which the particle moves. Now, for a fluid, the probability distribution for the particle's position $\phi(q, t)$ satisfies the diffusion equation

$$\frac{\partial \phi(q, t)}{\partial t} = \nabla \cdot \left[D \nabla \phi(q, t) \right],$$ (9.6)

where D is the diffusivity and quantifies the rate of movement of particles within the fluid from a high concentration gradient to a low concentration gradient, essentially the drag in diffusion. For our example, diffusivity is given by

$$D = \frac{k_B T}{m \gamma},$$ (9.7)

where k_B is Boltzmann's constant and T is the temperature of the fluid surrounding the particle. It is often assumed that $\xi(t)$ changes much faster than the timescale over which the particle's position changes. It is possible to show that this diffusivity is consistent with $\xi(t)$ being a Gaussian probability distribution that satisfies

$$\langle \xi(t) \rangle = 0$$ (9.8)
$$\langle \xi(t)\xi(t') \rangle = 2mk_B T\gamma \delta(t - t'),$$ (9.9)

where t' is some time later than t and $\delta(t - t')$ is the Dirac delta function. Eq. (9.8) is a direct result of the timescales over which these stochastic forces act, with the average force equal to zero. Eq. (9.9) relates the stochastic forces felt by the particle at different points in time. The factor $2mK_B T\gamma$ gives the maximum force felt by the particle, while the Dirac delta distribution tells us that the forces acting at different points in time are not related and are therefore memory-less. In the next section, we shall see that functions of the form of Eq. (9.9) play a key role in the determining systems' dynamics by characterising the interactions between a system and its environment. For a one-dimensional particle moving in a potential $V(q)$, the equation of motion becomes

$$\ddot{q}(t) + \gamma \dot{q}(t') + V'(q) = \frac{\xi(t)}{m}.$$ (9.10)

This generalises our first example of the damped pendulum and also includes an external stochastic driving force due to the fluid within which the particle interacts. Generalisation to three dimensions and allowing for a damping that changes with time yields

$$m\ddot{q} + m \int_{-\infty}^{t} \gamma(t - t')\dot{q}(t') \, dt' + \nabla V(q) = \xi(t),$$ (9.11)

which is the form of the Langevin equation we will need to use as a reference point in our analysis of open quantum systems in the next section[1].

---

1 The integral comes from the fact that we have replaced the damping rate of $\gamma \delta(t - t')$ with $\gamma(t - t')$, that is to say that it was always there. Using $f(x) = \int_{-\infty}^{\infty} \delta(x - x')f(x') \, dx'$, we see that

$$\gamma \dot{q}(t) = \int_{-\infty}^{t} \gamma \delta(t - t')\dot{q}(t') \, dt'.$$

9.2.1 Brownian Motion from Hamiltonian Mechanics

In this section, we will derive a quantum version of Eq. (9.11) from a Hamiltonian perspective. The reason we do this is to once more highlight the similarities of quantum and classical physics wherever possible. To avoid repetition, our argument will be quantum mechanical and made in the Heisenberg picture. Before going into details, we would like to emphasise that there are almost equivalent arguments to be made to derive the classical Eq. (9.11) as those presented here. This is because of (i) the similarity in form of

$$\frac{\mathrm{d}A}{\mathrm{d}t} = \{A, \mathcal{H}\} + \frac{\partial A}{\partial t} \tag{9.12}$$

and

$$\frac{\mathrm{d}\hat{A}_H}{\mathrm{d}t} = \frac{1}{i\hbar}[\hat{A}_H \hat{H}_H] + \frac{\partial \hat{A}_H}{\partial t}, \tag{9.13}$$

and (ii) much of the argument that follows relies on the algebra of the commutators. We already know that this algebra is shared with the Poisson bracket. The main differences in the argument is that quantum systems compose as tensors rather than Cartesian products, and that operators do not commute. As such, the fact that we will recover a quantum Langevin equation that is very similar in form to its classical counterpart is not as surprising as it may at first appear (recall that Ehrenfest's theorem is equivalent to taking the expectation value of the Heisenberg equations of motion).

> In this section only, we will adopt the following notation for working in the Heisenberg picture: for compactness we will not use the subscript H; rather, by $\hat{A}(t)$ we mean $\hat{A}_H(t)$ and by $\hat{A}$ we mean $\hat{A}_H(0)$, which is the same as the Schrödinger picture $\hat{A}$.

Our starting point is the Hamiltonian of the whole system plus environment. As an origin of fluctuation and loss, we will model the environment as an infinite set of oscillators at thermal equilibrium (a bath). The total Hamiltonian is then

$$\hat{H} = \hat{H}_S + \hat{H}_B + \hat{H}_I, \tag{9.14}$$

where $\hat{H}_S$ is the Hamiltonian of the system, $\hat{H}_B$ is the Hamiltonian of the bath and $\hat{H}_I$ is the interaction between the two. The bath comprises oscillators with mass m_n and natural frequency ω_n, and its Hamiltonian is

$$\hat{H}_B = \sum_n \frac{\hat{p}_n^2}{2m_n} + \frac{1}{2} m_n \omega_n^2 \hat{q}_n^2, \tag{9.15}$$

where $\hat{p}_n$ and $\hat{q}_n$ are the momentum and position of the nth oscillator within the bath. The Hamiltonian for the particle, or the system, is given by:

$$\hat{H}_S = \frac{\hat{p}^2}{2m} + V(\hat{q}) - \hat{q}^2 \sum_n \frac{\lambda_n^2}{2m_n \omega_n^2}, \tag{9.16}$$

where $\hat{p}$ and $\hat{q}$ are the momentum and position operators of the particle, with m and ω denoting the particle's mass and natural frequency, respectively. The first two terms are the usual kinetic and potential energies. The final term represents a renormalisation of the system's potential; this term is necessary to ensure that the potential minimum of the system remains unchanged after the system couples to the bath (we will not see the full explanation of this until page 300). Here, λ_n is the coupling constant for each oscillator, or bath mode, to the system. As Brownian motion is the manifestation of fluctuation effects of the environment on the system, the interaction Hamiltonian will couple the position of the particle to the position of each bath mode. It is given by

$$\hat{H}_I = -\hat{q} \sum_n \lambda_n \hat{q}_n = -\hat{q} \otimes \hat{B}, \tag{9.17}$$

where we have defined the bath operator

$$\hat{B} = \sum_n \lambda_n \hat{q}_n, \tag{9.18}$$

which combines all the couplings and positions of the bath modes into one net effect. We will see later that the bath operator plays a key role in determining how and to what degree information is lost by the system.

Using the Heisenberg equations of motion, Eq. (9.13), on each of the operators $\hat{p}, \hat{q}, \hat{p}_n$, and $\hat{q}_n$ gives

$$\frac{d\hat{q}(t)}{dt} = \frac{\hat{p}(t)}{m} \tag{9.19}$$

$$\frac{d\hat{p}(t)}{dt} = -V'(\hat{q}) + \sum_n \lambda_n \hat{q}_n(t) + \hat{q}(t) \sum_n \frac{\lambda_n^2}{m_n \omega_n^2} \tag{9.20}$$

$$\frac{d\hat{q}_n(t)}{dt} = \frac{\hat{p}_n(t)}{m_n} \tag{9.21}$$

$$\frac{d\hat{p}_n(t)}{dt} = \lambda_n \hat{q}(t) - m_n \omega_n \hat{q}_n(t), \tag{9.22}$$

where we have used the commutation relation $[\hat{q}_i, \hat{p}_j] = \delta_{ij} i \hbar$. Differentiating Eq. (9.21) with respect to time and substituting into Eq. (9.22), we find

$$\frac{d^2\hat{q}_n(t)}{dt^2} + \omega_n^2 \hat{q}_n(t) = -\frac{\lambda_n}{m_n} \hat{q}_n(t). \tag{9.23}$$

We note that the operators $\hat{q}_n(t)$ and $\hat{q}(t)$ describe the positions of different subsystems and are therefore different objects. Eq. (9.23) is therefore an inhomogeneous second-order differential equation with the solution

$$\hat{q}_n(t) = \hat{q}_n(0) \cos(\omega_n t) + \frac{\hat{p}_n(0)}{m_n \omega_n} \sin(\omega_n t) + \frac{\lambda_n}{m_n \omega_n} \int_0^t \sin(\omega_n(t - t'))\hat{q}(t') \, dt'. \tag{9.24}$$

Integrating by parts gives the familiar form [29],

$$\hat{q}_n(t) = \hat{q}_n(0) \cos(\omega_n t) + \frac{\hat{p}_n(0)}{m_n \omega_n} \sin(\omega_n t) + \frac{\lambda_n}{m_n \omega_n^2} (\hat{q}(t) - \hat{q}(0) \cos(\omega_n t))$$

$$- \frac{\lambda_n}{m_n \omega_n^2} \int_0^t \cos(\omega_n(t - t'))\dot{\hat{q}}(t') \, dt', \tag{9.25}$$

where $\dot{\hat{q}}(t')$ is the velocity of the particle. This can be substituted into the equation that results when differentiating Eq. (9.19) and substituting the result into Eq. (9.20)

$$m\frac{d^2\hat{q}(t)}{dt^2} + V'(\hat{q}) - \sum_n \lambda_n \hat{q}_n(t) - \hat{q}(t) \sum_n \frac{\lambda_n^2}{m_n \omega_n^2} = 0 \tag{9.26}$$

to give

$$m\frac{d^2\hat{q}(t)}{dt^2} + m\int_0^t \sum_n \frac{\lambda_n^2}{m_n m\omega_n^2}\cos(\omega_n(t-t'))\dot{\hat{q}}(t')\,dt' + V'(\hat{q})$$

$$= -\hat{q}(t)\sum_n m_n\omega_n^2 + \sum_n \lambda_n\left(\hat{q}_n(0)\cos(\omega_n t) + \frac{\hat{p}_n(0)}{m_n\omega_n}\sin(\omega_n t)\right) \tag{9.27}$$

$$+ \frac{\lambda_n}{m_n\omega_n^2}(\hat{q}(t) - \hat{q}(0)\cos(\omega_n t))\Bigg).$$

The first term in this equation gives the resultant force on the particle, while the third term gives the force due to the potential within which the particle sits. The second term bears resemblance to the frictional force seen in Eq. (9.11). Comparing coefficients between these integral terms, we can define the quantum equivalent to the damping rate,

$$\gamma(t-t') = \sum_n \frac{\lambda_n^2}{m_n m\omega_n^2}\cos(\omega_n(t-t')), \tag{9.28}$$

where we use

$$\gamma(t) = \sum_n \frac{\lambda_n^2}{m_n m\omega_n^2}\cos(\omega_n t) \tag{9.29}$$

to define the stochastic environmental forces,

$$\hat{\xi}(t) = \sum_n \lambda_n\left(\hat{q}_n(0)\cos(\omega_n t) + \frac{\hat{p}_n(0)}{m_n\omega_n}\sin(\omega_n t)\right) - m\hat{q}(0)\gamma(t) \tag{9.30}$$

to obtain the Quantum Langevin equation

$$m\ddot{\hat{q}}(t) + m\int_0^t \gamma(t-t')\dot{\hat{q}}(t')\,dt' + V'(\hat{q}) = \hat{\xi}(t). \tag{9.31}$$

This gives us a 'suitable' model for a Quantum particle subject to Brownian motion since it satisfies the correspondence principle: that classical dynamics are reproduced in the limit of large quantum numbers.

9.3 Master Equations

Derived in the Heisenberg picture, the Langevin equation may be used to describe the dynamics of observables and their expectation values. We may, however, desire a Schrödinger picture view and be concerned about the dynamics of the quantum state itself. In this case we take an alternative approach, as discussed in the introduction, and seek to model the dynamics of the reduced density operator of the system by a so-called master equation. We begin our analysis with the Liouville–von Neumann equation,

$$\frac{d\hat{\rho}}{dt} = -\frac{i}{\hbar}[\hat{H}, \hat{\rho}(t)], \tag{9.32}$$

where $\hat{\rho}$ is the density operator of the whole system and the environment and $\hat{H}$ is the total Hamiltonian. For the density operator to be physically meaningful, it must remain Hermitian, positive semi-definite, and normalised so that $\text{Tr}\{\hat{\rho}(t)\} = 1$ (pure states also satisfy $\text{Tr}\{\hat{\rho}^2(t)\} = 1$). We will use these conditions as constraints to inform us on the suitability of candidate models for open systems.

Assuming that it will not be possible to solve the Liouville–von Neumann equation for the system as a whole, we will seek to find approximations to $\hat{\rho}_S(t) = \text{Tr}_E\,\hat{\rho}(t)$ – the dynamics of the density operator of the system, tracing out all environmental components. As we are interested in the change of the state of the system-of-interest due to its interactions with the environment, it makes sense to begin our analysis by moving to the interaction picture. We can remove some of the complexity in this dynamical equation by moving into the interaction picture. In this way, we can analyse those dynamics of the state of the system-of-interest that are due only to the contributions from the interaction of the system with its environment. Assuming that the system and bath Hamiltonians are time-independent, the interaction picture Hamiltonian and density operator are

$$\tilde{H}_I(t) = \exp\left(\frac{it(\hat{H}_S + \hat{H}_B)}{\hbar}\right)\hat{H}_I \exp\left(\frac{-it(\hat{H}_S + \hat{H}_B)}{\hbar}\right) \tag{9.33}$$

$$\tilde{\rho}(t) = \exp\left(\frac{it(\hat{H}_S + \hat{H}_B)}{\hbar}\right)\hat{\rho}(t) \exp\left(\frac{-it(\hat{H}_S + \hat{H}_B)}{\hbar}\right). \tag{9.34}$$

We have replaced the hat with a tilde to denote operators in the interaction picture. As the system and bath Hamiltonians are time-independent, the Liouville–von Neumann equation becomes

$$\frac{d\tilde{\rho}(t)}{dt} = -\frac{i}{\hbar}\left[\tilde{H}_I, \tilde{\rho}(t)\right]. \tag{9.35}$$

The interaction picture will, by avoiding explicit mention of system and bath Hamiltonians, allow us to develop our discussion in a way that is general to any open system.

> If either the system or bath Hamiltonian is time-dependent, the following analysis should be repeated with either the time-dependent components included in $\hat{H}_I$ (which is not unreasonable, as an external drive can be considered an interaction) or the interaction picture should be defined using the evolution operator, $\hat{U}(t)$ for $\hat{H}_S(t) + \hat{H}_B(t)$. As $\dot{\tilde{\rho}} = \dot{\hat{U}}\hat{\rho}\hat{U}^\dagger + \hat{U}\dot{\hat{\rho}}\hat{U}^\dagger + \hat{U}\hat{\rho}\dot{\hat{U}}^\dagger = -\frac{i}{\hbar}\left[\tilde{H}_I, \tilde{\rho}(t)\right] + \dot{\hat{U}}\hat{\rho}\hat{U}^\dagger + \hat{U}\hat{\rho}\dot{\hat{U}}^\dagger$, there will be modifications to the following analysis in either approach. It is almost always the case that this is not considered when master equation methods are applied to time-dependent open systems.

The next step of the analysis is to use the mathematical trick of integrating Eq. (9.35) to obtain

$$\int_0^t ds\frac{d\tilde{\rho}(s)}{ds} = -\frac{i}{\hbar}\int_0^t ds[\tilde{H}_I(s), \tilde{\rho}(s)]$$

$$\Rightarrow \tilde{\rho}(t) = \tilde{\rho}(0) + \frac{i}{\hbar}\int_0^t ds[\tilde{H}_I(s), \tilde{\rho}(s)] \tag{9.36}$$

(where s is a dummy time variable) and then substituting the result back into the right-hand side of Eq. (9.35) to yield

$$\frac{d\tilde{\rho}(t)}{dt} = -\frac{i}{\hbar}[\tilde{H}_I(t), \tilde{\rho}(0)] - \frac{1}{\hbar^2}\int_0^t ds[\tilde{H}_I(t), [\tilde{H}_I(s), \tilde{\rho}(s)]], \tag{9.37}$$

where s has the limits of $s \in (0, t)$. Here, $s = 0$ marks the moment of interaction between system and environment and $s = t$ marks the evolution time for the system. It is possible, and sometimes desirable, to repeat this process of integration and back-substitution. However, this is not a level of detail we wish to consider on a first treatment of the subject. So far, our analysis is exact but not particularly useful. In order to make progress, we will have to make a number of assumptions and approximations. The derivation of the Redfield formalism contains many of the approximations common to most master equation derivations and is covered next.

9.3.1 The Redfield Master Equations

The important thing about the equation above is that it will enable us to make some useful approximations to the dynamics of $\hat{\rho}_S(t) = \mathrm{Tr}_E \hat{\rho}(t)$. For this reason, we note that we can always write the full density operator in the following form:

$$\tilde{\rho}(t) = \tilde{\rho}_S(t) \otimes \tilde{\rho}_B(t) + \tilde{\rho}_I(t), \qquad (9.38)$$

where the first term is a tensor product of the reduced density operators for the system and environment, while the $\tilde{\rho}_I$ contains all of the correlations between the two. It is important to note that while $\tilde{\rho}_S$ and $\tilde{\rho}_B$ are density operators (Hermitian, positive semi-definite, and normalised) this is not true of $\tilde{\rho}_I$. We may assert that at $t = 0$ the system-of-interest is in a pure state and so $\tilde{\rho}_I(0) = 0$ (this could be arranged by making a projective measurement of the system).

We will now argue that if the bath is sufficiently large, $\tilde{\rho}_I(t) \approx 0$ for all time. A classical analogy to this approximation is the introduction of an ice cube into a warm, stirred, bath: at the instant the ice cube enters the bath, the bath's average temperature will decrease; however, as it is big and at equilibrium this will only result in a small, rapid, change to the bath's state, it will quickly relax into a new equilibrium state. The ice cube, however, will suffer a constant and significant change to its state until it is fully melted, an equilibrium state it takes much longer to reach. It will do this without developing any significant correlations with the bath. The vastness of the environment, which comprises many components, each with a range of frequencies, means it has a spectral width that is far greater than that of the system; since the spectral width is a measure of how much energy is transferred per unit time from bath to system, it can be quantified using a decay rate, γ_B. The rate at which energy is transferred from bath to system is inversely proportional to the time it takes the bath to reach steady state; thus, we can assert the relationship

$$\gamma_B \sim \frac{1}{\tau_B}. \qquad (9.39)$$

Specifics of how to determine this rate can be found in more advanced treatments such as [41, 61]. We denote the relaxation rates of the system and the bath by τ_S and τ_B, respectively. Our argument means that we can assume $\tau_B \ll \tau_S$. This assumption means that any significant correlation between system and bath is only very short-lived compared to the overall dynamics of the system. For this reason, we assert the approximation that the elements of the matrix $\tilde{\rho}_I$ are approximately zero for most of the system's evolution. As a consequence, we can approximate the state of the total system by

$$\tilde{\rho}(t) \approx \tilde{\rho}_S(t) \otimes \tilde{\rho}_B(t). \qquad (9.40)$$

This means that the dynamics generated by master equations reliant on this approximation may not produce meaningful predictions in the short time limit. Potentially, one might liken this to the transient response of a classical systems (which, e.g. is often removed when studying chaotic dynamics). Care should therefore be taken when interpreting the dynamics of open quantum systems early in the system's evolution (and, when we come to the last section of this chapter, also their unravellings).

To make progress in simplifying Eq. (9.37), we shall take this approximation one step further. Returning to our analogy of the ice cube in a bathtub, we can assume that for the majority of the system's evolution, the bath remains in thermal equilibrium. This means that the environmental density operator remains approximately constant in time and we may assume that the total density operator is

$$\tilde{\rho}(t) \approx \tilde{\rho}_S(t) \otimes \hat{\rho}_B. \tag{9.41}$$

Substituting our approximated density operator into Eq. (9.37) we obtain

$$\frac{d\tilde{\rho}_S(t) \otimes \tilde{\rho}_B}{dt} = -\frac{i}{\hbar}[\tilde{H}_I(t), \tilde{\rho}_S(0) \otimes \hat{\rho}_B] - \frac{1}{\hbar^2} \int_0^t ds[\tilde{H}_I(t), [\tilde{H}_I(s), \tilde{\rho}_S(s) \otimes \hat{\rho}_B]]. \tag{9.42}$$

Tracing out the environment gives

$$\frac{d\tilde{\rho}_S(t)}{dt} = -\frac{i}{\hbar}\mathrm{Tr}_B[\tilde{H}_I(t), \tilde{\rho}_S(0) \otimes \hat{\rho}_B]$$

$$- \frac{1}{\hbar^2} \int_0^t ds\, \mathrm{Tr}_B[\tilde{H}_I(t), [\tilde{H}_I(s), \tilde{\rho}_S(s) \otimes \hat{\rho}_B]]. \tag{9.43}$$

Recall that in the derivation of the quantum Langevin equation, we coupled the bath to the system by position in order to manifest environment fluctuation effects. In this treatment we can be somewhat more general but we will still assume a linear interaction between system and bath operators $\hat{X}_S$ and $\hat{Y}_B$,

$$\hat{H}_I = \hat{X}_S \otimes \hat{Y}_B.$$

As we saw in quantum Brownian motion, we can assume that the average value of the bath operator $\hat{Y}_B$ is zero. That is to say,

$$\mathrm{Tr}_B\left\{\hat{Y}_B\right\} = 0. \tag{9.44}$$

If the interaction between a bath and its environment is not linear and $\mathrm{Tr}_B\left\{\hat{Y}_B\right\} \neq 0$ then the derivation of the master equation needs to take this into account at this stage of the analysis. In the interaction picture, the interaction Hamiltonian carries the form

$$\tilde{H}_I(t) = \tilde{X}_S(t) \otimes \tilde{Y}_B(t),$$

but it can be shown quite simply, using the cyclic nature of trace and the unitary nature of the evolution operator, that if Eq. (9.44) holds, then

$$\mathrm{Tr}_B\left\{\tilde{Y}_B\right\} = \mathrm{Tr}_B\left\{\hat{Y}_B\right\} = 0. \tag{9.45}$$

Thus, the first term on the right-hand side of Eq. (9.43),

$$
\begin{aligned}
\mathrm{Tr}_B\left[\tilde{H}_I(t)\tilde{\rho}_S(0)\otimes\hat{\rho}_B\right] &= \mathrm{Tr}_B\left[\tilde{X}_S(t)\otimes\tilde{Y}_B(t)\tilde{\rho}_S(0)\otimes\hat{\rho}_B\right] \\
&= \tilde{X}_S(t)\tilde{\rho}_S(0)\mathrm{Tr}_B\left\{\tilde{Y}_B(t)\hat{\rho}_B\right\} \\
&\quad - \tilde{\rho}_S(0)\tilde{X}_S(t)\mathrm{Tr}_B\left\{\hat{\rho}_B\tilde{Y}_B(t)\right\} \\
&= \left[\tilde{X}_S(t)\tilde{\rho}(0)\right]\mathrm{Tr}_B\left\{\hat{\rho}_B\tilde{Y}_B(t)\right\} \\
&= 0.
\end{aligned}
$$

Exercise 9.1 Show that $(\hat{A},\hat{B})\overset{\text{def}}{=}\mathrm{Tr}\left[\hat{A}^\dagger\hat{B}\right]$ is an inner product (this one is named after Frobenius). Use the Cauchy–Schwartz inequality to conclude $\mathrm{Tr}_B\left[\hat{\rho}_B\tilde{Y}_B(t)\right]=0$. *Hint:* Use the inner product to define the Frobenius norm and then conclude that $0\le\left(\mathrm{Tr}\left[\hat{A}^\dagger\hat{B}\right]\right)^2\le$ $\mathrm{Tr}\left[\hat{A}^2\right]\mathrm{Tr}\left[\hat{B}^2\right]\le\left(\mathrm{Tr}\left[\hat{A}\right]\right)^2\left(\mathrm{Tr}\left[\hat{B}\right]\right)^2.$ ∎

We therefore find that Eq. (9.43) can be approximated by

$$
\frac{d\tilde{\rho}_S(t)}{dt} = -\frac{1}{\hbar^2}\int_0^t ds\,\mathrm{Tr}_B[\tilde{H}_I(t),[\tilde{H}_I(s),\tilde{\rho}_S(s)\otimes\hat{\rho}_B]]. \tag{9.46}
$$

It is at this point that there is a choice of the level of approximation and concomitant derivations of master equations. A common, analytically solvable regime is one where it is assumed that the system is memory-less, and its dynamics depend only on its current state. This is known as the Markovian approximation. Non-Markovian master equations is a subtle and difficult topic, and for this reason we shall restrict our discussion to the Markovian case. Hence, we constrain the reduced density matrix of the system such that

$$
\tilde{\rho}_S(s) \to \tilde{\rho}_S(t). \tag{9.47}
$$

Due to our dismissal of short-term dynamics around the time of immediate interaction, we only consider a time after the initial correlations have ceased to impact on the system. We can therefore write the evolution as the difference between the total evolution time t and this initial interaction time interval, known as the correlation time τ:

$$
s = t - \tau. \tag{9.48}
$$

Since

$$
ds = -d\tau,\ \lim_{s\to t}\tau = 0,\ \text{and}\ \lim_{s\to 0}\tau = \infty, \tag{9.49}
$$

this changes the limits of integration and gives

$$
\frac{d\tilde{\rho}_S(t)}{dt} = -\frac{1}{\hbar^2}\int_0^\infty d\tau\,\mathrm{Tr}_B[\tilde{H}_I(t),[\tilde{H}_I(t-\tau),\tilde{\rho}_S(t)\otimes\hat{\rho}_B]]. \tag{9.50}
$$

It is important to note that Eq. (9.50) is only suitable when we consider long time-scales in comparison to the correlation time τ, that is to say that we must extend our upper limit to ∞ in order to be able to neglect the dynamics associated to the initial correlations.

The last step is to transform back into the Schrödinger picture so that Eq. (9.50) takes the form:

$$\frac{d}{dt}\hat{U}^{\dagger}(t)\hat{\rho}_S(t)\hat{U}(t) = \tag{9.51}$$

$$-\frac{1}{\hbar^2}\int_0^{\infty} d\tau\, \text{Tr}_B\left\{\left[\hat{U}^{\dagger}(t)\hat{H}_I\hat{U}(t), \left[\hat{U}^{\dagger}(t)\tilde{H}_I(-\tau)\hat{U}(t), \hat{U}^{\dagger}(t)\tilde{\rho}_S(t)\hat{U}(t) \otimes \hat{\rho}_B\right]\right]\right\}, \tag{9.52}$$

where we recall the evolution operator $\hat{U}(t) = \exp(-i[\hat{H}_S + \hat{H}_B]t/\hbar)$, and we note once more that the bath remains in equilibrium and remains unchanged. We can reduce Eq. (9.50) using a property of nested commutators, namely that

$$\left[e^{\hat{\mu}}\hat{A}e^{-\hat{\mu}}, \left[e^{\hat{\mu}}\hat{B}e^{-\hat{\mu}}, \left[e^{\hat{\mu}}\hat{C}e^{-\hat{\mu}}\right]\right]\right] = e^{\hat{\mu}}[\hat{A}, [\hat{B}, \hat{C}]]e^{-\hat{\mu}}, \tag{9.53}$$

where the general property of unitary operators, $\hat{U}\hat{U}^{\dagger} = \hat{U}^{\dagger}\hat{U} = \hat{I}$, has been used. The above calculations show that Eq. (9.50) can be reduced to:

$$\frac{d}{dt}\hat{U}^{\dagger}(t)\hat{\rho}_S(t)\hat{U}(t) =$$
$$-\frac{1}{\hbar^2}\int_0^{\infty} d\tau\, \text{Tr}_B\left\{\hat{U}^{\dagger}(t)\left[\hat{H}_I, \left[\hat{H}_I(-\tau), \hat{\rho}_S(t) \otimes \hat{\rho}_B\right]\right]\hat{U}(t)\right\}. \tag{9.54}$$

Expansion of the left-hand side of this equation yields:

$$\frac{d}{dt}\hat{U}^{\dagger}\hat{\rho}_S\hat{U} = \frac{d\hat{U}^{\dagger}}{dt}\hat{\rho}_S\hat{U} + \hat{U}^{\dagger}\frac{d\hat{\rho}_S}{dt}\hat{U} + \hat{U}^{\dagger}\hat{\rho}_S\frac{d\hat{U}}{dt}$$

$$= \left(\frac{i}{\hbar}[\hat{H}_S + \hat{H}_B]\right)\hat{U}^{\dagger}\hat{\rho}_S\hat{U} + \hat{U}^{\dagger}\frac{d\hat{\rho}_S}{dt}\hat{U} + \hat{U}^{\dagger}\hat{\rho}_S\left(-\frac{i}{\hbar}[\hat{H}_S + \hat{H}_B]\right)\hat{U}$$

$$= \hat{U}^{\dagger}\frac{d\hat{\rho}_S}{dt}\hat{U} + \frac{i}{\hbar}\hat{U}^{\dagger}[\hat{H}_S + \hat{H}_B\hat{\rho}_S]\hat{U}$$

$$= \hat{U}^{\dagger}\frac{d\hat{\rho}_S}{dt}\hat{U} + \frac{i}{\hbar}\hat{U}^{\dagger}[\hat{H}_S\hat{\rho}_S]\hat{U}, \tag{9.55}$$

where we have used the fact that the evolution operator is a function of $\hat{H}_S + \hat{H}_B$ and so must commute with this operator. We have also made use of $\hat{\rho}_S$ commuting with $\hat{H}_B$, leaving the commutator of $[\hat{H}_S, \hat{\rho}_S]$ last term. Substitution back into Eq. (9.54) brings us to

$$\hat{U}^{\dagger}(t)\frac{d\hat{\rho}_S(t)}{dt}\hat{U}(t) = -\frac{i}{\hbar}\hat{U}^{\dagger}(t)[\hat{H}_S, \hat{\rho}_S(t)]\hat{U}(t)$$
$$-\frac{1}{\hbar^2}\int_0^{\infty} d\tau\, \text{Tr}_B\left\{\hat{U}^{\dagger}(t)\left[\hat{H}_I, \left[\hat{H}_I(-\tau), \hat{\rho}_S(t) \otimes \hat{\rho}_B\right]\right]\hat{U}(t)\right\}. \tag{9.56}$$

Pre-multiplying by $\hat{U}(t)$ and post-multiplying by $\hat{U}^{\dagger}(t)$, and using the fact that the evolution operator is unitary, finally yields our first key result:

The Redfield master equation, which satisfies the Born (weak coupling) and Markov (no, or short, memory) approximations, is:

$$\frac{d\hat{\rho}_S(t)}{dt} = -\frac{i}{\hbar}[\hat{H}_S\hat{\rho}_S(t)] - \frac{1}{\hbar^2}\int_0^{\infty} d\tau\, \text{Tr}_B\left[\hat{H}_I\left[\hat{H}_I(-\tau)\hat{\rho}_S(t) \otimes \hat{\rho}_B\right]\right]. \tag{9.57}$$

The first term on the right-hand side describes the free evolution of the system, as it would evolve if left alone without any environmental interference, while the second term is known as the dissipator and describes the environmental influence on the system's dynamics. It is often denoted by $\mathcal{K}\hat{\rho}_S(t)$.

The Redfield equation is defined in terms of the system's Hamiltonian, density operator, interaction Hamiltonian, and initial bath density operator (albeit with the last two traced out). In the absence of more information and nice properties that might make this a solvable equation, it is rather challenging to make use of. In the next section, we will use the Redfield equation to derive the Caldeira–Leggett Model. This takes the form of a master equation entirely in the terms of the system.

Solutions to both the Redfield equation and the Caldeira–Leggett Model can fail to preserve the required properties for the density matrix: positivity, hermiticity, and normalisation. However, a type of Markovian master equations, which guarantees that these constraints are met, does exist: the set of master equations of Lindblad type. We will show how one may obtain a Lindblad equation from the Redfield equation, using an example system, and discuss the strengths and limitations in doing so.

9.3.2 The Caldeira–Leggett Model

We now combine the ideas of the previous two sections. We will use the fact that representing the source of random fluctuations by a bath of harmonic oscillators (Eqs. 9.15 and 9.17) led to a quantum Langevin equation with an appropriate classical limit. This result justifies the use of such an environment as the bath to be used in the Redfield equation. Our ambition in this section is to use the Redfield equation to obtain a master equation that approximates the dynamics associated to a quantum system, subject to a Brownian motion that contains only terms for the system itself. In our derivation of the quantum Langevin equation, we defined a linear coupling between the position of the particle and each bath mode. The Redfield equation, however, consists of a correlation-time-dependent interaction, which can be thought of as a delayed interaction by time τ. We can define this τ-dependent interaction through

$$\hat{H}_I(-\tau) = -\hat{q}(-\tau)\sum_n \lambda_n \hat{q}_n(-\tau) = -\hat{q}(-\tau)\hat{B}(-\tau), \tag{9.58}$$

where once again $\hat{B}$ is the bath operator. The dissipator of the Redfield equation can then be written as

$$\mathcal{K}[\hat{\rho}_S] = -\frac{1}{\hbar^2}\int_0^\infty d\tau\, \mathrm{Tr}_B\left[\hat{q}\hat{B},\left[\hat{q}(-\tau)\hat{B}(-\tau),\hat{\rho}_S(t)\otimes\hat{\rho}_B\right]\right], \tag{9.59}$$

where $\hat{q} = \hat{q}(0)$ and $\hat{B} = \hat{B}(0)$ represent the position and bath operators at time $t = 0$ (i.e. the Schrödinger picture operators). The next step is to use the partial trace to separate the operators in each of the system and bath space. We first must expand out the integrand in Eq. (9.59) to obtain

$$\begin{aligned}\mathrm{Tr}_B[\hat{q}\hat{B},[\hat{q}(-\tau)\hat{B}(-\tau),\hat{\rho}_S(t)\otimes\hat{\rho}_B]] = \mathrm{Tr}_B\,\big\{&\hat{q}\hat{B}\hat{q}(-\tau)\hat{B}(-\tau)\hat{\rho}_S(t)\otimes\hat{\rho}_B \\ &- \hat{q}\hat{B}\hat{\rho}_S(t)\otimes\hat{\rho}_B\hat{q}(-\tau)\hat{B}(-\tau) \\ &- \hat{q}(-\tau)\hat{B}(-\tau)\hat{\rho}_S(t)\otimes\hat{\rho}_B\hat{q}\hat{B} \\ &+ \hat{\rho}_S(t)\otimes\hat{\rho}_B\hat{q}(-\tau)\hat{B}(-\tau)\hat{q}\hat{B}\big\}.\end{aligned} \tag{9.60}$$

The partial trace then allows us to separate the system and bath operators, giving

$$
\hat{q}\hat{q}(-\tau)\hat{\rho}_S(t)\, \mathrm{Tr}_B\left\{\hat{B}\hat{B}(-\tau)\hat{\rho}_B\right\} - \hat{q}\hat{\rho}_S(t)\hat{q}(-\tau)\, \mathrm{Tr}_B\left\{\hat{B}\hat{\rho}_B\hat{B}(-\tau)\right\}
$$
$$
-\hat{q}(-\tau)\hat{\rho}_S(t)\hat{q}\, \mathrm{Tr}_B\left\{\hat{B}(-\tau)\hat{\rho}_B\hat{B}\right\} + \hat{\rho}_S(t)\hat{q}(-\tau)\hat{q}\, \mathrm{Tr}_B\left\{\hat{\rho}_B\hat{B}(-\tau)\hat{B}\right\}. \tag{9.61}
$$

We note again that the partial trace is cyclic and use $\left\langle\hat{A}\right\rangle_B = \mathrm{Tr}_B\{\hat{A}\hat{\rho}_B\}$ to write the bath terms as expectation values over the bath. Doing so, gives a dissipator of

$$
\mathcal{K}[\hat{\rho}_S] = -\frac{1}{\hbar^2}\int_0^\infty d\tau\left((\hat{q}\hat{q}(-\tau)\hat{\rho}_S(t) - \hat{q}(-\tau)\hat{q}\hat{\rho}_S(t))\left\langle\hat{B}\hat{B}(-\tau)\right\rangle_B\right.
$$
$$
\left. - (\hat{q}\hat{\rho}_S(t)\hat{q}(-\tau) - \hat{\rho}_S(t)\hat{q}(-\tau)\hat{q})\left\langle\hat{B}(-\tau)\hat{B}\right\rangle_B\right). \tag{9.62}
$$

The dissipator can be further simplified by writing it in commutator form; we break the terms up in such a way that we obtain nested commutators in the system space, multiplied by an anti-commutator (denoted by $\{\cdot,\cdot\}_+$) in the bath space and vice versa. This gives the dissipator

$$
\mathcal{K}[\hat{\rho}_S] = -\frac{1}{\hbar^2}\int_0^\infty d\tau\frac{1}{2}\left(\left\langle[\hat{B},\hat{B}(-\tau)]\right\rangle_B\left[\hat{q},\{\hat{q}(-\tau),\hat{\rho}_S(t)\}_+\right]\right.
$$
$$
\left. + \left\langle\{\hat{B},\hat{B}(-\tau)\}_+\right\rangle_B\left[\hat{q},[\hat{q}(-\tau),\hat{\rho}_S(t)]\right]\right). \tag{9.63}
$$

The terms acting in the bath space in Eq. (9.63) are sums and differences of two important quantities, the bath correlation functions. A correlation function quantifies how strongly related two properties are. The correlation between two operators can be calculated through

$$
C_{AB} = \left\langle\hat{A}\hat{B}\right\rangle - \left\langle\hat{A}\right\rangle\left\langle\hat{B}\right\rangle. \tag{9.64}
$$

If the expectation value of one of these operators is zero, as it is for the bath operator, the correlation function is given by

$$
C_{AB} = \left\langle\hat{A}\hat{B}\right\rangle. \tag{9.65}
$$

Note: we can more clearly understand how correlated two things are if we define

$$
\varrho = C_{AB}/\sigma_A\sigma_B
$$

where $\sigma_A = \sqrt{\left\langle\hat{A}^2\right\rangle - \left\langle\hat{A}\right\rangle^2}$ and $\sigma_B = \sqrt{\left\langle\hat{B}^2\right\rangle - \left\langle\hat{B}\right\rangle^2}$. Then ϱ can take on values from $+1$ to -1. If $\varrho = 1$, there is perfect correlation between the two distributions, and perfect anti-correlation if $\varrho = -1$.

A Time-Correlation Function describes for how long a given property of a system persists until it is averaged out by (microscopic) motions of said system,

$$
\left\langle\hat{A}(t)\hat{A}(t')\right\rangle = \left\langle\hat{A}(t-t')\hat{A}(0)\right\rangle.
$$

We can now see that the bath terms in Eq. (9.63) describe the correlation between bath operators before and after the initial interaction with the system. We therefore define the bath correlation function $C_1(\tau)$, and its Hermitian conjugate, $C_2(\tau)$ through

$$
C_1(\tau) = \left\langle\hat{B}\hat{B}(-\tau)\right\rangle_B \tag{9.66}
$$
$$
C_2(\tau) = \left\langle\hat{B}(-\tau)\hat{B}\right\rangle_B. \tag{9.67}
$$

To find out what these functions look like, we will make use of the harmonic oscillator nature of the bath and expand these expressions using the annihilation and creation operators of each oscillator in the bath at both times $t = 0$ and $t = -\tau$. Doing so for $C_1(\tau)$ gives

$$C_1(\tau) = \sum_n \lambda_n^2 \langle \hat{q}_n \hat{q}_n(-\tau) \rangle_B$$

$$= \sum_n \frac{\hbar \lambda_n^2}{2m_n \omega_n} \left\langle \left(\hat{a}_n + \hat{a}_n^\dagger \right) \left(\hat{a}_n(-\tau) + \hat{a}_n^\dagger(-\tau) \right) \right\rangle_B. \tag{9.68}$$

Note that as we assume the bath modes are uncorrelated, $\langle \hat{q}_n \hat{q}_m(\tau) \rangle = 0$ when $n \neq m$. The time-dependent terms in Eq. (9.64) can be written as a series of expanded exponential terms using the Hadamard lemma

$$e^{\hat{X}} \hat{Y} e^{-\hat{X}} = \hat{Y} + [\hat{X}, \hat{Y}] + \frac{1}{2!} [\hat{X}, [\hat{X}, \hat{Y}]] + \frac{1}{3!} [\hat{X}, [\hat{X}, [\hat{X}, \hat{Y}]]] + \cdots; \tag{9.69}$$

applying this to the τ-dependent ladder operators in the interaction picture yields

$$\hat{a}_n^\dagger(-\tau) = e^{-i\omega_n \tau \hat{n}} \hat{a}_n^\dagger e^{i\omega_n \tau \hat{n}} = \hat{a}_n^\dagger e^{i\omega_n \tau}$$

$$\hat{a}_n(-\tau) = e^{-i\omega_n \tau \hat{n}} \hat{a}_n e^{i\omega_n \tau \hat{n}} = \hat{a}_n e^{-i\omega_n \tau}. \tag{9.70}$$

By substitution into Eq. (9.68) we find the correlation function can be written as

$$C_1(\tau) = \sum_n \frac{\hbar \lambda_n^2}{2m_n \omega_n} \left\langle \hat{a}_n^2 e^{-i\omega_n \tau} + \hat{a}_n \hat{a}_n^\dagger e^{i\omega_n \tau} + \hat{a}_n^\dagger \hat{a}_n e^{-i\omega_n \tau} + \hat{a}_n^{\dagger 2} e^{i\omega_n \tau} \right\rangle_B. \tag{9.71}$$

It is at this point that we inspect the nature of each term in Eq. (9.71). We recall that the bath remains in thermal equilibrium and therefore exists as a statistical mixture of states. The action of any ladder operator on the bath density matrix will move the non-zero matrix elements from the diagonal, returning a trace, and therefore an expectation value, of zero. The only actions that will preserve the trace of the bath density matrix are those from 'single-photon events', where a ladder operator is paired with its conjugate. This means that the only terms we consider in Eq. (9.71) are the middle two. These two terms can be written in terms of the bath number operators, which tell us the number of excitations in the bath. Using

$$\hat{a}_n^\dagger \hat{a}_n = \hat{n}_n$$

$$\hat{a}_n \hat{a}_n^\dagger = \hat{n}_n + \hat{\mathbb{1}}$$

as well as Euler's formula to expand and recombine our terms, we obtain

$$C_1(\tau) = \sum_n \frac{\hbar \lambda_n^2}{2m_n \omega_n} \left\langle \left(2\hat{n}_n + \hat{\mathbb{1}} \right) \right\rangle_B (\cos(\omega_n \tau) - i \sin(\omega_n \tau)). \tag{9.72}$$

Recall that the bath is modelled as an infinite set of harmonic oscillators. The average number of excitations for each oscillator, $\langle \hat{n}_n \rangle$, at a given temperature can be found by use of Bose–Einstein distribution

$$N = \langle \hat{n}_n \rangle_B = \frac{1}{\exp\left(\dfrac{\hbar \omega_n}{k_B T} \right) - 1}. \tag{9.73}$$

After some rearrangement we can write this as

$$\left\langle 2\hat{n}_n + \hat{1} \right\rangle_B = \frac{1 + \exp\left(\dfrac{\hbar\omega_n}{k_B T}\right)}{\exp\left(\dfrac{\hbar\omega_n}{k_B T}\right) - 1} \tag{9.74}$$

$$= \frac{\exp\left(-\dfrac{\hbar\omega_n}{2k_B T}\right) + \exp\left(\dfrac{\hbar\omega_n}{2k_B T}\right)}{\exp\left(\dfrac{\hbar\omega_n}{2k_B T}\right) - \exp\left(-\dfrac{\hbar\omega_n}{2k_B T}\right)} \tag{9.75}$$

$$= \coth\left(\frac{\hbar\omega_n}{2k_B T}\right). \tag{9.76}$$

The first correlation function can finally be written

$$C_1(\tau) = \sum_n \frac{\hbar\lambda_n^2}{2m_n\omega_n}\left(\coth\left(\frac{\hbar\omega_n}{2k_B T}\right)\cos(\omega_n\tau) - \mathrm{i}\sin(\omega_n\tau)\right) \tag{9.77}$$

with its Hermitian conjugate given by

$$C_2(\tau) = \sum_n \frac{\hbar\lambda_n^2}{2m_n\omega_n}\left(\coth\left(\frac{\hbar\omega_n}{2k_B T}\right)\cos(\omega_n\tau) + \mathrm{i}\sin(\omega_n\tau)\right). \tag{9.78}$$

Combining Eq. (9.77) and Eq. (9.78) in linear combinations yields two new quantities

$$C_1(\tau) - C_2(\tau) = \left\langle [\hat{B}, \hat{B}(-\tau)] \right\rangle_B$$

$$= -2\,\mathrm{i}\,\hbar \sum_n \frac{\lambda_n^2}{2m_n\omega_n}\sin(\omega_n\tau)$$

$$= -\mathrm{i}\,D(\tau)$$

$$C_1(\tau) + C_2(\tau) = \left\langle \{\hat{B}, \hat{B}(-\tau)\}_+ \right\rangle_B \tag{9.79}$$

$$= 2\hbar \sum_n \frac{\lambda_n^2}{2m_n\omega_n}\coth\left(\frac{\hbar\omega_n}{2k_B T}\right)\cos(\omega_n\tau)$$

$$= D_1(-\tau).$$

The first quantity in Eq. (9.79), $D(\tau)$, is known as the dissipation kernel while the second quantity, $D_1(-\tau)$, is termed the noise kernel, each named after the nature of their influence on the system. The fact that the first term is purely imaginary makes us think of loss (as imaginary numbers can be used to model loss in classical systems). Thus, we can think of each of the terms (or modes) in the first power series as representing loss at different frequencies at a rate that depends on the coupling to the system. The noise kernel term is real and represents the net effect of lots of fluctuations, again at different frequencies, and at a rate that depends on the coupling λ_n to the system. Each kernel describes the distribution of power transferred between different frequency components of the bath and the system. For this reason, we can understand that these are quantities that we will be able to relate to the spectral density of the bath. We will return to this point very shortly, but before we do so, we will finish arranging the dissipator into the form we shall need later. Substitution of the kernels $D(\tau)$ and $D_1(\tau)$ into the dissipator gives

$$\mathcal{K}[\hat{\rho}_S] = \frac{1}{\hbar^2} \int_0^\infty d\tau \left(\frac{i}{2} D(\tau)[\hat{q}, \{\hat{q}(-\tau), \hat{\rho}_S(t)\}] - \frac{1}{2} D_1(-\tau)[\hat{q}, [\hat{q}(-\tau), \hat{\rho}_S(t)]] \right). \quad (9.80)$$

The next step in obtaining the master equation is evaluating the integrals in the dissipator; this is done by rewriting the position of the system at time τ

$$\hat{q}(-\tau) = e^{-i\hat{H}_s\tau/\hbar} \hat{q} e^{i\hat{H}_s\tau/\hbar}. \quad (9.81)$$

Using the Hadamard lemma, Eq. (9.69), we can truncate to the first order in τ to obtain a position of

$$\hat{q}(-\tau) \approx \hat{q} - \frac{\hat{p}}{m}\tau, \quad (9.82)$$

where $\hat{q} = \hat{q}(0)$ and $\hat{p} = \hat{p}(0)$ are the position and momentum operators at $t = 0$. This finally gives us a dissipator of

$$\mathcal{K}[\hat{\rho}_S] = \frac{i\,[\hat{q}^2, \hat{\rho}]}{2\hbar^2} \int_0^\infty d\tau\, D(\tau) - \frac{i\,[\hat{q}, \{\hat{p}, \hat{\rho}\}]}{2m\hbar^2} \int_0^\infty d\tau\tau D(\tau)$$
$$- \frac{[\hat{q}, [\hat{q}, \hat{\rho}]]}{2\hbar^2} \int_0^\infty d\tau D_1(\tau) + \frac{[\hat{q}, [\hat{p}, \hat{\rho}]]}{2m\hbar^2} \int_0^\infty d\tau\tau D_1(\tau). \quad (9.83)$$

In order to evaluate the integrals, we define the dissipation and noise kernels in terms of another physical property of the bath – the *spectral density*.

9.3.2.1 Spectral Density

We have already alluded to the possible connection between the noise and dissipation kernels and the spectral density of the environment. In this section, we will make that connection more explicit. The bath contains very many oscillators, each contributing power at its own frequency. Recall that we are considering the case of a Brownian motion at the moment, so we will assume that the frequencies of the environmental fluctuations are higher than the natural frequencies of the system it is weakly coupled to. With so many small but rapid degrees of freedom of the environment, we expect the system to be affected by their net effect. Hence, we can think about the average behaviour, and the power it can transfer to the system, at each of the frequency modes of which it is comprised. For any signal, the spectral density refers to the average energy distribution per unit time as a function of frequency. Let us start by clarifying what we specifically mean by the spectral density of a bath as a whole. To do this, we start with the interaction Hamiltonian of the system and the bath,

$$\hat{H}_I = \hat{q} \sum_n \lambda_n \hat{q}_n. \quad (9.84)$$

The coupling constant provides us with the strength of the coupling between the system and each mode of the bath. Although the coupling constant refers to each mode, we require a property that represents the bath as a whole, allowing us to describe the dissipation and noise in our bath through the respective kernels we have already discussed. To reiterate, the spectral density gives the amount of energy transferred between system and bath per unit time, per frequency (or bath mode); it is a frequency distribution telling us how much each mode impacts on the system. Since frequency and time are (classically) conjugate variables, the spectral density and correlation function are considered Fourier transform pairs of one another. For most bath correlation functions, as in Eq. (9.77), one can approximate the function as a series of exponentials [69],

$$C(\tau) \approx \sum_n^\infty c_n e^{i\,\omega_n \tau} \tag{9.85}$$

which generally allows the spectral density to be described by

$$J(\omega) = \sum_n \frac{\lambda_n^2}{2m_n\omega_n} \delta(\omega - \omega_n), \tag{9.86}$$

where, as a reminder, λ_n is the coupling between each bath mode and the system, while m_n and ω_n are the respective mass and frequency of each mode. Note that each mode can be considered an impulse that carries the dimensions of spectral irradiance (energy per second per frequency). When we consider a bath with very large numbers of degrees of freedom, instead of a set of impulses, we can consider the spectral density to converge to a smooth function. This is termed the effective spectral density. An illustrative example of a spectral and effective spectral density is shown in Figure 9.1.

Returning to Eq. (9.79), we see that it is possible to write the dissipation and noise kernels in terms of the spectral density through

$$D(\tau) = 2\hbar \int_0^\infty d\omega\, J(\omega) \sin(\omega\tau)$$

$$D_1(\tau) = 2\hbar \int_0^\infty d\omega\, J(\omega) \coth\left(\frac{\hbar\omega}{2k_BT}\right) \cos(\omega\tau). \tag{9.87}$$

The analysis can be greatly simplified if we replace a specific spectral density with its corresponding effective spectral density. A commonly used effective spectral density has a linear

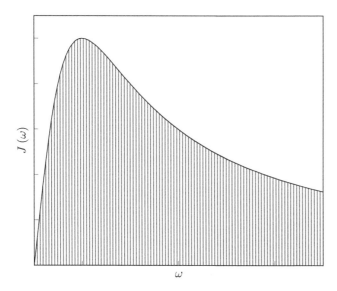

Figure 9.1 Diagram showing the difference between the spectral density (dark grey) and the effective spectral density (thick black). The spectral density is a collection of impulses, each with a weight depending on the coupling and mass of each oscillator in the bath.

dependence on frequency at low frequencies, otherwise known as Ohmic behaviour, and cuts out at higher frequencies. An example of such a function is given by

$$J(\omega) \approx \frac{2m\gamma\omega\Omega^2}{\pi(\Omega^2 + \omega^2)}, \tag{9.88}$$

where m is the mass of the system, and γ is the damping rate which describes the rate at which the system loses information to its surroundings. The quantity Ω is known as the cut-off frequency and it accounts for the renormalisation of the system's potential due to high-frequency modes - some of the energy associated to these higher frequencies is already accounted for in the system potential and therefore needs to be reduced in the spectral density of the bath to avoid double counting. The function that creates this cut-off in this example is the term

$$\frac{\Omega^2}{\Omega^2 + \omega^2}. \tag{9.89}$$

This is known as a Lorentz–Drude cut-off, but it is important to note that there are a number of other cut-off functions available that renormalise the bath in a way that suits the system and bath that is being considered.

Note that there exist many other types of spectral densities. The Lorentz–Drude spectral density is simply an extension of the Ohmic spectral density $J_0(\omega) = \frac{2m\gamma\omega}{\pi}$, whilst other cut-off functions can also be used. An alternative is the exponential cut-off, which is largely interchangeable with the Lorentz–Drude but can sometimes be easier to manipulate. The comparisons between spectral densities and their applicability will be discussed later when we scrutinise the approximations introduced in this chapter.

Making use of the spectral density above allows us to rewrite, and finally evaluate, the integrals in Eq. (9.83). Note that in doing this we make use of the fact that

$$\int_0^\infty \sin(\omega\tau)\, d\tau = \lim_{a \to 0} \int_0^\infty e^{-a\tau} \sin(\omega\tau)\, d\tau = \frac{1}{\omega} \tag{9.90}$$

to obtain

$$\frac{\hat{q}^2}{2\hbar} \int_0^\infty D(\tau)\, d\tau = \frac{\hat{q}^2}{2\hbar} 2\hbar \int_0^\infty d\omega \frac{J(\omega)}{\omega} = \hat{q}^2 \sum_n \frac{\lambda^2}{2m_n \omega_n^2}, \tag{9.91}$$

when this term is moved into the Hamiltonian it cancels with the Lamb shift term we introduced in Eq. (9.16) on page 286. The evaluation of the remaining integrals in Eq. (9.83) is covered widely in the literature [8] and requires Cauchy's residue theorem and the assumption that the cut-off frequency Ω is significantly larger than the frequency range considered. As this is not a course on integral calculus, we will not include those details here. We discuss types and limits of cut-offs, and the implications of other assumptions and approximations, later in the chapter. Finally, we arrive at the *Caldeira–Leggett equation* [10]:

$$\frac{d}{dt}\hat{\rho}_S(t) = -\frac{i}{\hbar}\left[\hat{H}_S, \hat{\rho}_S(t)\right] - \frac{i\gamma}{\hbar}\left[\hat{q}, \{\hat{p}, \hat{\rho}(t)\}\right] - \frac{2m\gamma k_B T}{\hbar^2}\left[\hat{q}, \left[\hat{q}, \hat{\rho}_S(t)\right]\right]. \tag{9.92}$$

The first term on the right-hand side describes the free evolution of the system as if there were no environmental influence, while the second and third terms account for the environment through dissipation and environmental noise, respectively.

The Caldeira–Leggett equation achieves the objective of obtaining a master equation in terms of system variables. As already noted, its solution is not guaranteed to meet the requirements of a density operator for all time. In the next section, we derive master equations (of so-called Lindblad form) whose solutions are guaranteed to be admissible as density operators.

9.3.3 Master Equations of Lindblad Form

In the previous section, we obtained the Caldeira–Leggett master equation that is used to approximate the quantum Brownian motion of a quantum particle subject to environmental influence. The issue with this equation, as we will see in this section, is that it does not guarantee physical dynamics. We will therefore seek to recast it into a form that gives us physicality, the Lindblad form. In order to understand what we mean by 'Lindblad form', we first retrace our steps to the general evolution of an open quantum system.

Let us consider the density matrix at time $t = 0$ that exists in a product state between system and environmental density matrices, that is to say

$$\rho(0) = \hat{\rho}_S(0) \otimes \hat{\rho}_B. \tag{9.93}$$

We have seen that we can obtain the initial reduced density matrix simply by tracing out the environment,

$$\hat{\rho}_S(0) = \mathrm{Tr}_B\{\hat{\rho}_S(0) \otimes \hat{\rho}_B\}. \tag{9.94}$$

We can also obtain the reduced matrix at some other point in time by applying the evolution operator to the initial density matrix and then tracing out the environment at a later point in time,

$$\hat{\rho}_S(t) = \mathrm{Tr}_B\{\hat{U}(t)(\hat{\rho}_S(0) \otimes \hat{\rho}_B)\hat{U}^\dagger(t)\}. \tag{9.95}$$

Note that since we assume an environment in equilibrium, the time evolution is only considered to be active in the open system's subspace.

Our challenge of modelling only the system's time evolution comes in creating an operation that can take us from the initial reduced density matrix and give us the density matrix at a later point in time without having to consider the system and environment as a whole,

$$\hat{\rho}_S(t) = V(t)\hat{\rho}_S(0). \tag{9.96}$$

We call $V(t)$ a dynamical map. It takes the non-unitary nature of environmental influence into account when evolving the system through time. In order for such an operation to be considered a dynamical map, it must be completely positive and trace preserving. The operations must also apply such that

$$V(t_1)V(t_2) = V(t_1 + t_2) \tag{9.97}$$

which shows that obtaining the state of the system at different points in time simply requires the continuous action of these dynamical maps. The aforementioned combination of dynamical maps is known as a *dynamical semigroup* [50].

Given a quantum dynamical semigroup (a continuous combination of dynamical maps), it can be shown that there exists a linear map known as the generator, $\mathcal{L}$, of the semigroup. The generator allows us to represent the semigroup in exponential form:

$$V(t) = \exp\{\mathcal{L}t\}. \tag{9.98}$$

Note here that we have defined $V(t)$ as a semigroup where $t = t_1 + t_2 + \cdots$.

Writing the reduced density matrix of the system $\hat{\rho}_S(t)$ in terms of $\mathcal{L}$ and differentiating with respect to time gives us the differential equation

$$\frac{d}{dt}\hat{\rho}_S(t) = \mathcal{L}\hat{\rho}_S(t) \tag{9.99}$$

This is known as the Markovian master equation, where $\mathcal{L}$ is the Liouvillian super-operator.

Exercise 9.2 Using the definition of $\hat{\rho}_S(t)$ in Eq. (9.96), show that the we obtain the differential equation given above. ∎

In order for the generator to appropriately describe the dynamical semigroup, the generator must take the form

$$\mathcal{L}\hat{\rho}_S(t) = -\frac{i}{\hbar}\left[\hat{H}, \hat{\rho}_S(t)\right] + \sum_j \hat{L}_j \hat{\rho}_S(t)\hat{L}_j^\dagger - \frac{1}{2}\left\{\hat{L}_j^\dagger \hat{L}_j, \hat{\rho}_S(t)\right\}_+, \tag{9.100}$$

where once again the first term on the right-hand side describes the unitary free evolution of the system, while the second is the dissipator. The operators $\hat{L}$ are known as *Lindblad operators* and give a physical representation of the environmental influence on the evolution of the open system. It is important to note that the Hamiltonian described in the above equation is not solely the system Hamiltonian but may include environmental terms that alter the free evolution of the system, giving an *effective Hamiltonian*; this will be discussed in more detail later.

In short, to be able to appropriately model the evolution of an open quantum system, one must be able to construct a master equation of Lindblad type. This will now be the focus of this section, where we will attempt to transform the Caldeira–Leggett equation into a Lindblad master equation.

As discussed previously, it is our goal to represent the quantum Brownian particle with an equation of Lindblad type. A Lindblad operator can be written as a linear combination of Hermitian operators, with complex coefficients. As an example, with the Caldeira–Leggett equation in mind, we will consider a linear combination of two Hermitian operators $\hat{C}$ and $\hat{D}$ with complex coefficients α and β, respectively,

$$\hat{L} = \alpha\hat{C} + \beta\hat{D}. \tag{9.101}$$

Substitution of our chosen Lindblad into the master equation Eq. (9.100) gives

$$\mathcal{L}\hat{\rho}_S(t) = -\frac{i}{\hbar}\left[\hat{H}, \hat{\rho}_S(t)\right] \tag{9.102}$$

$$+ (\alpha\hat{C} + \beta\hat{D})\hat{\rho}_S(t)(\alpha^*\hat{C} + \beta^*\hat{D}) \tag{9.103}$$

$$-\frac{1}{2}(\alpha^*\hat{C} + \beta^*\hat{D})(\alpha\hat{C} + \beta\hat{D})\hat{\rho}_S(t) \tag{9.104}$$

$$-\frac{1}{2}\hat{\rho}_S(t)(\alpha^*\hat{C} + \beta^*\hat{D})(\alpha\hat{C} + \beta\hat{D}). \tag{9.105}$$

Expanding the terms, we obtain

$$\mathcal{L}\hat{\rho}_S(t) = -\frac{i}{\hbar}\left[\hat{H}, \hat{\rho}_S(t)\right]$$

$$+ |\alpha|^2 \left(\hat{C}\hat{\rho}_S(t)\hat{C} - \frac{1}{2}\hat{C}\hat{C}\hat{\rho}_S(t) - \frac{1}{2}\hat{\rho}_S(t)\hat{C}\hat{C}\right)$$

$$+ \alpha\beta^* \left(\hat{C}\hat{\rho}_S(t)\hat{D} - \frac{1}{2}\hat{D}\hat{C}\hat{\rho}_S(t) - \frac{1}{2}\hat{\rho}_S(t)\hat{D}\hat{C}\right)$$

$$+ \alpha^*\beta \left(\hat{D}\hat{\rho}_S(t)\hat{C} - \frac{1}{2}\hat{C}\hat{D}\hat{\rho}_S(t) - \frac{1}{2}\hat{\rho}_S(t)\hat{C}\hat{D}\right)$$

$$+ |\beta|^2 \left(\hat{D}\hat{\rho}_S(t)\hat{D} - \frac{1}{2}\hat{D}\hat{D}\hat{\rho}_S(t) - \frac{1}{2}\hat{\rho}_S(t)\hat{D}\hat{D}\right), \tag{9.106}$$

which can be written in the contracted form

$$\mathcal{L}\hat{\rho}_S(t) = -\frac{i}{\hbar}\left[\hat{H}, \hat{\rho}_S(t)\right] + \sum_{ij} a_{ij}\left(\hat{F}_i\hat{\rho}_S(t)\hat{F}_j - \frac{1}{2}\left\{\hat{F}_j^\dagger\hat{F}_i, \hat{\rho}_S(t)\right\}_+\right) \tag{9.107}$$

with $\hat{F}_1 = \hat{C}$ and $\hat{F}_2 = \hat{D}$ and coefficients $a_{11} = |\alpha|^2$, $a_{12} = \alpha\beta^*$, $a_{21} = \alpha^*\beta$, $a_{22} = |\beta|^2$. The coefficients a_{ij} form the elements of the coefficient matrix

$$(a_{ij}) = \begin{pmatrix} |\alpha|^2 & \alpha\beta^* \\ \alpha^*\beta & |\beta|^2 \end{pmatrix}, \tag{9.108}$$

which must be positive semi-definite for the reduced density to maintain its properties and the generator to be considered physical for all time [50]. We can see that the diagonal elements of the coefficient matrix are real, while the off diagonals are generally complex. If we consider the case where $\alpha\beta^* = \mu + iv$, in which μ and v are real numbers, it is possible to write the generator as

$$\mathcal{L}\hat{\rho}_S(t) = -\frac{i}{\hbar}\left[\hat{H}, \hat{\rho}_S(t)\right]$$

$$- \frac{|\alpha|^2}{2}\left[\hat{C}, \left[\hat{C}, \hat{\rho}_S(t)\right]\right]$$

$$+ \frac{iv}{2}\left[\hat{C}, \{\hat{D}, \hat{\rho}_S(t)\}_+\right] - \frac{\mu}{2}\left[\hat{C}, \left[\hat{D}, \hat{\rho}_S(t)\right]\right] \tag{9.109}$$

$$- \frac{iv}{2}\left[\hat{D}, \{\hat{C}, \hat{\rho}_S(t)\}_+\right] - \frac{\mu}{2}\left[\hat{D}, \left[\hat{C}, \hat{\rho}_S(t)\right]\right]$$

$$- \frac{|\beta|^2}{2}\left[\hat{D}, \left[\hat{D}, \hat{\rho}_S(t)\right]\right],$$

which has the coefficient matrix

$$(a_{ij}) = \begin{pmatrix} |\alpha|^2 & \mu + iv \\ \mu - iv & |\beta|^2 \end{pmatrix}. \tag{9.110}$$

Note that real coefficients correspond to a commutator nested inside a commutator, while imaginary coefficients correspond to an anti-commutator nested inside a commutator. Also note that the signs of the double commutator terms are reversed while the anti-commutator terms are not. We now have a way of constructing the coefficient matrix from the commutator form of the master equation.

Exercise 9.3   Expand the commutator terms in Eq. (9.109) to verify that it is equivalent to Eq. (9.107). ∎

Let us return to the Caldeira–Leggett equation,

$$\frac{d}{dt}\hat{\rho}_S(t) = -\frac{i}{\hbar}\left[\hat{H}_S, \hat{\rho}_S(t)\right] - \frac{i\gamma}{\hbar}\left[\hat{q}, \{\hat{p}, \hat{\rho}_S(t)\}_+\right]$$
$$- \frac{2m\gamma k_B T}{\hbar^2}\left[\hat{q}, [\hat{q}, \hat{\rho}_S(t)]\right].$$

(9.111)

We can see that some of the double commutator terms seen in the Lindblad equation are also present here; this can be highlighted further by splitting the dissipation term in two and using the nested anti-commutator extension to Eq. (1.13)

$$\left[\hat{A}, \{\hat{B}, \hat{C}\}_+\right] + \left[\hat{C}, \{\hat{A}, \hat{B}\}_+\right] + \left[\hat{B}, \{\hat{C}, \hat{A}\}_+\right] = 0$$

(9.112)

to obtain

$$\frac{d}{dt}\hat{\rho}_S(t) = -\frac{i}{\hbar}\left[\hat{H}_S, \hat{\rho}_S(t)\right] - \frac{i\gamma}{2\hbar}\left[\{\hat{q}, \hat{p}\}_+, \hat{\rho}_S(t)\right]$$
$$- \frac{i\gamma}{2\hbar}\left[\hat{q}, \{\hat{p}, \hat{\rho}_S(t)\}_+\right] + \frac{i\gamma}{2\hbar}\left[\hat{p}, \{\hat{q}, \hat{\rho}_S(t)\}_+\right]$$
$$- \frac{2m\gamma k_B T}{\hbar^2}\left[\hat{q}, [\hat{q}, \hat{\rho}_S(t)]\right].$$

(9.113)

The second term is considered an effective Hamiltonian term and is often absorbed in the free evolution of the system. We will look into the importance of this term in our next section because it has physical significance. The next three terms now take a very similar form to that of the Lindblad equation. Comparing coefficients with Eq. (9.109), using $\hat{C} = \hat{q}$ and $\hat{D} = \hat{p}$, we see that

$$|\alpha|^2 = \frac{4mk_B T\gamma}{\hbar^2}$$

$$\mu = 0$$

$$i\nu = -\frac{i\gamma}{\hbar}$$

$$|\beta|^2 = 0.$$

The coefficient matrix for this equation is thus given by

$$(a_{ij}) = \begin{pmatrix} \frac{4m\gamma k_B T}{\hbar^2} & -\frac{i\gamma}{\hbar} \\ \frac{i\gamma}{\hbar} & 0 \end{pmatrix},$$

(9.114)

where we can see that the bottom right element is zero, since there is no $\left[\hat{p}, [\hat{p}, \hat{\rho}_S(t)]\right]$ term in the equation.

The determinant of this particular coefficient matrix is negative, which shows that the Caldeira–Leggett equation is not a suitable master equation as it is; this is often fixed with a minimally invasive adjustment that is made to ensure the matrix remains positive. The term to be introduced must ensure that the determinant of the coefficient matrix is at least zero. The resulting adjustment gives a master equation of

$$\frac{d}{dt}\hat{\rho}_S(t) = -\frac{i}{\hbar}\left[\hat{H}_S, \hat{\rho}_S(t)\right] - \frac{i\gamma}{2\hbar}\left[\{\hat{q}, \hat{p}\}_+, \hat{\rho}_S(t)\right]$$

$$-\frac{i\gamma}{2\hbar}\left[\hat{q}, \{\hat{p}, \hat{\rho}_S(t)\}_+\right] + \frac{i\gamma}{2\hbar}\left[\hat{p}, \{\hat{q}, \hat{\rho}_S(t)\}_+\right]$$

$$-\frac{2m\gamma k_B T}{\hbar^2}\left[\hat{q}, \left[\hat{q}, \hat{\rho}_S(t)\right]\right]$$

$$-\frac{\gamma}{8mk_B T}\left[\hat{p}, \left[\hat{p}, \hat{\rho}_S(t)\right]\right],$$

(9.115)

which can be simplified into the Lindblad equation

$$\frac{d}{dt}\hat{\rho}_S(t) = -\frac{i}{\hbar}\left[\hat{H}, \hat{\rho}_S(t)\right] + \hat{L}\hat{\rho}_S(t)\hat{L}^{\dagger} - \frac{1}{2}\left\{\hat{L}^{\dagger}\hat{L}, \hat{\rho}_S(t)\right\}_+$$

(9.116)

with the one Lindblad

$$\hat{L} = \sqrt{\frac{4m\gamma k_B T}{\hbar^2}}\hat{q} + i\sqrt{\frac{\gamma}{4mk_B T}}\hat{p}$$

(9.117)

and the effective Hamiltonian

$$\hat{H} = \hat{H}_S + \frac{\gamma}{2}\{\hat{q}, \hat{p}\}_+.$$

(9.118)

Note that $\{\hat{q}, \hat{p}\}_+ \propto \hat{a}^2 + \hat{a}^{\dagger 2}$ has a squeezing effect and no classical analogue, and that it is a quantum mechanical correction to the Hamiltonian. As we will discuss later in this chapter and again in Chapter 10, this quantum correction may be needed to correctly achieve a system's classical correspondence limit.

We can always rewrite the Lindblad in Eq. (9.117) in terms of annihilation and creation operators such that

$$\hat{L} = \sqrt{2\gamma}\left[\left(\sqrt{\frac{k_B T}{\hbar\omega_0}} + \sqrt{\frac{\hbar\omega_0}{16k_B T}}\right)\hat{a} + \left(\sqrt{\frac{k_B T}{\hbar\omega_0}} - \sqrt{\frac{\hbar\omega_0}{16k_B T}}\right)\hat{a}^{\dagger}\right],$$

(9.119)

where ω_0 is the natural frequency of the system. With the Lindblad in this form we see that when $k_B T \sim \hbar\omega_0/4$ the Lindblad effectively becomes $\hat{L} \sim \sqrt{2\gamma}\hat{a}$, and we see an equivalence between Lindblad Eq. (9.117) and the one we are about to derive.

The addition of the $\left[\hat{p}, \left[\hat{p}, \hat{\rho}_S(t)\right]\right]$ term to the Caldeira–Leggett equation presents complications because we require it to be minimally invasive. For this adjustment to be considered suitably small, we must therefore operate in the high-temperature limit. Specifically, the thermal energy of the system and bath must be much greater than the rate of loss from the system to the bath. Since the damping rate γ is related to the coupling strength λ_n through the spectral density, this suggests that the model is also only valid for a weak coupling. In these limits, the Caldeira–Leggett equations in both original and Lindblad form become equivalent and thus provide, at least, a physical model for the stochastic evolution of a quantum particle with an environment.

9.3.4 Low-temperature Regime

In our consideration of the Quantum Brownian particle, we assumed a high-temperature analysis. Specifically, we assumed that the bath was a thermal reservoir at some finite temperature. There are, however, many models in the literature that couple systems to a bath at zero (or very low) temperature. There are subtle differences in obtaining the Lindblad equation from the Hamiltonians of the system and bath in this limit. The first key difference comes in at the definition of the dissipation and noise kernels, as seen in Eq. (9.87). The dissipation kernel has no temperature dependence and so remains unchanged while we make use of the limit $\lim_{x\to\infty} \coth(x) = 1$ to produce

$$D(\tau) = 2\hbar \int_0^\infty d\omega\, J(\omega) \sin(\omega\tau)$$

$$D_1(\tau) = 2\hbar \int_0^\infty d\omega\, J(\omega) \cos(\omega\tau). \tag{9.120}$$

Making use of the spectral density, Eq. (9.88), the dissipation kernel is evaluated in the same way as in the previous sections, while the noise kernel is not. In order to evaluate the noise kernel by means of Cauchy's residue theorem, we consider the sinusoidal part of it to be fast oscillating, allowing us to characterise the kernel by the frequency $\omega_0/2$ [31], and approximate the noise kernel by

$$D_1(\tau) \approx \hbar\omega_0 \int_0^\infty d\omega\, \frac{J(\omega)}{\omega} \cos\omega\tau. \tag{9.121}$$

The integrals are then evaluated as before, and we obtain the master equation

$$\frac{d}{dt}\hat{\rho}_S(t) = -\frac{i}{\hbar}\left[\hat{H}_S, \hat{\rho}_S(t)\right] + \frac{m\gamma\Omega}{\hbar} \int_0^\infty d\tau \left(i\,\Omega e^{-\Omega|\tau|}\left[\hat{q}, \{\hat{q}(-\tau), \hat{\rho}_S(t)\}_+\right]\right. \tag{9.122}$$

$$\left. -\frac{\omega_0}{2}e^{-\Omega|\tau|}\left[\hat{q}, \left[\hat{q}(-\tau), \hat{\rho}_S(t)\right]\right]\right). \tag{9.123}$$

We have seen that we may write $\hat{q}(-\tau)$ using Hadamard lemma such that

$$\hat{q}(-\tau) = \sum_n \frac{\tau^n}{n!} A_n[\hat{q}] \tag{9.124}$$

with

$$A_0[\hat{q}] = \hat{q}$$

$$A_1[\hat{q}] = \left[-\frac{i\hat{H}_S}{\hbar}, \hat{q}\right] \tag{9.125}$$

$$A_2[\hat{q}] = \left[-\frac{i\hat{H}_S}{\hbar}, \left[-\frac{i\hat{H}_S}{\hbar}, \hat{q}\right]\right] \tag{9.126}$$

and so on. We can therefore write the master equation in terms of this series to obtain

$$\frac{d}{dt}\hat{\rho}_S(t) = -\frac{i}{\hbar}\left[\hat{H}_S, \hat{\rho}_S(t)\right] + \frac{m\gamma\Omega}{\hbar}\int_0^\infty d\tau \sum_n \frac{\tau^n}{n!} e^{-\Omega|\tau|}\left(i\Omega\left[\hat{q}, \{A_n[\hat{q}], \hat{\rho}_S(t)\}_+\right]\right.$$

$$\left. - \frac{\omega_0}{2}\left[\hat{q}, [A_n[\hat{q}], \hat{\rho}_S(t)]\right]\right). \quad (9.127)$$

It is at this point that we can make use of the gamma function relation

$$\int_0^\infty \tau^n e^{-\Omega|\tau|} d\tau = \frac{\Gamma(n+1)}{\Omega^{n+1}} = \frac{n!}{\Omega^{n+1}} \quad (9.128)$$

and substitute this into the master equation to obtain

$$\frac{d}{dt}\hat{\rho}_S(t) = -\frac{i}{\hbar}\left[\hat{H}_S, \hat{\rho}_S(t)\right] + \frac{m\gamma\Omega}{\hbar}\sum_n \frac{1}{\Omega^{n+1}}\left(i\Omega\left[\hat{q}, \{A_n[\hat{q}], \hat{\rho}_S(t)\}_+\right]\right.$$

$$\left. - \frac{\omega_0}{2}\left[\hat{q}, [A_n[\hat{q}], \hat{\rho}_S(t)]\right]\right). \quad (9.129)$$

As in the Caldeira–Leggett equation, the series $\sum_n \frac{1}{\Omega^{n+1}}A_n[\hat{q}]$ must be truncated. Doing so at the first order, where we use

$$\Omega\sum_n \frac{1}{\Omega^{n+1}}A_n[\hat{q}] \approx A_0 + \frac{A_1}{\Omega} = \hat{q} - \frac{\hat{p}}{m\Omega}, \quad (9.130)$$

gives us

$$\frac{d}{dt}\hat{\rho}_S(t) = -\frac{i}{\hbar}\left[\hat{H}_S, \hat{\rho}_S(t)\right] + \frac{i\,m\gamma\Omega}{\hbar}\left[\hat{q}^2, \hat{\rho}_S(t)\right] - \frac{i\gamma}{\hbar}\left[\hat{q}, \{\hat{p}, \hat{\rho}_S(t)\}_+\right]$$

$$- \frac{\gamma m\omega_0}{2\hbar}\left[\hat{q}, [\hat{q}, \hat{\rho}_S(t)]\right] - \frac{\gamma}{2\hbar}\frac{\omega_0}{\Omega}\left[\hat{q}, [\hat{p}, \hat{\rho}_S(t)]\right]. \quad (9.131)$$

The first term cancels with the Lamb shift in the system Hamiltonian, as Eq. (9.91) did with the Lamb shift in Eq. (9.16), while the next three form the dissipator.

> Evaluating the integral in this way gives a relationship between the damping rate and the coupling strength of each bath mode, at low temperatures, through
>
> $$\gamma = \frac{1}{m\Omega}\sum_n \frac{\lambda_n^2}{2m_n\omega_n^2}. \quad (9.132)$$

Considering the coefficient matrix of the above equation, we see that this can be put into Lindblad form with the addition of the term

$$a_{22} = \frac{\gamma}{m\hbar\omega_0}\left(1 + \frac{\omega_0^2}{4\Omega^2}\right), \quad (9.133)$$

which then gives the Lindblad

$$\hat{L} = \sqrt{\frac{\gamma m\omega_0}{\hbar}}\hat{q} + \sqrt{\frac{\gamma}{m\omega_0\hbar}}\left(i - \frac{\omega_0}{2\Omega}\right)\hat{p}. \quad (9.134)$$

If we now assume a high cut-off limit, as in the Caldeira–Leggett equation, we find that we obtain the Lindblad

$$\lim_{\Omega \to \infty} \hat{L} = \sqrt{\frac{\gamma m \omega_0}{\hbar}} \hat{q} + i \sqrt{\frac{\gamma}{m \omega_0 \hbar}} \hat{p} = \sqrt{2\gamma} \hat{a}, \tag{9.135}$$

where $\hat{a}$ is the annihilation operator. We saw that in the Caldeira–Leggett model, the combination of $\hat{q}$ and $\hat{p}$ operators give rise to Lindblads proportional to both annihilation and creation operators. The prevalence of low-temperature, high-cut-off, approximations explains the wide, almost default, use of Lindblad operators proportional to the annihilation operator. The difference between the Lindblad equations in each regime, however, is that the additional term in the high-temperature case can be considered minimally invasive, whereas in the low-temperature regime the additional term is significant. Although the equation guarantees validity of the master equation solutions, there are, as we will discuss next, implications that arise from using Lindblads of this form.

9.3.5 Lindblads: Strengths and Weaknesses

This section has covered an introduction to the modelling of a quantum system, open to a decohering environment, as a quantum Brownian motion, and at this stage it is useful to summarise the issues that arise. In particular, we should evaluate whether models using the higher-order Lindblad approach are able to correct the problems associated with the simple Caldeira–Leggett type case; whether a hierarchy of models of increasing accuracy can be generated, and whether such models are likely to be useful for engineering quantum systems. In general:

- The Lindblad form of Master Equation has the very attractive properties that it is Markovian, and hence its representation of the effect of the environment is time invariant, and that the density operator $\hat{\rho}(t)$ generated by such an equation is guaranteed to be physical $(\mathrm{Tr}(\hat{\rho}(t)) = 1$, its eigenvalues are non-negative at all times). However, it only represents a desired form and neither indicates what the Lindblad elements, which describe environmental interaction, should be, nor how a Lindblad master equation may be reached through a natural microscopic analysis. It is notably the only form of semigroup equation which has this property.
- The Caldeira–Leggett form of the Master Equation, guided by the classical equation for Brownian motion, allows the environmental interaction to be included explicitly, but is generally not of Lindblad form.
- Attempts to bring a Caldeira–Leggett type of Master Equation into Lindblad form require the adding of additional artificial terms, the movement of terms from the environment into the system Hamiltonian, and the prospect of cancelling terms, such as the Lamb shift and squeezing, which have physical significance.
- The time-dynamics of the resultant models have questionable reliability, making unclear their utility for describing system decoherence, which will limit how well quantum systems can be characterised [30].

The conclusion is that one is left with an unattractive choice. Either our equation is not guaranteed to produce physical solutions under all circumstances or, as we shall now go one to explore in some detail, it is contaminated by the addition of artificial terms whose impact on the dynamics can neither be guaranteed to be small, nor to be based in physics.

9.3.6 Effective Hamiltonians

One of the most important things to observe from the above discussion of the derivation of Lindblad master equations, is that terms that can be considered as modifications to the Hamiltonian arise. They are Hamiltonian-like because they occur in terms of the form $-\frac{i}{\hbar}\left[\hat{H}_X,\hat{\rho}_S\right]$ where $\hat{H}_X$ is some operator arising from the environment. We can see from the Liouville–von Neumann equation that these play the same role as the system Hamiltonian in generating dynamics. In our specific example of transforming the Caldeira–Leggett equation into a Lindblad equation, this involved taking a term out of the dissipator and combining it with the free evolution to provide the effective Hamiltonian

$$\hat{H} = \hat{H}_S + \frac{\gamma}{2}\{\hat{q},\hat{p}\}_+ = \hat{H}_S + \hat{H}_{qp}, \tag{9.136}$$

where $\hat{H}_{qp} = \frac{\gamma}{2}\{\hat{q},\hat{p}\}_+$. There is much discussion over whether this extra term should be omitted from the master equation, while a lot of literature includes the term without discussing its origin. In quantum to classical transition, as decoherence models such as the Quantum Boltzmann Machine describe, it is important to preserve correspondence; this means that classical dynamics should be reproducible from a quantum mechanical framework.

Understanding the quantum to classical transition is challenging, and we shall postpone its discussion until Chapter 10. For now, let us concern ourselves only with the dynamics of expectations of observables for a simple system and see what implications the addition of $\hat{H}_{qp}$ to the system Hamiltonian have. The dynamics of expectation values can be explored using the equation

$$\frac{d}{dt}\langle\hat{A}\rangle = \frac{d}{dt}\text{Tr}\left\{\hat{A}\hat{\rho}_S(t)\right\} \tag{9.137}$$

$$= \text{Tr}\left\{\frac{d\hat{A}}{dt}\hat{\rho}_S(t)\right\} + \text{Tr}\left\{\hat{A}\frac{d\hat{\rho}_S(t)}{dt}\right\}. \tag{9.138}$$

For an operator with no explicit dependence on time this reduces to

$$\frac{d}{dt}\langle\hat{A}\rangle = \text{Tr}\left\{\hat{A}\frac{d\hat{\rho}_S(t)}{dt}\right\}. \tag{9.139}$$

The time-derivative of the reduced density matrix of the system gives us the generator of the master equation. Substituting this in gives

$$\frac{d}{dt}\langle\hat{A}\rangle = \text{Tr}\left\{\hat{A}\left(-\frac{i}{\hbar}\left[\hat{H},\hat{\rho}_S(t)\right] + \mathcal{K}[\hat{\rho}_S(t)]\right)\right\}, \tag{9.140}$$

where we recall that $\mathcal{K}[\hat{\rho}_S(t)]$ is the dissipator. Expanding out the commutator in the first term on the right-hand side (free evolution) and using the cyclic property of trace allows us to write

$$\frac{d}{dt}\langle\hat{A}\rangle = -\frac{i}{\hbar}\text{Tr}\left\{\left[\hat{H},\hat{\rho}\right]\hat{A}\right\} + \text{Tr}\left\{\mathcal{K}[\hat{\rho}]\hat{A}\right\}. \tag{9.141}$$

If we now assume that the dissipator is in the Lindblad form as in Eq. (9.100), Eq. (9.141) becomes

$$\frac{d}{dt} \langle \hat{A} \rangle = -\frac{i}{\hbar} \langle [\hat{A}, \hat{H}] \rangle + \frac{1}{2} \left\{ \left\langle \left[\hat{L}^\dagger, \hat{A} \right] \hat{L} \right\rangle + \left\langle \hat{L}^\dagger \left[\hat{A}, \hat{L} \right] \right\rangle \right\}, \tag{9.142}$$

which is Ehrenfest's theorem with a correction.

Let us consider the specific example of the damped harmonic oscillator. At low temperatures, the Lindblad $\hat{L} = \sqrt{2\gamma}\hat{a}$, where $\hat{a}$ is the annihilation operator. The evolution of the expectation values of the system's position and momentum are given by:

$$\frac{d}{dt} \langle \hat{q} \rangle = -\frac{i}{\hbar} \langle [\hat{q}, \hat{H}] \rangle + \gamma \left\{ \left\langle \left[\hat{a}^\dagger, \hat{q} \right] \hat{a} \right\rangle + \left\langle \hat{a}^\dagger \left[\hat{q}, \hat{a} \right] \right\rangle \right\} \tag{9.143}$$

$$\frac{d}{dt} \langle \hat{p} \rangle = -\frac{i}{\hbar} \langle [\hat{p}, \hat{H}] \rangle + \gamma \left\{ \left\langle \left[\hat{a}^\dagger, \hat{p} \right] \hat{a} \right\rangle + \left\langle \hat{a}^\dagger \left[\hat{p}, \hat{a} \right] \right\rangle \right\}. \tag{9.144}$$

To see the impact of $\hat{H}_{qp}$, we will consider the dynamics of the system with and without this term in the Hamiltonian. We will begin by omitting this term. The Hamiltonian is therefore given by

$$\hat{H} = \hbar\omega \left(\hat{n} + \frac{1}{2} \right). \tag{9.145}$$

Substituting these expressions into the equations of motion for $\langle \hat{q} \rangle$ and $\langle \hat{p} \rangle$ yields

$$\frac{d}{dt} \langle \hat{q} \rangle = \frac{1}{m} \langle \hat{p} \rangle - \gamma \langle \hat{q} \rangle \tag{9.146}$$

$$\frac{d}{dt} \langle \hat{p} \rangle = -m\omega^2 \langle \hat{q} \rangle - \gamma \langle \hat{p} \rangle. \tag{9.147}$$

Eliminating $\langle \hat{p} \rangle$ from the above equations gives

$$\frac{d^2}{dt^2} \langle \hat{q} \rangle + 2\gamma \frac{d}{dt} \langle \hat{q} \rangle + (\omega^2 + \gamma^2) \langle \hat{q} \rangle = 0. \tag{9.148}$$

The solution to this equation is given by

$$\langle \hat{q} \rangle (t) = \left\langle \hat{q}_+ \right\rangle e^{i\omega t} e^{-\gamma t} + \langle \hat{q}_- \rangle e^{-i\omega t} e^{-\gamma t}, \tag{9.149}$$

which describes an oscillator with decaying oscillations of frequency ω at a decay rate γ; this is *not in agreement with the classical result*, since a damped oscillator will also undergo a frequency shift that is dependent on the level of damping applied to it.

Repeating the above analysis but now using the effective Hamiltonian

$$\hat{H} = \hbar\omega \left(\hat{n} + \frac{1}{2} \right) + \frac{\gamma}{2} \{\hat{q}, \hat{p}\}_+ \tag{9.150}$$

gives the result

$$\langle \hat{q} \rangle (t) = \left\langle \hat{q}_+ \right\rangle e^{i\omega' t} e^{-\gamma t} + \langle \hat{q}_- \rangle e^{-i\omega' t} e^{-\gamma t} \tag{9.151}$$

where $\omega' = \omega\sqrt{1 - \gamma^2/\omega^2}$ is the frequency of oscillation, in agreement with classical predictions.

Exercise 9.4 Using the Hamiltonian Eqs. 9.150, show 9.151 ∎

The above analysis shows that the dissipation term in the Caldeira–Leggett equation does not only describe the decay of a quantum system, but also the change in natural frequency due to external influence. It is therefore crucial to keep this term when considering the evolution of open quantum systems.

9.4 Master Equation Approximations and Their Implications

In this section, we discuss four key approximations used in the previous analysis. We will briefly discuss the suitability of each approximation, with an emphasis on the strengths and weaknesses of each. There is no set way to model an open quantum system, but there are certain regimes we can work in to simplify the problem. It is therefore important to know the appropriateness and limitations of each of these regimes, and what might be done to improve the validity of any master equation we derive. We split our discussion of the topic into two sections. In this section we make rather general points, to outline the main principles that should be considered. In the next section, we analyse a specific example to show how such considerations may be accounted for.

9.4.1 The Baker Campbell Hausdorff Approximations

To obtain the integral form of our dissipator we make use of the Hadamard lemma that arises from the Baker Campbell Hausdorff formula. We saw in Section 9.3.2 that the position of the system after the initial correlation time, $\hat{q}(-\tau)$, can be written

$$\hat{q}(-\tau) = e^{-i\hat{H}_S\tau/\hbar} \hat{q} e^{i\hat{H}_S\tau/\hbar}$$

$$= \hat{q} + \left(-\frac{i\tau}{\hbar}\right)[\hat{H}_S, \hat{q}] + \frac{1}{2!}\left(-\frac{i\tau}{\hbar}\right)^2 [\hat{H}_S, [\hat{H}_S, \hat{q}]]$$

$$+ \frac{1}{3!}\left(-\frac{i}{\hbar}\right)^3 [\hat{H}_S, [\hat{H}_S, [\hat{H}_S, \hat{q}]]] + \cdots. \tag{9.152}$$

When we consider a Hamiltonian of the form

$$\hat{H}_S = \frac{\hat{p}^2}{2m} + V(\hat{q}), \tag{9.153}$$

as we did for quantum Brownian motion, we said that τ was sufficiently small to truncate the series to first order, obtaining

$$\hat{q}(-\tau) \approx \hat{q} - \frac{\hat{p}\tau}{m}$$

in the process. But what if τ is not as small as we assumed? Alternatively, what if the higher-order commutators $[\hat{H}_s, [\ldots, [\hat{H}_S, \hat{q}]]]$ do not converge nicely? We might then want to investigate higher-order terms of the Hadamard lemma. For example, taking the series to second order yields

$$\hat{q}(-\tau) \approx \hat{q} - \frac{\hat{p}\tau}{m} - \frac{\tau^2 V'(\hat{q})}{m}, \tag{9.154}$$

where $V'(\hat{q})$ is the derivative of $V(\hat{q})$ with respect to $\hat{q}$, while the third order gives

$$\hat{q}(-\tau) \approx \hat{q} - \frac{\hat{p}\tau}{m} - \frac{\tau^2 V'(\hat{q})}{m} + \frac{\tau^3}{2m^2}\{\hat{p}, V''(\hat{q})\}_+, \tag{9.155}$$

where $V''(\hat{q})$ is the second-order derivative of the potential with respect to the position. It is quite evident at this point that consideration of higher-order terms will bring in more complexity due to the non-commutative nature of the conjugate observables, $\hat{q}$ and $\hat{p}$.

The nice example of a system comprising a harmonic oscillator with the potential

$$V(\hat{q}) = \frac{1}{2}m\omega^2\hat{q}^2$$

gives rise to a series that need not be truncated, and we obtain the exact solution of

$$\hat{q}(-\tau) = \left(\hat{q} - \frac{\hat{p}\tau}{m}\right)\cos(\omega_0\tau), \tag{9.156}$$

where ω_0 is the natural frequency of the oscillator. This result tells us that for the Caldeira–Leggett equation to be appropriate for the Harmonic oscillator, we require

$$\tau \ll \frac{1}{\omega_0} \tag{9.157}$$

or that the correlations between the system and bath have settled long before the system's wave function has completed one cycle. What happens if the system-of-interest is not a harmonic oscillator is a question we shall return to later.

Exercise 9.5 By inserting the potential for the harmonic oscillator, show that the position after some correlation time $\hat{q}(-\tau)$ is given by Eq. (9.156). ∎

9.4.2 The Born Approximation

The Born approximation requires a bath relaxation time that is far greater than that of the system. The assumption is that the correlations between system and bath are negligible, and the bath's density operator is constant in time. This allows a simplification of the overall density operator, leading to easy separability in order to trace over the environment and study the dynamics of the system only. Limitations therefore exist to the strength of coupling (this argument would only be true in the limit of weak coupling) and the spectral width ('size' one might say) of the bath relative to the system. However, it is far less clear if the approximations are compatible with composite quantum systems sharing environments, small environments, or strong coupling limits. This has a more subtle consequence of necessitating that the initial interaction between the system and the bath is discontinuous; it is 'switched on' instantaneously as there is no interaction term describing properly the initial interaction and transients, and there is no time dependence in the bath term, meaning that we do not allow for a more gradual or sophisticated interaction. A common issue with Born–Markov, or Redfield, Master Equations is that they often show non-physical time dynamics at the very start of an interaction. This is often given as a reason to both discard the time dynamics as unreliable, and also to pursue a Lindblad form – with the additional terms that entails. However, when one considers that the Born approximation primarily impacts the realism and continuity of the initial interaction, it is perhaps unsurprising that this is where we see the most issues. One might, therefore, wonder if it would be wise to avoid making the approximation at all – however, that would make the problem of modelling open quantum systems significantly less tractable. A different approach is attempting to retain the separable structure provided by the Born approximation, but postulate a more complex system – one that includes 'inseparable' elements of the environment due to the interaction. These would be low-frequency correlation terms that do not meet the Born approximation's requirement that the bath is able to respond quickly enough to follow any changes in the system whilst remaining in equilibrium. We will return to this discussion on page 375 in Chapter 11.

9.4.3 Choice of Spectral Density

In the previous section, we introduced the spectral density as the Fourier transform of the correlation function, giving us the distribution of modes of different frequencies during the initial correlation between the system and the bath. We saw that the general expression for the spectral density was given by

$$J(\omega) = \sum_n \frac{\lambda_n^2}{2m_n\omega_n}\delta(\omega - \omega_n). \tag{9.158}$$

When considering a bath with infinite degrees of freedom, the gap between neighbouring frequencies reduces to an infinitesimal amount, replacing our spectral distribution with a smooth function of ω.

In general, the smooth or *effective* spectral density can be defined in terms of a frequency-dependent coupling strength function $g(\omega)$ [56]

$$J(\omega) = \pi g^2(\omega),$$

the exact form of which is constrained such that it must be dimensionally consistent with the correlation function from which it is constructed, having units of energy per unit area. In the previous section we used an Ohmic spectral density of

$$J(\omega) = \frac{2m\gamma\omega}{\pi},$$

where we characterise the relative strength of system-bath coupling through the damping rate γ. We can see the Ohmic spectral density plotted as a solid line in Figure 9.2. We see that this form of spectral density suggests that power increases indefinitely with frequency, which in the high-frequency limit is not physical. This is because the maximum interaction should be centred around some critical mode frequency, above which the environment shows diminishing influence over the system (or some other non-linear function of frequency informed by the physics of the environment). To remedy this issue, we usually introduce a cut-off function dependent on some cut-off frequency that renormalises the spectral density. In the analysis shown in Section 9.3.2.1 we made use of the Lorentz–Drude cut-off function

$$X_L(\omega, \Omega) = \frac{\Omega^2}{\Omega^2 + \omega^2}, \tag{9.159}$$

where Ω was the cut-off frequency; this is depicted in Figure 9.2 as the dashed line and shows a drop-off past this critical value. We are, however, not limited to using a Lorentz–Drude cut-off. Another popular choice of cut-off function is the exponential cut-off:

$$X_e(\omega, \Omega) = \exp\left(-\frac{\omega}{\Omega}\right), \tag{9.160}$$

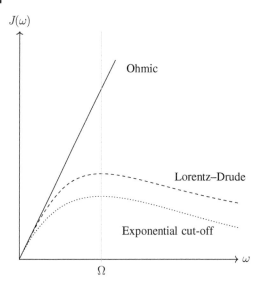

$J(\omega)$

Ohmic

Lorentz–Drude

Exponential cut-off

ω

Ω

Figure 9.2 Plot comparing (effective) spectral densities with and without different cut-off functions. The solid line is the Ohmic spectral density, which grows linearly with frequency. The dashed line shows the spectral density with a Lorentz–Drude function, while the dotted line has an exponential cut-off. Note that the general shapes of the Ohmic and Lorentz–Drude cut-off function are very similar; both can often serve the same purpose.

which is demonstrated by the dotted line in Figure 9.2. We see that both the Lorentz–Drude and exponential cut-off functions show similar behaviour and reasonably interchangeable.

> What is important to note is that all three spectral density functions show the same trend in the low-frequency regime. It is only for higher frequencies the general behaviours begin to diverge.

The relationship between spectral density and frequency need not be linear. Figure 9.3 shows three spectral density relationships for the case where $J(\omega) \propto \omega^s$ with $s < 1$ (sub-Ohmic), $s = 1$ (Ohmic), $s > 1$ (super-Ohmic). As we saw previously, the spectral density is the half-sided Fourier transform of the bath correlation function. The dependencies within the bath correlations, such as coupling strength and temperature, will all play a part in determining the dominant frequency range within the bath. Typically, since the spectral density is a measure of the amount of energy dissipated at a given frequency range, the choice of spectral density depends on the amount of damping at a lower temperature. For the sub-Ohmic case, low-energy excitations (frequencies), dominate, and so damping (dissipation) increases with reduced temperature. For the super-Ohmic case, the dominant excitations are at high energy, and so a reduction in temperature leads to a constant damping in this low-energy range [90]. What spectral density to use in a particular situation is a matter of experience. As an analogy, consider the process we would follow to accurately model the power loss of a real rather than an ideal resistor in a classical electrical circuit. Here we would characterise that component by measuring the current and voltage properties of the component for use in circuit models. The modelling of open quantum systems can benefit from a similar approach. If we could measure the environment and characterise its power spectrum we may, in an analogous way, include that in the details of the model. However, when considering a general case, or even just for convenience, we may make assumptions and approximations to the spectral density that we believe are suitable for the application at hand.

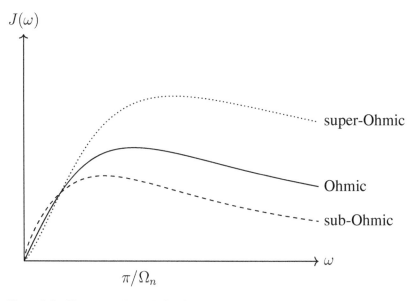

Figure 9.3 Plot comparing sub-Ohmic (dashed), Ohmic (solid), and super-Ohmic (dotted) spectral densities.

9.4.4 High Cut-off Limit

Previously, we found the integral form of the master equation for a quantum Brownian particle. We saw that making use of the dissipation and noise kernels allowed us to write

$$\mathcal{K}[\hat{\rho}_S] = \frac{i\,[\hat{q}^2, \hat{\rho}]}{2\hbar^2} \int_0^\infty d\tau D(\tau) - \frac{i\,[\hat{q}, \{\hat{p}, \hat{\rho}\}]}{2m\hbar^2} \int_0^\infty d\tau \tau D(\tau)$$

$$- \frac{[\hat{q}, [\hat{q}, \hat{\rho}]]}{2\hbar^2} \int_0^\infty d\tau D_1(\tau) + \frac{[\hat{q}, [\hat{p}, \hat{\rho}]]}{2m\hbar^2} \int_0^\infty d\tau \tau D_1(\tau).$$

We noticed that the first term gives rise to the Lamb shift, renormalising the system's potential. The second and third terms give rise to the dissipation and noise terms, respectively, while the fourth does not appear in the Caldeira–Leggett equation. The reason for this omission comes from the assumption that $\Omega \to \infty$, which we will now discuss in more detail. Evaluation of Eq. (9.87), using the spectral density defined in Eq. (9.88), yields an explicit form for the noise kernel by use of residues and the series expansion of

$$\coth\left(\frac{\hbar\omega}{2k_BT}\right) = \frac{2k_BT}{\hbar} \sum_{n=-\infty}^{\infty} \frac{\omega}{\omega^2 + v_n^2}, \tag{9.161}$$

where $v_n = 2\pi n k_B T$ denote the Matsubara frequencies, which are the poles of the Bose–Einstein distribution used to describe the average number of excitations within the bath. Combining the three expressions then yields

$$D_1(\tau) = 2\hbar \int_0^\infty d\omega\, J(\omega) \coth\left(\frac{\hbar\omega}{2k_BT}\right) \cos(\omega\tau)$$

$$= 2\hbar \int_0^\infty d\omega \frac{2m\gamma\omega\Omega^2}{\pi(\Omega^2 + \omega^2)} \frac{2k_BT}{\hbar} \sum_{n=-\infty}^{\infty} \frac{\omega}{\omega^2 + v_n^2} \cos(\omega\tau)$$

$$
= \frac{8m\gamma k_B T \Omega^2}{\pi} \int_0^\infty d\omega \frac{\omega}{\Omega^2 + \omega^2} \sum_{n=-\infty}^\infty \frac{\omega}{\omega^2 + v_n^2} \cos(\omega\tau)
$$

$$
= 4m\gamma k_B T \Omega^2 \sum_{n=-\infty}^\infty \frac{\Omega e^{-\Omega\tau} - |v_n| e^{-|v_n|\tau}}{\Omega^2 - |v_n|^2}. \tag{9.162}
$$

The fourth term in Eq. (9.83) therefore evaluates to

$$
\frac{[\hat{q},[\hat{p},\hat{\rho}]]}{2m\hbar^2} \int_0^\infty d\tau\tau D_1(\tau) = \frac{2\gamma k_B T}{\Omega\hbar^2} [\hat{q},[\hat{p},\hat{\rho}]] \sum_{n=-\infty}^\infty \frac{1 - \Omega/|v_n|}{1 - |v_n|^2/\Omega^2} \tag{9.163}
$$

$$
\approx \frac{2\gamma k_B T}{\Omega\hbar^2} [\hat{q},[\hat{p},\hat{\rho}]], \tag{9.164}
$$

where we constrain the Matsubara frequencies such that the thermal energy and cut-off energy are similar in magnitude, $k_B T \gtrsim \hbar\Omega$. To find the significance of this term, we must compare it to the third term in Eq. (9.83). The momentum of the particle roughly approximates to $\hat{p} \approx m\omega_0\hat{q}$, and so we see that the fourth term differs to this by a factor ω_0/Ω. For the Caldeira–Leggett model it is assumed that the cut-off frequency of the bath is substantially large enough for us to neglect this term, that is to say that environmental noise has a stronger dependence on position than on momentum in this case.

We must note that removal of this term from the Caldeira–Leggett equation requires a high cut-off limit. If the bath being considered does not possess a high cut-off frequency in comparison to the system's natural frequency, then one must retain the term and construct a new Lindblad operator that takes this additional noise into account, as seen in Section 9.3.4.

9.5 A Master Equation Derivation Example

In the previous section, we highlighted the key assumptions and approximations made when using the master equation framework for a general system. In this section, we will give some specific context for those issues by considering a physical example. One of the most commonly used examples of applying open systems methods is the harmonic oscillator. The example offers little opportunity to explore the more subtle sides of deriving master equations. Here, we choose to apply the methods described in the preceding section to a non-trivial system, namely the Superconducting Quantum Interference Device (SQUID), and we will discuss the potential problems that may arise in modelling these devices. This model has been chosen over other models, mainly because of our familiarity with the system, and also because its analysis is not trivial.

The SQUID is a superconducting loop containing a weak link or *Josephson junction*. For a proper introduction to the physics of this fascinating device, see any good textbook on superconductivity. When subjected to an external magnetic flux, Φ_x, a supercurrent will flow. The SQUID can therefore be modelled as an LC circuit with some quantum tunnelling through the weak link:

$$
\hat{H}'_S = \frac{\hat{Q}^2}{2C} + \frac{(\hat{\Phi} - \Phi_x)^2}{2L} - \hbar\nu\cos\left(\frac{2\pi\hat{\Phi}}{\Phi_0}\right), \tag{9.165}
$$

where C and L denote the effective capacitance and inductance of the SQUID, and $h\nu$ gives the Josephson energy (we also define $\omega_0 = 1/\sqrt{LC}$ as the resonant frequency of the system when $\nu = 0$). The canonical flux and charge operators share the same commutation relation $[\hat{\Phi}, \hat{Q}] = i\hbar$ as position and momentum, respectively. If we measure flux and charge, we could in principle obtain values over the entire real number line. As they share the same commutation relation and eigenvalues, these operators are therefore mathematically equivalent to position and momentum. A useful trick at this point is to shift the external flux dependence into the cosine part of the Hamiltonian. This is because it means that the SQUID Hamiltonian is of the form of a harmonic oscillator plus a cosine term. Such translation is achieved by applying the displacement operator[2] $\hat{D}(\Phi_x) = \exp(-i\,\hat{Q}\Phi_x\,\hbar)$ as a similarity transformation, and yields

$$\hat{H}_S = \hat{D}^{\dagger}(\Phi_x)\hat{H}'_S\hat{D}(\Phi_x) = \frac{\hat{Q}^2}{2C} + \frac{\hat{\Phi}^2}{2L} - h\nu\cos\left(\frac{2\pi}{\Phi_0}(\hat{\Phi} + \Phi_x)\right). \tag{9.166}$$

It is the cosine part of this Hamiltonian that makes the SQUID non-trivial and a great example for exploring the robustness of the tools and techniques discussed previously in this chapter. To help see this, we show in Figure 9.4 the non-linear dependence of the first three energy eigenvalues of $\hat{H}_S$ on the external flux.

As an electromagnetic device, the SQUID will couple inductively and capacitively to resonant modes around it. While both couplings will be present in any real system, circuit designs can be used to trade one off against the other. In any case, we will treat the usually assumed dominant inductive coupling first. We will later introduce the capacitive coupling to show what difference this makes to the analysis.

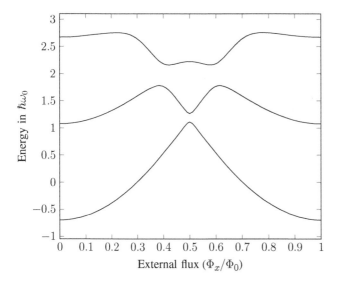

Figure 9.4 The first three energy eigenvalues for a SQUID ring with circuit parameters $\hbar\omega_0 = 0.043\Phi_0^2/L$ and $\hbar\nu = 0.07\Phi_0^2/L$ (expressing energy in terms of Φ_0^2/L is just one of a number of standard choices for natural units).

---

2 While not pertinent to the current discussion, it is important to be pedantic in the way the external flux depends on time – it is not as simple as a coordinate transformation in classical physics.

9.5.1 Inductive Coupling

We will begin exploring the case of the SQUID inductively coupled to its environment by taking a similar approach to the one used when considering Quantum Brownian Motion. We coupled the position of the Brownian particle with each of the modes in its environment, comprising a bath of harmonic oscillators. In cavity quantum electrodynamics, it is often the case that cavity modes are represented by a set of harmonic oscillators. As such, it is reasonable for us to model the SQUID's environment as a bath of oscillators (LC circuits) to which the SQUID is inductively coupled. This gives us the Hamiltonians:

$$\hat{H}_S = \frac{\hat{Q}^2}{2C} + \frac{\hat{\Phi}^2}{2L} - \hbar v \cos\left(\frac{2\pi}{\Phi_0}(\hat{\Phi} + \Phi_x)\right) \tag{9.167}$$

$$\hat{H}_B = \sum_n \frac{\hat{Q}_n^2}{2C_n} + \frac{\hat{\Phi}_n^2}{2L_n} \tag{9.168}$$

$$\hat{H}_I = -\hat{\Phi}\sum_n \kappa_n \hat{\Phi}_n = -\hat{\Phi}\hat{B}, \tag{9.169}$$

where we have once again defined the bath operator $\hat{B}_\Phi = \sum_n \kappa_n \hat{\Phi}_n$. Applying the techniques outlined in Section 9.3.2 (the analysis at this stage is unaltered), we arrive at the Redfield equation

$$
\begin{aligned}
\frac{d\hat{\rho}_S(t)}{dt} &= -\frac{i}{\hbar}[\hat{H}_S, \hat{\rho}_S(t)] \\
&+ \frac{1}{\hbar^2} \int_0^\infty d\tau \left(\frac{i}{2} D(\tau)[\hat{\Phi}, \{\hat{\Phi}(-\tau), \hat{\rho}_S(t)\}] \right. \\
&\left. - \frac{1}{2} D_1(-\tau)[\hat{\Phi}, [\hat{\Phi}(-\tau), \hat{\rho}_S(t)]] \right),
\end{aligned}
\tag{9.170}
$$

where once again the dissipation and noise kernels can be written in terms of the spectral density using Eq. (9.87):

$$D(\tau) = 2\hbar \int_0^\infty d\omega \, J(\omega) \sin(\omega\tau)$$

$$D_1(-\tau) = 2\hbar \int_0^\infty d\omega \, J(\omega) \coth\left(\frac{\hbar\omega}{2k_B T}\right) \cos(\omega\tau).$$

The kernels are evaluated in the low-temperature regime, as seen in Section 9.3.4, as is required if we are to consider a SQUID system. The approximated noise kernel is therefore given by

$$D_1(\tau) = \frac{\omega_0}{2} \coth\left(\frac{\hbar\omega_0}{2k_B T}\right) \int_0^\infty d\omega \frac{J(\omega)}{\omega} \cos(\omega\tau). \tag{9.171}$$

As we can consider the damping of the circuit (even though it is superconducting) as a form of electrical resistance, we assume an Ohmic bath with a Lorentz–Drude cut-off. This choice of cut-off function has been made because of its resemblance to impedance in Josephson circuits [51, 82]. Noting the analogy between system mass for the Brownian particle and SQUID capacitance, we have

$$J(\omega) = \frac{2C\gamma}{\pi} \omega \frac{\Omega^2}{\Omega^2 + \omega^2}. \tag{9.172}$$

Thus, we see that in the mid-low-temperature regime, the dissipation and noise kernels, $D(\tau)$ and $D_1(-\tau)$, may be written, respectively, as [19]:

$$D(\tau) = 2C\gamma\hbar\Omega^2\, e^{-\Omega|\tau|}\, \mathrm{sgn}\tau$$

$$D_1(-\tau) = C\hbar\gamma\Omega\omega_0 \coth\left(\frac{\hbar\omega_0}{4k_BT}\right) e^{-\Omega|\tau|},$$

where all the terms hold their previous definitions. As in Section 9.3.4, the noise kernel reduces in the low-temperature limit to[3]

$$D_1(-\tau) = C\hbar\gamma\Omega\omega_0\, e^{-\Omega|\tau|}. \tag{9.173}$$

As we saw in Section 9.3.2, we require a way of representing the correlation-time-dependent operator $\hat{\Phi}(-\tau)$; we do this in the same manner as in Section 9.3.4, by expanding as a power series of τ

$$\hat{\Phi}(-\tau) = \sum_n A_n[\hat{\Phi}]\tau^n, \tag{9.174}$$

where the functional $A_n[\hat{\Phi}]$ is found by equating powers of τ from the Hadamard lemma $\hat{\Phi}(-\tau) = e^{-i\hat{H}_S\tau/\hbar}\hat{\Phi}e^{i\hat{H}_S\tau/\hbar}$, i.e.:

$$\hat{\Phi}(-\tau) = \hat{\Phi} + \tau\left[-\frac{i\hat{H}_S}{\hbar},\hat{\Phi}\right] + \frac{\tau^2}{2!}\left[-\frac{i\hat{H}_S}{\hbar},\left[-\frac{i\hat{H}_S}{\hbar},\hat{\Phi}\right]\right]$$

$$+ \cdots + \frac{\tau^n}{n!}\left[-\frac{i\hat{H}_S}{\hbar},\ldots,\left[-\frac{i\hat{H}_S}{\hbar},\hat{\Phi}\right]\right]. \tag{9.175}$$

As discussed in Section 9.4.1, a harmonic oscillator possesses a series where each term $A_n[\hat{q}]$ is proportional to either the position or momentum operator, with pre-factors which add to give trigonometric terms [58]. The same cannot be said for the SQUID because of the trigonometric nature of the Josephson junction term in the Hamiltonian, the series grows in complexity as the order is increased. For this reason we have not been able to evaluate $\hat{\Phi}(-\tau)$ analytically and find it necessary to truncate the series in Eq. (9.174)[4]. Substituting Eq. (9.174) into the expressions for the dissipator of Eq. (9.170) yields the pre-Lindblad master equation:

$$\frac{d\hat{\rho}_S(t)}{dt} = -\frac{i}{\hbar}[\hat{H}_S,\hat{\rho}_S(t)] + \frac{iC\gamma\Omega}{\hbar}\left[\hat{\Phi},\left\{\sum_n\frac{n!}{\Omega^n}A_n[\hat{\Phi}],\hat{\rho}_S(t)\right\}_+\right]$$

$$-\frac{C\hbar\gamma\omega_0}{2\hbar}\left[\hat{\Phi},\left[\sum_n\frac{n!}{\Omega^n}A_n[\hat{\Phi}],\hat{\rho}_S(t)\right]\right], \tag{9.176}$$

where the identities for the dissipation and noise terms:

$$\frac{i}{2\hbar^2}\int_0^\infty d\tau D(\tau)\hat{\Phi}(-\tau) = \sum_n\frac{iC\gamma\Omega}{\hbar}\frac{n!}{\Omega^n}A_n[\hat{\Phi}]$$

$$-\frac{1}{2\hbar^2}\int_0^\infty d\tau D_1(-\tau)\hat{\Phi}(-\tau) = -\frac{C\hbar\gamma\omega_0}{2\hbar}\sum_n\frac{n!}{\Omega^n}A_n[\hat{\Phi}] \tag{9.177}$$

---

3 The simplification of the noise kernel at low temperatures has been questioned and discussed [31, 95], but is necessary here in order to follow the standard approach outlined in previous sections of this chapter.
4 If the system were to possess a time-dependent external flux $\Phi_x(t)$, the evolution operator would no longer be $e^{-i\hat{H}_S\tau/\hbar}$, and the series would grow significantly in complexity; hence, the method we present here may not be applicable.

have been used, alongside the identity

$$\Omega^{n+1} \int_0^\infty d\tau \tau^n e^{-\Omega\tau} = n!. \tag{9.178}$$

A first-order truncation of Eq. (9.174) means that the summations in Eq. (9.176) can be simplified to

$$\sum_n \frac{n!}{\Omega^n} A_n \approx A_0 + \frac{1}{\Omega} A_1 = \hat{\Phi} - \frac{\hat{Q}}{\Omega C}. \tag{9.179}$$

This in turn means that Eq. (9.176) yields the first-order master equation [19]:

$$
\frac{d\hat{\rho}_S(t)}{dt} = -\frac{i}{\hbar}[\hat{H}_S, \hat{\rho}_S(t)] + \overbrace{\frac{i\,C\gamma\Omega}{\hbar}[\hat{\Phi}^2, \hat{\rho}_S(t)]}^{\text{renormalises } L} - \overbrace{\frac{i\gamma}{\hbar}[\hat{\Phi}, \{\hat{Q}, \hat{\rho}_S(t)\}]}^{\text{dissipation term}}
$$

$$
- \frac{C\omega_0\gamma}{2\hbar} \left(\underbrace{[\hat{\Phi}, [\hat{\Phi}, \hat{\rho}_S(t)]]}_{\text{noise term}} - \underbrace{\frac{1}{\Omega C}[\hat{\Phi}, [\hat{Q}, \hat{\rho}_S(t)]]}_{\text{first-order cut-off}} \right). \tag{9.180}
$$

The first term again describes the free evolution of the SQUID, while the third and fourth terms, labelled dissipation and noise, have the same effect as the dissipation and noise terms we saw previously. We can also see that the final term in Eq. (9.180) vanishes in the limit of a high cut-off frequency. The nature of this approximation was discussed in Section 9.4.4, but this is not applicable to condensed matter systems, so it must be retained.

Also note that the second term in Eq. (9.180) is simply a renormalisation of the potential, or more specifically a shift in the SQUID inductance [16, 68] by a factor

$$\lambda = \frac{2\Omega\gamma}{\omega_0^2(1 + 2\Omega\gamma/\omega_0^2)},$$

as was the case in quantum Brownian motion. This term can therefore be considered a correction to the Hamiltonian of the SQUID. Again, as was the case for the Caldeira–Leggett equation, Eq. (9.180) is not Lindblad form and cannot guarantee physical dynamics. We therefore follow the same process of introducing a 'minimally invasive' $[\hat{Q}, [\hat{Q}, \hat{\rho}_S(t)]]$ term (remember that charge and momentum are analogous). Doing so, provides a master equation of Lindblad type

$$\frac{d\hat{\rho}_S(t)}{dt} = -\frac{i}{\hbar}[\hat{H}, \hat{\rho}_S(t)] + \frac{1}{2}\left([\hat{L}, \hat{\rho}_S(t)\hat{L}^\dagger] + [\hat{L}\hat{\rho}_S(t), \hat{L}^\dagger]\right) \tag{9.181}$$

with

$$\hat{H} = \frac{\hat{Q}^2}{2C} + \frac{\hat{\Phi}^2}{2L} - \hbar v \cos\left(\frac{2\pi}{\Phi_0}(\hat{\Phi} + \Phi_x)\right) + \frac{\gamma}{2}\left(\hat{\Phi}\hat{Q} + \hat{Q}\hat{\Phi}\right)$$

$$
\tag{9.182}
$$

$$\hat{L} = \sqrt{\frac{C\omega_0\gamma}{\hbar}}\,\hat{\Phi} + \left(i - \frac{\omega_0}{2\Omega}\right)\sqrt{\frac{\gamma}{Ch\omega_0}}\,\hat{Q}.$$

Note that in the high cut-off limit, as with the quantum Brownian case in the low-temperature regime, the Lindblad reduces to the scaled annihilation operator $\hat{L} = \sqrt{2\gamma}\hat{a}$. When not working in the high cut-off limit, our Lindblad can be written in terms of ladder operators; doing so gives

$$\hat{L} = \sqrt{2\gamma}\left[\left(1 - \frac{i\omega_0}{4\Omega}\right)\hat{a} + \frac{i\omega_0}{4\Omega}\hat{a}^\dagger\right]. \tag{9.183}$$

This will only be physically reasonable only when $\omega_0 \ll \Omega$.

The Lindblad master equation, when the series expansion $\hat{\Phi}(-\tau)$ is truncated at first order, looks very similar to the harmonic oscillator example. We cannot guarantee, however, that this is appropriate, since higher-order terms may play a significant role. If we now consider a second-order truncation of Eq. (9.174), we find that

$$\sum_n \frac{n!}{\Omega^n}A_n \approx \hat{\Phi} - \frac{\hat{Q}}{\Omega C} - \frac{\omega^2}{\Omega^2}\left(\hat{\Phi} + \frac{2\pi\hbar\nu L}{\Phi_0}\sin\left(\frac{2\pi}{\Phi_0}(\hat{\Phi} + \Phi_x)\right)\right), \tag{9.184}$$

where the external flux dependence, originating from the non-linear SQUID potential, can be seen to enter the dissipator for the first time. Substituting Eq. (9.184) into Eq. (9.177) then allows Eq. (9.170) to be rewritten as:

$$
\begin{aligned}
\frac{d\hat{\rho}_S}{dt} = &-\frac{i}{\hbar}[\hat{H}_S, \hat{\rho}_S(t)] + \frac{i\gamma\Omega C}{\hbar}\left(\overbrace{\left(1 - \frac{\omega_0^2}{\Omega^2}\right)[\hat{\Phi}^2, \hat{\rho}_S(t)]}^{\text{renormalises } L} - \overbrace{\frac{1}{\Omega C}[\hat{\Phi}, \{\hat{Q}, \hat{\rho}_S(t)\}]}^{\text{1}^{\text{st}}\text{-order dissipation}}\right. \\
&\left. - \underbrace{\frac{2\pi\hbar\nu L}{\Phi_0}\frac{\omega_0^2}{\Omega^2}\left[\hat{\Phi}, \left\{\sin\left(\frac{2\pi}{\Phi_0}(\hat{\Phi} + \Phi_x)\right), \hat{\rho}_S(t)\right\}\right]}_{\text{2}^{\text{nd}}\text{-order dissipation}}\right) \\
&- \frac{\gamma\omega_0 C}{2\hbar}\left(\underbrace{\left(1 - \frac{\omega_0^2}{\Omega^2}\right)[\hat{\Phi}, [\hat{\Phi}, \hat{\rho}_S(t)]]}_{\text{1}^{\text{st}}\text{ and 2}^{\text{nd}}\text{-order noise}} - \underbrace{\frac{1}{\Omega C}[\hat{\Phi}, [\hat{Q}, \hat{\rho}_S(t)]]}_{\text{1}^{\text{st}}\text{-order cut-off}}\right. \\
&\left. - \underbrace{\frac{2\pi\hbar\nu L}{\Phi_0}\frac{\omega_0^2}{\Omega^2}\left[\hat{\Phi}, \left[\sin\left(\frac{2\pi}{\Phi_0}(\hat{\Phi} + \Phi_x)\right), \hat{\rho}_S(t)\right]\right]}_{\text{2}^{\text{nd}}\text{-order cut-off}}\right),
\end{aligned}
\tag{9.185}
$$

where the term

$$\lambda = \frac{2\gamma\Omega\left(1 - \frac{\omega_0^2}{\Omega^2}\right)}{\omega_0^2\left(1 + \frac{2\gamma\Omega}{\omega_0^2}\left(1 - \frac{\omega_0^2}{\Omega^2}\right)\right)}$$

describes a renormalisation of the SQUID inductance with higher-order corrections.

The complexity of the master equation has increased substantially, and as higher-order truncations are considered, will only further increase. Again, this equation is not of Lindblad form, and drawing from the tools used in Section 9.3.3, we are in search of two Lindblad operators that have the form

$$\hat{L}_1 = \alpha_1\hat{\Phi} + \epsilon_1\hat{Q} \tag{9.186}$$

$$\hat{L}_2 = \alpha_2\hat{\Phi} + \epsilon_2\sin\left(\frac{2\pi}{\Phi_0}(\hat{\Phi} + \Phi_x)\right), \tag{9.187}$$

where $\alpha_1, \alpha_2, \epsilon_1, \epsilon_2$ are scalable constants. The issue with this task is that there is no unique way to split the terms up to assure a Lindblad equation, and we have some subjective freedom of choice. Each Lindblad has a corresponding coefficient matrix that requires insertion of a minimally invasive correction, but the definition of *minimally invasive* becomes blurred at this stage. One way of combating this is by introducing a weighting that dictates how much of each term contributes to each Lindblad, but this is a purely artificial insertion with no physical justification. Although we guarantee the required properties associated to the density matrix, we lose focus on what the model is actually trying to describe. Therefore, there appears to be a false security in transforming this equation into Lindblad type.

One could argue that we can reduce the number of corrections if we can find a way of naturally producing terms such as $[\hat{Q}, [\hat{Q}, \hat{\rho}_S(t)]]$. The first-order noise term carries exactly the same form and is a direct result of the inductive coupling between the SQUID and its environment. A natural progression is to assume that a degree of capacitive coupling will give rise to such terms without the need to introduce corrections; this is our next step.

9.5.2 Introducing Capacitive Coupling

We begin by assuming a SQUID that is coupled to a system of oscillators in the same way as before, except now we introduce some additional capacitive coupling. The Hamiltonians that describe this system-bath combination will then be given by

$$\hat{H}_S = \frac{\hat{Q}^2}{2C} + \frac{\hat{\Phi}^2}{2L} - \hbar v \cos\left(\frac{2\pi}{\Phi_0}(\hat{\Phi} + \Phi_x)\right) \tag{9.188}$$

$$\hat{H}_B = \sum_n \frac{\hat{Q}_n^2}{2C_n} + \frac{\hat{\Phi}_n^2}{2L_n} \tag{9.189}$$

$$\hat{H}_I = -\left(\hat{\Phi}\sum_n \kappa_n\hat{\Phi}_n + \hat{Q}\sum_n \eta_n\hat{Q}_n\right) = -\left(\hat{\Phi}\hat{B}_\Phi + \hat{Q}\hat{B}_Q\right), \tag{9.190}$$

where η_n is the capacitive coupling strength, and we have defined bath operators

$$\hat{B}_\phi = \sum_n \kappa_n \hat{\Phi}_n = \sum_n \sqrt{\frac{\kappa_n^2 \hbar}{2 C_n \omega_n}} \left(\hat{a}_n + \hat{a}_n^\dagger \right) \tag{9.191}$$

$$\hat{B}_Q = \sum_n \eta_n \hat{Q}_n = -i \sum_n \sqrt{\frac{\eta_n^2 \hbar C_n \omega_n}{2}} \left(\hat{a}_n - \hat{a}_n^\dagger \right) \tag{9.192}$$

in terms of the ladder operators $\hat{a}_n$ and $\hat{a}_n^\dagger$ once more. Using the same approach as in Section 9.3.2, with admittedly a few more steps, one arrives at a dissipator of

$$\begin{aligned}
\mathcal{K} = \frac{1}{2\hbar^2} \int_0^\infty d\tau & \left(\left\langle \left[\hat{B}_\Phi, \hat{B}_\Phi(-\tau) \right] \right\rangle_B \left[\hat{\Phi}, \left\{ \hat{\Phi}(-\tau), \hat{\rho}_S(t) \right\}_+ \right] \right. \\
& + \left\langle \left\{ \hat{B}_\Phi, \hat{B}_\Phi(-\tau) \right\}_+ \right\rangle_B \left[\hat{\Phi}, \left[\hat{\Phi}(-\tau), \hat{\rho}_S(t) \right] \right] \\
& + \left\langle \left[\hat{B}_Q, \hat{B}_Q(-\tau) \right] \right\rangle_B \left[\hat{Q}, \left\{ \hat{Q}(-\tau), \hat{\rho}_S(t) \right\}_+ \right] \\
& + \left\langle \left\{ \hat{B}_Q, \hat{B}_Q(-\tau) \right\}_+ \right\rangle_B \left[\hat{Q}, \left[\hat{Q}(-\tau), \hat{\rho}_S(t) \right] \right] \\
& + \left\langle \left[\hat{B}_\Phi, \hat{B}_Q(-\tau) \right] \right\rangle_B \left[\hat{\Phi}, \left\{ \hat{Q}(-\tau), \hat{\rho}_S(t) \right\}_+ \right] \\
& + \left\langle \left\{ \hat{B}_\Phi, \hat{B}_Q(-\tau) \right\}_+ \right\rangle_B \left[\hat{\Phi}, \left[\hat{Q}(-\tau), \hat{\rho}_S(t) \right] \right] \\
& + \left\langle \left[\hat{B}_Q, \hat{B}_\Phi(-\tau) \right] \right\rangle_B \left[\hat{Q}, \left\{ \hat{\Phi}(-\tau), \hat{\rho}_S(t) \right\}_+ \right] \\
& \left. + \left\langle \left\{ \hat{B}_Q, \hat{B}_\Phi(-\tau) \right\}_+ \right\rangle_B \left[\hat{Q}, \left[\hat{\Phi}(-\tau), \hat{\rho}_S(t) \right] \right] \right). \tag{9.193}
\end{aligned}$$

Note that we now have additional kernels that need evaluating. We can define our dissipation kernels such that

$$\begin{aligned}
i D_{\Phi\Phi}(\tau) &= \left\langle \left[\hat{B}_\Phi, \hat{B}_\Phi(-\tau) \right] \right\rangle_B \\
i D_{QQ}(\tau) &= \left\langle \left[\hat{B}_Q, \hat{B}_Q(-\tau) \right] \right\rangle_B \\
i D_{\Phi Q}(\tau) &= \left\langle \left[\hat{B}_\Phi, \hat{B}_Q(-\tau) \right] \right\rangle_B \\
i D_{Q\Phi}(\tau) &= \left\langle \left[\hat{B}_Q, \hat{B}_\Phi(-\tau) \right] \right\rangle_B,
\end{aligned} \tag{9.194}$$

and noise kernels such that

$$\begin{aligned}
D_{1\Phi\Phi}(\tau) &= \left\langle \left\{ \hat{B}_\Phi, \hat{B}_\Phi(-\tau) \right\}_+ \right\rangle_B \\
D_{1QQ}(\tau) &= \left\langle \left\{ \hat{B}_Q, \hat{B}_Q(-\tau) \right\}_+ \right\rangle_B \\
D_{1\Phi Q}(\tau) &= \left\langle \left\{ \hat{B}_\Phi, \hat{B}_Q(-\tau) \right\}_+ \right\rangle_B \\
D_{1Q\Phi}(\tau) &= \left\langle \left\{ \hat{B}_Q, \hat{B}_\Phi(-\tau) \right\}_+ \right\rangle_B,
\end{aligned} \tag{9.195}$$

using definitions of the effective spectral densities associated to each set of correlation functions, assuming Ohmic damping once more,

$$\begin{aligned}
J_{\Phi\Phi}(\omega) &= \frac{2C\gamma_{\Phi\Phi}}{\pi} \omega \frac{\Omega^2}{\omega^2 + \Omega^2} \\
J_{QQ}(\omega) &= \frac{2L\gamma_{QQ}}{\pi} \omega \frac{\Omega^2}{\omega^2 + \Omega^2} \\
J_{\Phi Q}(\omega) &= \frac{2\gamma_{\Phi Q}}{\pi\omega_0} \omega \frac{\Omega^2}{\omega^2 + \Omega^2} = J_{Q\Phi}(\omega),
\end{aligned} \tag{9.196}$$

where γ_{ij} is the damping rate associated to each effective spectral density. Recall that the spectral density and correlation function form a Fourier transform pair, so these damping rates quantify the correlation between flux and charge within the bath before and after the initial interaction. Inspection of each spectral density and how they relate to their corresponding correlation functions allows us to define the mixed damping rate in terms of the inductive and capacitive damping rates through

$$\gamma_{\Phi Q} = \sqrt{\gamma_{\Phi\Phi}\gamma_{QQ}}. \tag{9.197}$$

We can then write the capacitive damping rate γ_{QQ} in terms of the inductive damping rate $\gamma_{\Phi\Phi}$, using a relative coupling strength g,

$$\gamma_{QQ} = g^2\gamma_{\Phi\Phi} = g^2\gamma \tag{9.198}$$

and evaluate the correlation time-dependent operators $\hat{\Phi}(-\tau)$ and $\hat{Q}(-\tau)$ to the first order in the Hadamard expansion. We then find that the master equation can be written, rather dauntingly, as [20]

$$\frac{d\hat{\rho}_S(t)}{dt} = -\frac{i}{\hbar}\left[\hat{H}_S + \hat{H}_c, \hat{\rho}_S(t)\right]$$

$$-\frac{i\gamma}{\hbar}(1 + g^2 - g)\left[\hat{\Phi}, \{\hat{Q}, \hat{\rho}_S(t)\}_+\right] - \frac{i\gamma}{\hbar}\left(g^2 - \frac{g}{2}\right)\left[\{\hat{\Phi}, \hat{Q}\}_+, \hat{\rho}_S(t)\right]$$

$$-\frac{C\omega_0\gamma}{\hbar}\left(g + \frac{1}{2}\right)\left[\hat{\Phi}, [\hat{\Phi}, \hat{\rho}_S(t)]\right] - \frac{\omega_0\gamma}{2\hbar\Omega}(1 - g^2)\left[\hat{\Phi}, [\hat{Q}, \hat{\rho}_S(t)]\right]$$

$$-\frac{\gamma}{C\hbar\omega_0}\left(g + \frac{g^2}{2}\right)\left[\hat{Q}, [\hat{Q}\hat{\rho}_S(t)]\right]$$

$$+\frac{i g\gamma}{2\hbar L\Omega}\left[\hat{\Phi}, \left\{I_cL\sin\left(\frac{2\pi}{\Phi_0}(\hat{\Phi} + \Phi_x)\right), \hat{\rho}_S(t)\right\}_+\right]$$

$$-\frac{C\omega_0 g\gamma}{\hbar}\left[\hat{\Phi}, \left[I_cL\sin\left(\frac{2\pi}{\Phi_0}(\hat{\Phi} + \Phi_x)\right), \hat{\rho}_S(t)\right]\right]$$

$$+\frac{i g^2\gamma}{\hbar}\left[\hat{Q}, \left\{I_cL\sin\left(\frac{2\pi}{\Phi_0}(\hat{\Phi} + \Phi_x)\right), \hat{\rho}_S(t)\right\}_+\right]$$

$$-\frac{g^2\gamma\omega_0}{2\hbar\Omega}\left[\hat{Q}, \left[I_cL\sin\left(\frac{2\pi}{\Phi_0}(\hat{\Phi} + \Phi_x)\right), \hat{\rho}_S(t)\right]\right], \tag{9.199}$$

where

$$\hat{H}_C = \left(C\Omega\gamma + \frac{g\gamma}{2L\Omega}\right)\hat{\Phi}^2 + \left(L\omega g^2\gamma + \frac{g\gamma}{2C\Omega}\right)\hat{Q}^2 \tag{9.200}$$

is responsible for the re-normalisation of the SQUID's inductance and capacitance. We can clearly see that we are a long way away from the simple master equation for the quantum Brownian particle that we saw at the beginning of this chapter. A first-order analysis that incorporates parasitic capacitance still possesses more complexity than the second-order inductive model. As with the previous cases, this equation needs to be cast into Lindblad form. The different terms can all be restructured in the way outlined in Section 9.3.3, but this time we have a three-dimensional coefficient matrix, and the additional term required to guarantee a physical dynamical map is non-trivial. We find that casting this equation into Lindblad form, and guaranteeing the positivity of the map, automatically constrains the coupling ratio g, which further highlights the drawbacks of using Lindblad equations. The effective Hamiltonians produced in the 'Lindbladification' process are given by

$$\hat{H}_{\Phi Q} = \left(3g^2 - 2 + 1\right) \frac{\gamma}{2} \{\hat{\Phi}\hat{Q}\}_+ \tag{9.201}$$

$$\hat{H}_{\Phi S} = \frac{g}{L\Omega} \frac{\gamma}{2} I_c L \hat{\Phi} \sin\left(\frac{2\pi}{\Phi_0}(\hat{\Phi} + \Phi_x)\right) \tag{9.202}$$

$$\hat{H}_{QS} = g^2 I_c L \frac{\gamma}{2} \left\{ \hat{Q}, \sin\left(\frac{2\pi}{\Phi_0}(\hat{\Phi} + \Phi_x)\right) \right\}_+ + \frac{\gamma}{2} g^2 \frac{\pi I_c L}{\Phi_0 \Omega} \cos\left(\frac{2\pi}{\Phi_0}(\hat{\Phi} + \Phi_x)\right), \tag{9.203}$$

which have been obtained in the same manner as $\hat{H}_{qp}$ in Section 9.3.3.

We see that in constructing a Lindblad equation, we obtain many more effective Hamiltonian terms. Each term impacts the system that we are modelling, and it begs the question: have we altered the model too much in our goal to obtain a physically viable solution? Or are these real and physically meaningful renormalisations of the system's energy that we should account for? If the latter is the case, what does that mean for device characterisation? That is, does any experiment only probe the effective (or dressed) Hamiltonian of the system-of-interest so that we may never be able to directly access the parameters of the bare Hamiltonian? This is particularly important for engineering applications where accurate device characterisation may be of crucial importance for the integration of quantum components in complex systems (a topic we will return to in Chapter 11).

The questions raised by applying this master equation framework to a non-trivial system such as the SQUID will also be pertinent to many other systems. We also see that the Hamiltonian terms that arise in obtaining the master equation are not just renormalisations in the system's parameters, but also alter the functional form of the Hamiltonian. The crucial observation here is that open quantum systems are subtle and hard to understand. The methods of master equations may be useful for gaining a qualitative understanding of a system, but their validity for ubiquitous use in engineering applications is far from decided. Because of the emergence and importance of engineering quantum technologies, capabilities in modelling and simulation of open systems will be required (especially if robust model-based systems-engineering tools are to be developed); we will expand this point next.

9.5.3 Where Does the Model Break Down?

The preceding example of a SQUID coupled to an Ohmic bath acts as a nice example that demonstrates that, when considering any system more complex than a harmonic oscillator, many issues arise in deriving a suitable master equation. These include:

- The manipulations required to create a Lindblad form master equation may be substantial and subjective. They become more involved rather than less, as a hierarchy using Eq. (9.175) is developed. In order to achieve Lindblad form, more terms must subsequently be added to both the master equation and to the system Hamiltonian. These terms are arbitrary and invasive (as in they may both renormalise and change the functional form of the Hamiltonian) at all orders.

- Higher-order expansions necessitate the inclusion of additional assumptions in order to generate analytic expressions for the master equation, such as confining the system to the zero-temperature limit.
- The introduction of a more physically meaningful coupling, by including the important effects of capacitive coupling, introduces even greater arbitrariness in the resultant Lindblad model.
- It is unclear what the resultant models actually describe. Although they generate physically admissible density operators outputs for any set of input parameters, the structure of assumptions on which they are built, as well as the ability to arbitrarily and significantly change their outputs through non-physical parameter adjustment, cast ambiguity on their ties and usefulness for realistic modelling.

The conclusion, for any system other than the harmonic oscillator, is that one is left with an unattractive choice. Either the master equation is not guaranteed to produce physical solutions under all circumstances, or it is modified by the addition of artificial terms whose impact on the dynamics can neither be guaranteed to be small, nor to be based in physics. It may be that experiment might be able to resolve those issues, but in the absence of a general framework, the verification and validation of any open systems model might have to be done on a device-by-device and a system-by-system basis.

If, indeed, the price to pay for a density operator with guaranteed physicality is too high, then perhaps, in fact, the benefit is too small. A primary criticism levelled against non-Lindblad master equations is that their dynamics can be non-physical (in particular, a failure to conserve trace and positive eigenvalues), during their initial dynamics [8, 31, 32, 63, 80], in some regions of the overall parameter space. To some extent this is a consequence of the Markov approximation, and the extension of taking the integral limits in time to infinity, which will lose some transient effects. The consequence of this is heightened by the tendency to discontinuously switch on interaction between an uncorrelated system and bath, resulting in sudden initial system dynamics — the existing literature demonstrating non-physicality of non-Lindblad master equations typically highlight behaviour close to $t = 0$, where these effects are most significant (and where Lindblad dynamics are unlikely to be valid at all). The idea of having a model with good predictive (but not necessarily physical) capability within a well-specified, but not universal, predictive range is normal in engineering. The generation of models suitable for engineering open systems may therefore need to be more guided by the requirements of the technical ambition rather than physical modelling needs.

9.6 Unravelling the Master Equation

Prerequisite Material: In this section we will refer to noise a lot. The subject of noise is a subtle one. As part of this discussion, we will refer to stochastic processes that are a way of modelling random phenomena. Also note that we will write differential equations which include noise sources. Solving such equations with due mathematical rigour is an advanced subject in its own right and requires calculus such as that of Itô or Stratonovich. As we have already covered numerics using Euler methods, it is perhaps easiest to think of the equations we will use as defining Euler increments. Provided the step size is small enough, this view will be enough to understand the contents of this section. The interested reader, and especially those seeking to apply any modelling of random phenomena in earnest, would benefit from a study of stochastic calculus.

We began this chapter by considering the addition of classical Brownian motion to a dynamical system. We showed that the thermal fluctuations and environment effects on a particle lead to effective damping. We imagine that for each experiment, there is one set of random functions that influence the dynamics of the particle. In some circumstances (lots of very small fluctuations), we can consider an average effect and derive an effective deterministic equation of motion. However, this is not always the case, and sometimes we can only predict the time evolution of the probability density function deterministically. The Fokker–Planck and Kolmogorov equations are examples of this approach. We will not give details here, but suffice it to say that, just like the Liouville equation, they describe the dynamics of a system's probability density function, and contain drift and diffusion terms that describe the average effect of the stochastic fluctuations from environmental noise sources. They are in effect classical versions of quantum master equations. We have argued forwards from the Langevin equation to this position. We can, however, reverse this argument and ask instead: *what are the set of trajectories that could have produced a given average behaviour.* For example, given solutions to the Fokker–Planck equation, which Langevin equations exist, whose average probability density function generated over all trajectories is the same? The nature of fluctuations can take many forms, e.g. shot noise, Gaussian noise, white noise, thermal noise, and $1/f$-noise, each with their own very different characteristics. There may be a Langevin equation for each of the different noise sources that give the same average behaviour, but where each of the individual physical trajectories for each type of source looks very different. In order to understand the behaviour we might see in a real experiment, we may therefore want to perform an *unravelling* of the average behaviour into individual trajectories, for which we model noise of the kind that is present in the real system. In this way, we would compare the characteristic predictions of our theory with the behaviour we see in experiments. Unlike in a deterministic model, we would never expect to see exact matches between theoretical and experimental trajectories, only an agreement of qualitative behaviour or average properties such as a signal's power spectrum. Another, very different, use of the unravelling approach is as a useful computational trick, if solving for the probability density function directly is too computationally hard (it is an example of a Monte Carlo method).

We can consider the master equation as describing the average behaviour of a quantum system over many separate experiments. In the same vein as the above discussion, we might also be interested in knowing what the individual trajectories of pure state vectors of a quantum system subject to an environment might look like. Just as in the classical case, we would need to impose the convergence criterion that the average solution agrees with the predictions of the master equation. Specifically, we would require that for N trajectories $\{|\psi_n(t)\rangle\}$,

$\hat{\rho}(t) = \frac{1}{N}\sum_n |\psi_n(t)\rangle \langle\psi_n(t)|$. Note that each of these solutions, $|\psi_n\rangle$, is termed a realisation of the underlying stochastic process.

There are in fact an infinite number of ways to do this. Here we will describe two that we believe to be the most important, namely *quantum state diffusion* (QSD) and *quantum jumps*. We have chosen these examples, not only because they are amongst the most common unravellings of the master equation, but also because they are very different in their form. The former models a continuous noise process, and the latter a discontinuous one. Despite this substantial difference, they do indeed average over many realisations to the density matrix solution of the master equation. Note that we will not provide anything more than the key features of each of these models, as this is a field of study in its own right and there exist excellent resources for those who wish to gain an in-depth understanding of this fascinating approach.

Before going into details, we note that there are two main ways to understand the unravelling of the master equation. The first is to take a realist standpoint and argue that our model of the environment represents in some way the effects of the presence of a real environment on the system, and that each quantum trajectory is some realisation of what might happen to the quantum state in a real experiment[5]. In the other extreme of positivism, one can also consider any environment as a sink of information. That is, that the environment will take quantum information from the system and the state of the environment will therefore change. For many, but not all, environments unravellings of the master equation tend to show localisation of the quantum state which is consistent with this line of argument. One can therefore argue that each unravelling of the master equation is simply a way to mathematically model different measurement processes [94]. The two perspectives are not quite as mutually exclusive as one might at first think, as '*When the diffusion takes place as a result of an interaction of the system with its environment, it is immaterial whether that environment happens to include measuring apparatus, or some more general environment. Almost any interaction with an environment produces localization, with different environments producing different kinds of localization. Measurement is nothing special*' [65]. We will talk in more depth about such foundational questions in the next chapter – but for now it suffices to note that the situation is not as simple as such arguments imply.

The first unravelling we consider is QSD. Here, the environment is considered to affect the quantum state of the system-of-interest continuously, like a Brownian particle. The description is given in the Schrödinger picture. As the state vector is defined over the complex numbers, it seems reasonable to expect that the introduction of noise through a stochastic (random number) variable will also be complex. Just as with a Brownian motion, we would expect the mean distance traversed in the complex plane to be zero, and the variance to be proportional to the time. If we denote by $d\xi$ a stochastic complex

---

5 These models predict dynamics in terms of pure states. Hence, no accommodation is made for the entanglement that would happen between system and environment in a real physical system. In addition, the models are nonlinear and do not respect the principle of superposition. Hence, some care should be taken when thinking about these models from a realist perspective.

(so-called Wiener) increment, then the implication of assuming Brownian motion-type behaviour is that $\overline{d\xi^2} = \overline{d\xi} = 0$ and $\overline{d\xi d\xi^*} = dt$. After some work, one can show that the evolution of the state vector $|\psi\rangle$ is given by the increment (Itô equation)

$$|d\psi\rangle = -\frac{i}{\hbar}\hat{H}|\psi\rangle\,dt + \sum_j \left[\langle\hat{L}_j^\dagger\rangle\hat{L}_j - \frac{1}{2}\hat{L}_j^\dagger\hat{L}_j - \frac{1}{2}\langle\hat{L}\rangle_j^\dagger\langle\hat{L}_j\rangle\right]|\psi\rangle\,dt$$

$$+ \sum_j \left[\hat{L}_j - \langle\hat{L}_j\rangle\right]|\psi\rangle\,d\xi, \tag{9.204}$$

where $\hat{L}_i$ are the Lindblad operators from the master equation and dt is the time increment. The first term on the right-hand side is simply the normal dynamics from the Schrödinger equation. The second and third terms modify the dynamics in a way that is analogous to the respective drift and diffusion (or fluctuation) terms of the Fokker–Planck equation. As such, we see that QSD does look very much like a quantum analogue to Brownian motion. Concluding that it is also possible to interpret QSD in a way that has any connection to the measurement perspective may seem a little tenuous. If one considers that a bath of harmonic oscillators could represent a finite quality factor resonator, QSD can be thought of as a set-up that is in some ways similar to oscillator-based measurement such as the homodyne detection principles used by, e.g. lock-in amplifiers. In either case, QSD has found great utility in understanding open quantum systems, especially in the context of trying to understand the quantum-to-classical transition. These are topics we will revisit in the next chapter.

Our other example of unravelling is that of quantum jumps [11]. This model, being discontinuous, is very different from QSD. Here, we model an environment whose effect is to absorb whole 'photons' from the system-of-interest. The noise process for this reflects the discrete nature of photon absorption. The stochastic increment that reflects this process should share the statistics that we would expect based on observation. It therefore seems reasonable to assume that the increment, dN_j, is a Poissonian noise process where $dN_j\,dN_k = \delta_{jk}dN_j\,dN_j\,dt = 0$ (i.e. the jumps are uncorrelated and instantaneous) and $\overline{dN_j} = \langle\hat{L}_j^\dagger\hat{L}_j\rangle\,dt$. The last term indicates that jumps occur randomly at a rate that is determined by $\langle\hat{L}_j^\dagger\hat{L}_j\rangle$. In the frequently occurring case that a Lindblad is proportional to the annihilation operator, that $\langle\hat{a}^\dagger\hat{a}_j\rangle = \langle\hat{n}\rangle$ expresses the idea that the chance of detecting a photon is proportional to the expected number of photons in the state. The pure state stochastic evolution equation for quantum jumps is then given by

$$|d\psi\rangle = -\frac{i}{\hbar}\hat{H}|\psi\rangle\,dt - \frac{1}{2}\sum_j \left[\hat{L}_j^\dagger\hat{L}_j - \langle\hat{L}_j^\dagger\hat{L}_j\rangle\right]|\psi\rangle\,dt$$

$$+ \sum_j \left[\frac{\hat{L}_j}{\sqrt{\langle\hat{L}_j^\dagger\hat{L}_j\rangle}} - 1\right]|\psi\rangle\,dN_j. \tag{9.205}$$

Unlike QSD, the connection to a measurement process here is very clear and takes the form of photon counting. That is, an environment that absorbs whole photons and a photon measurement device are in effect equivalent concepts. The only practical difference is that in the latter case, one assumes the presence of some readout that is able to count when a jump has been made. It is also interesting to note that the behaviour of each can mimic the other in an appropriate limit. As to the foundational difference between the two, we will leave that until the next chapter, where we discuss such topics in some depth.

10

Foundations: Measurement and the Quantum-to-Classical Transition

> *"SOCRATES: Or suppose that we differ about magnitudes, do we not quickly end the differences by measuring?"*
>
> Plato – early Socratic dialogues

10.1 Introduction

In this chapter we present a discussion on two related topics that have generated lengthy (and sometimes heated) debates since the conception of quantum physics, and to which no consensus yet exists. Get any two physicists together in a social environment and, given sufficient time, there is a fair chance that they will at some point start a discussion on one or both of these topics. There is also a fair chance that they will find something to disagree on. There is even a fair chance that neither will concede much, if any, ground. Some will ardently defend specific interpretations of quantum mechanics and it seems the case that all of us, if not staunch irrational fans of a particular view, will harbour an irrational dislike of some. Perhaps ironically, and for full disclosure, my[1] irrational dislike is for the one named after Everett (it seems to me that it fails Occam's razor – it is an irrational dislike, as there is no reason to assume *a priori* that this principle of parsimony should apply). Most books on quantum mechanics will advocate (implicitly if not explicitly) in favour of one interpretation only. We have so far tried to minimise the metaphysics of interpreting quantum mechanics and rather tried to focus our discussion in terms of what the theory predicts for the outcomes of experiments. That is, as empirical scientists, we have tried to be as positivistic as we can be.

   The scientific process is never truly positivist and, from time to time, it is necessary to consider the ontology and metaphysics of a theory. But this is the case only in so far as it helps us form tests of our current theory or helps us extend or modify an existing theory to be a new and better theory. It is not easy to articulate the full scientific process, which is not one limited to induction, logic, and reason alone but also includes a good bit of creative guesswork. A detailed discussion of philosophy and the methodology of science is

---

1 As a somewhat personal set of reflections I have chosen to write this chapter in the first person.

Quantum Mechanics: From Analytical Mechanics to Quantum Mechanics, Simulation, Foundations, and Engineering,
First Edition. Mark Julian Everitt, Kieran Niels Bjergstrom and Stephen Neil Alexander Duffus.
© 2024 John Wiley & Sons Ltd. Published 2024 by John Wiley & Sons Ltd.
Companion website: www.wiley.com/go/everitt/quantum

beyond the scope of this book, but if you are keen to read around, please be aware that some philosophers of science have presented lengthy discussions of quantum mechanics, especially on the topics of this chapter. Such works can make interesting and valuable reading but it is also the case that the understanding of quantum mechanics can be somewhat limited or prejudiced by a specific interpretation. The one thing we can say, is that we must always return to reproducible, testable, and falsifiable experiments (positivists rejoice).[2]

For this reason, we are going to attempt a discussion of the foundations of quantum mechanics that tries to avoid any unnecessary metaphysical debate. Our approach follows a judicious application of Newton's Flaming Laser Sword, which follows '*Newtonian insistence on ensuring that any statement is testable by observation (or has logical consequences which are so testable)*' [1], but bringing in some ontological considerations where doing so is useful. We will try, without being overly pedantic, to focus our discussion on understanding the questions at hand and what needs to be done to resolve those questions. Before getting into specifics, we therefore start by listing some of the diverse forms in which classical and quantum mechanics can be presented as well as some features of each formulation:

		State	Dynamics	Linear	Local	Projection
Classical	Lagrangian	$q_i, \dot{q}_i$	Eq. (2.7)	N	Y[a]	N
	Hamiltonian	q_i, p_i	Eq. (2.20)	N	Y	N
	KVN	$|\psi(t)\rangle$	Eq. (2.48)	Y	Y	Sort of
	Louivillian	$\rho(q, p)$	Eq. (2.48)	Y	Y	Sort of
Quantum	Path integral	$q_i, \dot{q}_i$	Eq. (4.88)	Y	N	Sort of
	Heisenberg	$|\psi(0)\rangle$	Eq. (4.21)	Y	N	Y
	Schrödinger	$|\psi(t)\rangle$	Eq. (3.8)	Y	N	Y
	Phase space	$W(q, p)$	Eq. (4.52)	Y	N	Y

a) Least action may be a global view but it is still a local theory.

To try to make clear the open nature of the measurement problem and quantum-to-classical transition, we present the remaining discussion as a Socratic dialogue. We hope doing this will help stimulate a critical analysis of our arguments and make more clear any implicit assumptions we make along the way. Following Lakatos, who used this method to great effect,

> 'one has to purge one's mind from perverted illusions, one has to learn how to see and how to define correctly what one sees'.
>
> Imre Lakatos – Proofs and Refutations

---

2 Ideas such as these can be quite subtle and, for the sake of brevity, we have no room to discuss them here. Superficially, falsifiability is the capacity to logically contradict a statement by some empirical test. That said, we still use quantum mechanics even though it fails to predict how very massive objects evolve and we will use general relativity even though it fails to predict the structure of matter. This might lead one to ask, how right is each theory and how much does their falsification matter? For those interested in understanding these ideas in more detail, *The Logic of Scientific Discovery* by Popper and *The Methodology of Scientific Research Programmes* by Latakos et al. are good starting points. I would suggest reading these and then forming your own view on the '*Duhem–Quine thesis*'.

We have similarly sought to use the Socratic method as one of hypothesis elimination and to guide the discussion to some natural conclusions. That said, it is easy to miss hidden assumptions, so we leave as a challenge to the reader to see what other destinations may be possible. Feynman once said: *'I think I can safely say that nobody understands quantum mechanics'*. Often this is taken out of context, as much of quantum mechanics is understandable. Hopefully, our introduction to the different formulations of quantum mechanics and their connection to various forms of classical mechanics has helped make clear those parts of quantum theory that are as understandable as their classical counterparts. Feynman's statement does however carry weight for the current topic, as there have been many different attempts to interpret quantum mechanics, to resolve the measurement problem, and understand the theory's classical correspondence – none of which have been universally accepted. It would be safe to say that nobody fully understands quantum measurement and the quantum-to-classical transition. It is very much our intention not to be definitive in our treatment of this topic, but rather to use this chapter to generate thought and discussion. With this in mind, let us progress to the dialogue.

10.2 The Measurement Problem

Question: So what is this measurement problem anyway?

Reply: It is the problem that the rules of measurement in quantum mechanics appear to be fundamentally different from the rules of evolution. This is especially the case in the Schrödinger picture. Here, when a quantum system is evolving without being measured, the Schrödinger equation or some equivalent one in Heisenberg, phase space, or path integral formulation governs the system's dynamics. On measurement, the standard postulates of quantum mechanics in this formalism state:

M1 When an observable is measured, the only possible outcome of that measurement is an eigenvalue of the observable.

M2 The probability of measuring a given eigenvalue is the expectation value of the current state of the system with the projection operator into the subspace spanned by all the eigenvectors of that eigenvalue.

M3 On measurement the state of the system is projected into a normalised eigenstate of the system, associated with the measured eigenvalue (note that if the eigenvalue is degenerate, the state is projected into the subspace spanned by eigenvectors associated with that eigenvalue).

This is fundamentally inconsistent with its dynamical evolution, i.e. measurement has different rules from dynamics. Things are less problematic in the Heisenberg picture where the state is only ever the initial knowledge of the system. Here, measurement simply updates that information. Either case is unlike classical mechanics where, in theory, we can model any measurement device with the same rules as the system-of-interest by widening our model to include the measurement device itself. While we have no right to assume that the rules of nature will be consistent and elegant, it is very hard to be comfortable with this state-of-affairs. More specifically, the measurement problem is the *if* or the *how* or the *why* of the quantum-state of the system collapsing into an eigenstate of the observable quantity being measured.

Question: It sounds like ontology may be important here. Do we not need to discuss interpretations of quantum mechanics to be able to have this discussion?

Reply: As outlined in the introduction, no, or at least, not in any great depth. We will allude to some of them, but it is worth explaining why we will not be going into any great detail before proceeding further. Firstly, there are many works entirely dedicated to the ontological and epistemological issues in quantum mechanics, and a full discussion cannot fit within a single chapter of a book. Then there is the case that many philosophical discussions such as this are very entertaining but often depart from the strict remit of the scientific method. If different interpretations of any physical theory do not lead to experimentally distinguishable outcomes, then they are as good as each other (even if one personally dislikes one, it remains a viable candidate).

There is also the danger of thinking beyond the scope of validity of quantum mechanics. For example, many of the thought experiments used to illustrate interpretive arguments involve systems large enough for gravity to be important. As quantum theory does not even pretend to be valid in this domain, any such example has the danger of imposing classical (or at least non-quantum) thinking in an inappropriate way (and we may make similar category errors when later using open systems arguments – but we will try to point these out as we go). Interpretations therefore start to be important only when one specifically seeks to change the theory itself – such as might be the case if one wanted to modify the measurement axioms of quantum mechanics. It is rather the hope that out of this discussion we will establish some scientific basis for being able to have more rigorous metaphysical discussions with your friends and colleagues (or even yourself).

Question: Do we at least need to establish if quantum mechanics is fundamentally nondeterministic of nature? After all, are not many of these interpretations an attempt to refute or establish the fundamental probabilistic nature of quantum mechanics and are therefore tied directly to the measurement problem itself?

Reply: Some are indeed, but if one really wants to reject probabilistic interpretations, one is beholden to replace the existing measurement axioms with something else in sympathy with the correspondence principle that is deterministic. That is, the new theory (for it will be a new theory) must produce the same experimental predictions as standard quantum mechanics in the domain where standard quantum mechanics is known to be good. But before we consider this aspect of the measurement problem, I have a question for you: In terms of the experimental part of the scientific method, is classical mechanics not also a probabilistic theory?

Question: In Chapter 2 you talked a little about the fact that we can only measure the state of any system within some degree of accuracy and precision. This observation leads us to the fact that, for experimental purposes at least, in principle we should only ever represent a system by its probability density function (PDF). So is this the same?

Reply: It certainly has something in common with the Born Rules (M1 and M2), and we predict the future state of the system from this in a probabilistic model such as the dynamics of the PDF determined by the Liouville equation. If there is any non-linearity or damping, the evolution of the probability density may be highly non-trivial in nature (recall our Duffing oscillator example in Chapter 2). In this respect, both quantum and classical mechanics are probabilistic. In quantum mechanics there is the dynamics of, e.g. a state vector (Schrödinger picture), the observables (Heisenberg picture), or a phase space distribution (Wigner–Weyl picture). But the same can be said of some of the classical pictures (Lagrangian, Hamiltonian, Liouvillian, Koopman–von Neumann [KVN]). We could loosely describe all of the quantum and some of the classical as wave or ray theories – there is no 'particle' in them. This is why we choose not to discuss the wave/particle duality, as the closest we can come to the classical notion or a point particle is the Gaussian wave packet of a coherent state. The difference is that in classical mechanics there is an underlying deterministic model for a point particle (least action, Lagrangian, or Hamiltonian dynamics).

Question: I am not sure that I agree with this perspective; I feel that in some ways, classical mechanics is not as probabilistic as quantum mechanics. For now we can at least agree that there is a fundamental difference in that a point particle cannot be associated with the quantum description of a system, whereas in classical physics this is possible. I would have presented my argument differently though. The commutativity of classical physics means that, in theory, a valid limit of the PDF for the classical system is a Dirac delta-function (the 'point' particle). In quantum mechanics, as a corollary of the Heisenberg uncertainty principle, this is not allowed as $\Delta \hat{A} \Delta \hat{B} \geq \frac{1}{2} \left| \langle [\hat{A}, \hat{B}] \rangle \right|$. In this way, the point particle idea of classical physics is not in conflict with the probabilistic one; it is simply a limiting case – and this limit does not exist for the quantum description. Does this mean that we are doomed to fail in any attempt to resolve a classical correspondence limit?

Reply: Not necessarily. Remember that the correspondence limit only talks about both models producing the same experimental predictions. Together with any measurement process being compatible with that dynamics (indeed the measurement process itself might even be the cause of growing the classical correspondence in the first place [23]). In any case, because classical mechanics is practically (if not fundamentally) probabilistic, questions on the fundamental probabilistic nature of quantum mechanics may be irresolvable. Those on the measurement process itself and of achieving a correspondence limit may turn out to be tractable (even if a modification of the axioms of quantum mechanics is required).

Question: Let us turn now to the third measurement axiom. In Chapter 2 you talked a little about a 'collapse' of the classical PDF. On measurement of this system we will determine where, from within the PDF, the system is, and the new state of the system will be the PDF associated with the outcome of the measurement. In this way the PDF instantly 'collapses', based on the outcome of the measurement. So, is this the same as M3?

Reply: It is not at all the same. Let us now expand on the classical situation. Our discussion will (i) make clear the classical measurement process, (ii) make clear how very different the quantum measurement process is, and (iii) set the scene for later discussions on possible resolutions of the measurement problem.

Question: Collapse of the wave-function and collapse of the PDF seem to have a lot in common – how do they differ?

Reply: As we gain a better knowledge of a classical system, we are simply excluding probabilities from its PDF. If, for example, we look at the PDF for the Duffing oscillator in Chapter 2 on page 52, we see that there are positions for which multiple momenta are defined. If, classically, we were just to measure the position of the oscillator with some sufficient degree of accuracy, at such an approximate time and position the resulting PDF would be a number of disjoint blobs in momentum (likely with a Gaussian profile over position). In quantum mechanics we are told that on measurement of an observable (here $\hat{q}$), we can only measure an eigenvalue (q_0, say). After the measurement, the system is in an eigenstate $|q_0\rangle$ of $\hat{q}$, whose position representation wave-function is $\psi(q) = \langle q|q_0\rangle = \delta(q - q_0)$. This state is delocalised over all of momentum, and the outcome is shockingly different to the classically 'equivalent' process.

Question: So, if we take a probabilistic view of the quantum state, the problem is not so much that the probability distribution suddenly 'collapses' – it is what it collapses into?

Reply: If we can only make measurements every so often in the classical picture. On measurement, we simply refine the knowledge of the PDF. In the quantum situation, the state of the system is projected into an eigenstate (so long as that eigenstate was a component of the initial quantum state) of that observable (with some ambiguity only if there is some degeneracy associated with the eigenvalue being measured).

It is worth noting again that the conceptual difficulty of collapsing the wave-function is really only an issue associated with the dynamics of the state, such as in the Schrödinger picture. In the Heisenberg picture the state vector is only ever the last state measured and does not have any dynamics of its own (it is the observables that evolve). In this way the Heisenberg picture is conceptually less problematic than the Schrödinger picture with respect to the measurement axioms.

10.3 Refining the Idea of Measurement

Question: The above discussion all seems rather abstract. I am also not sure that what they say happens, really is what happens experimentally. We have discussed measurement of eigenvalues of operators without ever saying anything about the way that they are actually measured. As physics is an empirical science, do we not have an obligation to try to describe this process?

Reply: In undergraduate classical physics, outside the laboratory, it is unusual to talk about measurement in great detail. Granted that almost all physics students are aware that the finite resistance of a voltage or current meter is sometimes important to account for, but I cannot think of many other ubiquitous examples. Interestingly, this is an area where engineering disciplines often make use of more detail in their modelling than we do in physics. In applications such as feedback and control or device characterisation as well as disciplines such as model-based systems engineering (MBSE), the behaviour of a sensor *operating* in a system can be crucially important and may well form part of the model and design process. Such models may be physical, or they may be abstracted data-driven 'sensor models', or even a combination of the two, depending on the use-case. An example might be the sensors that form part of the lidar systems of an autonomous vehicle. If they change behaviour due to any environmental condition, set of inputs, or being in a particular state then it is crucial this is understood to ensure agreement between the systems self-perception and the reality of the world around it. I think it is fair to say that in most physics research such detailed consideration of the behaviour of measurement devices is not often practised. This is usually fine because we are normally able to arrange our apparatus such that the measurement component is negligibly effected by such considerations.

Question: So measuring a classical system may be more subtle than at first glance. Can we be clear about what the core measurement model should be?

Reply: The simplest implicit position is that a classical system evolves on some trajectory and then a device like a camera (e.g. with a graticule) or oscilloscope can monitor a component of the dynamics without substantially disturbing the system. The measurement device must either directly or indirectly, by some amplification process, produce some record that can be experienced at a human level. That record should also be a response that we can convert into numerical values which we can relate in some known way to the canonical coordinates and momentum, and use to check the predictions of the model.

Question: That clearly does not cover the whole gamut of what is needed. But it is already somewhat complicated. It is not clear to me that, if an amplification stage is needed, under what conditions should that be considered part of the measurement device?

Reply: It might be tempting to say that we can make this separation if the amplifier's effect on the measurement system is negligible. However, that is an oversimplification for any measurement process. From the system-of-interest to human interpreted outcomes there is always a chain of physical systems that form the experiment as a whole. Preparing the way for a later discussion on quantum measurement, in classical microelectronics, there may be quite a few stages between 'measuring' a voltage or current at such scales and getting a human interpretable output. This is, loosely speaking, a classical analogue of the von Neumann chain, where for some set-ups it may be somewhat ambiguous exactly which components in this chain actually form the 'measurement device' as a single identifiable subsystem with clear system boundaries. Having alluded to the von Neumann chain, I feel we need to jump ahead of our argument for one second to make clear that we will be avoiding any discussion of mind-consciousness in our discussions of quantum measurement. While entertaining, those arguments are redundant as all such arguments can be replaced by simpler ones. Furthermore, to avoid contradictions, the logical conclusion of such a train of thought is a form of solipsism. Such arguments are untestable and to my mind purely metaphysical. Thus, they fall outside of the remit of science (anyone who does assert mind-conciseness is needed for quantum measurement should also consider applying Hitchens's razor). Quantum mechanics is, in any case, known not to be valid at the scale of the human brain (as to whether or not quantum mechanics plays a role in mind-consciousness itself, it is worth reading the relevant monographs by Roger Penrose).

Question: We are in danger of getting off topic. Is it possible to form a single statement of what a measurement is, or are there different kinds of measurements?

Reply: Well, there is one trait common to all measurement processes: it is always a comparative process that can numerically relate an unknown quantity to a known quantity (e.g. the current through a wire to the position of a needle against a fixed scale). Note that even photographic film can be covered by this definition because the amount that the film is developed at a certain position can be converted into a numeric value. Indeed, to quote:

> 'Measurement of magnitudes is, in its most general sense, any method by which a unique and reciprocal correspondence is established between all or some of the magnitudes of a kind and all or some of the numbers, integral, rational, or real, as the case may be. (It might be thought that complex numbers ought to be included; but what can only be measured by complex numbers is in fact always an aggregate of magnitudes of different kinds, not a single magnitude.) In this general sense, measurement demands some one–one relation between the numbers and magnitudes in question – a relation which may be direct or indirect, important or trivial, according to circumstances'.
>
> Bertrand Russell [76]

Question: In statements such as this, is there not an implicit assumption that some property or collection of properties exist independently of the system being measured?

Reply: Perhaps not. Later we will discuss an example of measuring a quantum system whose classical limit exhibits dissipative chaos. An argument there is that this limit is achieved through the act of measurement itself, that the measurement of the quantum system may be thought to grow the non-linear classical dynamics outcome that is measured – that the system's dynamics would be totally different and linear without the measurement device being present. Nevertheless, in the classical limit, we can assign **unique** and **reciprocal** numbers to the magnitude being measured (in this case the magnitude of the phase space coordinate ($\langle \hat{q} \rangle , \langle \hat{p} \rangle$) from the origin).

Question: Ok, so let us agree for now that we need that a 'unique and reciprocal correspondence is established between all or some of the magnitudes of a kind and all or some of the numbers'. You still have not answered my question – are there different kinds of measurement?

Reply: There are several categories I can think of; they are: weak versus strong (where the measurement device may affect the classical system being measured); with or without noticeable back-action (where the system being measured affects the measuring device non-negligibly); and direct versus indirect measurement (where we measure the thing we are directly interested in or, if that is not possible, some other quantity that we can relate in some way to that magnitude). The usual postulates of quantum measurement understood from this perspective would therefore fall into the category of strong, direct measurement without noticeable back action.

Reply: Before proceeding, can you give an example of a type of measurement that is not relevant to quantum systems?

Question: What about an inertial or global positioning system (GPS) navigation system located inside a supertanker? Here the device is smaller than the system being measured and it gives an example of direct versus indirect measurement. Within each navigation system, direct measurements of, e.g. a gyroscope are being made. The measurements are not of the coordinates of the ship itself but of some components within the device, which, in its turn, is attached to the ship. The state of a few of the degrees of freedom of the ship is then inferred from this measurement record. It is worth noting that the ship is a composite and (hopefully) rigid body which enables the use of multiple measurement devices in its 'measurement'. What this example does highlight is that there are different 'size' (measurement device):(measured object) ratios that determine effective measurement models.

Question: What about continuous measurement?

Reply: This is something we take for granted. After all, human senses such as sight operate in this way (at least to our level of perception).

Question: What we actually get out of an experiment will be some record from an experiment. This might be the needle moving on a galvanometer. This most certainly disturbs the system been measured, but from the needle's position we might have a continuously updated record of a voltage.

To me, a measurement record results from setting up physical systems arranged in such a way that the measurement stage will change its recordable state (e.g. the needle on a galvanometer) depending on the state of the system we wish to measure. As stated before, it must do this in a way that can be meaningfully interpreted by a human. It is this change of state of the measurement apparatus (MA) that forms the record. Is there a quantum version of this interpretation of measurement?

Reply: Yes, there is. Given the sensitive nature of quantum states, it is not quite as straightforward as the classical counterpart. As we alluded to above, in some engineering applications, model-based measurement techniques combine recorded experimental data and simulations of the underlying physical system estimates of the state of a system. To make sense of a quantum version of continuous measurement, the concept of model-based measurement is of great utility. We will discuss one such example set-up in the next section.[3]

10.4 My First Foray into Model-based Measurement

In this section, we discuss a toy model-based measurement of a quantum device.[4] I have chosen one from my own research as (i) this is a close example of quantum measurement to the classical moving coil galvanometer – both of which enable us to understand the concept of the measurement process from a simple model-based approach; (ii) I am much more comfortable being critical of my own work than that of others (in this case work done for my DPhil from the mid-1990s). The group I joined (led by Terry Clark) had been working on a way to measure a macroscopic quantum object (QO) called a superconducting

---

3 We use the term 'model-based measurement' in this book in hindsight; at the time the work was done we had not encountered such terminology, but we would have used it, as it makes a useful link between physics and engineering perspectives.
4 A 'toy model' is a simplified model where all but the most important details have been removed in order to explain a specific mechanism of operation using the fewest assumptions possible.

quantum interference device (SQUID) since the early 1980s, when Alan Widom first put forward a model for superconducting quantum electrodynamics on a macroscopic scale [91] (and later by Leggett [48]).[5] As we discussed in the previous chapter, the SQUID ring is a device that comprises a 'thick' ring of superconducting material where, at some point in the ring, there is an area of weak superconductivity – this is called the weak-link or Josephson junction (JJ). The weak-link provides a potential barrier where the superconducting condensate can quantum mechanically interfere with itself. The ring Hamiltonian is

$$\hat{H}_s(\Phi_x) = \frac{\hat{Q}_s^2}{2C_s} + \frac{(\hat{\Phi}_s - \Phi_x)^2}{2L_s} + \hbar v \cos \frac{\hat{\Phi}_s}{\Phi_0},$$

where the charge $\hat{Q}_s$ and flux $\hat{\Phi}_s$ operators are analogous to momentum and position with $[\hat{\Phi}_s, \hat{Q}_s] = i\hbar$. Here, C is capacitance, L is inductance, and $\hbar v$ quantifies the energy associated with the self-interference of the condensate. Crucially, $\Phi_0 = h/2e$, the so-called flux quantum, arises from a topological and not a canonical quantisation condition (we will return to an important consequence of this later).

As an example of a device that is asserted to be behaving quantum mechanically at a macroscopic scale, it is natural to ask if we can measure this object continuously in a way that is analogous to the measurement of classical systems. The idea is to form a continuous classical record (indirectly) of the quantum state by weakly interacting an LC oscillator with the quantum system. The usual set up the group used is shown in Figure 10.1. The LC oscillator is referred to as a tank circuit because of its ability to store energy. The fact that the SQUID is a macroscopic object is key, because it means that its important parameters, and

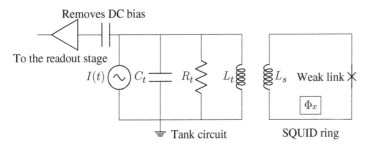

Figure 10.1 Schematic of a SQUID ring coupled to an *LCR* oscillator (tank circuit). The SQUID is the quantum system, where the ring has an inductance L_s, and the JJ brings with it a capacitance, C_s, and a tunnelling term whose energy is proportional to $\cos \hat{\Phi}_s/\Phi_0$ where $\Phi_0 = h/2e$. The SQUID is inductively coupled to the tank circuit, where the inductance $L_t \gg L_s$ and capacitance $C_t \gg C_s$, and which is assumed to be operating as a classical device. The tank circuit is capacitively coupled to an amplifier stage in order to prevent unwanted direct current (DC) biases entering the system. Also note that the SQUID may be subject to an external magnetic field, denoted by Φ_x.

---

5 Given how widely SQUIDs are now finding application in quantum technologies, it is odd to reflect that at that time their quantum nature was not only not well accepted, there were a significant number of people who were emotionally opposed to the idea. I recall some quite hostile, unpleasant, and unscientific arguments that SQUIDs 'must be classical' that made the superconductivity community an environment somewhat toxic to work in. Happily, that seems to be something that is no longer such a problem, and the community has become much more amicable.

those of the measurement device, are of more comparable magnitudes than, e.g. electrons entering a Geiger counter, which makes modelling much more tractable. An electron is an intangible object that we can understand only by the effects it has on the objects it interacts with (including measurement devices). For this reason, we have no intuition of what an electron actually is beyond the mathematical/Platonist ideals of it that we derive through modelling its behaviour. As a macroscopic object, a SQUID is something we can in some way directly partially experience (e.g. hold). Other objects such as nano-mechanical resonators take this idea even further, and the fact that these appear to exhibit quantum phenomena is truly fascinating. Because of this notion of scale, it seems reasonable that such objects might form a good basis for testing the measurement problem.

In this model, the tank circuit is the 'larger' or 'heavier' object (i.e. the one with the larger inductance and capacitance) but still within a few orders of magnitude of the SQUID circuit. So, unlike the example of the galvanometer or voltmeter, the back-action can be very significant, and the circuit set-up and mode of operation are important considerations for controlling measurement disturbance. For this reason, the inductive coupling was kept weak, and the tank circuit typically operated in one of two modes. In the first, the tank circuit was driven at small amplitudes and its frequency response analysed. In the second, the frequency was fixed and 'slow', and the tank circuit was driven at increasingly large amplitudes to examine its voltage–current (VI) characteristic. Both were considered as potential candidate alternatives to the normal projective measurement process, an example of weak measurement [59].

The challenge of understanding quantum circuits such as this one, as well as their interactions with each other and classical MA, is what drew me to research in the first place – the fact that we have yet to properly resolve how such systems really work remains a fascination for me. In any case, this is the conceptually simplest non-trivial system, which is also likely to be modellable, that I can think of. My view is that if one is to understand quantum measurement, and potentially improve quantum theory so that the measurement postulates are no longer needed,[6] it is only through simple modellable systems such as this that we will be able to make any progress.

Question: Given your previous quote from Bertrand Russell about measurement of magnitude, and the need for a unique and reciprocal correspondence, is the tank circuit really a measurement device?

Reply: I would say that it is, but as part of a the whole set-up including amplification and readout (say on a spectrum analyser). The idea is that the tank circuit is classical enough to allow the rest of the circuitry to be treated in terms of classical electronics. Strictly speaking, our model should really include every single element quantum mechanically, but we do have to draw the line somewhere – as to where we should do so is known as the von-Neumann chain (although, as previously mentioned, we stop short of extrapolating as far as consciousness). We understand the tank circuit as a measurement device operating in the same way as it would, if the SQUID were replaced by a classical LC circuit (harmonic oscillator). By understanding how the coupled circuits work, we can establish facts about the

---

6 Interestingly, it is only when thinking in terms of the Schrödinger picture that it seems essential to me that we do this; when I'm thinking in the Heisenberg picture I do not feel that way at all.

frequency or amplitude of oscillation on the *LC* circuit by monitoring the tank circuit (e.g. if the tank circuit has no detectable oscillation, then the *LC* circuit must be approximately at rest).

Question: You assert that it is necessary to model the measurement device to understand the process of measurement. How did/do you do this?

Reply: When I joined the group, we used an analogue of the Born–Oppenheimer approximation, where the tank circuit's dynamics were given by

$$C_t \frac{d^2\Phi_t}{dt^2} + \frac{1}{R_t}\frac{d\Phi_t}{dt} + \frac{1}{L_t^*}\Phi_t = I(t) + \mu \left\langle\left\langle \hat{I}(\mu\Phi_t + \Phi_x)\right\rangle\right\rangle, \tag{10.1}$$

where $\mu = M/L_t$, $L_t^* = L_t(1 - M^2/L_tL_s)$, and M is the mutual inductance. Here, $\left\langle\left\langle \hat{I}(\mu\Phi_t + \Phi_x)\right\rangle\right\rangle$ is the time average of a screening current operator defined as $\hat{I}(\Phi_x) = -\partial\hat{H}_s/\partial\Phi_x = (\hat{\Phi}_s - \Phi_x)/L_s$ for some chosen state (usually an energy eigenstate). The idea was to understand the different effects that different states would have on the tank circuit dynamics, intending to use the results of this modelling to subsequently infer, in a given experiment, the state of the SQUID through either the frequency response or *VI* characteristics of the tank circuit. The issue with this model is that it was implicitly assumed that the SQUID was in some fixed state, and this was usually an energy eigenstate. As such, it was not really a model of a weak-measurement process at all. This is because it assumes there is in place the so-called quantum Zeno effect, a phenomenon where continuous strong measurements freeze the quantum state of a system.

My contribution, which at the time I thought was more substantial than I do now, was to remove the time averaging and solve the Schrödinger and tank circuit equations as a single set of coupled differential equations, specifically:

$$C_t \frac{d^2\Phi_t(t)}{dt^2} + \frac{1}{R_t}\frac{d\Phi_t(t)}{dt} + \frac{1}{L_t^*}\Phi_t(t) = I(t) + \mu \left\langle \psi(t) \left| \hat{I}(\mu\Phi_t(t) + \Phi_x(t))\right| \psi(t)\right\rangle \tag{10.2}$$

$$\left[\frac{\hat{Q}_s^2}{2C_s} + \frac{(\hat{\Phi}_s - \mu\Phi_t(t) - \Phi_x(t))^2}{2L_s^*} + \hbar\nu \cos\frac{\hat{\Phi}_s}{\Phi_0}\right] |\psi(t)\rangle = i\hbar\frac{\partial}{\partial t}|\psi(t)\rangle, \tag{10.3}$$

where $L_s^* = L_s(1 - M^2/L_tL_s)$. At the time this was quite a computationally demanding set of equations to solve, as computations were done in the Heisenberg matrix representation and quite a few basis state were needed to ensure numerically accurate solutions (it took a further five years or so to be comfortable enough to have journal-quality publishable results [24]). This model would allow the quantum state of the SQUID to evolve and was consistent with expected physics. For example, in the adiabatic limit (slow-moving fields that approximate a DC bias), if the SQUID started in the ground state, it would remain there as expected. This produced predictions in agreement with experiments in this limit. At the time there was an additional motivation because experiments had shown that the interesting phenomenon of down-conversion by many orders of magnitude, from a microwave source, through the SQUID to a radio frequency tank circuit, would happen at specific points in external flux. We had been unable to explain this behaviour with any other model, and theory needed to catch up with experiment. The reason for introducing this model was primarily to understand if this might be a quantum process.

Question: So what did you show with this model?

Reply: The model indeed predicted that (i) energy could be down-converted by several orders of magnitude between an external microwave flux through the SQUID ring to the tank circuit; and (ii) this happened at points in the DC bias flux which correspond to avoided crossings of the SQUID energy eigenvalues.[7] This was consistent with the experimental results and indicated that the behaviour was explainable by the SQUID being excited at its avoided crossings and therefore could be quantum in origin.

Question: How can you be sure that there was no classical explanation of the same effect?

Reply: I cannot; what I can say is that I was unable to reproduce the same results using the semi-classical equivalent model of the SQUID (the so-called resistively shunted junction model) coupled to the same tank circuit equation of motion.

Question: So this seems really promising – at the very least this model may tell you that you can measure where some avoided crossings in the SQUID ring spectrum are and assert something about the underlying state?

Reply: Maybe, but I became quite uncomfortable with the *ad hoc* way that I mixed classical and quantum dynamics (recall, I earlier criticised the sort of thinking that mix together quantum and classical models, assuming everything will work out). The approach may still have value as a heuristic or phenomenological model and there may be limits where it is good enough to help with device design.

I have, however, lost confidence that this model can help me understand the measurement problem or the quantum-to-classical transition as the classical behaviour of the tank circuit has been assumed. To be a proper replacement for the actions of measurement, this model would need to produce predictions in complete agreement with the measurement actions of quantum mechanics in the appropriate limit. This would need to include the Born rule, projection, and observed experimental phenomena such as the quantum Zeno effect. Due to the nature of the coupling between the quantum and classical components of the model it is not clear how all of this could arise. To make progress I wanted something

---

7 An avoided crossing or anti-crossing refers to a property of eigenvalues of Hermitian operators such as the Hamiltonian. Consider a two-state Hamiltonian in its energy eigenbasis where the states are ordered in increasing eigenvalues *and* those eigenvalues can be continuously changed so that they become degenerate. If some perturbation is added that lifts this degeneracy it will do so in a way that does not alter the ordering of the eigenvalues and eigenvectors. That is, the eigenvalues cannot cross and so we say that this is an avoided or anti-crossing. The full definition is a little more complex and described more fully in [47], where it is observed that '*in a diatomic molecule, only terms of different symmetry can intersect, while the intersection of terms of like symmetry is impossible (E. Wigner and J. von Neumann 1929). If, as a result of some approximate calculation, we obtain two intersecting terms of the same symmetry, they are found to move apart on calculating the next approximation...*' and the result '*... is a general theorem of quantum mechanics; it holds for any case where the Hamiltonian contains some parameter and its eigenvalues are consequently functions of that parameter*'.

better – something where all devices are treated on the same quantum footing and the measurement device became classical of its own accord.

Question: So what needs to be done in order to address that question directly?

Reply: Fundamentally, we have to include the classical measurement device within our quantum model. Before we can address the question of measurement, we therefore really need to first understand what it means for a quantum device to behave classically. The other question we need to address is what sort of measurement devices we need to bring into such a model and how do we understand their behaviour as classical. For the tank circuit the classical output is the amplified current that is (classically) measured at the output stage. We may accept the tank circuit as a valid measurement device but it is not a typical one. In most cases the measurement device is not of a comparable scale to the QO. So let us now better understand the physics of some of the measurement processes.

10.5 Two Other Measurement Devices and Their Classical Limit

To understand measurement, we need to first understand what a measurement device is. Motivated by the (possibly unwise) desire to explain all physics within a single unified theory, we would want to model both the quantum system to be measured and the measuring device within the same framework. This is especially desirable to those motivated by a quantum realism perspective, and those that look at decoherence as a path to the quantum-to-classical transition. If a measurement device is to be brought into the quantum model, it must be operating in its classical limit so that a classical record can be made – otherwise it cannot be considered a measurement device (we will come to what that means in more detail later). The model must be sufficiently good to be able to explain the emergence of a measurement record that is in agreement with experiment (a click of a photomultiplier tube [PMT], a signal on an oscilloscope etc.). This view needs to cover all sorts of measurement devices – not just the tank circuit example of the previous section.

Question: PMTs are seen in many experiments in quantum optics at the end of a string of polarisers, mirrors, and other objects that manipulate the quantum states of light. As such, it is a rather common example of a measurement device and seems quite simple in its design. What are its main features?

Reply: Yes, see Figure 10.2 for a schematic of a typical device. The PMT is designed so that when it absorbs one (or more) photon(s), it will provide sufficient classical voltage and current to be recorded somehow. We expect the probability of it making a measurement to be proportional to the number of photons in the field. This means that a PMT may also make measurements of quantities such as position or momentum for states that are localised (such as coherent states). The nearer the state is to the vacuum state, the fewer measurements will be made, the further away, the more. After making a measurement, a PMT will need some time to reset itself before it is capable of making another measurement.

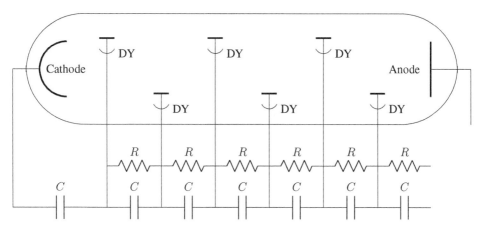

Figure 10.2 An approximate, and incomplete, schematic of a photomultiplier tube. At one end of the vacuum tube, the cathode is the source of the initial electron(s). Then there are a series of dynodes (DY), electrodes, each separated by approximately 100V, that use secondary emission as an electron multiplier. The Idea is that a photon expelling an electron from the cathode instigates an electron avalanche so that a substantial current can be detected at the anode.

Question: This indeed seems like a simple example of a classical measurement device, so can we start by understanding that first? The electron avalanche referred to in the caption seems to capture the process very well, doesn't it?

Reply: The explanation in the caption is very classical indeed and not the complete story. In many textbook diagrams of a PMT we would see an arrow indicating an incoming photon as a corpuscle. To indicate the avalanche effect, there would then be an increasing number of arrows down the dynode chain indicating more and more electrons – thus implied to be particles – until we see a large number 'hit' the anode. We might then sat that the arrival of all of these electrons generates a classical current that, e.g. triggers a speaker to click.

Question: That seems reasonable. Have you deliberately chosen not to include any of these typical annotations; are there specific reasons?

Reply: There are indeed, just as with our schematics for the Stern–Gerlach experiment. Beams or trajectories (except in a Feynman path integral way) are not the right way to think of any quantum particle. We know we have a source of photons, whatever that may be, and then some dynamics happens under quantum mechanical laws of motion. We would think that the emission of an electron from a cathode will be by the photoelectric effect and this is a quantum process. But we have not measured that event, so we can't strictly talk about it. As a quantum process all that happens is that there is some interaction between the Hamiltonians for the two systems, and it evolves some initial state into a superposition. What we can say is that we have a photon source – to make this conceptually easier, let us think of this as a two-level atom in its excited state coupled to a photon field as a harmonic oscillator (just like the Jaynes–Cummings model). An atom in free space cannot decay unless it is coupled to a system that can accept it – in this case the cathode. The interaction between the atom and one or more electron(s) in the cathode would be unitary.

If there were no dynodes and no further measurement, we might expect the dynamics to follow unitary Schrödinger evolution (such as that of Figure 8.2).

However, this is not everything there is present; there are the dynodes, and each element has a complex environment that may be enough to prevent such simple quantum dynamics from happening. Let us assume for now that the environment is not the issue. As we will find later, there are strong arguments suggesting that the environment does play a role in achieving a quantum-to-classical transition; hence, this is probably not true and may turn out to be a poor assumption (the resistors will decohere the state, maybe into something 'classical'). Nevertheless, from the standpoint of *reductio ad absurdum*, neglecting any potential effect of the environment does serve to illustrate the issue of trying to describe the operation of a PMT wholly within quantum theory. If we now include the first dynode in this analysis, we have a tripartite quantum system comprising an atom, a cathode, and a dynode. This three-part system must also evolve as some dynamical superposition of states, so this subsystem could not be considered to make a measurement. Continuing the same argument to include dynode after dynode until we get to the anode, it is not clear at what point this culminates in a current that can make the speaker go 'click'.

Question: I can see what you are trying to say here but you have definitely mixed classical and quantum ideas in that argument. The free electron model in a metal is after all quite a big approximation. Electrons in a real metal are interacting particles and the actual electron state will not be some sort of gas but rather an entangled and complicated state. This entanglement will not just be with other electrons but also with the lattice itself. What does it mean to excite an electron out of such a system? Also – how does this work within the standard measurement postulates? After all, the idea is that on measurement the system is projected into an eigenstate of an observable. What does that mean for a photon counter when there is an idea that the photon has been absorbed – are we in fact measuring the state of an atom?

Reply: These are questions I wish I could answer. One problem is that I cannot model a system as complex as this quantum mechanically – there is just too much going on. It may feel to you like this is a form of academic cowardice but all I can suggest is that we try to understand a simpler system.

Question: How about we consider Young's double-slit experiment, as this is one of the standard examples of quantum mechanics?

Reply: Yes, let's. As I recall, we have a source of electrons, for example. They are attenuated to such a level that they arrive at the detecting screen, or film, one at a time. By this I mean that we see the film develop in one localised spot at a time, as we never actually see an electron. The set-up is shown in Figure 10.3 together with an example probability density of where the electron might be detected on the film.[8] After enough events, the exposed areas of the film will begin to resemble such a distribution. This all seems rather straightforward, so what are the complications?

Reply: There are a few. First let us note that the measurement is a chemical reaction where the developed chemicals are localised to a certain part of the film. The chemical reaction

---

8 The interference pattern was generated using the pgf-interference package by Keno Wehr.

Figure 10.3 (a) The experimental set-up for Young's double-slit experiment, and (b) an example of the probability density function for the film being developed at a given position and each development event (note that the pattern is rotated by 90° relative to the top figure).

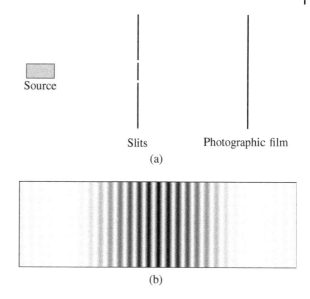

Source

Slits   Photographic film

(a)

(b)

itself is fundamentally quantum mechanical in nature, but the fact that the chemicals are in some ways localised in the film is rather classical. We would like to interpret this as a position measurement. However, the electron takes part in a chemical reaction and it is not clear that on measurement it is projected into an eigenstate of the position observable. The only way we would be able to do that would be to perform further experiments on the same electron immediately after the measurement. This is clearly not something that is possible. So while this experiment is consistent with measurement axioms M1 and M2, the last one remains undecidable.

Question: What about all these example experiments where one queries which hole the particle went through – if we have this knowledge then only single-slit patterns are observed?

Reply: Ahh, the delayed choice experiments, such as those proposed by John Wheeler, or the quantum eraser. Certainly these are conceptually very interesting but what most treatments do not do, is to include the presence of the extra components of the apparatus needed to perform the delayed choice within the system Hamiltonian itself or its accompanying environment. It seems natural to me that in order to properly understand the behaviour of such a set-up, this should strictly speaking be done. As decoherence seems to typically have a localising effect that can destroy entanglement, and any part of a physical apparatus will produce some decoherence, this should also be included in any such model.

Question: Without doing such complicated modelling is there any understanding we can extract from an experiment such as this?

Reply: I do not believe that we can – as any empirical test of the measurement hypothesis has background assumptions about the experiment and detector apparatus. We either include the apparatus in the model to make as explicit many of those assumptions as possible, or they remain implicit. If the system is 'measured' by a device recording an event in the way a film of PMT does, how do we know if it is projected into an eigenstate of the measured observable? The photon, electron, or whatever goes on to interact with the apparatus-measurement device–environment will do so in some complex way. The only way to understand if the usual axioms of quantum measurement naturally arise as the limit of some process (such as a PMT), is to model that process. We do know that, by the correspondence principle, whatever a model-based approach to measurement looks like, it must reproduce the measurement axioms for whatever scenario those axioms were seen to predict the observed behaviour.

Question: It's a bit off topic, but this discussion has prompted a question: what about the quantum Zeno effect – surely the idea that one can maintain a specific quantum state through measurement as a corollary of the postulates of measurement is proof enough?

Reply: The Zeno effect has been approximately realised in a number of systems but, to the best of my knowledge, it has only ever been observed to slow rather than freeze dynamics. As we shall later see, dissipation in coupled quantum systems can produce the same effect without the measurement axioms.

Question: I am beginning to think that the tank circuit was by far the simplest example of a measurement device, maybe we should return to considering that?

Reply: Yes, maybe it was. But before returning to that, or something similar, we still need to articulate what it means to achieve quantum-to-classical transition.

10.6   The Quantum-to-Classical Transition

This section is dedicated to trying to understand what it means for a quantum system to behave classically. Recasting our expression of the correspondence principle from Chapter 2 in a form specifically intended to address this question, we have

> 'To satisfy the correspondence principle, quantum mechanics must reproduce the predictions of classical physics in the domains where classical mechanics produces predictions in line with experiments'.

A standard textbook realisation of this principle is

> 'If a quantum system has a classical analogue, expectation values of operators behave, in the limit $\hbar \rightarrow 0$, like the corresponding classical quantities'.
>
> Merzbacher [60]

As already noted, others, such as Bohr, have argued that quantum systems become classical when quantum numbers describing the system are large. Some argue from a macro-realism perspective, which essentially states that systems become classical when they are large. Others have argued that scale is not enough, that it was complexity that made the limit happen. Yet another alternative view is that environmental decoherence can be the cause, including views such as environment-induced super-selection (einselection). Regardless, the approach, if it is truly to satisfy the correspondence principle, must reproduce predictions in line with classical mechanics where appropriate. It should do so using only the postulates of quantum mechanics applied appropriately. Specifically, in this limit, one might expect all the following to hold:

1. The phase space dynamics of the Wigner function to reproduce the Liouville equation or, more generally,

$$\frac{dA}{dt} = \{\{A, \mathcal{H}\}\} + \frac{\partial A}{\partial t} \longrightarrow \frac{dA}{dt} = \{A, \mathcal{H}\} + \frac{\partial A}{\partial t}.$$

This limit was kept in mind in our construction of phase space quantum mechanics.

2. The Schrödinger and Heisenberg pictures we would expect to produce models with some equivalence to KVN theory,

$$-\frac{i}{\hbar}\hat{H}|\psi\rangle = \frac{\partial}{\partial t}|\psi\rangle \longrightarrow -i\hat{L}|\psi\rangle = \frac{\partial}{\partial t}|\psi\rangle.$$

This may be reasonable, as KVN was constructed to be as a classical operator model to compare to the Schrödinger picture.

3. Additionally, we would want path integrals to reproduce the least action principle,

$$\sum_{\text{all }\Gamma} \exp\left[\frac{i}{\hbar}(\text{constant})\int_{\Gamma} dt\,\mathcal{L}\right] = K(a, b) \longrightarrow \delta\int_{t_i}^{t_f} dt\,\mathcal{L} = 0.$$

By inspection one can see that the phase varies the least near stationary action points. Thus, these are the greatest contributors to the Feynman sum over all actions (path integral) principal, and this limit also seems reasonable.

Just as with measurement, I would argue that as a minimum a model-based approach should be used to confirm that such limits are indeed achieved.

Question: As you mention multiple approaches, what are the issues with $\hbar \to 0$?

Reply: Limits can be funny things, as Michael Berry puts it,

> 'Biting into an apple and finding a maggot is unpleasant enough, but finding half a maggot is worse. Discovering one-third of a maggot would be more distressing still: The less you find, the more you might have eaten. Extrapolating to the limit, an encounter with no maggot at all should be the ultimate bad-apple experience. This remorseless logic fails, however, because the limit is singular: A very small maggot fraction ($f \ll 1$) is qualitatively different from no maggot. Limits in physics can be singular too—indeed they usually—are reflecting deep aspects of our scientific description of the world'.
>
> M Berry, 'Singular Limits', Physics Today Vol 55 (2002)

In quantum mechanics the limit $\hbar \to 0$ is a singular limit. It can be shown that neither this limit, Bohr's high energy limit, nor arguments of scale and complexity, on their own, are sufficient to achieve a classical limit.

Question: Really? Can you give an example? In that article you just quoted, following a Young's slit example, Michael Berry also says that, from this singular limit combined with decoherence, the combined *'concept governs the emergence of the classical from the quantum world in situations are more sophisticated than Young's, where chaos is involved'.* This seems hugely promising, is there anything wrong with it?

Reply: Again, I find it easier to criticise my own work than that of others. I have done quite a bit of work on decoherence as a path to the quantum-to-classical transition. There has been some success in this work but it all suffers from one flaw of reasoning because all models of the coherence contain implicit or explicit assumptions that mean the resulting dynamics are not equivalent to unitary quantum evolution.[9] With this substantial caveat in mind, let us see how far the approach can take us. What would you like me to address first?

Question: How about a reason to think that the idea might work?

Reply: A very elegant example that supports this can be found by following the example of [79], and many related works. The example system-of-interest is a Duffing oscillator whose classical damped driven dynamics provide a standard example of dissipative chaos. The approach to understanding the quantum-to-classical transition of this system is to use a quantum description of the system, where an environment is introduced to affect damping. This was modelled using quantum state diffusion, as the idea is to reproduce the dynamics of a single experiment, and a master equation approach would not do this.[10] The Duffing oscillator Hamiltonian is

$$\hat{H} = \frac{\hat{p}^2}{2} + \frac{\beta^2}{4}\hat{q}^4 - \frac{\hat{p}^2}{2} + \frac{\Gamma}{2}(\hat{p}\hat{q} + \hat{q}\hat{p}) + \frac{g}{\beta}\hat{q}\cos t,$$

and the damping was provided using a Lindblad, $\hat{L} = \sqrt{2\Gamma}\hat{a}$ – which effectively produces the velocity-dependent friction analogous to that seen in the classical system (see Chapter 9 for details of this model and an explanation of the Γ term). Here, β is equivalent to a scaling factor of $\hbar$ so that $\beta \to 0$ is the same as $\hbar \to 0$. One can clearly see that the Poinacré section consistent with the PDF of Figure 2.3 emerges in this limit.[11] Finding emergent

---

9 An interesting exercise is to revisit the open quantum systems chapter and identify all the assumptions made in deriving the master equation. If we accept decoherence as the path to the quantum-to-classical transition, each of these assumptions essentially becomes a new axiom of quantum mechanics. For example, in deriving master equations, there is always a point where one makes some judicious application of a Baker Campbell Hausdorff expansion. For all but the simplest of systems, this approximation may have some quite substantial implications – approximation becomes an effective ansatz (and different levels of truncation a different ansatz). Where such assumptions lead to models that produce good agreement with experimental observation, we might want to look for a post-quantum theory, in which such an effect is produced through a more physical explanation. At least master equations of the Lindblad form are linear and thus preserve their statistical interpretation. The situation is worse for unravellings of the master equation such as quantum jumps and quantum state diffusion – these models are no longer linear and do not satisfy the principle of superposition. As such, we need to treat all conclusions drawn from any of these open-systems models with due care, caution, and a healthy degree of scepticism. A post-quantum theory such as Penrose's musings on quantum gravity might themselves introduce slight violations of the principle of superposition. Hence, the fact that the open systems approaches do have some problems. With a post-quantum theory in mind, there remains real value in exploring them.
10 Following [40], we could instead have considered Wigner function dynamics and compared the outcome to the Liouville equation, but that would have been far more computationally demanding.
11 As $\beta \to 0$, the computations become more challenging. In the associated material you will find several example programs from reproducing these results that get increasingly more sophisticated in order to deal with this computational demand. This is also the reason that the different plots contain different numbers

chaotic behaviour in this limit is a good reason to believe that there is merit to the idea that decoherence is a path to the quantum-to-classical transition.

Question: Can you help me understand why this happens?

Reply: In unravelling the master equation using models such as quantum state diffusion and quantum jumps, the Lindblad required to model damping is the annihilation operator, as one effect of the Lindblad in unravellings of the master equation is to decohere the state towards an eigenstate of the Lindblad. For damping with this kind of Lindblad, the effect is one of localising the state of the system towards a coherent state, thus helping to realise the localisation needed to achieve the quantum-to-classical transition. It helps if we think in the phase space representation of the quantum system as the state approximates a state that is continually reshaped by the environment to be more Gaussian and localised with the Hamiltonian dynamics competing against this process so that the classical limit is only achieved in the limiting case $\hbar \to 0$.

Question: How about your claim that the limit $\hbar \to 0$ is insufficient even with decoherence?

Reply: My go-to system for testing many quantum ideas, including the correspondence principle, is the SQUID that we discussed before. If one tries to achieve the quantum-to-classical transition by taking $\hbar$ to zero, we find that there are real problems, as the canonical and topological Plank's constant prevents the limiting case from leading to a correspondence with the SQUID's 'classical' equation of motion (those derived from Hamilton's equations).

Question: I understand the 'classical' model is the so-called restively shunted (Josephson) junction (RSJ) model and that it includes damping. How did you model the quantum SQUID to reproduce RSJ dynamics?

Reply: Following exactly the same procedure as in the Duffing oscillator example above. In this case though, at the limit $\hbar \to 0$, the correspondence between the two models is not achieved. This is because the flux quantum, $\Phi_0 = h/2e$, that appears in the Joseph junction term and in the Hamiltonian arises from a topological quantisation condition arising from the fact that the SQUID is a physical ring. The terms where Plank's constant arise from commutation relations scale differently from the topological terms, and a meaningful limit is not achieved. We find there is another limit-based approach that works instead. Here we scale the capacitance and inductance, and sympathetically re-scaling some of the other parameters appropriately, we can leave the dimensionless form of the classical equations of motion unchanged. The findings of the model are therefore sympathetic with the macro-realistic 'bigger is more classical' argument, as, e.g. larger capacitance makes the ring behave more classically. In this limit we can get the two models to agree in terms of the behaviour of expectation values of observables. It led me to propose an alternative form of the correspondence principle:

---

of points. The different examples have been kept to show how a program can evolve with the attempted solution of a problem. In the working code-base the good parts (such as the more efficient calculation of the displacement operator) of this code would be migrated into the main body of the library.

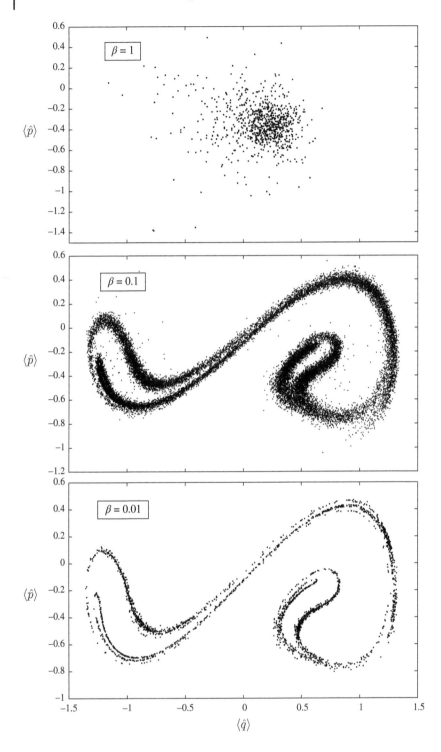

Figure 10.4 Example Poincaré sections for the quantum state diffusion evolution of the Duffing oscillator with $g = 0.3$ and $\Gamma = 0.125$. The classical limit is very well approximated for small β ($\hbar$ scaling factor), providing chaotic-like dynamics as expected from a quantum-to-classical transition.

Consider $\hbar$ fixed (it is) and scale the Hamiltonian, in such a way as to preserve the form of the classical phase space, so that when compared with the minimum area $\hbar/2$ in phase space:

1. the relative motion of the expectation values of any observable (generalised) coordinate (and hence the associated classical action) becomes large, and
2. the state vector is localised (in any representation); then, under these circumstances, expectation values of operators will behave like their classical counterparts [22].

Question: Are you happy with that expression?

Reply: Not at all. I have two objections to it that, at the very least, indicate that it is incomplete. These are (i) consider the simple harmonic oscillator in a high-energy coherent state. This would satisfy the above statement exactly. We know from our study of the Jaynes–Cummings model in the computing chapter that appropriately coupling a quantum harmonic oscillator to a two-level system will delocalise the coherent state into a Schrödinger's cat state. This contradicts the original assertion that the isolated harmonic oscillator was in its classical limit even though it satisfied the assertion (no matter how big it is); and (ii) for most systems and most initial states dynamics naturally leads to the spread of the quantum state. The explicit requirement for localisation is not achievable within unitary quantum dynamics as localisation requires the addition of things such as decoherence. The new dynamical model therefore has all the implicit and explicit assumptions made in its derivation as additional postulates of the theory. This expression is therefore not a closed statement within the axioms of quantum mechanics.

Question: Is there an alternative way of expressing the quantum-classical correspondence principle?

Reply: I think so. This followed a study with the system-of-interest being a driven two-level system. When driven by a classical field, this would display continuous Rabi oscillations. The idea was to replace the classical field with a quantum equivalent and see if it was possible to realise Rabi oscillation behaviour in the correspondence limit. The field was modelled using a driven damped quantum harmonic oscillator.[12] We used the Jaynes–Cummings model to study this system. Recall that without decoherence the spin exhibits collapse and revival behaviour (as we computed in Chapter 7). Once more, using decoherence as a key ingredient in achieving the quantum-classical transition, damping was applied to the field mode using quantum state diffusion. We then used the spin as a probe of the field behaviour.

We demonstrated that, providing the coupling was weak enough, we could transition between the two limits by turning on decoherence and drive in a controlled way [25].

Question: In this approach decoherence seems central to the result?

Reply: It is, and the same objection that was raised earlier is certainly valid for this model. The point is that the limit is reached when a test two-level system behaves in a certain way.

---

12 One could argue that the fact they also need driving with a classical field has only pushed the classical limit back. Nevertheless, the question as to whether or not the field could behave like a classical drive remains an interesting and pertinent one.

So long as the system is really a two-level system (and not some truncated approximation), they are the most quantum of objects.

Question: So, how do you express correspondence in this way?

Reply: How about the following?

> For those quantum systems with a classical analogue, when suitably coupled to a two level quantum system (the quantum probe[13]), the system is in its correspondence limit if the expectation values of the spin observables behave like the two-level system is coupled to the system's classical counterpart[14].

While the work inspiring this conjecture contained conceptual issues of decoherence, this expression is in-and-of-itself model independent. As such, it should work for quantum and post-quantum theories.

Question: Does it really need to be a two-level system – why not a three-level system?

Reply: Good question, it should be possible to generalise the statement, but my feeling is that this is probably its strongest form.

Question: How do you know if the probe is behaving like its classical counterpart and is this also linked to the measurement problem?

Reply: We have to measure it, as there is no way of verifying that any quantum system has achieved its correspondence limit without measuring something. In this case we have to measure the two-level system.

Question: So it appears we have come back to where we started, which is to ask what measurement is?

Reply: Yes, but at least we are now equipped with a way of expressing the correspondence limit for the measurement device. This opens up the possibility of considering a model-based approach to the measurement problem. What we need is a toy model of a MA coupled to a system-of-interest. In this model, all the objects should be described quantum mechanically, and at the MA realise their quantum-to-classical transition. If this can reproduce the actions of measurement, it may be that these are really only a limiting case of some underlying nearly quantum behaviour.[15]

10.7 A Model-based Approach to Quantum Measurement

In this section, we outline an idea based on work presented in [26]. The standard measurement axioms in the Schrödinger picture and Copenhagen interpretation seem at odds with

---

13 It it too much to call this a qrobe?

14 Note that this expression appears here in print for the first time. As such it has not been subjected to the peer-review process of the paper which motivated it.

15 The reason for stating 'nearly quantum' behaviour is that we will be using an open systems approach to realise the quantum-classical transition. As already mentioned in a previous footnote, this will contain assumptions beyond the core axioms of quantum mechanics. Those assumptions could be explained by some modification to quantum theory. For example, it may be that just a small amount of non-linearity, as might be realised through the introduction of gravity into a post-quantum model, could produce equivalent behaviours.

the rest of the theory. It is thus natural to ask if they are needed or can be replaced with something that seems less odd. The aim of this section is to present one alternative to the usual axioms of quantum measurement and explore some implications of pursuing such a programme of work. The idea of this discussion, even more than the previous ones, is to generate objections and stimulate discussion.[16]

The core idea is that we assume environmental decoherence is the valid path to achieving the quantum-to-classical transition. The environment is thus asserted to grow the classical-like dynamics of a probe circuit, and the combined probe-environment system is an apparatus suitable to act as a measurement device. This is a bit like the idea that a large internal resistance can make a galvanometer act as a voltmeter. Without the resistor the galvanometer is not a voltmeter, and without the environment the probe is also not a measurement device. This means that we can in principle model the classical-like MA only within the unitary evolution aspects of quantum theory. That is, the measurement axioms should arise as emergent phenomena resulting from the collective unitary dynamics. The Hamiltonian of the MA will be of the form

$$\hat{H}_{MA} = \hat{H}_{probe} + \hat{H}_{environment} + \hat{H}_{environment\ interaction},$$

where: $\hat{H}_{probe}$ is the Hamiltonian of the probe, whose corresponding classical limit we wish to achieve, without any damping, friction, or other loss; $\hat{H}_{environment}$ contains all the environmental damping effects; $\hat{H}_{environment\ interaction}$ contains all coupling energies between the environment and the probe. While one could in theory include in $\hat{H}_{environment}$ the rest of the universe, our assertion is that damping in itself will be enough to terminate the von Neumann chain and realise the quantum-to-classical transition. One then couples the QO to be measured to the MA, yielding a total Hamiltonian of the form:

$$\hat{H} = \hat{H}_{MA} + \hat{H}_{interaction} + \hat{H}_{QO},$$

where $\hat{H}_{interaction}$ is the coupling energy between the probe and the QO. For simplicity, it is assumed that the QO does not couple directly to the environment (in a real experiment, this can never be achieved). A schematic of the set-up is shown in Figure 10.5.

Strong supporters of such schemes might argue that if the environment is sufficiently complex, this procedure will be good enough to reproduce quantum measurement as an emergent phenomenon in-and-of itself. The fact is that unitary evolution can never actually realise stable localisation (excepting very specific scenarios, such as coherent states in a perfect simple harmonic oscillator). That said, if, over the complete timescale of an experiment, an approximation to localisation does occur, then the quantum-to-classical-transition could be argued to have been realised. In other words, the quantum environment is assumed to sufficiently suppress the delocalisation of the probe for long enough for it to effectively reproduce measurement behaviour. Unfortunately, to the best of my knowledge, actually modelling the dynamics of such a system is intractable, rendering this an untestable hypothesis. An application of Newton's Flaming Laser sword means that we should seek a testable hypothesis. The best we can do is to make some suitable approximations, such as to model the effect of the environment using open quantum systems methods, to estimate

---

16 I should note that the views expressed here are my own. I have different ways of interpreting this model from some of my collaborators. I have had many a fun conversation about these ideas and hope to have many more.

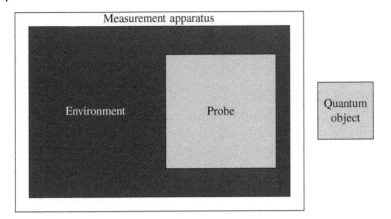

Figure 10.5 Schematic of the set-up of a minimal 'fully quantum mechanical' approach to model-based quantum measurement. Here, the MA is considered to be the probe, which is itself a QO, together with its environment. The MA is set up so that it is operating within its classical correspondence limit, and we therefore assume it can be coupled to the usual readout mechanisms available to any classical system.

the behaviour of such a model. Here, the fact that we are forced to use approximations in deriving an effective model means we should be careful not to overinterpret our findings. As previously observed, we have in our approximation process added to the axioms of quantum mechanics (and in some cases violated others – such as superposition). With this substantial caveat aside, we at least have a path to making modelling measurement tractable, even if there's a limit to what we can say about the foundations of the subject.

Question: The general idea seems reasonable and I can see that it would appeal, especially from a quantum realism perspective. Can you summarise what you did, and the findings?

Reply: The set-up was a probe that was a modified Duffing oscillator, and the QO a two-level system. We arranged the coupling, so one could argue that if the spin was in one of the eigenstates of $\hat{S}_z$, the probe potential would be a quartic function, but for the other eigenstate a double well potential (a physical realisable example would be a cross-Kerr type coupling). The idea was that with the MA undergoing periodic dynamics, the spin must be in one eigenstate, and when the MA was undergoing chaotic dynamics, the spin had to be in the other state. The reason we felt achieving chaotic dynamics in one case was important, was to support our assertion that the measurement device was operating in its correspondence limit[17]). We used both master equation and quantum state diffusion methods to model the system's dynamics (the first to get a statistical view and the second to mimic individual runs of an experiment).

A summary of conclusions that can be drawn from the findings are

(a) If $\hat{H}_{\text{interaction}}$ commutes with $\hat{H}_{QO}$, then the two-level system always ends up in one state of $\hat{S}_z$ or the other (this is analogous to measurement of compatible observables).

---

17 I had not at this time conceived of my previous articulation that correspondence limit is achieved when a two-level probe behaves as if it is coupled to the classical system. If I had, my approach and articulation would have taken this into account.

(b) The measurement device dynamics of expectation values are chaotic-like or periodic, depending on the measured state.

(c) Which final state is selected (mimicking collapse of the wave-function) is a consequence of environmentally induced symmetry breaking together with the localisation effects of decoherence.

(d) In this model, the preferred basis emerges as a natural consequence of coupling between the measurement device and the QO.

(e) Individual classical-like trajectories of an open quantum system can act as a record of the measurement of an individual qubit, and be shown to reproduce to good approximation the Born rule and Zeno effect.

(f) As the measurement took place, the two-level system entangled with the probe, but once the measurement was complete, they disentangled. So when there is a maximum of classical correlation between the dynamics of the probe and the measured state, the quantum correlations are minimised.

Question: What about measurement of incompatible observables?

Reply: There were some deviations from the Born rule and Zeno effect but as such behaviour might be expected in experiments, the findings remained reasonable. Moreover, it was possible to model a continuous measurement process, in which the two-level system would occasionally jump between the measured eigenstates, and the dynamics of the probe would be chaotic-like or periodic depending on the state measured. As such, it is my view that this approach, at the very least, provides a potential pathway into model-based measurements (something which may be useful for engineering complex quantum system measurements).

Question: Do you think that this scheme could, without the open systems approximations, provide an explanation of measurement as an emergent phenomenon?

Reply: The only hope for something tractable might be in a slightly modified post-quantum theory. For example, in the future we may have the computational ability to model a small environment. In a phase space approach to quantum mechanics, because of the way its formulated, one could potentially add a small amount of space-time curvature to the model to test Penrose's musings that gravity might be enough to realise measurement as an emergent process. To achieve success, one would have to model the full quantum set-up, such as the one proposed at the beginning of this section, and demonstrate a classical-like reproduction of a measurement by the probe/environment set-up. The probabilistic element of the system would be due to the environment starting in different configurations. In this way, the probabilistic interpretation of quantum mechanics would no longer be a fundamental one – rather one that just indicates ignorance of all the degree of freedom within a system (much like in the dynamics of a real chaotic system). This, at the very least, would open the gates for conversations about free will in a social environment.

Question: Is this the only way to understand your results?

Reply: No. One can interpret any environment as a form of measurement (as information loss of any kind can be considered a measurement if one can measure the change of state to which the loss goes). One could for example use quantum jumps to model the decoherence of the environment, and then interpret each jump as modelling photon detection (see, for example, [23]). In that work we showed that an analysis of the quantum jump record of just such an environment could evidence the chaotic-like behaviour of a Duffing oscillator in the limit $\hbar \to 0$. As an aside, on re-reading this paper many years after writing it, I was disappointed to note that I did not clearly convey one of its main foundational messages (an omission by me, as my co-authors may have a different perspective). What is not clear from the text is that the classical-like behaviour only emerges because the system is being measured. Without the decohering environment of the jumps measurement device, the dynamics would be unitary and linear, and there would be no chaos in the system. Somewhat like a physical realisation of Berkeley's *'The Objects of Sense exist only when they are perceived: The Trees therefore are in the Garden, or the Chairs in the Parlour, no longer than while there is some body by to perceive them. Upon shutting my Eyes all the Furniture in the Room is reduced to nothing, and barely upon opening them it is again created'.* in *A Treatise Concerning the Principles of Human Knowledge.* We could paraphrase and say: *'The classical limit exist only when objects are decohered/measured: the Duffing oscillator therefore is chaotic no longer than while there is some decoherence/measuring environment by which to record. Upon removing the apparatus, all dynamics are unitary, and only upon restoring them its classical-like trajectory is again created'.* Just because an environment can be thought of as a measuring device, this does not mean that all environments are interpreted in this way. So I am not saying that all classical mechanics arises through some MA – but it does seem that decoherence plays a central role in achieving the classical limit, and we cannot measure without introducing some decoherence. There are many interesting other ways of approaching this topic (such as *Quantum Measurement and Control* by Wiseman and Milburn) and relatively recent interpretations of quantum mechanics that merit investigation, e.g. Quantum Bayesianism. There is plenty of material out there, and the hope is that this section has provided sufficient background to act as a platform for you to engage with the literature in an informed way.

10.8 Questions for the Reader to Ponder

As with many discussions or foundations of quantum mechanics, this chapter has probably raised more questions than it has answered. The previous section, for example, outlined my current best guess at how measurement works within a realistic framework. However, currently unresolvable questions about the role of the environment and deviation from unitary quantum dynamics remain. At best the idea is incomplete and it would be too much to reject the normal Copenhagen interpretation of quantum mechanics based upon it. Instead of drawing conclusions, we shall therefore close this chapter by asking some more open questions for you to ponder for fun.

Question: Is pure state quantum mechanics (QM) any more meaningful than deterministic pure state classical mechanics (by this we mean, e.g. Hamilton's equations rather than Liouville's equation)? If it is not, should we only deal with purely probabilistic (density matrix or Wigner function) formulations of quantum mechanics?

Question: Can we argue that the theory of the nature of reality is any more probabilistic than, e.g. classical Hamiltonian mechanics?

Question: If environmental coherence is the path to the quantum-to-classical transition, might the probabilistic nature of quantum theory simply be an expression of ignorance of environmental degrees of freedom? How might the role of measurement in such an outcome effect this discussion?

Question: Consider a quantum coherent (i.e. pure state) composite and entangled system where each component is entangled with the others. Now also consider making those components spatially separated and consider one specific component that we shall call the object. Does it make sense to say that some of the quantum correlations between the object and the rest of the components are space-like (outside the objects light cone) and some are time-like (inside the objects light cone)? If so, would tests of quantum coherence on the object indicate decoherence even though there is none. That is, would any experiment that can be performed on the object, view the loss of information outside of its light cone as decoherence? Might such a process be a quantum version of the second law of thermodynamics? Are these questions well formulated and if not what would need to be done to make them scientific?

Question: Stueckelberg and Feynman once proposed that antimatter could be interpreted as a matter going backwards in time. If equal amounts of matter and antimatter were created at the Big Bang could this explain the matter antimatter asymmetry we see in our universe (there would be another antimatter universe evolving the other direction in time but evolving as if it were travelling forward in time)? Does this question violate Newton's flaming laser sword?

Question: Look up Milburn's ideas on intrinsic decorherence. What do you think about them?

Question: What does it take for a quantum system's environment to be considered a measurement device?

Question: Consider a universe comprising only a cricket ball and free space. Would this continue to behave like a cricket ball indefinitely (e.g. why would it not undergo unitary quantum dynamics)?

Question: There is one more axiom of quantum mechanics that we have not discussed but which is very important, namely the *symmetrisation postulate*. This states that systems containing identical particles will have one of two symmetry properties with respect to permutation of the particles. This is known as *exchange symmetry*. It is defined through the permutation operator $\hat{P}_{nm}$ – which acts to interchange the nth and mth particles in a system of particles. Mathematically, this simply interchanges the labels of their state spaces. If the state, $|\psi\rangle$, of the system is symmetric with regards to the permutation operator

$$\hat{P}_{nm}|\psi\rangle = |\psi\rangle,$$

we say that the particles are *Bosons*. If, on the other hand, the state of the system is anti-symmetric under permutation

$$\hat{P}_{nm}\,|\psi\rangle = -\,|\psi\rangle,$$

we say that the particles are *Fermions*. Experiment has only ever found that particles are either one kind or the other, which kind being dictated by the spin of the particle. The consequences are substantial and lead to the Pauli exclusion principle, Maxwell–Boltzmann and Bose–Einstein statistics as well as the structure and chemistry of atoms and molecules. The two questions for your consideration are:

- Without looking it up, can you show that two electrons where all quantum numbers apart from spin are the same can only be in the singlet state? Hint: apply the permutation operator to each of the singlet and triplet states. The fact that the subspace spanned by the triplet states is unavailable to these electrons is an example of the Pauli exclusion principle.
- Why do you think it is the case that this postulate sits apart from the other axioms of quantum mechanics?

11

Quantum Systems Engineering

"It is a rare mind indeed that can render the hitherto non-existent blindingly obvious. The cry 'I could have thought of that' is a very popular and misleading one, for the fact is that they didn't, and a very significant and revealing fact it is too."
Douglas Adams, Dirk Gently's Holistic Detective Agency

"Experience is the child of thought, and thought is the child of action."
Benjamin Disraeli – Vivian Grey

11.1 Introduction

This text, so far, has focused on developing an understanding of quantum mechanics – mathematically and conceptually – and on illustrating the practical approaches and methods that constitute '*doing physics*'; which, even in the theoretical and mathematical domains, requires a deft practitioner well skilled in varied approaches. One might be tempted to imagine physics as a pure, somewhat abstract, study of nature in pursuit of fundamental truths. Experience of performing experiments to test even simple theories to a high degree of certainty soon leads one to appreciate that this view of physics does not quite survive contact with reality. At its most practical, physics serves as the interface between theory, experimentation, and engineering; it is here that great technological innovation occurs, from semiconductors and transistors, to space technologies, to quantum technologies – which shall be our focus. Quantum technologies is a rapidly growing field, increasing in technical maturity and commercial relevance. It is premised on the deeply exciting notion that harnessing the unique phenomena of quantum physics within engineered devices and systems will realise both step-change improvements to existing areas of technology and entirely new paradigms.

Quantum Mechanics: From Analytical Mechanics to Quantum Mechanics, Simulation, Foundations, and Engineering, First Edition. Mark Julian Everitt, Kieran Niels Bjergstrom and Stephen Neil Alexander Duffus.
© 2024 John Wiley & Sons Ltd. Published 2024 by John Wiley & Sons Ltd.
Companion website: www.wiley.com/go/everitt/quantum

The notion of taking fundamental aspects of physics that we have never truly[1] used or controlled for technology and bringing those into the realm of engineering to see what we can create, has true novelty. More importantly, it has substantial potential, for example based on the impact we foresee quantum computing having on specific types of problems (from quantum simulations of chemistry for medicine and pharmaceuticals, to commonly touted cryptographic consequences, to improvements in optimisation that annealers may realise) and in the step-change improvements realised by quantum sensors. Although there is always a new leading edge of technology, it is very rare that this necessitates a fundamental evolution in the physics we can use for technology – those based on quantum physics are not merely doing the same things better; it is translating the phenomena we have discussed throughout this text, with all of its nuances, challenges, and potential, into the domain of engineering. That a quantum computer can sidestep the complexity class of solving certain mathematical problems is a clear indication of how deep a capability fully quantum devices present. We hope that our readers share our interest and enthusiasm in quantum technologies, and that some of you pursue its development – whether in academia, industry, or start-ups. For all of our readers, we hope this chapter presents an interesting perspective on what is needed to bridge quantum physics to engineering, and how the problems that most need to be addressed are traceable to open questions in the foundations of quantum mechanics.

We focus our discussion on the different engineering needs of quantum, as opposed to non-quantum, devices from a holistic perspective, why the nature of quantum systems results in specific challenges for developing new technologies, and some strategies to mitigate them. Our arguments will be considered through the lens of *systems engineering*, a discipline that has grown out of the early astrospace, aerospace, and transistor industries to now underpin the development and realisation of almost all complex integrated technologies, and that has strongly influenced software engineering and coding, which also grapple with the challenge of integrating many pieces into a complex system. The interested reader may relate some of the ideas discussed here with the philosophy behind good coding practice discussed in Chapter 8, and the wider challenges of modelling and simulating quantum systems that we have investigated in Chapters 9 and 10. For those avid physicists who perhaps see engineering as parallel to their interests, please do not think that this chapter only bears relevance to engineering and technology development; the underlying problems we discuss will relate back to fundamental problems in quantum mechanics, and the methodologies will have value for experiment and perhaps even theory.

It is very much the case that properties unique to quantum systems pose unique challenges to engineering technologies, and this will be the focus of our discussion. It is also the case that some of the ideas we discuss here transcend quantum mechanics and physics and apply equally to all science. One of the defining features of systems engineering is that we always start from the *problem we are trying to solve*, not a solution, technology, or

---

1 The distinction between Quantum '1.0' and '2.0' technologies has sometimes been made, with the former referring to technologies that depend on quantum mechanics without really manipulating and harnessing its unique facets, e.g. transistors or even photovoltaic cells, and also those that require quantum effects to be mitigated, such as electronic tunnelling prevention in modern chips which have feature sizes of just a few nanometres. Herein we consider quantum technologies to refer to those that control and utilise uniquely quantum effects, such as superposition and entanglement, to realise their intended function.

implementation we seek to promulgate; we build a quantum sensor because it is the best way to meet a need (and not because it is quantum *per se*). This 'need' translates into a set of system requirements that cover operational (core purpose), functional (what it must do), and non-functional (how we want it to be done/perform) requirements. From the outset this provides constraints and criteria against which we will test our solution in a process known as verification and validation (V&V). One crucial feature of this approach is that all requirements must be testable.

A requirements view in a scientific context thus leads us to a systematic and holistic approach to science, maybe expressed by Newton's Flaming Laser Sword (see page 332 in Chapter 10). The development of any new model or theory has requirements that lead to verification, validation, and acceptance criteria. Taking this perspective, later in the chapter we will consider as a concrete example what it means for a model of an open quantum system to be 'acceptable'. The other key aspect of systems thinking is that we consider the system as a whole. The quote by David Bohm on page 281 nicely sums this up for quantum systems, where the phenomenon of entanglement maximises the notion of 'interconnectedness'. Even in more simple classical systems there are examples of inseparable interconnectedness, such as the back action of a system on its measurement device (once more consider the humble moving coil galvanometer). Fundamentally, both physics and systems engineering take a systemic view of problems, systems, and solutions, and the methods found in systems engineering add to that systematic processes and structures to frame our thinking. In physics, our structured record of knowledge is often only a lab book (physical or electronic), in systems engineering this is extended to a structured set of tools and approaches that add traceability to what we do, maintain a record of knowledge, and fundamentally contribute to solving hard problems. In this way systems thinking and systems engineering methods can be of great benefit to the theorist and the experimental physicist, not only the technologist and engineer, as one might presume.

In this chapter, we will discuss some of the aspects of systems engineering that will be necessary for quantum technology development and useful to the physicist. Throughout, we will consider how fundamental aspects of quantum mechanics will affect and change systems engineering principles, altering how we need to consider engineering problems, prototyping, and technology translation from lab to product. We will start by setting the general scene before expanding on the specific challenges of quantum systems engineering.

11.2   What Is Systems Engineering?

11.2.1   An Overview

As physicists, we are familiar with the ideas of 'physics thinking', encapsulating ideas such as reductionism, through to Fermi problems. Systems engineering starts with a similar philosophical outlook, or world view, of systems thinking which seeks to understand complexity through a holistic perspective of a 'whole system'. This is codified by the discipline of systems engineering, which puts to practice systems thinking principles towards engineering complex, highly connected, systems. As we shall see, systems thinking and systems engineering are as much ancillaries to developing highly complex devices, as the scientific method is to doing physics – they form the method and discipline by which we assure engineering quality, expedience, and success for very complex and highly (inter)connected systems where performance, reliability, verifiability, and other factors have to be guaranteed. Systems engineering originated in Bell Laboratories in the late 1940s and matured significantly by NASA through the early space projects. As the field has matured, it has been formalised, aggregating best practice, methodologies, and analytic tools into a coherent engineering discipline.

In the most general sense, the structured process that systems engineering follows has four key stages:

1. *Requirements capture and analysis*: Systems engineering is always led by requirements. These are the 'needs' that we are addressing through an engineered solution. Whilst 'needs' may be informal and non-technical, we translate these into a set of system requirements that cover operational (core purpose), functional (what it must do), and non-functional (how we want it to be done/perform) requirements. These requirements are then matured and translated from the system-level down to a technical implementation through an iterative process that becomes more detailed as a project develops. Revealing the detailed requirements may be a matter of modelling, analysis, science, and engineering (figuring out a good 'solution'), but may also be managing external factors such as component availability through supply chains (semiconductor shortages, for example), or the customer not fully understanding what they need, requiring a more involved requirements capture process.

2. *Functional analysis and technology selection*: Functional analysis is part of the requirements process. It takes the system requirements and breaks them down into a functional structure that defines what the system needs to do without specifying any implementation. This enables technology selection, where one may comparatively evaluate different means of meeting each functional requirement, resulting in a top-level choice of technology. For example, it would be through a technology selection process that one would conclude that a quantum sensor was the right type of sensor to meet requirements.

3. *Architecting and design*: With a justified technology selection made, we continue to mature requirements whilst designing the system, until we have a detailed technical specification and requirements set alongside a full system design to implement. Architecture choices can be complex and involved (as we shall see in quantum), and there are often iterative aspects to this.

4. *Implementation, integration test, and V&V*: Lastly, we implement the designed system – usually in granular stages – and begin to test it against the requirements we have formulated. As components and subsystems succeed testing, we integrate the system and verify it meets requirements, iterating design and implementation if it does not. Eventually, we verify the whole system against the initial requirements, and the customer validates it against their needs.

Traditionally, these processes were captured and managed through documentation-heavy methods. Increasingly, software-enabled model-based systems engineering (MBSE) approaches are being adopted. Analysis of their advantages and disadvantages is beyond this chapter's scope; however, they emphasise modelling and simulation-led design, which is very similar to how we do physics, and very important to novel technology development.

An important benefit of MBSE for innovative technology development is that by starting with modelling, it is possible to quickly discover the things we do not know, and cannot adequately model or describe. These may be at a very high (functional) level, where we discover that we have a poor understanding of the physical relationships between aspects of our system, or how they would be practically implemented. Similarly, as our models increase in detail towards a technical design, we discover the specific technical aspects we do not know, and capture all of this within the MBSE process. Innovation involves many 'unknown unknowns', and MBSE can valuably convert these into 'known unknowns' before it is too late to address them.

11.2.2 Notes on Complexity and Interconnectedness

It is important to touch on the notion of complexity; a point of commonality amongst the technologies that drove early systems engineering was the need to integrate numerous, highly connected, technically sophisticated subsystems and components. Whether a fighter jet, spacecraft, electrical grid, or, today, a computer chip with billions of nano-metre transistors; in engineering these systems it was found that, as scale and connectivity increased, the performance and reliability of constituent components stopped providing a good indication of the performance and reliability of the system. Furthermore, intuitive notions such as the system performing best when each individual component performed best no longer held true – indeed, and as has also been shown for quantum systems – there were cases where 'good + good = bad'. This experience gave rise to a more formal notion of complexity: *'Complexity is the degree of difficulty in predicting the properties and behaviour of a system when the behaviour of the parts is known'* [89]. It is a consequence of the *integration* of elements, and a matter not only of *what* has been integrated, but *how* it has been organised and connected. If a system cannot be described as the simple (or complicated, as it were) combination of its parts, the implication is that it must be described and characterised as its systemic whole. Therein, systems engineering differentiates itself as a complement to, but not replacement of, the numerous fields of device engineering; its concern is with characterising and predicting system-level properties, and supporting the engineering and delivery of the whole. Today, systems engineering is applied, as standard practice, to the development of most complex products across industries and technical domains (often allocated 10–15% of a project budget, a substantial amount).

11.2.3 Industry Standard Systems Engineering

Its scope has been broadened and codified[2] to include everything from the specification of systems and their technological design, to maintenance, to supply chain analysis, and

---

2 The gamut of systems engineering processes are recorded in the *ISO/IEC/IEEE 15288 Systems and software engineering – System lifecycle processes* standard. This standard does not specify how to do systems engineering, but does specify the top-level processes it contains, i.e. its remit. This is an interesting mixture of concrete, technical, engineering, from system design to implementation and test, to broader notions of engineering management that facilitate the delivery of complex systems. For our purposes, we shall be considering (quantum) systems engineering from a technical perspective here, but for the reader interested in effectively undertaking complex R&D and product delivery, a full understanding of systems engineering would be of benefit.

to all areas that might fall in between – it manages and enables the *design for* a system's intended purpose and entire lifecycle. It serves (among other things) to analyse and improve design feasibility, track and record project progress, and develop objective criteria against which system and subsystem functionality can be verified and validated. Its ubiquity and proven usefulness in realising the current cutting edge of technology leaves little reason to expect that systems engineering methods will not also be standard practice in the future of quantum technologies development.

Whilst the Systems Engineering approach is typical for 'product' development at high *technology readiness levels (TRLs)*,[3] it is unusual in the context of low-TRL [52] research. Whether for experimental physics, early R&D, prototyping, or demonstration, it is obviously desirable, and in many cases necessary, that low-TRL lab-based experimental systems and devices reliably satisfy their operational requirements, whether for science or technology. Furthermore, establishing processes that make experimental development more traceable, inform and record design choices, and eventually improve translation to some next step (whether further experimentation, or towards mature technology development) – in combination improving research whilst minimising time and resource waste – brings benefit to low-TRL science. More recently these benefits are being realised, with methodologies for low-TRL systems engineering, including for quantum technologies development, showing benefit to device development and research outputs alike. We do not intend in this text to provide an encompassing description of systems engineering; instead, we aim to discuss why this is an interesting subject for quantum devices.

11.3 Quantum Systems Engineering

11.3.1 Motivation

Systems engineering, as a whole, has, as mentioned above, been developed for complex classical technologies. It has a broad canon of methodologies and tools, some general and some bespoke; in practice, systems engineering is always tailored to the task at hand, with appropriate tools and methods selected or developed. Quantum technologies development is still nascent; there are a few early technologies that have been translated from lab to product in a few domains,[4] but most remain in the lab, seeking device demonstration; there is not yet a mature notion of quantum engineering processes. For a variety of reasons, it is clear that quantum engineering has unique facets, owing to practical considerations and the unique properties of quantum states. Quantum states are fragile, not only easily destroyed, but fundamentally (and often irreversibly) changed through measurement, interaction, and environmental variations. Furthermore, interacting with quantum states implies entanglement, which affects concepts of system boundary definition, and introduces systemic

---

3 TRL is a standard engineering concept assessing the technological maturity of a device, system, or product. The notion of TRLs was originally developed by NASA for space technologies, and have since been adopted by various organisations with slightly different definitions but similar concepts of maturity. Consistently, TRLs span TRL 1 to TRL 9, where 1 is a hypothesis or theoretical notion, and 9 is a mature in-service implementation.

4 Quantum communications technology, for example, which is a front-runner in technical maturity and product development.

complexity. The ability to *model* systems is fundamental to systems engineering[5] but, as we have already discussed, is a particular challenge in quantum physics, especially if one is seeking a realistic and detailed description of quantum systems and their interaction. The modelling capabilities required for some aspects of systems engineering do not presently exist for quantum systems, and it is not obvious how they would be realised – at the very least the approaches developed for the study of physics seem to be a poor fit for engineering.

Lastly, systems engineering also encompasses processes for test, verification, and validation – assuring that a system is what is predicted and intended. In a classical context this can be challenging for so-called 'black box' systems, such as deep learning models, yet it is likely to be even worse for quantum systems, which cannot be easily interrogated and may be performing operations with no classical analogue and no obvious means of checking if an answer is 'right' – a quantum analogue to 'black-box' V&V is needed ('cat-box' as it were). A practical example of this is Feynman's original use case for quantum computation for simulation of quantum systems to controllably investigate aspects of physics; this will result in outputs that cannot be classically verified, and methods of delineating a true answer from the breadth of errors that might obscure it is a nuanced area of research. Together, these differences suggest that quantum engineering will have significant, unique aspects, and that systems processes may have to be fundamentally altered for quantum technology development; that *Quantum Systems Engineering* is not just systems engineering quantum technologies.

Considering that you may be inclined towards engineering quantum devices in the future, it is worth briefly exploring the 'what' and the 'why' of some of these differences. We emphasise *some* since, as hinted at earlier, quantum technology development sits at the forefront of knowledge. We do not yet have an overview of its similarities and differences to other fields of engineering, nor have we attempted to construct large and complexly integrated devices. When it comes to complexity and challenge, quantum computing is the high watermark of the current engineering of quantum devices. Whilst there are a variety of qubit types and quantum computing architectures, superconducting quantum interference device (SQUID)-based superconducting quantum computers are one of the primary variants; a system we have considered earlier. Chapter 9 should provide the reader with an appreciation of just how challenging modelling and engineering a controllable, high-fidelity, qubit array might be, requiring on-demand entangling and disentangling of SQUIDs without extraneous coupling, or excessive decoherence.

As qubit numbers grow, we are beginning to see how architectural design assists in this task, the notion of every qubit being able to entangle with every other is foregone in favour of architectures of qubit clusters, with full connectivity within each cluster, and structured connectivity between clusters. This engineering solution enables 'logical' qubits to be constructed from a more sophisticated, but also more easily realisable, architecture. While advanced, the art of the possible remains far from the very-large scale integration that has driven much of systems engineering development – there is certainly more to learn. Therefore, the points made in this chapter are also a reflection of the maturity of quantum technologies development, and the points of difference we highlight should neither be seen

---

5 Particularly the sub-discipline of MBSE.

as an exhaustive list nor the final word on the subject. With this in mind, let us consider a few key areas of difference between classical and quantum systems engineering processes.

11.3.2 Modelling and Simulation Challenges

11.3.2.1 What Is in a Model?

We have already discussed the theory and practicalities of modelling and simulating quantum systems in some detail, so we shall not reprise those discussions here, but just summarise a few key points. The role of modelling and simulation in device design and engineering, particularly for fundamental device engineering, is intuitively clear; in order to inform design, we must understand the consequences of architectural and engineering choices, and therefore be able to predict the properties (functionality, performance, reliability, manufacturability, etc.) of a system we propose. Therefore, we need to be able to describe quantum systems with a level of detail reflecting their 'real' implementation and behaviour in a non-idealised environment.

> The quality of a model for engineering is defined by its ability to very exactly reflect the behaviour of real engineered systems in use.

Whilst a good model for physics may work well for engineering, its purpose is quite fundamentally different. Models in physics are often descriptive as opposed to exact, they seek to highlight and identify phenomena of interest, and to represent a theoretical fabric by which truths of nature may be represented. This may sound overly philosophical, but it is important in that what constitutes a good model for physics is somewhat different to what constitutes a good model for engineering; to name one difference, in physics we often value generality – for example, that our theory of open quantum systems is valid at very low temperatures and higher temperatures alike, or that it is valid for both weak and strong coupling regimes, generality indicates that we are closer to a representation of nature as it is.

For engineering, the concern is primarily that one is able to model behaviour in these different parameter ranges very exactly; whether this is through one or several models, or fundamental, or phenomenological, is far less important so long as the model's validity can be well evidenced. In the semiconductor industry this approach is well exemplified by the Berkeley short-channel IGFET model (IGFET – insulated gate field-effect transistor) (BSIM) [13] family of models, which contains models for different classes of semiconductor devices, each assembled from numerous sub-models reflecting different parameter ranges and conditions. This pragmatic approach to modelling is less unified than what we often aim to achieve in physics, but provides the predictive capability necessary for engineering. Models with the qualities necessary to support quantum engineering are currently immature; in the near term this shall be a challenge for the quantum engineer, who will likely need to develop modelling and simulation approaches alongside experimentation and prototype device development. When looking at open quantum systems, we also saw how challenging it is to derive analytic expressions for quantum systems – even in very simple cases – and the numerous layers of assumptions and simplifications we apply in order to arrive at a mathematically and physically tractable scenario. This can be seen as a barrier to technology development; however, we may also interpret it as a requirement on the devices we design.

11.3.2.2 Design for Modelability

Systems engineering includes the notion of *design for…* methodologies, such as *design for…* reliability, manufacturability, integrability, testability,[6] and so forth. For quantum devices, we may need to consider an additional design requirements, such as: *design for modelability*. Conceptually, this invites a tautology, design for our ability to design, but pragmatically, quantum systems design may need to err towards what we know we can model and simulate, and attempts to tame what we cannot – such as any environment to which the system will couple as it decoheres. The notion of engineered environments can be taken quite far, and one could imagine a carefully designed environment meant to simplify system behaviour and maximise coherence times (e.g. by making use of two-photon absorption); however, it need not be this complex. A dilution refrigerator is also a type of engineered environment, with controllable low temperature and isolation from its 'external' environment by way of thermal insulation, potentially vibration damping, and electromagnetic (EM) field damping (e.g. mu-metal shielding or active compensation); this in and of itself may simplify the challenges of modelling a quantum device, how substantially, and to what valid degree, remains to be seen.

11.3.2.3 Hierarchical Modelling

So far, we have only considered low-level device engineering; however, modelling and simulation also serve abstract design – bringing it from the realm of the specialist engineer or physicist, to that of a high-level designer or systems integrator. This is achieved through *hierarchical modelling*, a hierarchy of models allowing system behaviour to be described realistically at different levels of abstraction. High-level models often represent a simplified view of system behaviour, and describe performance in terms of non-fundamental measures of effectiveness for the integrated system whole, whereas low-level models eventually reach the physical foundations of a device's operations. An example from the semiconductor industry is describing integrated-circuit performance in terms of timings, which can be related to high-level concepts such as clock cycles, but is a result of their specific semiconductor physics. It is also exemplified in software development and the nature of programming languages, including in the implementation hiding an intuitive hierarchy presented in a single code base (e.g. as in Chapter 8), and in the abstraction offered by high-level programming languages enabling someone with very limited architectural knowledge to write code, with libraries and the compiler hiding the details of the implementation and cascading instructions down to machine code – all without inhibiting design.

This has the important consequence that it makes the design and requirements processes accessible; a good hierarchy of models hides detail and complexity whilst retaining accuracy sufficient for system design to take place without having a total knowledge of system operation. Practically, this means a systems integrator can compare devices and make a technology selection, or write requirements for a system they need without understanding how exactly that system is made. This works down the design chain, distributing the

---

6 These are sometimes referred to as Systems Engineering '-ilities', for obvious reasons.

need for specialist expertise at low-level fundamental device engineering, whilst retaining a linked overarching model of device behaviour and performance at higher levels. As we briefly explored when looking at decoherence, we do not currently have a natural hierarchy of models of quantum systems. How to implement hierarchical models, and what choice of (fundamental or abstract) parameters to use, remains an open problem. The decoherence and dephasing times (T_1 and T_2) used in Quantum Computing are examples of the abstract parameters, we use today, but as more diverse quantum technologies are realised along with underpinning quantum components, one might imagine that this will expand.

There is also an immediately practical importance to hierarchical models; in the near term, they may be needed in order to realise computationally tractable models of complex quantum systems. If one considers a truly 'realistic' model of a quantum device, including all influences on the system from its environment, the effects of measurement, internal correlations, and potential feedback and control mechanisms, a very large Hilbert space will be required to describe the system. To an extent, this may be a matter of making surreptitious simplifications; however, if we then consider the integration of quantum components or subsystems in an overall device, we will again quickly run into issues of scale. The concept of quantum computers to simulate quantum states is a recognition of this challenge, and, for quantum computing, classical simulation on supercomputers currently enables a little over 50-qubit simulations. Modelling and simulation of quantum computers (which currently integrate over 100 qubits), or other technologies including relatively simple quantum systems, exceed the limits of classical simulation on conventional hardware. If hierarchical models can be used to simplify this at points of device integration, they may significantly aid in the realisation of early integrated devices.

11.3.2.4 An Aside on Standards

There are some surprisingly important reasons for establishing hierarchical models. One example is the development of appropriate taxonomies – which are usually hierarchical classification schemes that can be used to abstract entities into different layers of complexity. These are used to organise, index, and make connections between related things. Taxonomies can be sufficiently important in the development of a technology that they may be governed by international standards. The development of vocabulary and taxonomies for quantum technologies is one of many activities of standards agencies. Whilst it is beyond the scope of this chapter, design for standardisation and the role of standards organisations will be important to the success of many quantum technologies (from, e.g. specification of interconnects to the methodology of defining a quantum sensor profile). The nuance of quantum physics will be important in developing, understanding, and demonstrating conformance to many of the standards that will emerge as quantum technologies mature.

11.3.2.5 Extensibility

A concept closely related to model hierarchy is model *extensibility*: that basic models can easily be extended, or combined, to include different configurations, elements, or behaviours. At the simplest level, this could enable several components to be modelled together, based on their individual models, or for one to change the model of a component in an environment by simply adding or removing complementary environment models. This property of models is important to the classical engineering design process, and typical of low-complexity technologies, but it is unclear if this will be possible for quantum systems at all. Adding new elements to a system or environment ultimately means changing the system, environment, and/or interaction Hamiltonians, which quite fundamentally alters the system's description; if extensible models are to be realised, they are likely contingent on careful design for modelability. At the time of writing, there is significant ongoing research on extensible quantum computing architectures, searching to enable scalable many-qubit systems to be realised. These are contingent on the addition of qubits having a predictable, and ideally minimally negative, effect on overall system behaviour and performance, implying that a model for such systems will have some extensibility, at least to system scale. However, a broader notion of extensibility, encompassing environmental effects such as EM coupling and temperature – which would very fundamentally change the system environment and coupling – seems incongruous with the physics of quantum systems. For example, taking a flux-coupled SQUID and adding capacitive coupling changes the system description and parameterisation far more profoundly than an additive term in an extensible model could account for.

11.3.2.6 State of Play

It is clear from this discussion that modelling quantum systems is immature and has potentially fundamental differences to classical modelling. For the quantum engineer this will be a significant challenge, and work to develop better models should be undertaken alongside technology development (whether towards product realisation or for experimental science). To an extent this is a statement of our *position of knowledge* in quantum physics, and shortcomings in modelling relate closely to the various quantum engineering differences we will describe next; it is worth considering, throughout this chapter, how engineering challenges relate to our position of knowledge. It is easy to imagine that engineering, as a discipline or assembly of processes and methods, sits after physics, translating science into operational use with all the knowledge of the foundations – for quantum physics this could not be further from the truth, engineering here is the synthesis of theoretical and experimental physics with partially translated experience gained from systems integration and engineering in other domains. The challenges of quantum engineering stem from those of quantum mechanics, will be addressed by research that addresses both, and – as we shall also discuss later – will present solutions to the benefit of fundamental science and technology development alike. In the future, one might imagine a mature field of quantum engineering, complete with fundamental understanding, model hierarchies, design tools, and its own quantum engineering methods, potentially quite abstract from its underlying physics. Perhaps, by resolving the challenges highlighted in this chapter and more, unmentioned or unknown, this notion of quantum engineering can be reached – for now, though, the quantum engineer is the quantum physicist, engaged with theory and application, and pushing the frontiers of knowledge to realise new technology.

We shall now look at some, but not all, of the other areas of difference – many of the points shall be corollaries of the discussion above.

11.3.3 Reliability Engineering

A core element of systems engineering and technology development is *reliability engineering*; succinctly, this is the analysis of a system's ability to provide its core functions to an expected performance level for an expected duration or lifetime – conversely, it is the study of system failure, and therein of operational reliability. This ranges from the very simple (e.g. that any key on your keyboard, will, on average, last a minimum number of presses without breaking) to the tremendously complex (e.g. that a spacecraft can carry out a mission, or indeed that your smartphone, with its complex integration of components and subsystems, will work as expected for at least its warranty duration).

Reliability is not a deterministic notion – we cannot guarantee that something behaves perfectly and lasts so long, but we can describe how it can fail (its *failure modes*), the mechanisms for these failures, the probability distributions that describe them,[7] and how these failures interrelate – both in terms of their mutual correlation, and how they sit within a fault tree for a device with potential redundancies and mitigations. It is easy to conceptualise failure modes as something highly mechanical – a spring breaking, a rubber gasket becoming stiff, or something warping due to temperature – however, there are many more subtle and physically complex aspects of reliability.

Taking our running quasi-classical example of semiconductors, three important, physically complex, failure modes are thermally induced electromigration, electrostatic discharge over an integrated circuit,[8] and quantum tunnelling (noting that modern transistor circuits have feature sizes of 3 nm and very dense packing). We shall not unpack any of those here, the point is only that device reliability is nuanced, and for the engineer to predict and characterise it, a deep understanding of the underlying systems is necessary – to an extent the challenges in modelling and simulating quantum systems appear again.

Considering the above examples, it seems reasonable to expect that reliability engineering will be important to quantum technologies, just as it is elsewhere, and will be quite challenging due to issues of complexity and the need to mature system-level models. We should also consider if there will be unique aspects to reliability engineering quantum systems. The most likely difference is of failure modes. A classical device cannot fail due to decoherence, or errant entanglement; uncontrolled or unexpected quantum behaviours will manifest as failure modes, as shall improper state preparation and decoherence. Whilst these are new causes of failure, it remains to be seen how they are characterised from a reliability perspective. Decoherence, for example, may be quite a binary failure, whereby, if system coherence is above a threshold, it performs adequately (or is mitigated through error correction), and once below it effectively fails completely – whilst this is a uniquely quantum case, its effect on the system's reliability may be quite ordinary, simply described by a probability for discontinuous system failure (although calculating this probability may

7 For example, very often things have a 'bathtub' failure distribution, meaning that probability of failure is high when a device is new (typically manufacturing defects) and old (age-related failures and wear and tear), but constant and low for most of its operational life 'in the middle'.

8 This is particularly interesting since it is a complex failure mode; there are various ways to protect components and sub-circuits from electrostatic discharge; however, if you protect each component maximally the risk of electrostatic discharge causing critical damage to the integrated circuit actually increases. This is an example of 'good + good = bad'. Optimum reliability can only be achieved by optimising at the system level which, unintuitively, is not necessarily making the 'best' subcomponents – a practical example of complexity. We know from our understanding of open quantum systems that mixed, entangled, quantum systems lose separability (i.e. they are not fully described by the subsystems alone) and therefore are complex. Their failure, from a reliability perspective, is likely to be complex too, and require analysis and engineering at the system level.

be challenging) or a decoherence time (mean time to failure). Unintended entanglement, however, seems to be more unique; for a system with many controllably entangled quantum components, such as qubits in a quantum computer, it is reasonable to expect that coupling will be imperfect – there will be some unintentional coupling between quantum components. This may have direct consequences, such as decreasing the fidelity of qubit operations or inducing unexpected state changes (a qubit flip, as it were), and indirect consequences, such as hastening decoherence of the system. How this is described from a reliability engineering perspective is an open question; those who are interested may find value in comparing unintended entanglement with classical failure modes to see if there is anything similar – electrical cross-talk bears an intuitive similarity. It is left as an exercise to the reader to consider how far that similarity goes, and to formulate a justified opinion on whether errant entanglement is likely to be a unique type of failure from a reliability engineering perspective.

There may also be quantum consequences for seemingly classical failure modes. A tremendously common failure mode, for practically any device, is it is physically breaking – material fractures in the device due to thermal cycling, vibration, impact, wear and tear, and so forth. Superficially, we might dismiss this as being quite uninteresting to quantum, but let us consider what it might mean to a superconducting device such as a SQUID. Fundamentally, a SQUID is a superconductive ring interrupted by a Josephson junction, a weak link, made from a thin insulating layer (typically $< 1\,nm$). So what if, perhaps due to thermal effects from numerous cooling cycles, a very thin small crack were to form through one part of the SQUID ring? It may simply break the device and render it non-functional; however, under the right circumstances it might also behave as a Josephson junction – a weak link in a superconducting ring. This would certainly have changed the system's Hamiltonian in an unpredictable and unexpected way (it may change an radio frequency (RF) SQUID into a direct current (DC) SQUID); however, it may still be a coherent quantum system. Such a failure mode may result in unexpected quantum behaviours by fundamentally changing the system, without leading 'simply' to decoherence – a classical form of failure leading to a nuanced quantum failure mode.

Clearly there is a lot of nuance to quantum reliability engineering; research on reliability will necessarily be a part of mature technology realisation, from an engineering perspective there is not, nor should there be, any way around it. However, rather than waiting for it to become an engineering problem, characterising quantum reliability, experimentally inducing failure modes and cataloguing them, could be a fascinating area of empirical research in and of itself – after all, using experiments to discover unexpected behaviour has long been part of physical enquiry.

11.3.4 System-of-interest Boundaries

At various points in this text, we have referred to a 'system-of-interest' (SOI), and used the term intuitively; in systems engineering it has a specific meaning, closely related to system boundaries: the SOI is the system whose lifecycle is under consideration [42]. We shall not go into the broader notion of a system lifecycle here. It spans everything from an initial system concept, requirements analysis, through to design, development, test, V&V, deployment, and subsequently the system's deployed life (including factors such as repair and maintenance, upgrade, and eventually retirement, decommissioning, and potentially even recycling). This extends far beyond the scope of this chapter, and is covered well in standard systems engineering references, however, do recognise that we are discussing part of this lifecycle, spanning from design to validation, and therefore the notion of a SOI is relevant and important to delimiting what we are developing. Typically, a 'system' (which may be anything from a system of devices, a system-of-systems, to a whole device, to a component, depending on what is being developed) is split into a *narrower SOI*, a *wider SOI*, an *environment*, and a *wider environment*. These boundaries attempt to separate the system controlled, developed, and owned by the engineer (the narrower SOI) from the system into which this will be integrated and contribute to (the wider SOI). For example, the narrower SOI may be a qubit, a quantum measurement and read-out subsystem, or a cooling system, integrated into the wider SOI of a functioning quantum computer; if one is developing a stand-alone component, an encapsulated system, or something that cannot be broken down into constituent subsystems, it may be that the wider and narrower SOIs are the same.

This wider SOI will operate in an environment that directly influences it; this cannot be controlled by the engineer, but its properties may be mitigated (or utilised) in the system's design. The environment need not be a purely technical notion and can include human elements, such as usage, or regulatory elements, such as standards or compliance requirements. We shall touch on these, but in the context of our engineering discussion here, this mainly means the EM environment, temperature, vibrations, power supply, and so on. Lastly, the wider environment is a wider context that does not interact with the SOI at all but may influence the whole 'system lifecycle', and decisions made over its course (e.g. system disposal, or upgrade for future regulatory compliance); this is important to commercial systems development, but not to our analysis of quantum technologies, and therefore shall not be considered here.

So, the SOI definition provides the engineer with a means to structure the system elements being developed in order to define boundaries within a project, and within a technical system architecture. This defines what is *internal* to the system, and what is *external* to the system, influencing how the system is developed (where do you split it apart?), how these parts are brought together and integrated, and, therefore, also how the system is modelled and described. The difficulty in effectively delimiting SOI boundaries for classical technologies is increasing; in complex, highly connected, devices, identifying the SOI, and separating narrower and wider SOIs, is both a conceptual and engineering challenge. This is exacerbated for devices where connectivity changes, those that are intrinsically reconfigurable or serve many different and changing functions within a wider system.

Similar challenges may exist for quantum technologies; with the potential for quantum components to controllably or unintentionally entangle with the narrower and wider SOIs and the environment (potentially manifesting as decoherence, but maybe something more nuanced if the environment is quantum too, as might be the case for a quantum component's environment). The 'connectivity' of a quantum system has the potential to change significantly in operation. Considering our previous study on open quantum systems, if, for part of an operational cycle, a wider system must couple by entanglement to the narrow SOI (e.g. quantum memory, input–output (I/O) or readout, or an ancillary qubit array) in a manner that generates a non-separable quantum state, it is hard to argue physically that they are not now both part of the narrow SOI – if the system Hamiltonian has changed, so surely has our SOI. If one were to take the alternative approach of considering all operational conditions and bringing every wider system that we might entangle with into the narrow SOI, where would one stop? – and would this be practicable for engineering?

Fundamentally, this begets the question 'how do we define a quantum system's architecture?' We are used to being able to establish system boundaries and interfaces, based on material structure and connection – how do we box things? what plugs into what? – but is this adequate for quantum technology development, or does architecture need to fundamentally reflect types of connectivity, entanglement regimes, and the fundamental mutability of the system? This certainly requires more nuanced consideration than cutting the von Neumann chain at a convenient point and calling that a system boundary. Furthermore, if we were to deviate from traditional notions of a SOI towards a quantum analogue, couched in terms of quantum systems engineering and quantum architectures, how shall we treat our classical co-systems – for example, measurement devices that eventually provide a classical read-out? Perhaps that point of classical transition is one of the few system boundaries we can draw with some degree of confidence.

Before we move on from the SOI, it is worth raising a practical consideration: the SOI may be defined according to what we can most easily model, and, as we have already suggested, the quantum architecture developed according to modelability. An observation we may make from our discussion of open quantum systems in Chapter 9 is that the mathematical, especially analytic, tractability of our modelling problem depends on whether we choose to include aspects of the environment in the system Hamiltonian. Now, this may sound logically suspicious. Take as an example a SQUID coupled to a thermal bath; we know the Hamiltonian of our SQUID, and that of our bath – so for what reason would bringing part of the environment into the system Hamiltonian improve the modelling problem's tractability, unless by some mathematical trickery? The answer lies in the assumptions we make, particularly the Born assumption, which underpins the separability of environment and system, and the absence of back-action from the system to the environment due to the environment being vast relative to the system, and having a far shorter correlation time – ergo, we argue that they are effectively uncorrelated.

So, what it means to bring a term from the environment into the system Hamiltonian, is that this 'part' of the environment does not abide by our notions of separability and cannot be treated with this assumption – we split the environment into *separable* and *inseparable* environments, an addition to the SOI model put forward previously, and the latter part of an effective SOI described by an overall system Hamiltonian. Not only is this another way in which the SOI concept changes for quantum technologies, it may also provide us with a pragmatic method of SOI definition: if we can use the modelling process, particularly the choices we make to reach tractable problems, to define the effective system Hamiltonian for a quantum device – perhaps several of these for different operational regimes or environmental conditions – maybe this is a fundamental, bottom-up, method of SOI and quantum architecture definition. Therefore, the quantum SOI boundary is no longer just a logical delimiter, but an expression of quantum mechanical reality, and any reconfiguration of it implies a fundamentally different system. Not all SOI boundaries will be valid; and this analysis may reveal key engineering choices between bringing the inseparable environment into the SOI, or devising engineering solutions to remove it from the environment altogether.

An interesting corollary of this is that it implies that the prospective quantum systems engineer may need to have specific knowledge of modelling quantum systems in order to identify design configurations that are physical, and understand their consequences.

A question that has been posed by the growing quantum technologies and 'quantum-ready' industries is 'how much fundamental quantum physics does an engineer need to know to work with or on quantum technologies?'. The answer seems to be 'quite a lot', and, considering our discussion of the SOI, abstraction through hierarchical models is unlikely to be sufficient to change this.

11.3.5 Requirements Analysis

As alluded to in the introduction, a fundamental part of technology development is requirements capture and analysis; from a systems engineering perspective, it is the start of the development process, which should always be requirements led, and seek to deliver solutions to well-articulated problems.

As we have been doing up to this point in the chapter, we shall consider how the requirements process is affected by quantum. However, since requirements are so fundamental to technology development (and, as we shall highlight later in this chapter, research science), we will first take a moment to explain what requirements are, and how they are formalised. Let us start with a simple example of a requirement, and then go into some more detail. A current quantum technologies use case gaining significant attention is *position, navigation, and timing (PNT)*, looking to quantum to provide 'better'[9] PNT by using highly accurate quantum sensors, and negating reliance on satellite systems such as the global positioning system (GPS).

A fundamental requirement for a Quantum PNT system might be to provide 'GPS quality PNT without reliance on satellites or external positioning systems' – this requirement says nothing of *how*, but does specify *what* needs to be achieved in order to deliver a useful and valuable technology. This, of course, would not be the only requirement, but the important point is that technology development is, and should be, derived from a set of 'needs', articulated by end-users or stakeholders, for which we seek to develop a solution. The identification of these needs is termed *requirements capture*. These are then analysed, and translated into formal system requirements, which articulate what we aim to develop and provide us with a point of reference against which we can validate an eventual solution. Subsequently, a process of technical requirements analysis, solution analysis, architecting, and technical design translates these from high-level statements to low-level implementation; correspondingly, at each stage of development, the solution can be verified and validated against its corresponding requirements, from components, to subsystems, to an overall integrated solution demonstrated in its intended use environment.

---

9 We shall not go into what this means in great detail, since it is a nuanced discussion that requires some knowledge of technical subjects far outside this text's scope. Do note, though, that 'better' should be based on an analysis of needs, and codified in the requirements – and certainly not assumed. The most common assumption is that 'better' means 'more accurate'; this assumption is often made, and in our experience, rarely true in a straightforward sense.

If, at any of these development stages, the solution does not meet the corresponding requirements, we must iterate, identifying why it does not, and engineer a better solution. By following this process, and assuming the requirements set is realistic (both that it is logical, that the requirements are non-contradictory and do not contravene the laws of physics, and feasible, that meeting the requirements is within the art of the possible), the final solution can be verified against all the top-level requirements, and validated against end-user or stakeholder expectations.

Lastly, before we consider the quantum side of this, it is worth describing the types of requirements that can exist – there are various requirements models, but our view is that the Holistic Requirements Model [9] provides a clear structure to requirements well suited to technology development and science:

- *Operational requirements*: These are the high-level requirements that reflect the 'need' we are seeking to address, our earlier example of a technology providing 'GPS quality PNT without reliance on satellites or external positioning systems' would be an example of this. Another example might be a device that 'Allows for the accurate and verifiable simulation of molecular chemistry to enable drug discovery' – this might apply to a quantum computer. Note that neither of these examples say *how*, only *what*; high-level requirements statements should not specify a solution unless there is a unique reason to do so; they should be technology-agnostic, and the engineer should select the solution that best meets the needs (as opposed to any preconceptions).
- *Functional requirements*: If the operational requirements are the 'demands', the functional requirements are the 'solutions'. These are the specific functions a system must perform in order to meet the operational requirements. Maintaining the PNT example above, two of the functional requirements derived from the operational requirement may be to 'estimate changes in position' and 'provide accurate timing'. We spoke of translating requirements from high- to low-level statements. This requirement could, for example, be broken down into requirements to 'measure linear acceleration in three dimensions', 'measure rotational acceleration', and 'continuously integrate acceleration measurements to derive an estimate of motion'. This still has not implied a solution, but supposing we – for some good reason based on technical capability – selected a quantum accelerometer for this system, we might with some intermediary analysis[10] reach requirements to 'produce a trapped gas of atoms', 'cool the trapped atomic gas', and 'measure the acceleration of the cold-atom gas', amongst many other requirements. Evidently, even more detailed requirements could be derived from these, leading to the specific functional architecture for the system implemented by a concrete design.
- *Non-functional requirements*: As the name implies, non-functional requirements are those that do not describe a system function but do describe a property of the system or a requirement on its design. This is typically broken into three categories:
  - Non-functional *performance* requirements: These specify how well a system needs to perform, typically associated with a measurable quantity. When the system's performance is tested, it is these requirements that set the benchmark. One may note that the

---

10 For clarity, this example is substantially simplified, and should not be taken as a literal description of a cold-atom accelerometer.

example functional requirements above only specify *what* the system must do, not *how well*; a non-functional performance requirements would do this, for example specifying the navigational accuracy at the top-level, down to the temperature of the cold-atom cloud at the bottom-level.

- Non-functional *system* requirements: These describe properties of the system that constrain design, but are not necessarily specific to its functions. The physical properties of the system are the most obvious example, such as its size, weight, power (and potentially cost). System-level '-ility' requirements such as reliability also fall into this category, although they may be translated to, or result in, performance requirements or functional requirements allowing them to be satisfied.
- Non-functional *implementation* requirements: Lastly, these requirements place specific constraints on the implementation of a system or function. These can come from various sources, but most often capture compliance standards. Again, taking our PNT example, at the high-level this might be that the system implements standardised interfaces and data formats to communicate PNT information with other platform systems; conversely, a specific low-level implementation requirement might be that the atom source used in the cold-atom accelerometer complies with standards for radioactive sources.

Using this formalism to understand one's goals – in technology development or science – is invaluable. Furthermore, it is the starting point for creating a *record of knowledge*, a consistent means of documenting what you intend to do, for what reason, and how this has led you to a specific design decision and implementation.

We shall talk more about the value of maintaining a record of knowledge and applying methods, including requirements analysis, to scientific research a little later in this chapter; for now, let us return to considering whether quantum makes things different. Initially, one might think that requirements analysis will expand, in so much as there could be new requirements pertaining to quantum aspects of the system. However, stepping back, we must not forget that at the top-level, an operational requirement is not meant to specify implementation – it is meant to capture need. We do not set out with requirements that specify a quantum system, rather we develop a quantum system, if that is the best technology choice to meet the requirements.[11] Therefore, requirements capture should not change due to quantum – requirements analysis will need to expand thought to include the technical requirements associated with quantum implementations. This is not a profound change to the process, just an addition.

The only challenge that appears in the context of requirements analysis is in actually translating high-level requirements to a technical specification and design for a quantum solution – this relies on hierarchical modelling, allowing one to translate a requirement for position accuracy into, for example, the specifications for an atom source and magneto-optical trap (MOT), or even the entanglement regime used in a sensor array to improve accuracy and suppress noise.

---

11 Doing the opposite is often called creating 'solutions looking for problems'. Even if, as a scientist or technologist, your goal is to develop quantum technologies or something similarly specific, you should start by investigating requirements until you find one that justifies a quantum solution, as opposed to presuming the implementation and forging ahead without understanding the needs or requirements.

It seems that no matter where we look in technology development, the lack of sophisticated predictive models for open quantum systems rears its head as a major issue. The context of requirements analysis emphasises how serious this issue is, too; if we are even to assess the merits of applying quantum technologies to meet a set of needs, we require better models with a level of abstraction suited to high-level requirements analysis. In the lab, and as researchers, we can justify looking at quantum out of scientific interest, but if our goal is to establish a *technology pull* towards commercial quantum technologies, it is imperative that their advantages and disadvantages can be evaluated from an engineering perspective.

11.3.6 Test and Verification

When we introduced requirements, we also introduced the corollary notions of test, verification, and validation. These close the cycle of technology development: at the start we form requirements and translate these down levels of technical granularity, until we have a detailed specification to which we can design; in the middle, we do the developmental work to implement these designs in engineered devices or even experimental systems; and then, at the end, we test that our implementation meet our technical requirements at each layer of granularity and integration.

This last statement may not be entirely intuitive, so let us provide an example of the 'layers of granularity' at which test, verification, and validation occur. Starting at the 'lowest' level of granularity – that of the fundamental technical design and implementation – one first tests the disparate components that will eventually make up the system whole.

At this stage we would test components that we have designed, built, or even purchased externally, against the technical specifications derived from the requirements analysis. Let us continue using the cold-atom accelerometer example that we started to explore in the requirements subsection; we might test that the atom source produces atoms at the expected rate, that a laser, or a vacuum pump, meet their specification, or perhaps even that some mu metal shielding sufficiently attenuates external electromagnetic fields. For a system that relies more profoundly on quantum effects, such as a quantum computer, we might test the coherence time of a modular component such as a physical qubit. Practically, this might be a SQUID, which should by now be a device familiar to the reader, and we would be testing that its properties defined in its manufacture, the superconductor used, and its Josephson junction, meet the requirements we set out.

Clearly, these components are not yet a 'system' – we next assemble components to make subsystems. These will have corresponding requirements and architectures, and we would next verify that they satisfy their requirements. For example, we may now have a vacuum system (with pump, chamber, control electronics, and so forth), and may verify that it performs to specification. Similarly, we may verify (using some appropriate test apparatus) that the MOT for our cold-atom accelerometer traps a sufficient number of atoms in a sufficiently constrained volume, meeting the requirements associated with it. For a quantum computing example, we might now be testing a logical qubit subsystem, which, depending on architecture, may be several physical qubits with control circuitry and controllable means of coupling.

Once we have verified that our subsystems meet their requirements, we may move on to integrate them into assemblies or larger systems. Whilst, from a simplified perspective, subsystems can just be assembled to form a system, in reality subsystem integration tends to occur over several levels of assembly.

For example, we might integrate the vacuum system, MOT, and laser system in our cold-atom accelerometer to create a subsystem capable of creating a trapped atom cloud, and cooling it by laser cooling, the performance of which we test and verify against our requirements. Subsequently, this assembled subsystem (which might be considered the core sensor) may be integrated with the other constituent subsystems, including control electronics, power supplies, measurement systems, EM shielding and ruggedisation, and perhaps more sensor assemblies to create an entangled array, which together would make the cold-atom accelerometer system.

Verification can occur at various stages of integration; usually, it is wise to test integrated subsystems at the earliest possible point, making this a gradual and layered process. Whether for science or technology, it is rather unhelpful to have put everything together only to find it 'does not work', and have no clear means of understanding why.

This all leads to top-level V&V, which, whilst often placed in the same phrase, are two distinct processes. Verification is the use of a set of tests to verify that a system meets its requirements as written. For our cold-atom accelerometer, this may be that it measures rates of acceleration between well-defined magnitudes to a well-defined accuracy, that it can repeat measurements at some defined measurement rate, that it meets these performance requirements in a defined 'environment' (e.g. a temperature range, background EM environment, and a physical environment constituting factors such as vibrations), that it meets relevant standards, which may include things such as digital interfaces and data output, and so on – depending on the intended use-case there may be very many other requirements. If these tests are successful, we have verified that the system meets its requirements.

Validation is subtly different; returning to the notion of a system being designed to address key *needs*, validation is the end-user confirming that the system 'does what it is supposed to do' and meets the needs at hand. One would hope, with the right requirements, that verification naturally leads to validation – but requirements analysis is not always easy nor obvious, so this is a separate step.

For technology development, a significant issue is that end-users do not always understand the problems they face well enough to articulate 'needs' that can be reliably analysed to form requirements. Similarly, they do not always possess a technical understanding sufficiently deep to check that the top-level requirements are appropriate. This issue has a scientific analogue, too: a system that can be verified but not validated might be an experiment that has been correctly built, but does not actually enable the measurements or observations necessary to test the intended physical theory. This could be a consequence of shortcomings in planning or design, and a more formal systems approach should address this; however, it might also be a reality of working at the edge of knowledge.

In experiment design, the requirements translation may be an investigative 'guess' based on the theory and observations of the time, leaving open the possibility of designing experiments that we hope – for some justifiable reasons – will tell us more about the physics we are investigating, but that we cannot guarantee will 'validate' against their intended purposes. There is not necessarily an issue with this; what is important is that the postulates formed to 'bridge the gap' in knowledge are well recorded and well understood, as the experiment will also be testing those.

Whilst we used some quantum examples, the process above describes V&V as conventionally carried out in a classical context; as in the previous sections, let us consider how this might change for quantum technologies.

Validation may be the least affected part of this, since it is not the technical testing, but the acceptance of the technology by the end-user (often through a formal acceptance process). If we defer the issue of a technical test, and assume that the end-user trusts the tests we use to verify the system, there is no reason to believe that validation will change. However, as the contemporary discussion of artificial intelligence (AI) technology shows, establishing trust may be a real barrier to technology acceptance and validation. The tendency for AI systems to be 'black box' (i.e. systems where you can only observe the inputs and outputs, but not the internal processes) has proven to be a real barrier to their acceptance, and is now driving 'explainable' AI, which aims to create AI models with decision-making processes that can be extracted, recorded, and scrutinised.

This is a matter of trust, and whilst it has a human dimension, there is also a significant legal and regulatory one: How are these systems audited? What would it mean to use one in a safety-critical circumstance? What liability can you place on them? How would their fault (or not) be proven? So, a more nuanced position is that validation is unlikely to change for a quantum system that can be transparently verified and investigated. The meaning of this depends entirely on the complexity of the quantum system, in a sense 'how quantum' it is. A single quantum sensor or a simple device (a quantum random number generator is another example) may be verified and validated, based on measurement of its outputs through controlled tests – much as similar classical devices today.

A more complex measurement system that, for example, uses an array of entangled sensors to suppress noise may be treated similarly, but require more tests and scrutiny to ensure that the entanglement is well controlled and leads to the expected system behaviour and performance under all operational conditions – we accept that the system may fail through unexpected entanglement (a quantum failure mode), but thoroughly test that it is operating correctly, and we may even have ancillary sensors to compare against during operation to discover if the quantum sensor is behaving strangely. Therefore, V&V does not change, but we may have to adapt our design to ensure that our system is easily, and even continuously, testable. Conversely, a complex and deeply quantum system, such as a 'cat-box' quantum simulator, or a computer for drug discovery (or other molecular simulations), solving NP-Hard problems, or as a part of Quantum AI, is likely to face trust and validation challenges even greater than current AI algorithms and systems.

This is not a superficial issue, and it is grounded in the challenges of modelling complex quantum systems and of measuring quantum systems. The former means we lack a way to predict the behaviour of a complex quantum system across operational scenarios. If we were able to do this, we would have a nuanced way to test if a system is behaving as we expect according to theory – even if this is just as a black box – which would provide a type of verification. To add to the problem, we are likely to use quantum computers and simulators to analyse problems that exceed current computational capability and that have answers that are not necessarily easy to check.

The example of simulating quantum chemistry, which we have used several times in this chapter, is one such case; we may use quantum simulators on complex industrial chemistry problems, such as novel catalyst design or drug discovery, with no equivalent classical simulation to check against, and 'real' testing (experimentally in the lab) may require costly industrial processes. In these cases, it will be important to know if the quantum simulator can be believed on a case-by-case basis. Pragmatically, one approach to this may be by having a number of quantum simulators or computers, built from different architectures and qubit types, running the same simulation in order to see if they provide the same result. Furthermore, we might also combine the simulation of known and unknown chemical processes to have a virtual test of sorts – since they would both be within one quantum simulation, if the known part is simulated incorrectly, we know to distrust the output. Pragmatic solutions like these are part of *design for test*, which will be a key design philosophy, if we are to make complex quantum systems that can be verified. A counterpart to this may be *designing failure*, the worst failures are those that are deceptive – where a system produces a believable but untrue output. For a quantum system, a variety of things may cause this, from subtly incorrect state preparation to unintended entanglement or mixing, or faulty gate operations – if one could design a system that *fails to an obvious state*, and induce failures to demonstrate this, it may address part of the issues of verification.

Lastly, the nature of quantum measurement means that it is physically impossible to make quantum systems 'transparent' in a classical sense – we cannot measure the state of a quantum computer at each operation, check that it is what we expect, and continue on with our computation. So, even more so than the convolutional neural networks used in AI, complex quantum systems are non-deterministic black boxes that cannot be made completely transparent. However, supported by design for test and designed failure, it might be that surreptitious non-demolition measurements can be used to provide some indication that a quantum process is, or is not, working as intended without destroying the quantum state.

> Overall, much as our limitations in modelling quantum systems inhibit requirements analysis, it also stops us 'closing the loop' on system development through verification. Combined with the physical reality of quantum measurement, V&V of complex quantum technologies will depend on nuanced engineering, designing for test and failure, pragmatic and statistical approaches to verification, and eventually on advances in modelling.

11.3.7 Device Characterisation

Before we move on to some practical approaches, let us briefly consider device characterisation. Whilst not V&V, device characterisation is an important outcome of testing devices, playing an important role in the engineering design and manufacturing processes. It is the testing of a specific manufactured system, subsystem, or component to measure the parameters that indicate its quality, performance, or properties. The choice of these parameters will depend on the system at hand and could be any relevant, measurable, parameter (T_1 and T_2 times, resistances, power stability, temperature, physical properties such as hardness or elasticity – this notion is as applicable to a rubber gasket as to a CPU or quantum device).

This *parameter extraction* can inform numerous things: it can be part of verification; if the extracted parameters lie outside an acceptable range, the system fails verification, otherwise it passes; it can, with greater nuance, inform where a system can be used, a 'better' system may be used in a higher-end application, or a 'worse' one in a less demanding environment; it may be used for device (or wider system) calibration, e.g. a phone screen will be manufactured, characterised, and then calibrated for colour accuracy; and it may also be used for feedback and control, where a system needs to be dynamically monitored and adjusted, based on requiring a knowledge of its individual parameters to feed into appropriate models.

Feedback and control is particularly important, as this engineering approach, across electronics, underpins how we enable systems to couple, interact, and decouple in a predictable and controllable way. This requires models, but we shan't reprise that conversation – rather let us assume we have some suitable 'black box' model for the behaviour of our quantum system. However, it requires the parameters of this system from device characterisation in order to be applied. This may be a problem. Consider the concrete example of the SQUID as described in Section 9.5 on page 316. We see that the environment renormalises the system Hamiltonian, and, as per our earlier discussion on the SOI, from an open quantum system's perspective, an inseparable part of the environment may be included in the system Hamiltonian to form an effective Hamiltonian. This changes fundamental properties of the model of the system, such as its ground-state energy; these fundamental properties are likely to directly change the parameters we seek to measure and extract for device characterisation. The point is that a real physical SQUID in a real environment will have an energy that is renormalised by that environment in some way. As we say in Section 9.5, even simple open system models produce different environmental renormalisation at different levels of approximation. It is not clear that the model of the same SQUID, in the same environment, is able to converge to something that realistically and similarly models the renormalisation across levels of approximation, to ensure consistently accurate device characterisation.

If we can guarantee that the conditions, in which we perform device characterisation, exactly match those of the device in-use, then the parameter extraction will be a correct representation of the system; if not, for example if the electromagnetic-environment is subtly or significantly different, or the quantum system will be coupling to a wider system as part of its operation, we do not know if the parameter extraction is representative. Furthermore, noting our previous discussion on the extensibility of quantum models, we do not know, either, whether there is a straightforward 'correction' to the device characterisation that would extend its validity to different conditions, and, if so, what it might be.

In the near term, the solution to this is most likely to tightly engineer and control a quantum system's environment, and to partition complex quantum systems into relatively small, well-tested, subsystems that can be characterised as integrated assemblies, and that contain most of the sophisticated entangled operations. This is far from the grand vision of fully quantum devices with controllable entanglement between every element, but is a pragmatic and feasible path forwards whilst we take the many steps needed to mature engineering capability.

11.3.8 Model-based Systems Engineering

We began this section by considering the engineering perspective on modelling, how it relates to engineering processes, and where our current capability falls short. The consequences of this have been a recurring theme throughout the chapter. To close, let us consider them again, now with a deeper understanding of quantum systems engineering and its relationship with fundamental science.

In systems engineering, there is a paradigm termed MBSE. The premise of MBSE is to consistently apply a modelling and simulation approach to systems engineering processes, acknowledging that hierarchical and extensible models both inform systems engineering steps and decisions (such as requirements translation, technology selection, architecture and design, device characterisation, and established V&V criteria), and intrinsically capture much of our systems engineering knowledge – from how a system behaves (and therefore why we have made engineering decisions), to the environment we are considering, to the failure modes we model and mitigate. MBSE tends to seek simplicity. At a given level of abstraction, it promotes having a model that provides the least detail still sufficient for systems engineering, usually through abstract parameters and lumped models that hide detail at high levels of abstraction – an implicit requirement for model hierarchies. Unsurprisingly, it also demands extensibility. Since the approach fundamentally utilises models, not only to inform the systems engineering, but also to capture all relevant information about the system, there is an expectation that anything outside the conventional SOI, but important to system behaviour, can be modelled and typically 'added in' as necessary within an extensible framework. What exactly is needed beyond the SOI will depend entirely on the technology and the use-case, which is why extensibility is favoured over an enumeration of independent whole-system models. We can already see from this conceptual analysis that MBSE, whilst tightening our notion of what a good model should be capable of, and what models we need, does nothing to reduce the 'quantum challenge'. However, with our current understanding, we can go a bit further regarding how we might try to take an MBSE approach.

Attempts have been made to apply MBSE to quantum technologies development. An example is [44], which investigates an MBSE approach to developing quantum dot systems. Some engineering progress can be made by only using the time-independent Schrödinger equation to generate a model, a convenient simplification of the problem made possible by the technology's limited quantum complexity. Nevertheless, the paper notes that the 'implementation of the newly developed physical rules in the systems development and the employment of a combination of quantum technologies and the relevant classical feedback are still big challenges'.

It is clear that, even in simple or toy-box cases (e.g. as we investigated in Chapters 9 and 10), it is hard to reduce complexity by establishing a degree of model separability – between quantum elements (quantum-to-quantum coupling), but also between a single quantum element and its feedback, control, readout circuitry, and environment (which may include quantum-to-quantum and quantum-to-classical coupling). We have stated the necessity of hierarchical models to the systems engineering process, but with separability so hard to achieve, which paths are there towards realising them? We think developing models that can support systems engineering processes such as device characterisation and verification, not to mention frameworks like MBSE, will require an approach more nuanced and more sympathetic to engineering needs than those we have explored in this text.

To take an example, consider the simple model discussed on page 342. Whilst this provides a way to model interacting quantum and classical circuits, it also contains a number of unjustified and unvalidated assumptions, making it a naïve approach. Conversely, in

Section 10.7 on page 354, perhaps we go too far in the other direction when we develop a model-based approach to quantum measurement. The idea that we would have to model all circuitry within a quantum framework may be too computationally expensive and cumbersome to be of practical use. It might make much more sense (depending on our requirements) to try to model classical systems classically and quantum systems quantum mechanically, but within the same ontological framework. We might be able to do this in phase space by deforming the Moyal bracket in

$$\frac{dA}{dt} = \{\{A, \mathcal{H}\}\} + \frac{\partial A}{\partial t},$$

so that $\hbar \to 0$ for the classical parts of the system (see, for example, [3]), although this approach is also likely to be computationally challenging. An alternative might be found using the ideas in Ref. [7]. This work quantises Koopman–von Neumann theory to produce an elegant reformulation of quantum mechanics (where the Wigner function naturally arises as the joint position, momentum representation of the state vector). Furthermore, this creates a unified framework for modelling quantum and classical systems, including their interaction, together. It remains to be seen if this provides a pathway towards hierarchical models, or some compromise that is sufficient for systems engineering.

In general, it is interesting to consider what properties of a modelling approach would recommend it as a good pathway towards engineering. One, as we have just discussed, is the ability to evenhandedly model both quantum and classical elements and therefore a 'whole system' without an artificial separation of elements. Another may be to test if an approach provides the ability to identify when, where, or even if, it is possible to make a valid simplification of the model, eliminating a degree of complexity or unifying parameters – both of which would indicate pathways to a model hierarchy. And a third, which we explored in Section 9.5, is whether we can extract a physical parametrisation from a model that better reflects device characterisation. Currently, we lack a 'good' approach for any of these criteria, and we may find that a model, which meets one, fails to meet the others.

Let us put the above discussion in context and consider as an example the extent to which modelling might be applied to the specific case of a portable sensor. It is clear that a good model should, at the very least, include all the necessary feedback, control, and readout circuitry to predict its operation under ideal circumstances. It should also be clear from the previous discussion that to do this, environmental effects must be taken into account in a suitable way. But what about everything else that forms part of the entire system that the sensor might interact with when deployed? This might include considerations like the choice of power supply, as the performance of the sensor may depend on, e.g. the state, the ability to deliver current, or the noise characteristics of specific supplies. Our consideration may need to extend further than this component modelling. We may also need to take the device's likely operational environment and its potential uses into account in the model. For example, can our model include the effects of impacts, temperature changes, electromagnetic interference? If we expect our device to be used in extreme environments, such factors may be crucially important to consider in order to ensure verifiable and reliable operation. We would want, therefore, to accommodate these considerations in the model, to be able to engineer a design that is fit for purpose. Given the fragility of quantum states, it is not at all clear that a modelling abstraction that works under certain assumptions, will continue to be valid under other conditions. It seems unlikely to us that it will be sufficient to extrapolate from a lab-type model to a real-world scenario such as the one we have just outlined. The conceptual and practical challenges of wanting to perform modelling and simulation of systems with quantum components to meet the demands of MBSE are really quite exciting from a physics perspective.

While we have a tendency to seek ideal solutions in physics, in this case we may need to be careful not to let the great become the enemy of the good; whilst we make progress in the domain, we may seek different models for different purposes, satisfying the right acceptability criteria for their purpose, rather than only accepting one modelling approach that meets all our needs.

> In all cases, it is clear that to establish models for engineering applications, further work on the foundations of quantum mechanics is of crucial importance.

11.4 Concluding Remarks

We hope that it is clear that making progress in developing quantum technologies will require the development of new tools within a quantum systems engineering framework, and therefore, practically, will be in the hands of physicists for the foreseeable future.

In light of everything, we have discussed in this chapter, we would like for you to reflect on a final, practical, question: *How can systems engineering methods benefit the science I do today?* This is quite an unusual question, the imagined boundary between science and engineering means that formalised methodologies like systems engineering are rarely applied in physics, or at low TRL in general. For example, just for fun, one might like to consider what the formal requirements specification and the associated V&V criteria of a theory of quantum gravity might be? We mention throughout the notion of end-users or stakeholders driving 'needs' and requirements; we tend to assume that the stakeholder is extrinsic to our work, but it is not necessary for that to be the case. The stakeholder is whoever is defining why we do what we do, and most often for us as scientists that is ourselves; driven by our scientific interests, desire to realise innovation, and practical factors like our collaborators or grants. This is not a reason to avoid systems engineering, or to avoid a requirements-led approach. Considering our position as stakeholders in the physics we do, systems methods can give us clarity, improve methodology and, if used judiciously, accelerate our research and development.

It is fascinating that in order to achieve success in engineering, the most applied of subjects, we may both require and shed light on three of the longest open problems in the foundations of quantum physics, namely the measurement problem, understanding the von Neumann chain, and the quantum-to-classical transition.

Bibliography

1 M. Alder. Newton's flaming laser sword. *Philosophy Now*, 46:29–33, 2004. URL https://philosophynow.org/issues/46/Newtons_Flaming_Laser_Sword.

2 A. Alex, M. Kalus, A. Huckleberry, and J. von Delft. A numerical algorithm for the explicit calculation of SU(n) and SL(n,c)SL(n,c) clebsch–gordan coefficients. *Journal of Mathematical Physics*, 52(2):023507, 2011. doi: 10.1063/1.3521562.

3 M. Amin and M. A. Walton. Quantum-classical dynamical brackets. *Physical Review A*, 104:032216, 2021. doi: 10.1103/PhysRevA.104.032216. URL https://link.aps.org/doi/10.1103/PhysRevA.104.032216.

4 K. Beck. *Test Driven Development*. By Example (Addison-Wesley Signature). Addison-Wesley Longman, Amsterdam, 2002. ISBN 0321146530.

5 R. F. Bishop and A. Vourdas. Displaced and squeezed parity operator: Its role in classical mappings of quantum theories. *Physical Review A*, 50:4488–4501, 1994. doi: 10.1103/PhysRevA.50.4488. URL http://link.aps.org/doi/10.1103/PhysRevA.50.4488.

6 M. Błaszak and Z. Domański. Canonical quantization of classical mechanics in curvilinear coordinates. Invariant quantization procedure. *Annals of Physics*, 339:89–108, 2013. ISSN 0003-4916. doi: https://doi.org/10.1016/j.aop.2013.08.014. URL https://www.sciencedirect.com/science/article/pii/S0003491613001899.

7 D. I. Bondar, R. Cabrera, D. V. Zhdanov, and H. A. Rabitz. Wigner phase-space distribution as a wave function. *Physical Review A*, 88:052108, 2013. doi: 10.1103/PhysRevA.88.052108. URL https://link.aps.org/doi/10.1103/PhysRevA.88.052108.

8 H. P. Breuer and F. Petruccione. *The Theory of Open Quantum Systems*. OUP Oxford, 2007. ISBN 9780199213900. URL https://books.google.co.uk/books?id=DkcJPwAACAAJ.

9 S. Burge. Holistic requirements model. *The Systems Engineering Toolbox*, pages 1–15, 2006. URL www.burgehugheswalsh.co.uk/Uploaded/1/Documents/HRM-Tool-Box-V1.0.pdf.

10 A. O. Caldeira and A. J. Leggett. Quantum tunnelling in a dissipative system. *Annals of Physics*, 149(2):374–456, 1983.

11 H. J. Carmichael. *An Open Systems Approach to Quantum Optics*. Springer-Verlag, Berlin, Heidelberg, 1993.

12 C. Cohen-Tannoudji, B. Diu, and F. Laloe. *Quantum Mechanics*, Volumes 1 and 2. Wiley, 1991.

Quantum Mechanics: From Analytical Mechanics to Quantum Mechanics, Simulation, Foundations, and Engineering, First Edition. Mark Julian Everitt, Kieran Niels Bjergstrom and Stephen Neil Alexander Duffus.
© 2024 John Wiley & Sons Ltd. Published 2024 by John Wiley & Sons Ltd.
Companion website: www.wiley.com/go/everitt/quantum

13 C. K. Dabhi, S. S. Parihar, H. Agrawal, N. Paydavosi, T. H. Morshed, D. D. Lu, W. Yang, and others. BSIM4 4.8.1 MOSFET model. *Dept. Elect. Eng. Comput. Sci., Univ. California, Berkeley, CA, Tech. Rep*, 94720:185, 2017.

14 B. I. Davies, R. P. Rundle, V. M. Dwyer, J. H. Samson, T. Tilma, and M. J. Everitt. Visualizing spin degrees of freedom in atoms and molecules. *Physical Review A*, 100:042102, 2019. doi: 10.1103/PhysRevA.100.042102. URL https://link.aps.org/doi/10.1103/PhysRevA.100.042102.

15 B. S. DeWitt. Point transformations in quantum mechanics. *Physical Review*, 85:653–661, 1952. doi: 10.1103/PhysRev.85.653. URL https://link.aps.org/doi/10.1103/PhysRev.85.653.

16 J. Diggins, J. F. Ralph, T. P. Spiller, T. D. Clark, H. Prance, and R. J. Prance. Chaotic dynamics in the RF superconducting quantum-interference-device magnetometer: A coupled quantum-classical system. *Physical Review E*, 49:1854–1859, 1994. doi: 10.1103/PhysRevE.49.1854. URL https://link.aps.org/doi/10.1103/PhysRevE.49.1854.

17 P. A. M. Dirac. Bakerian lecture - the physical interpretation of quantum mechanics. *Proceedings of the Royal Society of London. Series A. Mathematical and Physical Sciences*, 180(980):1–40, 1942. doi: 10.1098/rspa.1942.0023. URL https://royalsocietypublishing.org/doi/abs/10.1098/rspa.1942.0023.

18 P. A. M. Dirac. *The Principles of Quantum Mechanics*. Clarendon Press, 1981. ISBN 9780198520115.

19 S. N. A. Duffus, K. N. Bjergstrom, V. M. Dwyer, J. H. Samson, T. P. Spiller, A. M. Zagoskin, W. J. Munro, K. Nemoto, and M. J. Everitt. Some implications of superconducting quantum interference to the application of master equations in engineering quantum technologies. *Physical Review B*, 94:064518, 2016. doi: 10.1103/PhysRevB.94.064518. URL http://link.aps.org/doi/10.1103/PhysRevB.94.064518.

20 S. N. A. Duffus, V. M. Dwyer, and M. J. Everitt. Open quantum systems, effective hamiltonians, and device characterization. *Physical Review B*, 96:134520, 2017. doi: 10.1103/PhysRevB.96.134520. URL https://link.aps.org/doi/10.1103/PhysRevB.96.134520.

21 P. Ehrenfest. Bemerkung über die angenäherte gültigkeit der klassischen mechanik innerhalb der quantenmechanik. *Zeitschrift für Physik*, 45(7):455–457, 1927. ISSN 0044-3328. doi: 10.1007/BF01329203. URL https://doi.org/10.1007/BF01329203.

22 M. J. Everitt. On the correspondence principle: Implications from a study of the nonlinear dynamics of a macroscopic quantum device. *New Journal of Physics*, 11(1):013014, 2009. doi: 10.1088/1367-2630/11/1/013014. URL https://doi.org/10.1088%2F1367-2630%2F11%2F1%2F013014.

23 M. J. Everitt, T. D. Clark, P. B. Stiffell, J. F. Ralph, A. R. Bulsara, and C. J. Harland. Signatures of chaoticlike and nonchaoticlike behavior in a nonlinear quantum oscillator through photon detection. *Physical Review E*, 72:066209, 2005. doi: 10.1103/PhysRevE.72.066209. URL https://link.aps.org/doi/10.1103/PhysRevE.72.066209.

24 M. J. Everitt, T. D. Clark, P. B. Stiffell, J. F. Ralph, and C. J. Harland. Energy downconversion between classical electromagnetic fields via a quantum mechanical squid ring. *Physical Review B*, 72:094509, 2005. doi: 10.1103/PhysRevB.72.094509. URL https://link.aps.org/doi/10.1103/PhysRevB.72.094509.

25 M. J. Everitt, W. J. Munro, and T. P. Spiller. Quantum-classical crossover of a field mode. *Physical Review A*, 79:032328, 2009. doi: 10.1103/PhysRevA.79.032328. URL https://link.aps.org/doi/10.1103/PhysRevA.79.032328.

26 M. J. Everitt, W. J. Munro, and T. P. Spiller. Quantum measurement with chaotic apparatus. *Physics Letters A*, 374(28):2809–2815, 2010. ISSN 0375-9601. doi: https://doi.org/10.1016/j.physleta.2010.05.006. URL https://www.sciencedirect.com/science/article/pii/S0375960110005700.

27 R. P. Feynman, R. B. Leighton, and M. Sands. *The Feynman Lectures on Physics: The Definitive Edition*, Volume 3. Pearson, 2009.

28 R. P. Feynman, A. R. Hibbs, and D. F. Styer. *Quantum Mechanics and Path Integrals*. Dover Books on Physics. Dover Publications, 2010. ISBN 9780486477220. URL https://books.google.co.uk/books?id=JkMuDAAAQBAJ.

29 G. W. Ford, J. T. Lewis, and R. F. O'Connell. Quantum Langevin equation. *Physical Review A*, 37:4419–4428, 1988. doi: 10.1103/PhysRevA.37.4419. URL https://link.aps.org/doi/10.1103/PhysRevA.37.4419.

30 J. Gambetta and S. Sheldon. Cramming more power into a quantum device, 2019. URL https://www.ibm.com/blogs/research/2019/03/power-quantum-device/.

31 S. Gao. Dissipative quantum dynamics with a Lindblad functional. *Physical Review Letters*, 79(17):3101, 1997.

32 C. Gardiner, P. Zoller, and P. Zoller. *Quantum Noise: A Handbook of Markovian and Non-Markovian Quantum Stochastic Methods with Applications to Quantum Optics*, Volume 56. Springer Science & Business Media, 2004.

33 C. Gerry and P. Knight. *Introductory Quantum Optics*. Cambridge University Press, 2004. ISBN 9781139453554. URL https://books.google.co.uk/books?id=MDwgAwAAQBAJ.

34 D. T. Gillespie. *A Quantum Mechanics Primer*. Open University Set Book. International Textbook Company, 1970. ISBN 9780700222902. URL https://books.google.co.uk/books?id=2XAsAAAAYAAJ.

35 R. J. Glauber. Coherent and incoherent states of the radiation field. *Physical Review*, 131(6):2766, 1963. doi: 10.1103/PhysRev.131.2766.

36 H. Goldstein, C. P. Poole, and J. L. Safko. *Classical Mechanics*. Addison Wesley, 2002. ISBN 9780201657029. URL https://books.google.co.uk/books?id=tJCuQgAACAAJ.

37 D. T. Greenwood. *Classical Dynamics*. Dover Books on Mathematics. Dover Publications, 1997. ISBN 9780486696904. URL https://books.google.co.uk/books?id=x7rj83I98yMC.

38 M. Grifoni and P. Hänggi. Driven quantum tunneling. *Physics Reports*, 304(5):229–354, 1998. ISSN 0370-1573. doi: https://doi.org/10.1016/S0370-1573(98)00022-2. URL https://www.sciencedirect.com/science/article/pii/S0370157398000222.

39 H. J. Groenewold. On the principles of elementary quantum mechanics. *Physica*, 12(7):405–460, 1946. ISSN 00318914. doi: 10.1016/S0031-8914(46)80059-4.

40 S. Habib, K. Shizume, and W. H. Zurek. Decoherence, Chaos, and the correspondence principle. *Physical Review Letters*, 80(20):4361–4365, 1998.

41 T. Haga, M. Nakagawa, R. Hamazaki, and M. Ueda. Liouvillian skin effect: Slowing down of relaxation processes without gap closing. *Physical Review Letters*, 127:070402, 2021. doi: 10.1103/PhysRevLett.127.070402. URL https://link.aps.org/doi/10.1103/PhysRevLett.127.070402.

42 ISO/IEC/IEEE. Systems and software engineering: System life cycle processes. 15288, 2015.

43 J. R. Johansson, P. D. Nation, and F. Nori. QuTiP: An open-source python framework for the dynamics of open quantum systems. *Computer Physics Communications*,

183(8):1760–1772, 2012. ISSN 0010-4655. doi: https://doi.org/10.1016/j.cpc.2012.02.021. URL https://www.sciencedirect.com/science/article/pii/S0010465512000835.

44 M. Karimaghaei, R. Cloutier, A. Khan, J. D. Richardson, and A.-V. Phan. A model-based systems engineering framework for quantum dot solar cells development. *Systems Engineering*, 26(3):279–290, 2023. doi: https://doi.org/10.1002/sys.21655. URL https://incose .onlinelibrary.wiley.com/doi/abs/10.1002/sys.21655.

45 S. Krämer, D. Plankensteiner, L. Ostermann, and H. Ritsch. QuantumOptics.jl: A Julia framework for simulating open quantum systems. *Computer Physics Communications*, 227:109–116, 2018. ISSN 0010-4655. doi: https://doi.org/10.1016/j.cpc.2018.02.004. URL https://www.sciencedirect.com/science/article/pii/S0010465518300328.

46 E. Kreyszig. *Introductory Functional Analysis with Applications*. Wiley Classics Library. Wiley, 1989. ISBN 9780471504597. URL https://books.google.co.uk/books? id=nZmpQgAACAAJ.

47 L. D. Landau, E. M. Lifshitz, J. B. Sykes, and J. S. Bell. *Mechanics*, Volume 1. Butterworth-Heinemann. Elsevier Science, 1976. ISBN 9780750628969. URL https://books.google.co.uk/books?id=e-xASAehg1sC.

48 A. J. Leggett. Macroscopic quantum systems and the quantum theory of measurement. *Progress of Theoretical Physics Supplement*, 69:80–100, 1980. ISSN 0375-9687. doi: 10 .1143/PTP.69.80. URL https://doi.org/10.1143/PTP.69.80.

49 R. L. Liboff. *Introductory Quantum Mechanics*. Addison-Wesley, 1987. ISBN 9780201122213. URL https://books.google.co.uk/books?id=FbIPAQAAMAAJ.

50 G. Lindblad. On the generators of quantum dynamical semigroups. *Communications in Mathematical Physics*, 48(2):119–130, 1976.

51 Y. Makhlin, G. Schön, and A. Shnirman. Dissipative effects in Josephson qubits. *Chemical Physics*, 296(2):315–324, 2004. ISSN 0301-0104. doi: https://doi.org/10.1016/ j.chemphys.2003.09.025. URL https://www.sciencedirect.com/science/article/pii/ S0301010403004981. The Spin-Boson Problem: From Electron Transfer to Quantum Computing … to the 60th Birthday of Professor Ulrich Weiss.

52 J. C. Mankins. Technology readiness levels. *White Paper, April*, 6, 1995.

53 R. C. Martin. *Agile Software Development: Principles, Patterns, and Practices*. Prentice Hall PTR, 2003. URL http://dl.acm.org/citation.cfm?id=515230.

54 R. C. Martin. *Clean Code: A Handbook of Agile Software Craftsmanship*. Pearson, 2008.

55 A. E. P. Martínez and L. M. Aguilar. Quantifying the hybrid entanglement of the Stern-Gerlach experiment using discrete reductions. *Physics Letters A*, 394:127200, 2021. ISSN 0375-9601. doi: https://doi.org/10.1016/j.physleta.2021.127200. URL https://www .sciencedirect.com/science/article/pii/S0375960121000645.

56 F. Mascherpa, A. Smirne, A. D. Somoza, P. Fernández-Acebal, S. Donadi, D. Tamascelli, S. F. Huelga, and M. B. Plenio. Optimized auxiliary oscillators for the simulation of general open quantum systems. *Physical Review A*, 101:052108, 2020. doi: 10.1103/PhysRevA .101.052108. URL https://link.aps.org/doi/10.1103/PhysRevA.101.052108.

57 S. Massar and P. Spindel. Uncertainty relation for the discrete Fourier transform. *Physical Review Letters*, 100:190401, 2008. doi: 10.1103/PhysRevLett.100.190401. URL https://link.aps.org/doi/10.1103/PhysRevLett.100.190401.

58 P. Massignan, A. Lampo, J. Wehr, and M. Lewenstein. Quantum Brownian motion with inhomogeneous damping and diffusion. *Physical Review A*, 91(3):033627, 2015.

59 M. B. Mensky. Quantum restrictions for continuous observation of an oscillator. *Physical Review D*, 20:384–387, 1979. doi: 10.1103/PhysRevD.20.384. URL https://link.aps.org/doi/10.1103/PhysRevD.20.384.

60 E. Merzbacher. *Quantum Mechanics*. John Wiley, New York, 2nd edition, 1970.

61 F. Minganti, A. Biella, N. Bartolo, and C. Ciuti. Spectral theory of Liouvillians for dissipative phase transitions. *Physical Review A*, 98:042118, 2018. doi: 10.1103/PhysRevA.98.042118. URL https://link.aps.org/doi/10.1103/PhysRevA.98.042118.

62 J. E. Moyal. Quantum mechanics as a statistical theory. *Proceedings of the Cambridge Philological Society*, 45:99, 1949.

63 W. J. Munro and C. W. Gardiner. Non-rotating-wave master equation. *Physical Review A*, 53(4):2633, 1996.

64 M. A. Nielsen and I. L. Chuang. *Quantum Computation and Quantum Information: 10th Anniversary Edition*. Cambridge University Press, 2010. doi: 10.1017/CBO9780511976667.

65 I. Percival. *Quantum State Diffusion*. Cambridge University Press, 2005. ISBN 9780521021203. URL https://books.google.co.uk/books?id=iejnHAAACAAJ.

66 G. Potel, F. Barranco, S. Cruz-Barrios, and J. Gómez-Camacho. Quantum mechanical description of Stern-Gerlach experiments. *Physical Review A*, 71:052106, 2005. doi: 10.1103/PhysRevA.71.052106. URL https://link.aps.org/doi/10.1103/PhysRevA.71.052106.

67 W. H. Press, S. A. Teukolsky, W. T. Vetterling, and B. P. Flannery. *Numerical Recipes in C*. Cambridge University Press, Cambridge, USA, 2nd edition, 1992.

68 J. F. Ralph, T. D. Clark, M. J. Everitt, and P. Stiffell. Nonlinear backreaction in a quantum mechanical SQUID. *Physical Review B*, 64:180504, 2001. doi: 10.1103/PhysRevB.64.180504. URL https://link.aps.org/doi/10.1103/PhysRevB.64.180504.

69 G. Ritschel and A. Eisfeld. Analytic representations of bath correlation functions for ohmic and superohmic spectral densities using simple poles. *The Journal of Chemical Physics*, 141(9):094101, 2014. doi: 10.1063/1.4893931. URL https://doi.org/10.1063%2F1.4893931.

70 R. P. Rundle and M. J. Everitt. An informationally complete Wigner function for the Tavis–Cummings model. *Journal of Computational Electronics*, 20(6):2180–2188, 2021. doi: 10.1007/s10825-021-01777-6. URL https://doi.org/10.1007/s10825-021-01777-6.

71 R. P. Rundle and M. J. Everitt. Overview of the phase space formulation of quantum mechanics with application to quantum technologies. *Advanced Quantum Technologies*, 4(6):2100016, 2021. doi: https://doi.org/10.1002/qute.202100016. URL https://onlinelibrary.wiley.com/doi/abs/10.1002/qute.202100016.

72 R. P. Rundle, B. I. Davies, V. M. Dwyer, T. Tilma, and M. J. Everitt. Visualization of correlations in hybrid discrete—continuous variable quantum systems. *Journal of Physics Communications*, 4(2):025002, 2020. doi: 10.1088/2399-6528/ab6fb6. URL https://doi.org/10.1088/2399-6528/ab6fb6.

73 R. P. Rundle, P. W. Mills, T. Tilma, J. H. Samson, and M. J. Everitt. Simple procedure for phase-space measurement and entanglement validation. *Physical Review A*, 96:022117, 2017. doi: 10.1103/PhysRevA.96.022117. URL https://link.aps.org/doi/10.1103/PhysRevA.96.022117.

74 R. P. Rundle, T. Tilma, J. H. Samson, V. M. Dwyer, R. F. Bishop, and M. J. Everitt. General approach to quantum mechanics as a statistical theory. *Physical Review A*,

99:012115, 2019. doi: 10.1103/PhysRevA.99.012115. URL https://link.aps.org/doi/10
.1103/PhysRevA.99.012115.

75 M. Rushka and J. K. Freericks. A completely algebraic solution of the simple harmonic
oscillator. *American Journal of Physics*, 88(11):976–985, 2020. doi: 10.1119/10.0001702.
URL https://doi.org/10.1119/10.0001702.

76 B. Russell. *The Principles of Mathematics*. W. W. Norton, 1903.

77 J. J. Sakurai. *Modern Quantum Mechanics (Revised Edition)*. Addison Wesley, 1st edition,
1993. ISBN 0201539292. URL http://www.worldcat.org/isbn/0201539292.

78 R. Schack and T. A. Brun. A C++ library using quantum trajectories to solve quantum master equations. *Computer Physics Communications*, 102(1):210–228, 1997. ISSN
0010-4655. doi: https://doi.org/10.1016/S0010-4655(97)00019-2. URL https://www
.sciencedirect.com/science/article/pii/S0010465597000192.

79 R. Schack, T. A. Brun, and I. C. Percival. Quantum state diffusion, localization and
computation. *Journal of Physics A: Mathematical and General*, 28(18):5401, 1995. doi:
10.1088/0305-4470/28/18/028. URL https://dx.doi.org/10.1088/0305-4470/28/18/028.

80 M. A. Schlosshauer. *Decoherence: And the Quantum-To-Classical Transition*. The Frontiers Collection. Springer, 2007. ISBN 9783540357735. URL https://books.google.co.uk/
books?id=1qrJUS5zNbEC.

81 E. Schrödinger. An undulatory theory of the mechanics of atoms and molecules. *Physical Review*, 28:1049–1070, 1926. doi: 10.1103/PhysRev.28.1049. URL https://link.aps.org/
doi/10.1103/PhysRev.28.1049.

82 A. Shnirman, Y. Makhlin, and G. Schön. Noise and decoherence in quantum two-level
systems. *Physica Scripta*, T102(1):147, 2002. doi: 10.1238/physica.topical.102a00147. URL
https://doi.org/10.1238%2Fphysica.topical.102a00147.

83 R. L. Stratonovich. On distributions in representation space. *Soviet Physics - JETP*,
31:1012, 1956.

84 L. Susskind and G. Hrabovsky. *Classical Mechanics: The Theoretical Minimum*. Theoretical Minimum. Penguin Books, 2014. ISBN 9780141976228. URL https://books.google.co
.uk/books?id=-WOCngEACAAJ.

85 S. M. Tan. A computational toolbox for quantum and atomic optics. *Journal of Optics B:
Quantum and Semiclassical Optics*, 1(4):424–432, 1999. doi: 10.1088/1464-4266/1/4/312.
URL https://doi.org/10.1088/1464-4266/1/4/312.

86 T. Tilma, M. J. Everitt, J. H. Samson, W. J. Munro, and K. Nemoto. Wigner functions
for arbitrary quantum systems. *Physical Review Letters*, 117:180401, 2016. doi: 10.1103/
PhysRevLett.117.180401. URL http://link.aps.org/doi/10.1103/PhysRevLett.117.180401.

87 K. Vogtmann, A. Weinstein, and V. I. Arnol'd. *Mathematical Methods of Classical Mechanics*. Graduate Texts in Mathematics. Springer, New York, 2013. ISBN
9781475720631. URL https://books.google.co.uk/books?id=5OQlBQAAQBAJ.

88 S. N. Walck. Learn physics by programming in Haskell. In J. L. Caldwell, P. K. F.
Hölzenspies, and P. Achten, editors, *Proceedings 3rd International Workshop on Trends
in Functional Programming in Education, TFPIE 2014, Soesterberg, The Netherlands, 25th
May 2014*, Volume 170 of *EPTCS*, pages 67–77, 2014. doi: 10.4204/EPTCS.170.5. URL
https://doi.org/10.4204/EPTCS.170.5.

89 W. Weaver. Science and complexity. *American Scientist*, 36(4):536–544, 1948.

90 U. Weiss. *Quantum Dissipative Systems*. Series in Modern Condensed Matter Physics. World Scientific, 2012. ISBN 9789814374910. URL https://books.google.co.uk/books? id=qgfuFZxvGKQC.

91 A. Widom. Quantum electrodynamic circuits at ultralow temperature. *Journal of Low Temperature Physics*, 37 (3–4):449–460, 1979. doi: 10.1007/bf00119200.

92 E. Wigner. On the quantum correction for thermodynamic equilibrium. *Physical Review*, 40:749–759, 1932. doi: 10.1103/PhysRev.40.749. URL https://link.aps.org/doi/10.1103/ PhysRev.40.749.

93 G. Wilson, D. A. Aruliah, C. T. Brown, N. P. C. Hong, M. Davis, R. T. Guy, S. H. D. Haddock, K. D. Huff, I. M. Mitchell, M. D. Plumbley, B. Waugh, E. P. White, and P. Wilson. Best practices for scientific computing. *PLoS Biology*, 12(1):1–7, 2014. doi: 10.1371/journal.pbio.1001745. URL https://doi.org/10.1371/journal.pbio.1001745.

94 H. M. Wiseman. Quantum trajectories and quantum measurement theory. *Quantum and Semiclassical Optics: Journal of the European Optical Society Part B*, 8(1):205, 1996. doi: 10.1088/1355-5111/8/1/015. URL https://dx.doi.org/10.1088/1355-5111/8/1/015.

95 H. M. Wiseman and W. J. Munro. Comment on "dissipative quantum dynamics with a lindblad functional". *Physical Review Letters*, 80(25):5702, 1998.

Index

Quantum Mechanics: From Analytical Mechanics to Quantum Mechanics, Simulation, Foundations, and Engineering,
First Edition. Mark Julian Everitt, Kieran Niels Bjergstrom and Stephen Neil Alexander Duffus.
© 2024 John Wiley & Sons Ltd. Published 2024 by John Wiley & Sons Ltd.
Companion website: www.wiley.com/go/everitt/quantum